NONLINEARITY AND CHAOS
IN ENGINEERING DYNAMICS

NONLINEARITY AND CHAOS IN ENGINEERING DYNAMICS

Edited by

J. M. T. Thompson
S. R. Bishop

Centre for Nonlinear Dynamics,
University College London, UK

JOHN WILEY & SONS
Chichester · New York · Brisbane · Toronto · Singapore

Other Wiley Editorial Offices

John Wiley & Sons, Inc., 605 Third Avenue,
New York, NY 10158-0012, USA

Jacaranda Wiley Ltd, 33 Park Road, Milton,
Queensland 4064, Australia

John Wiley & Sons (Canada) Ltd, 22 Worcester Road,
Rexdale, Ontario M9W 1L1, Canada

John Wiley & Sons (SEA) Pte Ltd, 37 Jalan Pemimpin #05-04,
Block B, Union Industrial Building, Singapore 2057

Library of Congress Cataloging-in-Publication Data

IUTAM Symposium (1993:University College London)
    Nonlinearity and chaos in engineering dynamics:IUTAM Symposium,
  UCL, July 1993/edited by J. M. T. Thompson, S. R. Bishop.
        p.   cm.
    Includes bibliographical references and index.
    ISBN 0 -471 -94458 -0
    1. Dynamics—Congresses.   2. Nonlinear theories—Congresses.
  3. Chaotic behavior in systems—Congresses.   I. Thompson, J. M. T.
  II. Bishop. S. R.   III. Title.
  TA352.I98   1993
  620.1'04—dc20

British Library Cataloguing in Publication Data

A catalogue record for this book is available from the British Library
ISBN 0 471 94458 0

Typeset $10\frac{1}{2}$/12pt Palatino by Thomson Press (India) Ltd, New Delhi
Printed and bound in Great Britain by Bookcraft (Bath) Ltd.

CONTENTS

PREFACE

Engineers and applied scientists are increasingly responding to the revolution in nonlinear dynamics which has uncovered the full complexity inherent in the equations of motion of macroscopic mechanical systems. The unpredictability implied by the chaotic motions and fractal basin boundaries of simple 'deterministic' systems is but one feature which nicely epitomizes the field. The ubiquitous new phenomena necessitate a change in emphasis away from the classical reliance on perturbation and averaging methods towards the use of computational techniques employing the powerful geometrical concepts developed by mathematicians.

This book gives a coherent overview of recent developments in the field, presenting in a structured fashion the latest research of leading international groups in theoretical and applied dynamics. Topics covered include: nonlinear problems of structural, mechanical, aerospace and naval engineering; topological and computer methods, cell mapping and global analysis; phenomenological studies of attractors, fractal basins, bifurcations and escape; control of chaos; experimental studies; piecewise linear, stick–slip and impacting systems; time series analysis, phase-space reconstruction, parametric identification; dynamics of cables, beams and structures under fluid loading; stochastic bifurcation and resonance under random excitation.

Particular emphasis is given to recent developments in knot theory, which can identify bifurcational precedences in driven oscillators. New ideas are presented on the control and potential use of chaos; and on the phase-space reconstruction of time-series data. Discontinuous systems, including, for example, sliding friction or impacts, are another important and developing area of study that is strongly represented. Other contributions present important experimental verifications of theoretical concepts; and novel engineering applications related to vehicle dynamics and the chaotic motions of machine tools.

These research contributions were presented by their authors at a Symposium on *Nonlinearity and Chaos in Engineering Dynamics*, sponsored by the International Union of Theoretical and Applied Mechanics (IUTAM). This was held on 19–23 July 1993, at University College London. As with all specialist IUTAM symposia, invitations to attend, and to present papers, were made by a Scientific Committee which was constituted as follows: J. M. T. Thompson (Chairman), S. Al-Athel (Saudi Arabia), S. T. Ariaratnam (Canada), S. Arimoto (Japan), D. H. van Campen (Netherlands), F. L. Chernousko (Russia), C. S. Hsu (USA), F. C. Moon (USA), W. Schiehlen (Germany), S. W. Shaw (USA), W. Szemplińska-Stupnicka (Poland) and H. Troger (Austria). The Symposium brought together a wide spectrum of theoretical and applied dynamicists, 78 participants from 23 countries; and promoted a vigorous exchange of ideas. The extensive discussions helped to establish, consolidate and direct the emerging body of topological, analytical and computational expertise that is needed to address the challenging practical problems of engineering dynamics. The proceedings began with an

Opening Address by Sir James Lighthill who was the President of IUTAM during his period as Provost of University College London; followed by the opening general lecture given by Philip Holmes of Cornell University. They ended with a comprehensive final discussion session which provided a valuable focus for identifying desirable developments and lines of future research.

Financial support for the Symposium was generously provided by IUTAM, and most major international publishers contributed to the display of scientific books and periodicals. The detailed organization was in the hands of the Local Steering Committee comprising the following researchers from the Centre for Nonlinear Dynamics and its Applications: S. R. Bishop (Chairman), M. E. Davies, S. Foale, P. G. Holborn, F. A. McRobie, J. Stark and J. M. T. Thompson. Thanks are also due to Derek Roberts, Provost of University College London, and to Jim Croll, Head of the Department of Civil and Environmental Engineering, for their encouragement and support; to Anne Power for her invaluable work on the local organization; and to Margaret Thompson for her cheerful and enthusiastic help with the social programme.

This book contains a specially written intoduction to the subject area (Chapter 1); the full texts of all the lectures presented at the Symposium; abstracts of the posters in Appendix I; the names and addresses of all participants in Appendix II; and a comprehensive index. We are pleased to have had it so efficiently and attractively produced by John Wiley & Sons Ltd.

Michael Thompson (Director)
Steven Bishop (Manager)
Centre for Nonlinear Dynamics and its Applications
Civil Engineering Building
University College London
Gower Street
London WC1E 6BT
UK

1 BASIC CONCEPTS OF NONLINEAR DYNAMICS

J. M. T. Thompson

1.1 INTRODUCTION

Like any new scientific discipline, the new geometrical theory of *nonlinear dynamics and chaos* has spawned a multitude of specialized concept and terminologies. These can be a major obstacle to applied scientists and engineers wishing to apply the powerful new methods in their own fields. To help overcome this, we provide here an overview that aims to highlight the central concepts and ideas that will be of particular importance in practical applications.

Recent books which the reader may find helpful are those of Guckenheimer and Holmes (1983), Thompson and Stewart (1986), Moon (1987), Arrowsmith and Place (1990) and Abraham and Shaw (1992). Collections of modern applications are edited by Schiehlen (1990), Thompson and Gray (1990), Kim and Stringer (1992), Thompson and Schiehlen (1992), and Mullin (1993).

1.2 DYNAMICAL SYSTEMS AND THE POINCARÉ SECTION

The general type of continuous dynamical system that will concern us here is described by an *autonomous* set of n first-order ordinary differential equations (ODEs),

$$\dot{x} = f(x) \tag{1}$$

giving a stationary *vector field* in the n-dimensional *phase space*, $\mathbb{R}^n$ say, spanned by the components of vector x. Non-autonomous equations, in which time, t, appears explicitly, can be rendered autonomous by identifying t as an extra phase coordinate governed by the dummy equation $\dot{t} = 1$. A driven mechanical oscillator can be put into the required form by identifying the velocity as a second phase coordinate and the time as a third. In a typical phase space the vectors vary smoothly with position, and trajectories are everywhere tangent to them. This leads naturally to the *Euler* time integration scheme. For a small time step, Δt,

Nonlinearity and Chaos in Engineering Dynamics
Edited by J. M. T. Thompson and S. R. Bishop,   © 1994 John Wiley & Sons Ltd

we can write $\Delta x = f(x)\Delta t$ allowing us to make a small finite step from point i to the next point $i + 1$ using

$$x_{i+1} = x_i + f(x_i)\Delta t. \tag{2}$$

An improvement of this basic scheme is the *Runge–Kutta* method which uses an Euler-type prediction followed by corrections to achieve higher-order accuracy. The trajectories fill the phase space to form a *phase portrait*. In a dissipative system this portrait will show the structure of the attractors and basins, and is sometimes called the attractor–basin phase portrait.

A discrete dynamical system is described by the iterated *map* (or *mapping*),

$$x_{i+1} = F(x_i). \tag{3}$$

The Euler time integration of a continuous system does, for example, generate a discrete system of this type. More globally, continuous dynamical systems on $\mathbb{R}^n$ are formally reduced to a mapping of dimension n-1 by the use of a Poincaré section.

Such a section transverse to the flow of a continuous system on $\mathbb{R}^n$ generates a Poincaré mapping on $\mathbb{R}^{n-1}$, taking a point in the surface of section to its image upon first return to the section. The value of such a mapping lies in the fact that it captures the (attractor-basin) dynamics of the system, and has the same general stability properties as the flow.

For a mechanical oscillator driven by a periodic excitation of period T, Poincaré sections can be defined, most simply, by the planes $t = iT$ where $i = 1, 2, 3, \ldots$. This corresponds to the *stroboscopic sampling* of the velocity and displacement. It should be emphasized, however, that alternative Poincaré sections are often advantageous: in impacting systems, for example, it can be useful to work with the 2D impact map whose coordinates are the phase and velocity sampled at impact (Foale and Bishop, 1992).

The Poincaré mapping of a smooth continuous dynamical system will typically be a *diffeomorphism*, namely a smooth differentiable one-to-one mapping with a unique and smooth differentiable inverse.

1.3   DIVERGENCE, DISSIPATION AND RECURRENT BEHAVIOUR

The non-crossing trajectories of a continuous system give a fluid-like *flow* in the phase space. Writing the set of n first-order ODEs in scalar form as

$$\dot{x}_i = f_i(x_j) \tag{4}$$

we have the important scalar divergence,

$$\mathrm{div}(x_i) = \partial f_1 / \partial x_1 + \partial f_2 / \partial x_2 + \cdots + \partial f_n / \partial x_n. \tag{5}$$

The rate of change of a small volume, V, of the phase 'fluid' is given by

$$\dot{V}(t)/V(t) = \mathrm{div}(x_i). \tag{6}$$

The analogous result for the two-dimensional mapping

$$x_{i+1} = G(x_i, y_i), \quad y_{i+1} = H(x_i, y_i) \tag{7}$$

gives us the ratio of small areas

$$A_{i+1}/A_i = D = (\partial G/\partial x)(\partial H/\partial y) - (\partial G/\partial y)(\partial H/\partial x) \tag{8}$$

where D is the Jacobian determinant.

A conservative, autonomous mechanical system with no energy dissipation is called a *Hamiltonian* system. Its equations of motion can be written in terms of the Hamiltonian function, $\mathscr{H}$, (numerically equal to the sum of the kinetic and potential energies) as

$$\dot{q}_i = \partial \mathscr{H}/\partial p_i, \quad \dot{p}_i = -\partial \mathscr{H}/\partial q_i \tag{9}$$

where q_i are the r generalized coordinates and p_i the generalized momenta. This canonical form shows immediately that a Hamiltonian system has an identically zero divergence function on $\mathbb{R}^{2r}$, this result being known as *Liouville's theorem*. The Hamiltonian flow is thus akin to that of an incompressible fluid. In the wider context of non-mechanical systems, not all systems with an identically zero divergence can be reduced to this classical canonical form: we can refer to these, more generally, as *volume-preserving* systems.

Dissipation of energy tends to give a negative divergence to the flow. Consider for example the driven oscillator

$$\ddot{x} + b(\dot{x}) + c(x) = F\sin(\omega t) \tag{10}$$

which we reduce to the first-order form

$$\dot{x} = y, \quad \dot{y} = -c(x) - b(y) + F\sin\theta, \quad \dot{\theta} = \omega \tag{11}$$

to obtain for the (x, y, θ) phase space the divergence function

$$\mathrm{div}(x, y, \theta) = -\mathrm{d}b/\mathrm{d}y = -b_y(y). \tag{12}$$

We see that the divergence is just a function of y, and governed only by the dissipation function $b(y)$: the sinusoidal forcing does not appear in it; nor does the restoring force, so that even in the vicinity of an unstable hilltop, with for example $c(x) = -x$, the sign of the divergence depends only on the form of $b(y)$. The (x, y, θ) phase space of such a periodically driven oscillator can be usefully viewed in the toroidal space $\mathbb{R}^2 \times S^1$, product of the plane $\mathbb{R}^2$ and the circle S^1.

We use the adjective *dissipative* to describe any system that does not have an identically zero divergence. Often a system so described will be totally dissipative, in the sense that the divergence function is everywhere negative. This would be the case with *Duffing's equation*, describing a driven oscillator with a cubic or polynomial restoring force, but with simple, positive linear damping corresponding to an energy sink. But we also encounter systems in which the phase space might have regimes of positive divergence, containing for example a repellor. This arises in the *van der Pol equation* of an oscillator with a nonlinear damping characteristic such that the autonomous system is capable of sustained self-excited oscillation in a limit cycle. The energy source for such behaviour is typically provided by a fluid flowing over an elastic structure. Similar results and subdivisions according to the divergence properties apply to iterated mappings, and we shall focus most of our attention on dissipative flows and maps.

In a phase space regime of negative divergence, a cloud or ensemble of starts will shrink asymptotically onto an attracting set of zero volume. Setting $\text{div}(x_i)$ equal to a constant, $-k$, in equation (6) gives, for example,

$$V(t) = V(0)\exp(-kt) \tag{13}$$

A typical start within this cloud will experience a *transient* before settling asymptotically onto a stable *steady-state* solution, called an *attractor*. Such a post-transient set can be a point attractor, a periodic or quasi-periodic attractor, or a chaotic attractor. Generically each attractor is entirely surrounded in phase space by its own basin of attraction. All transients initialized in a small neighbourhood around the attractor move back to it, making it *asymptotically stable* in the local sense of Lyapunov. All the above attractor types can appear alternatively as unstable steady states giving the saddles and repellors that we discuss later.

To distinguish these steady-state attractors, saddles and repellors from transients, geo-metrical dynamics uses the concept of a *recurrent state*. A particular state of a dynamical system is deemed recurrent if, after sufficient time, the system returns arbitrarily close to the state. The relaxation of the definition away from precise repetition is here used to embrace quasi-periodic and chaotic motion as recurrent. An ensemble of recurrent states linked together by a single trajectory constitutes recurrent behaviour. A further relaxation is to the *non-wandering* state which is one that has arbitrarily close states that return arbitrarily close. This generalization of a recurrent state (any recurrent state is non-wandering, but not vice versa) is needed to embrace a homoclinic orbit.

1.4   POINT, PERIODIC AND QUASI-PERIODIC ATTRACTORS

An equilibrium or fixed point, x_e, of (1) is characterized by

$$\dot{x} = f(x_e) = 0. \tag{14}$$

It can be stable or unstable, and is the first (trivial) form of recurrent behaviour. If it is asymptotically stable it is a *point attractor*.

A local stability analysis of any fixed point starts with the linearized equations describing small variations about the point. For a fixed point of a flow, stability hinges on the signs of the real parts of the eigenvalues. Typically the point will be hyperbolic (non-critical) with no zero real parts and we then have: the necessary and sufficient condition for stability is that all signs be negative; the necessary and sufficient condition for instability is that at least one sign be positive.

A fixed point of a flow (or map) that has all its linear eigenvalues in the stable or unstable domains is called *hyperbolic*. There are then no critical eigenvalues corresponding to neutral stability, and the phase portrait around the fixed point is structurally stable against perturbations of the system. (We should note that the term *hyperbolic point* is used differently in the literature on Hamiltonian systems to mean a saddle, near which trajectories follow a roughly hyperbolic shape.) Fixed points of a Hamiltonian system (or any volume-preserving system) can be at most neutrally stable with all local trajectories staying close, though not returning to, the point.

In the phase space of a flow, a closed orbit satisfying recurrence by returning precisely to its starting point after its periodic time T, is called a *periodic motion*. Such a motion (not normally sinusoidal) is called a *harmonic* oscillation if it has the same period as the driving or sampling: the more restrictive expression, *simple harmonic motion*, is reserved for sinusoidal behaviour. Correspondingly, a *subharmonic* oscillation of order n is a periodic oscillation with period n times that of the driving or sampling: so a harmonic oscillation can be thought of as a subharmonic of order $n = 1$.

If a periodic motion is asymptotically stable, it is called a *periodic attractor*, or *limit cycle*. The stability is best assessed using a Poincaré section in which the periodic motion will appear as a *fixed point* of the map (3) characterized by $x_{i+1} = x_i$. Such a fixed point, stable or unstable, is the simplest (trivial) form of recurrent behaviour for an iterated mapping. Its local stability hinges on the moduli of the linear eigenvalues. If the point is hyperbolic with no modulus equal to unity, we have: the necessary and sufficient condition for stability is that all moduli be less than unity; the necessary and sufficient condition for instability is that at least one modulus is greater than unity.

A periodic orbit in $\mathbb{R}^3$ forms a closed circuit, which might be a simple loop like a rubber band or might form a *knot*: a pair of such orbits might be linked, as in a chain. The uniqueness of the flow means that orbits cannot cross, so a knot or a link cannot be eliminated even under the variation of a control parameter. A knowledge of the knot-types and linkages of a set of periodic orbits is therefore useful (Ghrist and Holmes, 1993; McRobie and Thompson, 1993), and can in particular be used to establish constraints and precedences on possible bifurcations in driven oscillators (McRobie, 1992a). Two orbits cannot, for example, merge at a saddle-node fold bifurcation if one is knotted while the other is unknotted: neither can they merge if one, but not the other, is linked to a third orbit. To establish the orbit topology for a driven oscillator, the time history $x(t)$ can be interpreted as a *braid* diagram, as in Figure 1.1: the orbit structure in the full 3D phase space of $(x, \dot{x}, t)$ is easily deduced from it because when two time histories cross it is obvious which has the greater $\dot{x}$. Similar deductions cannot be made from the $(x, \dot{x})$ phase projection.

Consider a motion which has one periodic component of period T_1 and a second of period T_2. The first component repeats after time $t = nT_1$ while the second repeats after $t = mT_2$ where n and m are positive integers. If we can find a (lowest) time for which $t = T = nT_1 = mT_2$ then the composite motion is periodic with period T. If, however, T_1/T_2 is irrational (that is to say T_1 and T_2 are *incommensurate*) then no such time can be found: the composite motion never precisely repeats itself, and is declared *quasi-periodic*. Over a long time scale there will however be arbitrarily close repetition, allowing the motion to be declared recurrent. If the motion is asymptotically stable we have a *quasi-periodic attractor*. Quasi-periodic motions with just two incommensurate periods can be visualized as filling out a torus in a 3D phase space. In strongly coupled systems, quasi-periodic attractors with more than two incommensurate periods are unlikely to be observed experimentally because they are easily perturbed into chaotic attractors. The ratio of the periods, T_1/T_2, which measures the average number of orbits of period T_2 during one orbit of period T_1, is called the *rotation number* (winding number).

Arnold horns (tongues) are tongue-like regimes in the control space of generic two-frequency oscillators in which two competing periodic components of the response are mode-locked into periodic motion. Outside these regimes there is drift, corresponding, for example, to quasi-periodic motion. The tongue boundaries are cyclic folds. This mode-locking can be illustrated in the *circle map*, an archetypal one-dimensional mapping of the circumference of a circle onto itself. A similar mapping of a toroidal surface is the *standard map*, used to delineate the transition from regular to chaotic motion in Hamiltonian systems.

Two linked orbits: $x = \cos t, \quad x = 0.5 \cos t, \mod 2\pi$

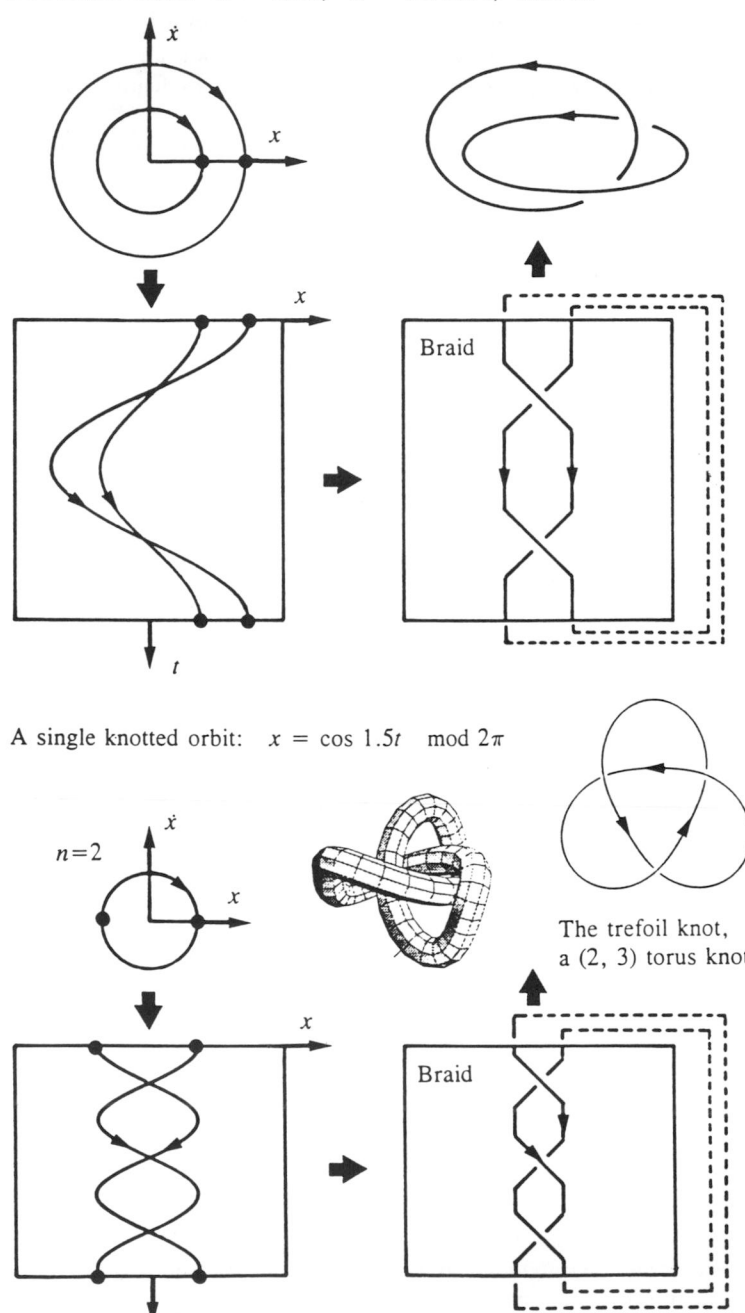

A single knotted orbit: $x = \cos 1.5t \mod 2\pi$

The trefoil knot,
a (2, 3) torus knot

Figure 1.1 (a) Imagining them to be steady-state responses of an oscillator driven at period 2π, the sketches show how a braid analysis of two simple sinusoidal wave-forms, $x(t)$, identifies them as linked orbits in the toroidal space $(x, \dot{x}, \theta)$, where θ is the phase angle. (b) In a similar way, a single subharmonic of order $n = 2$, with rotation number 2/3, is shown to be a (2, 3) torus knot (a trefoil) in $\mathbb{R}^2 \times S^1$.

1.5   SADDLES AND REPELLORS

A recurrent motion is unstable if any single adjacent motion moves out of its immediate phase-space neighbourhood. The *index of instability* of a regular hyperbolic state is the degree of instability based on the linear eigenvalues. For a fixed point of a flow it is the number of eigenvalues with a positive real part: for a fixed point of a map it is the number of eigenvalues with modulus greater than one.

If the index of instability is zero, the state is a stable attractor. If the index lies between zero and the phase dimension, we have an unstable *saddle* solution that repels in some directions but attracts in others. Such saddles, which can be fixed, periodic, quasi-periodic or, more generally, chaotic motions, are typical ingredients of dissipative phase spaces. Although, being unstable, they are not directly observable in a physical system, they play a key role in structuring the phase space. In particular, saddles with index one, having a single unstable direction, frequently lie in and organize basin boundaries.

If the index is equal to the phase dimension, we have a *repellor*, corresponding to fixed, periodic, quasi-periodic or chaotic motions. These repel all adjacent motions like a fluid source. Such a state can only be observed in a phase regime of positive divergence, associated for example with an energy source. It should be noted that saddles may become repellors if the dynamics is projected down to a lower dimension; on the centre manifold of a saddle-node bifurcation, for example, the saddle will appear as a one-dimensional repellor.

For a vector field on $\mathbb{R}^2$, the *Poincaré index* of a fixed point is the net rotation of the vector field direction along a nearby closed path encircling the point. The index of an attractor or repellor is $+1$, while the index of a saddle is -1. This index concept can be generalised to maps and to higher dimensional phase spaces. Topological theorems give global constraints on the number and type of fixed points based on the Poincaré index. This index, like the index of instability, is an example of a *discrete invariant*, whose integer value does not depend on the choice of phase space coordinates.

1.6   CHAOTIC MOTION AND ITS IDENTIFICATION

The last and most complex form of bounded post-transient recurrent behaviour in dissipative dynamics, after the point, periodic and quasi-periodic attractors, is the *chaotic attractor*. The first such attractor to be examined in detail arose in the *Lorenz equations*. These are an autonomous set of three first-order ODEs with simple nonlinearities, devised by Lorenz as a model of thermal convection in his atmospheric studies: they played a seminal role in the development of chaos theory.

The chaotic attractor exhibits a sensitive dependence on initial conditions, and, despite the absence of any stochastic forcing of the deterministic system, (1), a broad-band noisy power spectrum. Typically a fast contraction onto a sheet allows an exponential divergence of trajectories within the sheet, consistent with an overall volume contraction. The repeated folding of the sheet, as in the making of flaky pastry, gives the attractor a bounded fractal structure and induces a complex mixing of trajectories. The chaotic nature of an attractor can be established by examining the divergence of trajectories, quantified by Lyapunov exponents; its fractal dimension as assessed for example by the capacity dimension; and its bifurcational precursors such as a period doubling cascade.

In experimental and computational dynamics the term *strange attractor* is often used loosely and interchangeably with chaotic attractor. Strict mathematical definitions of a strange attractor are not always mutually consistent, but often demand that it should contain a transversal homoclinic orbit. Sometimes the term strange is taken to refer only to a fractal structure, which does not always imply sensitive dependence on initial conditions. *Chaos* is a generic term for the complex, seemingly irregular motions of deterministic systems characterized by a sensitive dependence on initial conditions and a broad-band spectrum. In dissipative dynamics it embraces chaotic attractors, saddles and repellors, as well as the chaotic transients observed in the vicinity of these recurrent states.

Chaotic saddles, which attract nearby trajectories in at least one direction and repel trajectories in at least one complementary direction, play an important role in the fractal basin boundaries of driven oscillators: they can collide with, and destroy, a chaotic attractor in a global bifurcation (Stewart, 1987). The recurrent *chaotic repellor* can be most easily analysed in a computer simulation by running time backwards so that it is transformed into a chaotic attractor. A *chaotic transient* is an irregular transient motion, observed in the vicinity of any of the chaotic recurrent states. It is often of particular importance when in the vicinity of a chaotic saddle which forms a fractal basin boundary: it can be of arbitrarily long duration, and leads eventually to an attractor of any type (point, periodic, etc.).

Under the slow variation of a control, a common route to a chaotic attractor is a cascade of *period-doubling* bifurcations, each of which is a super-critical flip generating a stable oscillation with twice the period of its precursor. In this classical cascade the bifurcations become more and more closely spaced, the ratio of successive parameter intervals tending in the limit to the Feigenbaum number $\delta_\infty = 4.669\ 20\ldots$. This number is universal (generic) in the sense that it arises in a very wide class of problems. The limiting ratio ensures that in such a *Feigenbaum cascade* the repeatedly doubled period quickly reaches infinity at an accumulation point in a finite parameter interval, after which a complex pattern of chaotic motions and periodic windows is observed.

The *logistic map*, a nonlinear one-dimensional mapping typically arising in population dynamics, forms a useful introduction to this period-doubling route to chaos. It is equivalent to the *quadratic map*, to which it can be transformed by a change of variable. A related 2D mapping, governed by two coupled difference equations with one quadratic nonlinearity, is the *Hénon map*: this can be seen as an embedding of the quadratic map, to which it degenerates as one of its parameters drops to zero.

The *Lyapunov exponents*, σ_i, of a trajectory measure the average long-term exponential rate of divergence of all adjacent motions based on

$$\sigma = \lim_{t \to \infty} (1/t)\ln[(\text{separation at } t)/(\text{separation at } 0)]. \qquad (15)$$

This limit must be taken only amongst trajectories whose final separation remains small. In the simplest case in which the fundamental trajectory is the trivial fixed point of the 3D linear(ized) flow $\dot{x}_i = \lambda_i x_i (i = 1$ to $3)$ with the ordered eigenvalues $\lambda_1 > \lambda_2 > \lambda_3$, starts on the unit sphere will be transported by the flow into an ellipsoid with principal semi-axes of lengths $\exp(\lambda_i t)$. The Lyapunov exponents are here simply the eigenvalues, $\sigma_i = \lambda_i$, with the sum, $\sigma_1 + \sigma_2 + \sigma_3$ equal to the divergence of the flow. This result holds for more general situations, so in a totally dissipative system for which the volume divergence is negative, the sum of the Lyapunov exponents of any trajectory will be negative, with however no such restriction on any single exponent. If the maximum exponent, σ_1, is positive then *some* adjacent trajectories will diverge from the fundamental; and there will be a sensitive dependence on initial conditions (albeit a trivial one for the above fixed point for which $\sigma_1 > 0$ implies instability). A positive Lyapunov exponent in a *bounded attractor* is a sign of chaotic motion.

Symbolic dynamics is the theory and manipulation of sequences of symbols, used for example in exploring the invariant sets of the *Smale horseshoe*, a 2D mapping introduced by Smale to explore the folding and mixing actions of chaotic motion. A simple example is the shifting of the decimal point in the bi-infinite sequence

$$\ldots\ldots010010.1110101011\ldots\ldots \tag{16}$$

This arises in the 1D mapping $x_{i+1} = 2x_i(\text{mod } 1)$, which describes, under a change of variable, the fully developed chaos at the end of the logistic map. Symbolic dynamics should not be confused with similar expressions describing automated algebraic manipulations on a computer.

A common diagnostic test used to establish the existence of a chaotic attractor relies on a proof of its fractal properties, to which we now turn.

1.7 FRACTALS AND THEIR DIMENSIONS

A *fractal*, or fractal set, has fine detail on all scales, the term being coined by Mandelbrot from the Latin *fractus*, meaning broken. There is as yet no agreed mathematical definition. Falconer (1990) suggests that the term is best used loosely for sets, often defined by a simple recursion, that have properties such as: fine structure, with detail on arbitrarily small scales; local and global irregularity which cannot be described by traditional geometry; self-similarity, perhaps approximate or statistical; a non-integer fractal dimension.

A simple example of a fractal is the *Cantor set*, namely an infinitely desiccated set such as the triadic Cantor set. The latter is generated by removing the middle third from the unit interval, and then progressively removing the middle thirds from the remaining intervals until in the limit we are left with just an infinity of disconnected points. Since the total length of line removed is

$$(1/3) + 2(1/3)^2 + 4(1/3)^3 + 8(1/3)^4 + \cdots = 1 \tag{17}$$

the length or measure of the remaining points is zero. The final set of points has the attributes of a fractal, with every magnification revealing more detail, which in this simple example is always self-similar. Because of their distribution in space, the points can be assigned a fractal dimension that we shall now evaluate.

The *fractal dimension* of an infinite set of points in $\mathbb{R}^n$ is a measure, non-integer and less than n, of the extent to which the points fill the space. Of the numerous definitions for determining such a measure, one of the most useful is the capacity dimension. A more mathematical measure is the *Hausdorff dimension*, which is sometimes numerically equal to the capacity dimension. A third is the *information dimension*, a kind of non-integer dimension in which points in a dynamical invariant set are weighted according to their relative probability of occurrence in a typical long trajectory.

The *capacity dimension* (box dimension) is based on the covering of a set, as we shall illustrate using cubes of side ε to cover an object in $\mathbb{R}^3$. The number of cubes, $N(\varepsilon)$, needed to cover a point is 1, so $N(\varepsilon)$ here scales as ε^0. The number to cover a line of length L is of order L/ε, scaling as ε^{-1}, while the number to cover a surface of area A is of order A/ε^2, scaling as ε^{-2}. We thus define the capacity dimension, d, such that $N(\varepsilon) \propto \varepsilon^{-d}$, giving

$$d \equiv \lim_{\varepsilon \to 0} \log N(\varepsilon)/\log(1/\varepsilon). \tag{18}$$

Applied to the triadic Cantor set in a plane, we see that the unit interval is covered by one square of unit side ($N = 1$, $\varepsilon = 1$): after the first middle-third removal we can cover with two squares of side one-third ($N = 2$, $\varepsilon = 1/3$): followed by $N = 2^2$, $\varepsilon = (1/3)^2$, with the general result $N = 2^i$, $(1/\varepsilon) = 3^i$. Using these differently sized squares to cover the *final* limiting set, we evaluate its capacity dimension as

$$d = \log 2/\log 3 = 0.630\,92\ldots . \tag{19}$$

This is a non-integer result between that of a point (0) and a line (1). This figure is also obtained for the triadic set if we evaluate its Hausdorff dimension but such agreement is not guaranteed for more complex sets. The capacity dimension of a set F can be alternatively assessed by determining: the smallest number of closed balls of radius ε that cover F; the smallest number of cubes of side ε that cover F; the number of ε-mesh cubes that intersect F; the smallest number of sets of diameter at most ε that cover F; the largest number of disjoint balls of radius ε with centres in F (Falconer, 1990, p. 41).

The capacity dimension (like a Lyapunov exponent) is an example of a *continuous invariant*, whose numerical value from the continuum of real numbers does not depend on the choice of phase space coordinates.

1.8   BASINS AND THEIR BOUNDARIES

The phase space of a dissipative system on $\mathbb{R}^n$ decomposes into basins of attraction bordered and separated from each other by $(n - 1)$-dimensional basin boundaries. Trajectories initialized within a basin tend asymptotically, as $t \to +\infty$, to the attractor lying within the basin. Trajectories initialized precisely on a basin boundary (*separatrix*) often flow towards a saddle solution which attracts within the boundary but repels across it. Such a saddle can be located numerically using the *straddle-orbit technique*: once found, the complete boundary can be located by running time back-wards on a computer from starts close to the saddle.

Trajectories that flow or map towards a saddle as $t \to +\infty$, approaching it tangentially along the stable incoming eigenvectors, are called the *stable manifolds* or *insets*. Similarly, trajectories asymptotic to a saddle as $t \to -\infty$ are called the *unstable manifolds* or *outsets*. Insets and outsets are known collectively as *outstructures*, and being invariant with time they are *invariant manifolds*. The insets are particularly important in organizing a phase portrait, and often form basin boundaries. They can be located computationally by running time backwards from close to the saddle. In a 2D flow, one start displaced a small distance along an incoming eigenvector will be enough to locate the corresponding inset. In a 2D map, a ladder of starts along the eigenvector will be needed to fill out the whole manifold: a complete ladder, forming a *fundamental neighbourhood* in which the bottom rung maps to the top rung, ensures that the inset is properly filled with computed points.

In a 2D dissipative map, a basin boundary formed by a saddle inset will become fractal when, after a homoclinic tangency, the saddle develops a homoclinic tangle (McDonald *et al.*, 1985). A non-integer fractal dimension can then be estimated for this *fractal basin boundary.* Such fractal boundaries are often observed in driven oscillators, where they can give rise to the phenomenon of basin erosion, implying a sudden loss of integrity of a system in a noisy environment (Thompson, 1989; Thompson and Soliman, 1990). Practical applications of this arise in ship capsize studies (Thompson *et al.*, 1990, 1992).

1.9 CONTROLS, STRUCTURAL STABILITY AND BIFURCATION

Parameters, μ, in the functions governing a dynamical system

$$\dot{x} = f(x, \mu) \tag{20}$$

are understood to remain constant during the dynamical motions of the system. They are, however, imagined to be under the control of an external agent who can give them very slow quasi-static variation, thereby driving the system slowly from one parameter regime to another; in this way bifurcations can be realised. An example would be a human controller slowly varying a throttle or rheostat setting in a laboratory experiment. The conceptual multi-dimensional space whose coordinates are the control parameters is the *control space.* The conceptual $n + m$ dimensional space spanned by the n phase coordinates and the m control parameters is called the *phase-control space.* Under one control ($m = 1$), paths of attractors will be observable in this space: under two controls, these become attractor surfaces. Computer techniques for following paths in phase-control space, and bifurcation arcs in control space are described by Parker and Chua (1989) and Foale and Thompson (1991).

Because the parameters of a physical system are never known precisely, and may be subject to small variation, it can be argued that a good mathematical model should have a phase portrait that is robust against small changes in the model itself. Small changes of the parameters and functions of the model should not qualitatively change the topology of the phase portrait. A phase portrait that is robust in this way is declared *structurally stable*, and the system is sometimes called a *coarse system* following the terminology of Andronov. Structurally unstable portraits, encountered under the variation of a control parameter, signal a bifurcation, and an extension of the concept of structural stability to the phase-control space gives us the idea of co-dimension described below. A precise phase-space definition of structural stability appropriate to chaotic and homoclinic structures is not yet agreed upon.

A *bifurcation* is a qualitative change in the topology of the attractor-basin phase portrait, realisable under the quasi-static variation of a single control parameter μ as it crosses a critical value μ_c. The phase portrait at μ_c is structurally unstable. Local bifurcations, restricted to a small neighbourhood of phase space, involve the creation, destruction and splitting of attractors. Global bifurcations involve connections between the outstructures of saddles, often producing abrupt changes in the structure of the attractors and/or their basins.

Generic bifurcations observed under variation of one control lie on surfaces of dimension $m - 1$, and are said to have co-dimension one. A bifurcational event in phase-control space is declared structurally stable if the whole event is topologically robust against additional perturbations of the system. In the manner popularised by *catastrophe theory*, bifurcations that are not structurally stable under one control can sometimes be made structurally stable (unfolded) by embedding in a higher dimensional control space. A typical example of this is the symmetry-breaking imperfection needed to *unfold* a symmetric static pitch-fork into the co-dimension-two cusp. The *codimension, c,* of a bifurcation is the number of control parameters needed to achieve this structural stability.

1.10 LOCAL BIFURCATIONS OF EQUILIBRIA AND CYCLES

A system under increasing μ can lose its stability at a local bifurcation when a fundamental primary path (of fixed points or cycles) becomes unstable at $\mu = \mu_c$. This can happen either when the path reaches a maximum value of the control parameter at a *fold* or when the path bifurcates and throws off a secondary path. In the latter case, the form and stability of the secondary path gives us the following three-fold classification. At a *super-critical* bifurcation the continuous fundamental path loses its stability as it intersects a stable secondary path that only exists for $\mu > \mu_c$. At a *sub-critical* bifurcation the fundamental path loses stability as it intersects an unstable secondary path that only exists for $\mu < \mu_c$. The third form, that arises in the paths of fixed points, is the *trans-critical* bifurcation. Here the fundamental path destabilizes as it intersects a continuous secondary path of fixed points which exists at sub- and super-critical values of μ, being unstable for $\mu < \mu_c$ and stable for $\mu > \mu_c$. This *exchange of stability* was discussed by Poincaré, and is the asymmetric bifurcation encountered in the elastic stability of solids and structures: it is essentially a pathological form of the saddle-node.

A local bifurcation involving the loss of stability of an attractor will have one or more critical eigenvalues for which the response is linearly neutral, being neither attracting nor repelling within the linear approximation. All the essential nonlinear bifurcational behaviour can be viewed in the reduced space defined by a curved *centre manifold* originating from the associated critical eigenspace. Similar reductions are achieved by the elimination of passive coordinates, the Fredholm alternative, and the Lyapunov–Schmidt scheme.

A local bifurcation is observed on an equilibrium path of a flow as the real part of a linear eigenvalue passes through zero. If a real eigenvalue passes through zero, we have a *saddle-node* bifurcation. This typically manifests itself as a *fold*, the simplest $c = 1$ bifurcation in which a control-phase path of fixed points reaches an extreme value, say a maximum, μ_c, of a control parameter μ. As μ is increased towards μ_c a saddle and a node coalesce parabolically; there is locally no fixed point for $\mu > \mu_c$, forcing the system to jump dynamically to a distant, unrelated attractor. In the presence of symmetry, or other constraints, the saddle-node can however manifest itself as a sub-, super- or trans-critical bifurcation in which a secondary equilibrium path bifurcates from a monotonically increasing fundamental path. These pathological bifurcating forms are common in elastic stability theory, and are rendered structurally stable by the addition of a second symmetry-breaking control parameter, typically an imperfection.

If the real part of a complex conjugate pair of flow eigenvalues passes through zero, we have the *Hopf* bifurcation in which a secondary path of cycles bifurcates from the monotonic equilibrium path as it loses its stability. Depending on the sign of a nonlinear coefficient, this secondary path can be stable with $\mu > \mu_c$, giving us the super-critical Hopf bifurcation: conversely, it can be unstable with $\mu < \mu_c$, giving the sub-critical form. The Hopf bifurcation has $c = 1$, being structurally stable under one control: it cannot be destroyed by a symmetry-breaking imperfection, as can the symmetric static bifurcations.

A local bifurcation can analogously be observed on a path of fixed points of a map, representing perhaps a path of cycles in a stroboscopically sampled flow. A loss of stability will now correspond to a mapping eigenvalue passing through the unit circle. If a real eigenvalue passes through $+ 1$ we have a saddle-node, which typically manifests itself as a *cyclic fold* with $c = 1$ and features similar to the equilibrium fold, generating in particular an inevitable dynamic jump to a remote attractor. If a real eigenvalue passes through $- 1$ we have a $c = 1$ *flip* at which a secodary path of period-doubled fixed points bifurcates off the monotonic fundamental path as it loses its stability. Depending on the sign of a nonlinear coefficient, this secondary path can be stable with $\mu > \mu_c$, giving us the super-critical flip: conversely, it can be unstable with $\mu < \mu_c$, giving the sub-critical form. A flip bifurcation in the Poincaré mapping of a flow generates secondary cycles with twice the period of the fundamental cycle. If a complex conjugate pair of eigenvalues penetrates the unit circle we have the *Neimark* bifurcation. This is the mapping equivalent of the Hopf bifurcation, and is often called a *secondary Hopf* bifurcation. Quasi-periodic motions on a torus typically bifurcate off the fundamental path of cycles, in either super-critical or sub-critical form.

1.11 HOMOCLINIC AND HETEROCLINIC GLOBAL BIFURCATIONS

The adjective *global* is used to describe bifurcations and other phenomena that are not essentially described in a local region of phase space. Global bifurcations often involve homoclinic and heteroclinic connections between the outstructures of saddles, producing changes in the basin structure of the phase portrait: they are important features governing the dynamics of driven oscillators (Thompson and McRobie, 1993). Temporal *intermittency* is a complex steady-state motion involving sudden irregular switching from one form of behaviour to another, which is typically observed close to certain types of global bifurcation.

In a 2D flow the outset of a saddle can return as its inset, forming a *homoclinic connection*, also called a *saddle connection*. In a dissipative system such a connection is structurally unstable: it is typically encountered under the sweep of a single control, μ, identifying it as a $c - 1$ global bifurcation. The introduction of periodic forcing typically thickens this single event into a train of events in a 2D map that starts with a homoclinic tangency and gives rise to a homoclinic tangle. On the other hand, higher-dimensional flows also commonly exhibit homoclinic connections, in which an outset lies inside an inset (or vice versa) at a single parameter value.

In 3D flows a structurally-unstable homoclinic connection may involve a fixed point with a 1D outset lying inside a 2D inset with complex conjugate eigenvalues. This is the *Shilnikov* homoclinic connection, named after Shilnikov who showed that if the real eigenvalue is larger in magnitude than the real part of the complex eigenvalues, the existence of persistent horseshoes can be inferred for a nearby interval of parameters.

In a 2D dissipative map the outset of a saddle can touch its inset in a *homoclinic tangency*, a global $c = 1$ bifurcation often encountered in the Poincaré mapping of a driven oscillator as illustrated in Figure 1.2. This sketch is based on our extensive studies of the *escape equation*,

$$\ddot{x} + \beta\dot{x} + x - x^2 = F\sin\omega t \tag{21}$$

governing the escape from the generic cubic potential well

$$v(x) = (1/2)x^2 - (1/3)x^3. \tag{22}$$

As a simple consequence of the forward and backward mapping along the manifolds, when they touch once, they must touch simultaneously an infinite number of times. Infinite sets of stable periodic orbits, called *Newhouse sinks*, were shown by Newhouse to exist at certain parameter values close to those at which such a homoclinic tangency occurs. Sweeping μ

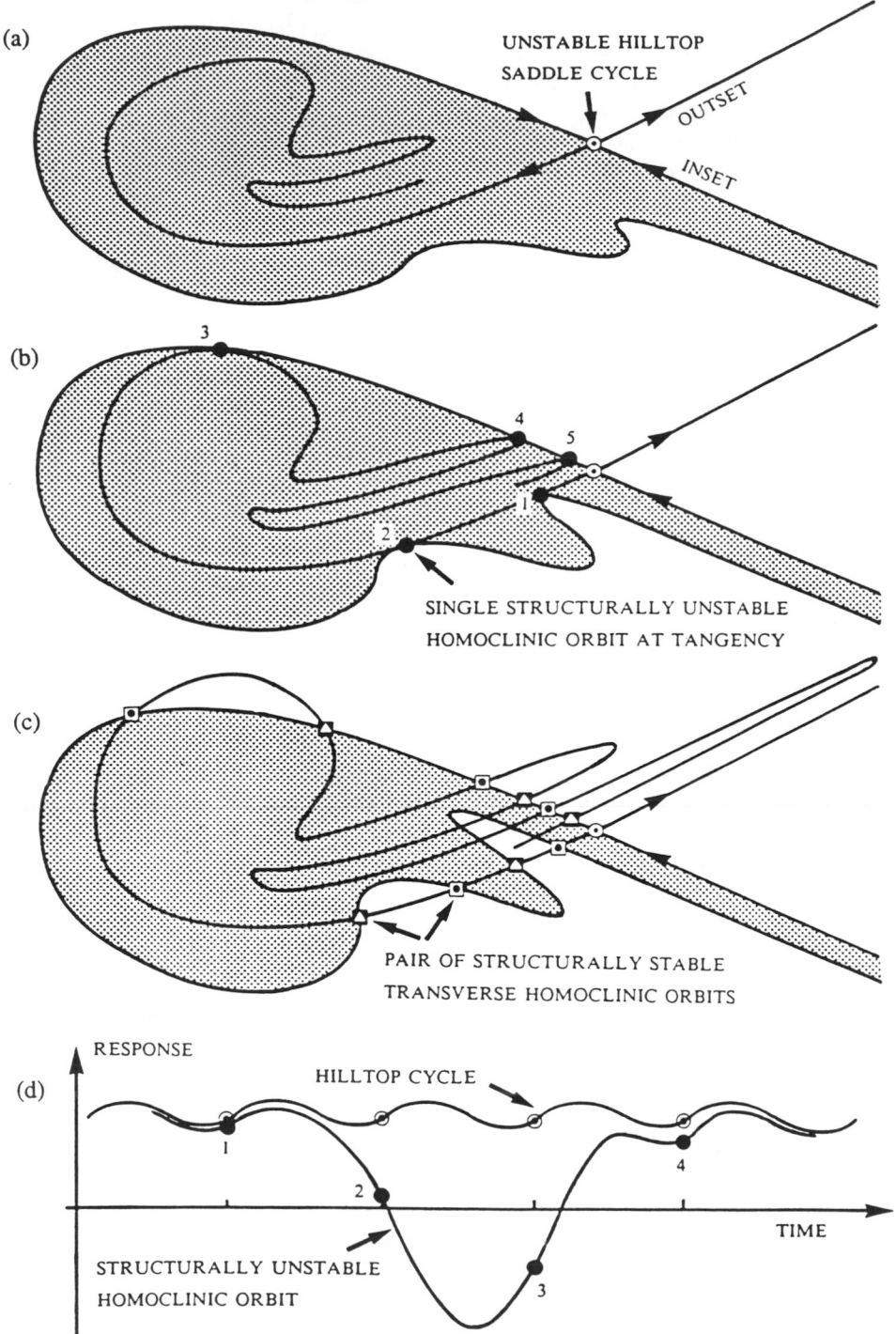

(a) UNSTABLE HILLTOP
SADDLE CYCLE

OUTSET

INSET

(b)

3

4
5
1
2

SINGLE STRUCTURALLY UNSTABLE
HOMOCLINIC ORBIT AT TANGENCY

(c)

PAIR OF STRUCTURALLY STABLE
TRANSVERSE HOMOCLINIC ORBITS

(d) RESPONSE

HILLTOP CYCLE

1

2

4

3

TIME

STRUCTURALLY UNSTABLE
HOMOCLINIC ORBIT

Figure 1.2 The initial homoclinic tangency of the inset and outset of the hill-top saddle cycle of the escape equation (21) in a stroboscopic Poincaré section. Three sketches before, at and after the tangency show how the grey safe basin acquires a fractal boundary. The lower picture shows the time history of the homoclinic orbit (McRobie, 1992b).

(a) <u>BASIC SIGNATURE AFTER INITIAL HOMOCLINIC TANGENCY</u>

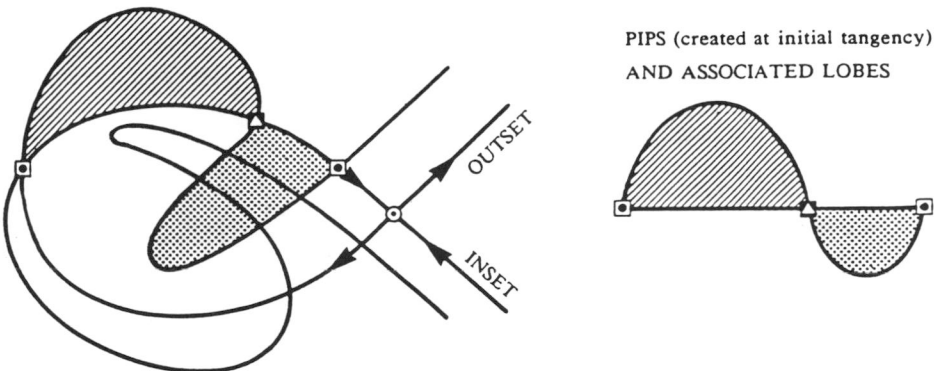

PIPS (created at initial tangency)
AND ASSOCIATED LOBES

(b) <u>NEW SIGNATURE AFTER INTERNAL HOMOCLINIC TANGENCY</u>

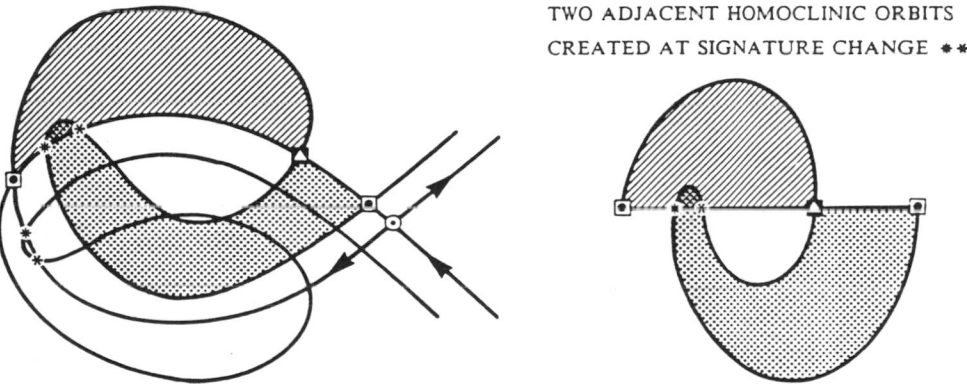

TWO ADJACENT HOMOCLINIC ORBITS
CREATED AT SIGNATURE CHANGE ✱✱

(c) <u>THE TIME HISTORY OF THE SIGNATURE CHANGE ORBIT</u>

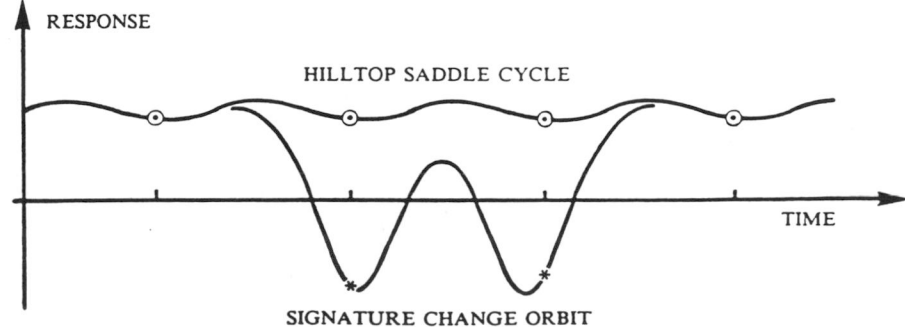

Figure 1.3 A schematic diagram showing the first period-one Birkhoff signature change of the escape equation (21) in two stroboscopic Poincaré sections. The lower picture shows a time history of the signature-change orbit (McRobie, 1992b).

16

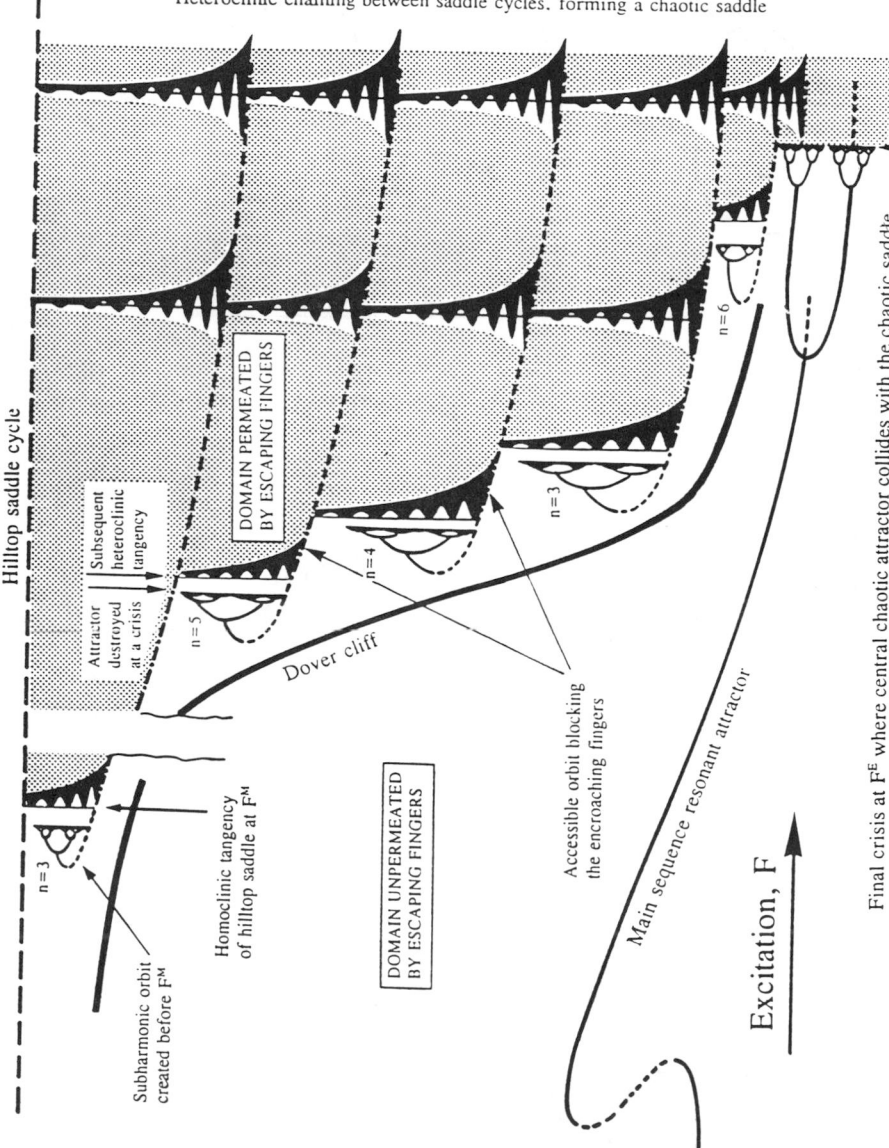

Figure 1.4 A schematic diagram, showing the series of basin implosions, associated with discontinuous changes of the accessible orbit, during the Dover cliff erosion process. The diagram is based on numerical results of Lansbury *et al.* (1992) for the twin-well Duffing oscillator.

beyond the first tangency generates a *homoclinic tangle* in which the inset and outset of the saddle will have an infinite number of structurally stable transverse intersections. This implies complex dynamics, associated with the formation of a horseshoe: and the inset, which may represent a basin boundary, will now have a fractal nature.

The *lobes* of such a tangle are amenable to analysis, and can provide useful information about the global dynamics of a system (McRobie and Thompson, 1991, 1992). As the tangle develops under further increase of μ, new tangencies between the inset and outset will herald changes in the *Birkhoff signature* of the manifolds (Figure 1.3): one such change can be used to predict the loss of stability of the chaotic attractor governing the escape from a potential well (McRobie, 1992b).

Melnikov methods can estimate the parameter values at which homoclinic and heteroclinic tangencies occur. Using a perturbation from an underlying integrable system (often a Hamiltonian flow) for which an analytical closed-form solution is available, the Melnikov function provides a measure of the distance between the relevant inset and outset. For driven oscillators, the Melnikov method can be viewed as a simple energy balance, in which the energy lost due to damping is equated to the energy input from the sinusoidal forcing, as the system is assumed to move around the saddle connection of the unperturbed Hamiltonian orbit (McRobie and Thompson, 1994).

In a 2D flow the outset of a saddle can be the inset of a second remote saddle forming a *heteroclinic (saddle) connection*, which in a dissipative system is another example of a global $c = 1$ bifurcation. Higher-dimensional flows also commonly exhibit heteroclinic connections, in which an outset lies inside an inset (or vice versa) at a single parameter value. Addition of periodic forcing to the 2D dissipative flow typically thickens the distinct connection into a train of events in a 2D map that starts with a $c - 1$ *heteroclinic tangency*, at which the manifolds touch simultaneously an infinite number of times. Sweeping μ beyond the first tangency generates a *heteroclinic tangle* in which the inset and outset have an infinite number of structurally stable transverse intersections. Unlike a homoclinic tangle, this is not by itself sufficient to generate fractal geometry.

A heteroclinic tangency has been shown to play a key role in the rapid *Dover cliff* erosion of fractal basins in driven oscillators (Soliman and Thompson, 1992c). The detailed mechanics of the erosion process (Figure 1.4) involve a growing *heteroclinic chain* between a succession of *accessible* subharmonic orbits (Grebogi *et al.*, 1987) as mapped out for the twin-well Duffing oscillator by Lansbury *et al.* (1992).

1.12   BIFURCATIONAL CLASSIFICATION

We can classify the $c = 1$ bifurcations of attractors according to the response that would be observed physically if a control parameter were slowly varied to sweep the system through the bifurcation in the direction that generates instability or increased complexity (Thompson *et al.*, 1994). *Subtle* (i.e. continuous) bifurcations are the local super-critical bifurcations with the continuous growth of a new attractor path. They are safe with no fast dynamic jump or instantaneous enlargement of the attracting set. They are determinate with a single outcome, and generate no hysteresis on reversal of the control sweep. The *catastrophic-explosive* bifurcations (our first sub-division of the discontinuous bifurcations) are global events with a sudden, instantaneous enlargement of the attracting set, with however no jump to a remote disconnected attractor. They are determinate in outcome, and are reversible with no

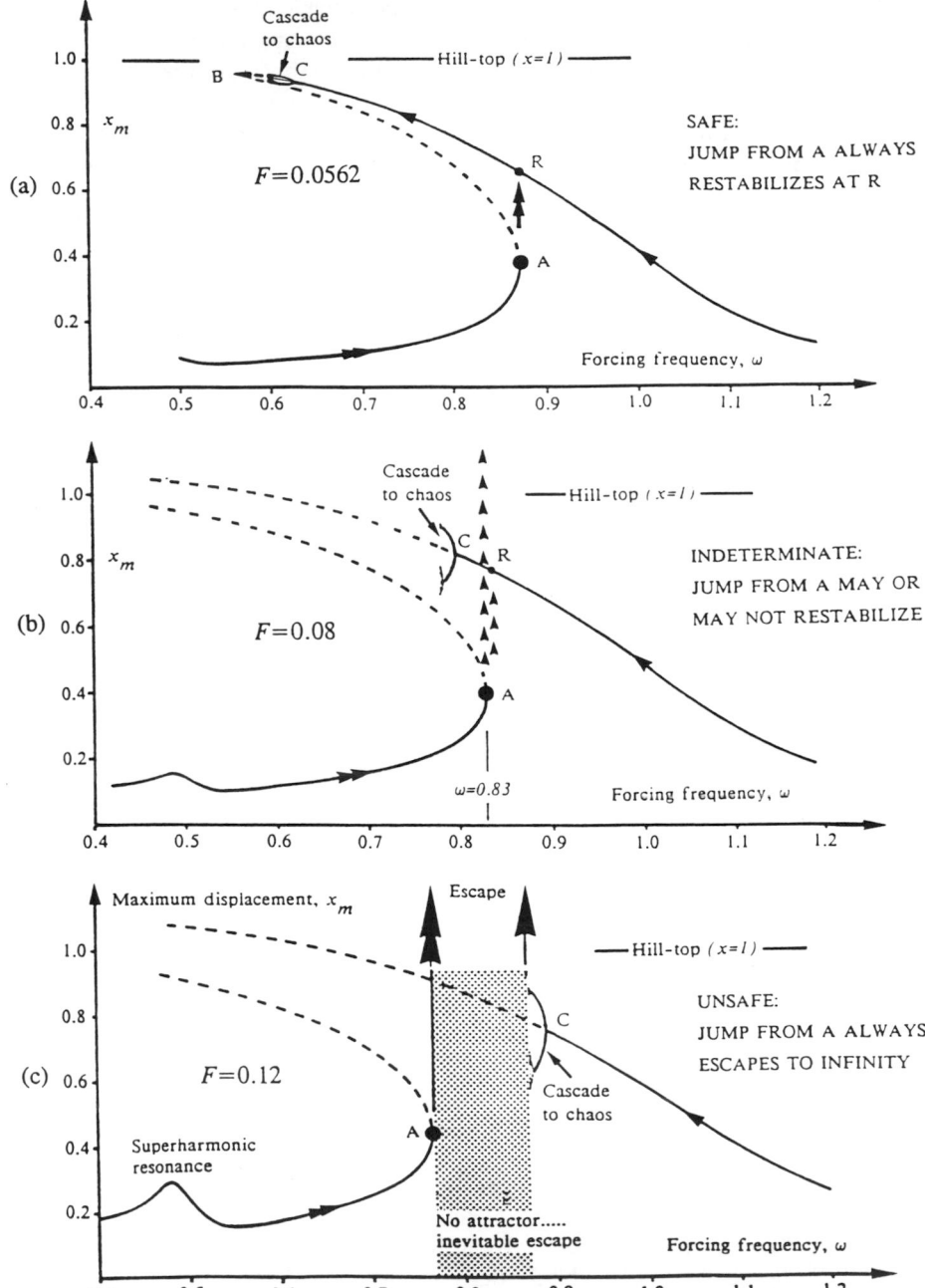

Figure 1.5 Safe indeterminate and unsafe jumps to resonance for the escape equation (21) at $\beta = 0.1$. Stable steady-state responses are indicated by solid curves, unstable responses by broken curves. Time histories corresponding to the indeterminate jump of the centre diagram are given by Thompson and Soliman (1991).

hysteresis. The *catastrophic-dangerous* bifurcations (our second sub-division) have a blue-sky disappearance of the initial attractor giving a dangerous fast dynamic jump to a distant unrelated attractor: examples are the sub-critical local bifurcations and the folds. They can be determinate or indeterminate in outcome, depending on the global topology. The original attractor is not reinstated on reversal of the control sweep. Two such bifurcations typically generate a *hysteresis loop*, observed for example during frequency sweeps through nonlinear resonance.

The subtle and explosive bifurcations have an outcome that is insensitive to small details such as the rate of the control sweep, small perturbations, and noise. The outcome is therefore predictable, and they are classified as *determinate bifurcations.* In contrast, when a system is swept through a catastrophic-dangerous bifurcation there is an inevitable jump to a remote uncorrelated attractor. On the point of instability, the system may find itself well inside the basin of a single distant attractor, to which it must inevitably jump even in the presence of small perturbations: the jump is then determinate. Conversely, the system may find itself precisely on a (smooth or fractal) basin boundary: there is then a sensitive choice concerning to which of the two (or more) remote attractors the system will jump. This choice will depend delicately on the precise rate of sweeping, perturbations and noise, making the outcome unpredictable from a practical point of view: we then have an *indeterminate bifurcation.* These have recently been shown to be typical and common features in the resonance response behaviour of driven oscillators under both direct (Thompson and Soliman, 1991; Stewart and Ueda, 1991; Thompson, 1992) and parametric (Soliman and Thompson, 1992a, b) excitation. An example is shown in Figure 1.5(b) in which the jump to resonance from the saddle-node fold, A, may lead to escape from the potential well, or may restabilize onto the available harmonic attractor R: a third possible outcome, arising from the fractal basin structure, is a restabilization onto a co-existing subharmonic attractor of order $n = 3$ as illustrated in figure 5 of Thompson and Soliman (1991).

ACKNOWLEDGEMENTS

I would like to acknowledge the award of a *Senior Fellowship* by the Science and Engineering Research Council of the UK. Some parts of this article are based on a tutorial glossary of terms (Thompson and Stewart, 1993), and I am deeply indebted to Bruce Stewart for his many contributions.

REFERENCES

Abraham, R. H. and Shaw, C. D. (1992) *Dynamics: The Geometry of Behaviour*, Redwood City, Addison-Wesley.

Arrowsmith, D. K. and Place, C. M. (1990) *An Introduction to Dynamical Systems*, Cambridge, Cambridge University Press.

Falconer, K. (1990) *Fractal Geometry*, Chichester, Wiley.

Foale, S. and Bishop, S. R. (1992) Dynamical complexities of forced impacting systems, *Phil. Trans. R. Soc.* **A338**, 547–556.

Foale, S. and Thompson, J. M. T. (1991) Geometrical concepts and computational techniques of nonlinear dynamics, *Computer Methods in Applied Mechanics and Engng*, **89**, 381–394.

Ghrist, R. and Holmes, P. (1993) Knots and orbit genealogies in three dimensional flows. In: *Bifurcations and Periodic Orbits of Vector Fields*, NATO ASI Series, Amsterdam, Kluwer.

Grebogi, C., Ott, E. and Yorke, J. A. (1987) Basin boundary metamorphoses: changes in accessible boundary orbits, *Physica D*, **24**, 243–262.

Guckenheimer, J. and Holmes, P. (1983) *Nonlinear Oscillations, Dynamical Systems and Bifurcations of Vector Fields*, New York, Springer.

Kim, J. H. and Stringer, J. (eds) (1992) *Applied Chaos*, New York, Wiley.

Lansbury, A. N., Thompson, J. M. T. and Stewart, H. B. (1992) Basin erosion in the twin-well Duffing oscillator: two distinct bifurcation scenarios, *Int. J. Bifn & Chaos*, **2**, 505–532.

McDonald, S. W., Grebogi, C., Ott, E. and Yorke, J. A. (1985) Fractal basin boundaries, *Physica D*, **17**, 125–153.

McRobie, F. A. (1992a) Bifurcational precedences in the braids of periodic orbits of spiral 3-shoes in driven oscillators, *Proc. R. Soc.* **A438**, 545–569.

McRobie, F. A. (1992b) Birkhoff signature change: a criterion for the instability of chaotic resonance, *Phil. Trans. R. Soc.* **A338**, 557–568.

McRobie, F. A. and Thompson, J. M. T. (1991) Lobe dynamics and the escape from a potential well, *Proc. R. Soc.* **A435**, 659–672.

McRobie, F. A. and Thompson, J. M. T. (1992) Invariant sets of planar diffeomorphisms in nonlinear vibrations, *Proc. R. Soc.* **A436**, 427–448.

McRobie, F. A. and Thompson, J. M. T. (1993) Braids and knots in driven oscillators, *Int. J. Bifn & Chaos*, **3**, 1343–1361.

McRobie, F. A. and Thompson, J. M. T. (1994) Criteria for escape phenomena in driven oscillators using Melnikov-like energy estimates, *Nonlinear Dynamics*, in press.

Moon, F. C. (1987) *Chaotic Vibrations*, New York, Wiley.

Mullin, T. (ed.) (1993) *The Nature of Chaos*, Oxford, Oxford University Press.

Parker, T. S. and Chua, L. O. (1989) *Practical Numerical Algorithms for Chaotic Systems*, New York, Springer.

Schiehlen, W. (ed.) (1990) *Nonlinear Dynamics in Engineering Systems*, Proc. IUTAM Symposium, Stuttgart, August 1989, Berlin, Springer.

Soliman, M. S. and Thompson, J. M. T. (1992a) Indeterminate sub-critical bifurcations in parametric resonance, *Proc. R. Soc.* **A438**, 511–518.

Soliman, M. S. and Thompson, J. M. T. (1992b) Indeterminate trans-critical bifurcations in parametrically excited systems, *Proc. R. Soc.* **A439**, 601–610.

Soliman, M. S. and Thompson, J. M. T. (1992c) Global dynamics underlying sharp basin erosion in nonlinear driven oscillators, *Physical Review* **A45**, 3425–3431.

Stewart, H. B. (1987) A chaotic saddle catastrophe in forced oscillators. In: *Dynamical Systems Approaches to Nonlinear Problems in Systems and Circuits*, Salam, F. and Levi, M. (eds.), Philadelphia, SIAM.

Stewart, H. B. and Ueda, Y. (1991) Catastrophes with indeterminate outcome, *Proc. R. Soc.* **A432**, 113–123.

Thompson, J. M. T. (1989) Chaotic phenomena triggering the escape from a potential well, *Proc. R. Soc.* **A421**, 195–225.

Thompson, J. M. T. (1992) Global unpredictability in nonlinear dynamics: capture, dispersal and the indeterminate bifurcations, *Physica D*, **58**, 260–272.

Thompson, J. M. T. and Gray, P. (eds) (1990) *Chaos and Dynamical Complexity in the Physical Sciences*, First Theme Issue, *Phil. Trans. R. Soc.* **A332**, 49–186.

Thompson, J. M. T. and McRobie, F. A. (1993) Indeterminate bifurcations and the global dynamics of driven oscillators, Plenary lecture, *First European Nonlinear Oscillations Conference, Hamburg, August 1993*, Berlin, Akademie Verlag.

Thompson, J. M. T., Rainey, R. C. T. and Soliman, M.S. (1990) Ship stability criteria based on chaotic transients from incursive fractals, *Phil. Trans. R. Soc.*, **A332**, 149–167.

Thompson, J. M. T., Rainey, R. C. T. and Soliman, M.S. (1992) Mechanics of ship capsize under direct and parametric wave excitation, *Phil. Trans. R. Soc.* **A338**, 471–490.

Thompson, J. M. T. and Schiehlen, W. (eds) (1992) *Nonlinear Dynamics of Engineering Systems*, Theme Issue, *Phil. Trans. R. Soc.* **A338**, 451–568.

Thompson, J. M. T. and Soliman, M. S. (1990) Fractal control boundaries of driven oscillators and their relevance to safe engineering design, *Proc. R. Soc.* **A428**, 1–13.

Thompson, J. M. T. and Soliman, M. S. (1991) Indeterminate jumps to resonance from a tangled saddle-node bifurcation, *Proc. R. Soc.* **A432**, 101–111.

Thompson, J. M. T. and Stewart, H. B. (1986) *Nonlinear Dynamics and Chaos*, Chichester, Wiley.

Thompson, J. M. T. and Stewart, H. B. (1993) A tutorial glossary of geometrical dynamics, *Int. J. Bifn & Chaos*, **3**, 223–239.

Thompson, J. M. T., Stewart, H. B. and Ueda, Y. (1994) Safe, explosive and dangerous bifurcations in dissipative dynamical systems, *Physical Review*, **E49**, 1019–1027.

J. M. T. Thompson, *Centre for Nonlinear Dynamics, University College London, Gower Street, London WC1E 6BT, UK*

PART I
Experiments

It was in 1963 that Lorenz first noticed that a set of deterministic equations can exhibit solutions that do not settle down to a periodic or equilibrium state but rather wander randomly within a bounded region. Prior to this, though some may have noted a strange response to a particular system, nobody had put the mathematics and the numerical simulations together to provide firm evidence and so, more than likely, irregular responses were seen as a fault or quirk and not a true solution of the problem. Since that time many researchers have made this discovery for themselves with any number of different equations – so much so that today, although it is still interesting and informative to find regions of chaos within a given system of equations, this behaviour is of no great surprise to anyone. Smale (1967) and Feigenbaum (1978) added significant theoretical insight into the complex behaviour of nonlinear systems but this is not to say that all the important mathematical advances have already been made.

A natural question often asked is whether this chaotic motion is *only* due to the choice of mathematical model, or is our macroscopic engineering world really chaotic? The only way to fully resolve this question is either to design and construct physical systems which can behave chaotically, or alternatively to view and monitor an existing system and identify chaos. Notwithstanding chaotic motions there is still much that we do not fully understand about the behaviour of nonlinear systems. In order to improve that knowledge sophisticated experimental set-ups need to be designed to confirm or promote corresponding theoretical concepts.

In the first chapter in this section *Moon* reveals the chaotic and fractal characteristics generated when metal is cut by a lathe. It is proposed that the chaotic vibrations of the cutting tool produce surface records that reveal the fractal type of scaling that we have become used to from complex maps similar to those of Mandelbrot (1982). In the second chapter *Nayfeh* and his co-workers use a parametrically excited cantilever beam to perform very careful investigations of the transfer of energy between different modes within a weakly nonlinear system. Significantly, analytical models are proposed to explain interactions between modes.

In the chapter by *Bayly et al.* the application of theoretical advances in nonlinear dynamics to an approximation of a physical system allows estimates of stability zones to be determined. That is to say, the experiments can be guided by the theory and numerical simulations to achieve a desirable state. Finally, in the chapter by *Cusumano and Kimble* the complex basin structure as a result of a homoclinic bifurcation is observed in a magneto-mechanical oscillator.

REFERENCES

Feigenbaum, M. J. (1978) Quantitative universality for a class of nonlinear transformations, *J. Stat. Phys.*, **19**, 25–52.

Lorenz, E. (1963) Deterministic non-periodic flows, *J. Atmos. Sci.*, **20**, 130–141.

Mandelbrot, B. B. (1982) *The Fractal Geometry of Nature*, Freeman, San Francisco.

Smale, S. (1967) Differentiable dynamical systems, *Bull. Am. Math. Soc.*, **73**, 747–817.

2 CHAOTIC DYNAMICS AND FRACTALS IN MATERIAL REMOVAL PROCESSES

F. C. Moon

Experimental and numerical studies are presented on the nonlinear dynamics of metal cutting. The data suggest that under certain conditions material removal processes will exhibit chaotic dynamics. Signal process analysis involves reconstructed phase-space orbits, probability density function and fractal dimension calculation in the phase space. The experiments show a fractal dimension less than three for turning of aluminium on a small lathe. The nonlinear model involved both time delay effects and velocity-dependent friction forces in the cutting process. Numerical simulation shows a loss of stability with increase of depth of cut and a transition to quasi-periodic dynamics. These data are the beginning of a study to demonstrate the relation between chaotic and fractal dynamics in the material removal process and the spatially complex or fractal nature of the surface topography of the workpiece.

2.1   INTRODUCTION

Research into the understanding of the cutting process in machining has been an active field of study for many decades yet tool chatter remains a problem. Early models were based on linear stability analysis (e.g. Tobias 1965), although nonlinear effects were recognized as important to the dynamics. Recent studies have begun to recognize the role of nonlinearities (Tlusty and Ismail 1981), but only a few have incorporated more up-to-date research resulting from the study of chaotic vibrations, fractal mathematics, and modern nonlinear dynamical systems.

It is the thesis of this study that deterministic self-generating random-like vibrations or chaos in the cutting process will produce surface records that reveal fractal-type scaling of the finished surface geometry, and conversely, that measurement of the apparently random properties of the surface geometry will reveal clues to the underlying nature of the cutting process and overall machine-tool dynamics. This chapter is the first in our effort to establish this link between nonlinear tool dynamics and surface quality of the workpiece.

Nonlinearity and Chaos in Engineering Dynamics
Edited by J. M. T. Thompson and S. R. Bishop,   © 1994 John Wiley & Sons Ltd

The understanding of material-cutting dynamics involves four aspects: (i) the physics of the material removal process, (ii) modelling of the machine-tool dynamics, (iii) understanding the coupled dynamics of the machine tool and the deformation process, and (iv) relating the material removal dynamics to the resultant surface properties and topography of the workpiece.

Theories and models to understand cutting processes of metals have been evolving for nearly half a century. The early work of Merchant (1945) established a shear plane model with an averaged friction law for the interaction of the chip and the tool face (see Figure 2.1). More recent models use a finite-width shear zone along with strain-hardening plasticity including temperature effects (see e.g. Hastings *et al.* 1980).

The study of machine-tool chatter has had a long history as evidenced by the number of research papers and monographs such as Tobias (1965), Welbourn and Smith (1970), Weck (1978), Tlusty (1985), and Chiriacescu (1990). These studies have long recognized the self-excited nature of machine-tool chatter and many dynamics models have been developed. Most of these models are linearized, and periodic motions are assumed to occur. However, experimental evidence in the very papers and books which proposed these theories often show non-steady oscillations and downright chaotic-looking vibrations. In spite of this the linear theory has been useful to develop a stability theory to predict the stable and unstable operating range of machine tools as a function of cutting speed, depth of cut, and other parameters.

Recently nonlinear analyses of machine-tool chatter have shown the possibility of chaotic vibrations. Grabec (1986, 1988) used a two-degree-of-freedom dynamic model with a friction nonlinearity. Lin and Weng (1991) have used a nonlinear model, based on the work of Tobias (1965) and Tlusty and Ismail (1981), to perform numerical simulation that exhibits chaotic dynamics. These models are based on a basic paradigm shown in Figure 2.1 with one or two degrees of elastic deformation, and a workpiece whose free surface has a profile proportional to the tool motion at an earlier time. This dynamic cutting force model has two important terms: a velocity-dependent term and a displacement-dependent term with time lag.

Differential equations with delay are known to exhibit dynamic instabilities. A mathematical treatment of such equations is given by Stépán (1989) who includes a study of equations of

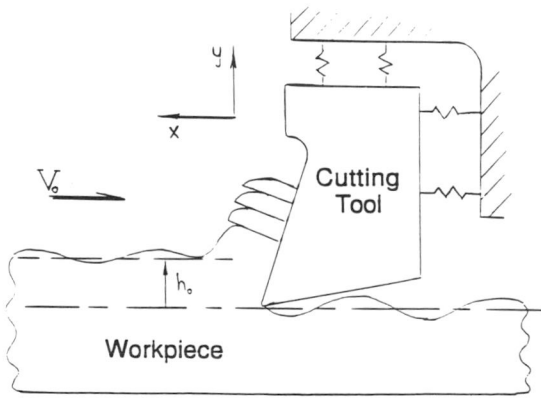

Figure 2.1 Sketch of orthogonal cutting process.

machine-tool chatter with delay. In addition to the single-delay time model of Tobias, Stépán presents a model with a cutting force with distributed time delay. This model is proposed to account for the distribution of chip forces on the tool face.

The nonlinear studies of Grabec (1986, 1988) and Lin and Weng (1991) employ two-degree-of-freedom models for the tool dynamics. Lin and Weng also use a multiple regenerative effect with several time-delay terms to account for the tool leaving the workpiece. The loss of contact between the tool and workpiece was first modelled by Tlusty and Ismail (1981).

Aside from the numerical studies of these cutting dynamics models, models with delay differential equations have received little study from the nonlinear dynamics community, especially as regards the global dynamical behaviour (see e.g. Plaut and Hsieh 1987, Ueda *et al.* 1993). The monograph of Stépán (1989) is largely concerned with stability questions of the linearized delay equation. Thus, the nature of the bifurcation and routes to chaotic behaviour are not readily understood. Nor is it known if there are multiple solutions dependent on initial conditions or whether such solutions have fractal or smooth basins of attraction. The search for the routes to chaotic dynamics in such dynamical models of machine-tool chatter will also permit one to begin to determine criteria for unpredictable machine-tool vibration.

Studies of friction-induced chaos are related to cutting dynamics and may be found in the work of Feeny and Moon (1989, 1994) and Popp and Stelter (1990). Analytical, numerical and experimental studies in these papers all show evidence of chaotic or unpredictable behaviour.

2.2   FRACTALS AND MATERIAL REMOVAL PROCESSES

Fractals is the mathematics of geometric and mathematical objects with many length scales and patterns. Fractals in dynamical systems are now well known in strange attractors and fractal basin boundaries in phase space. Both machine-made and natural objects have also been found to have fractal-like surfaces. There are now well accepted numerical tools for measuring the so-called fractal dimension (see e.g. Feder 1988, Moon 1992).

In physical systems with material surfaces such as those created by geological forces (see e.g. Turcotte 1992) or by fracturing a solid by impact (e.g. Mandelbrot *et al.* 1984), surface features have been measured which exhibit fractal-like behaviour over several length scales. In another work on fractured surfaces from the University of Texas (Fineberg *et al.* 1991), fractal-like surfaces were observed as a result of dynamic impact.

The proposition that machine-made surfaces have fractal distribution has recently received experimental verification in several studies. Gagnepain and Roques-Carmes (1986) in France have used an optical profilometer to measure the distribution of altitude scales and obtained a value of the fractal dimension of $D = 2.70$ for a polyethylene-terephthalate (PET) film surface.

Majumdar and Tien (1990) have used a contact diamond stylus to measure surface profiles on lapped, ground and turned stainless-steel surfaces, with the latter measuring fractal dimensions of $D = 1.5$ to 1.7. They use the Weierstrass–Mandelbrot functions to characterize an isotropic rough surface. Also, Majumdar and Bhushan (1991) have used a fractal surface model to predict elasto-plastic contact forces between rough surfaces and have found good agreement with experiments. Recently Srinivasan and Wood (1992) have proposed using fractals in geometric tolerancing. These authors refer to the observation of these fractal surfaces as randomly generated. However, one of the theses of our research is that nonlinear deterministic material removal processes can create such random-like surfaces. That is, there is an intimate link between the nonlinear dynamics of material removal and the surface statistics and quality (Scott 1989).

2.3   DESCRIPTION OF EXPERIMENTAL STUDIES

Experiments on chaotic cutting dynamics were carried out on a small bench-top lathe using metals, plastics and ceramics. Chaotic dynamics can easily be observed by affixing a transducer to a tool bit or tool holder in the form of a strain gauge, accelerometer, or load cell. One indirect method which was also used was measurement of the voltage between the tool and the workpiece (Postnikov 1978). An attempt was made to measure tool-temperature time history with an infrared sensor, but the chip motion interfered with the measurement.

A sketch of the experimental set-up is shown in Figure 2.2. The lathe is a German-built Präzi machining centre with turning speeds from 300 to 1200 rev./min. Workpiece diameters ranged from 20 mm to 40 mm. Thus cutting speeds ranged from 25 to 140 m/min. Two different tool bits were used, one of conventional tool steel and the other a standard 'C6' carbide tool-bit insert with a 0.38 mm radius. Much of the data presented in this chapter was obtained using an aluminium alloy workpiece. Thus, the lower cutting speed of 25 m/min was too low by industrial practice guidelines and was subject to a cutting phenomenon called built-up edge'. The higher cutting speed was within accepted practice standards. For most of the aluminium tests no lubrication was used.

Time histories of the tool-bit motion were recorded on a Nickolet digital oscilloscope and transferred to a Micro Vax computer system for data processing such as phase-plane reconstruction and fractal-dimension calculations. Other signal-processing techniques such as Fourier transforms, probability density function and autocorrelation function were performed with a dual channel Hewlett-Packard signal analyser.

Classical tool-chatter theory assumes a periodic vibration. However, as these experiments show, the motion does not always appear to be periodic (Figure 2.3). One method for presenting the dynamic data that has emerged from modern nonlinear dynamics is to use a

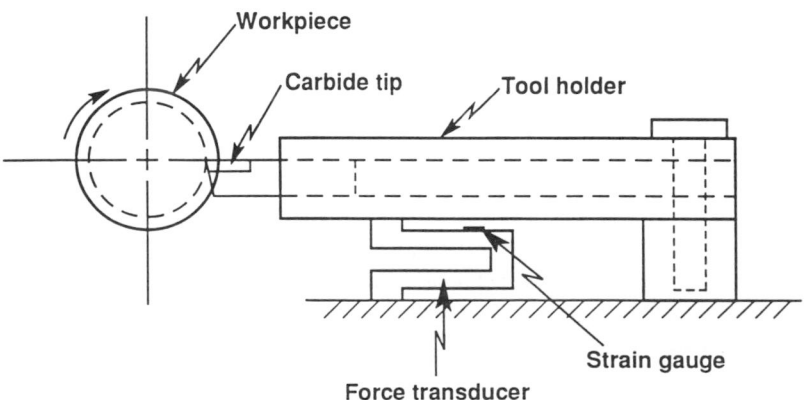

Figure 2.2   Sketch of cutting tool and force transducer.

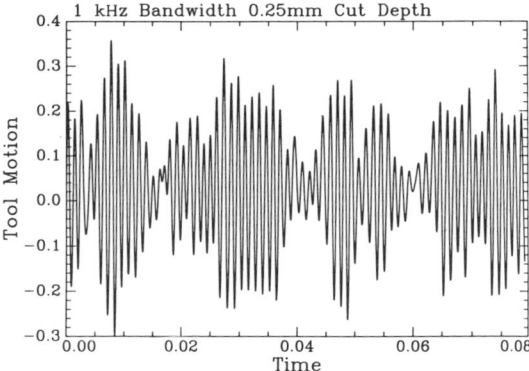

Figure 2.3   Experimental time history of tool-chatter vibrations for aluminium alloy at a cutting speed of 120 m/min.

pseudo-phase-space or embedding-space technique (see e.g. Moon 1992). For example, a two-dimensional (2D) phase space can be generated by plotting the signal $x(t)$ versus the delayed signal $x(t - \tau)$. To create a three-dimensional phase space, one uses $x(t)$, $x(t - \tau)$ and $x(t - 2\tau)$. Here τ represents a delay time that is not commensurate with any of the natural frequencies in the dynamics. In this representation, a periodic signal should show up as a closed curve (Figure 2.4a). However, in the experimental data in Figure 2.4b, the phase space exhibits a tangled trajectory which gives some evidence of chaotic motion. These data are for an aluminium alloy cut at a low speed of 25 m/min.

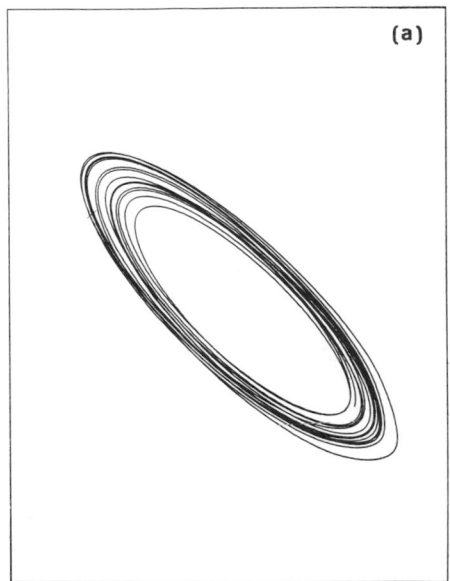
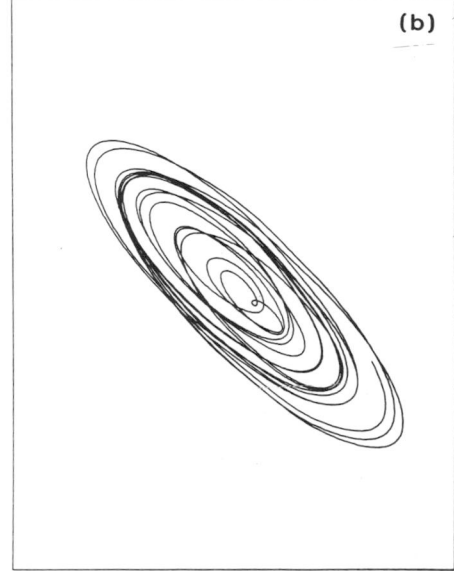

Figure 2.4   Reconstructed phase space, $|x(t), x(t - \tau), x(t - 2\tau)|$, for cutting dynamics for aluminium alloy at 120/min. (a) 0.46 mm depth of cut; (b) 0.25 mm depth of cut.

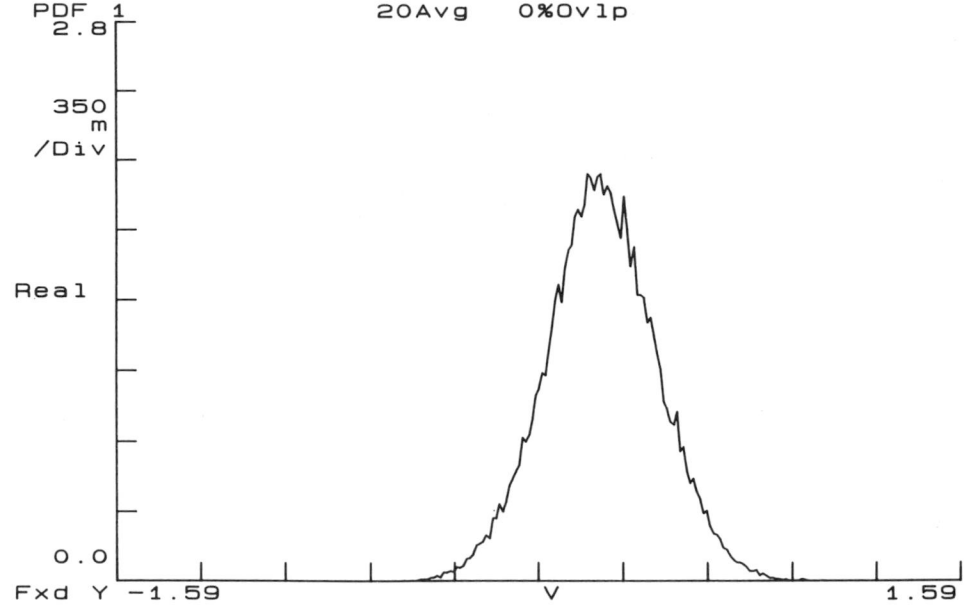

Figure 2.5   Probability density function for tool vibrations: aluminium alloy, 120 m/mm, 0.39 mm depth of cut.

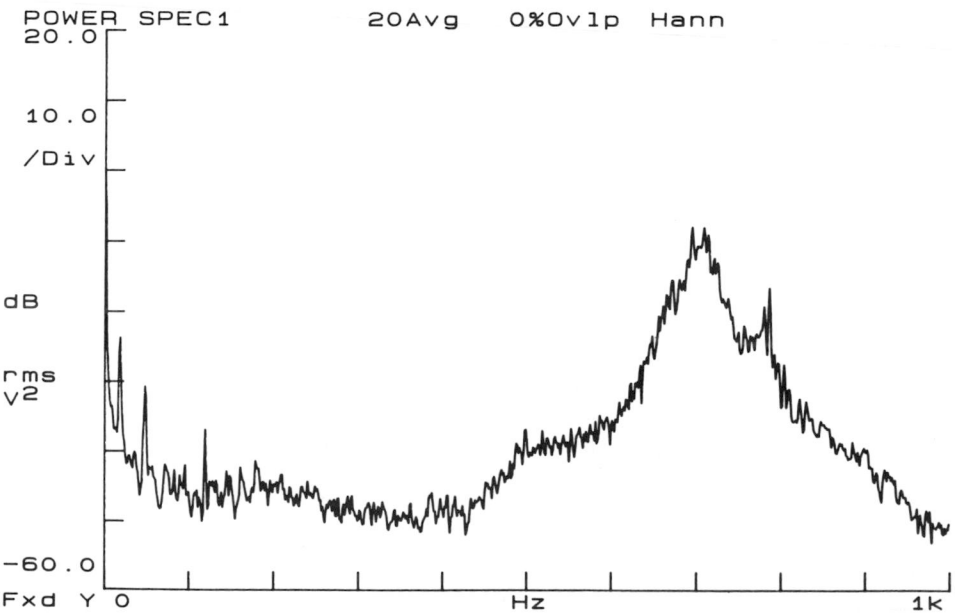

Figure 2.6   Fourier transform, tool dynamics during cutting of aluminium alloy at 120 m/min, 1.53 mm depth of cut.

Another modern measure of nonlinear chaotic dynamics is the use of a probability distribution function (PDF). The periodic signal exhibits a double-spike distribution function. However, the PDF for chaotic-type cutting (Figure 2.5) shows a function more similar to a Gaussian distribution or random-like processes even when the system is deterministic. Also, the power spectral density of a Fourier transform of the tool motion shows a broad spectrum characteristic of chaotic dynamics (Figure 2.6). Another measure of the motion, the fractal dimension, is described below.

2.4   FRACTAL DIMENSION OF PHASE-SPACE ATTRACTOR

Fractal geometry of chaotic motions in the phase space are well known (see e.g. Moon 1992). The fractal properties of dynamical systems can be measured by looking at the dynamics in an embedding space:

$$(x(t), x(t - \tau), x(t - 2\tau), \ldots, x(t - m\tau)).$$

Evidence for fractal geometry in the phase space for the tool dynamics experiments is shown in Figures 2.7 and 2.8. In the calculation of fractal dimension of dynamical attractors a correlation function $C(\varepsilon)$ is used. The correlation function is found to be a function of the length scale ε in the form as $\varepsilon \to 0$,

$$C(\varepsilon) = C_0 \varepsilon^D.$$

$C(\varepsilon)$ measures the probability that two points on the trajectory will be found in a ball of size ε in the phase space. If the points of the trajectory completely fill some region of an n-dimensional space, then $D = n$. The data in Figures 2.7 and 2.8 show that as we increase the dimension of the embedding space, the fractal dimension reaches an asymptotic value less than three.

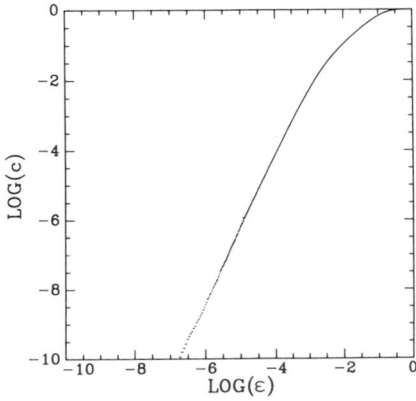

Figure 2.7   Fractal-dimension-calculation, $\log C(\varepsilon)$ v. $\log \varepsilon$ using a 1000 Hz bandwidth, aluminium alloy 120 m/min; embedding space dimension = 7.

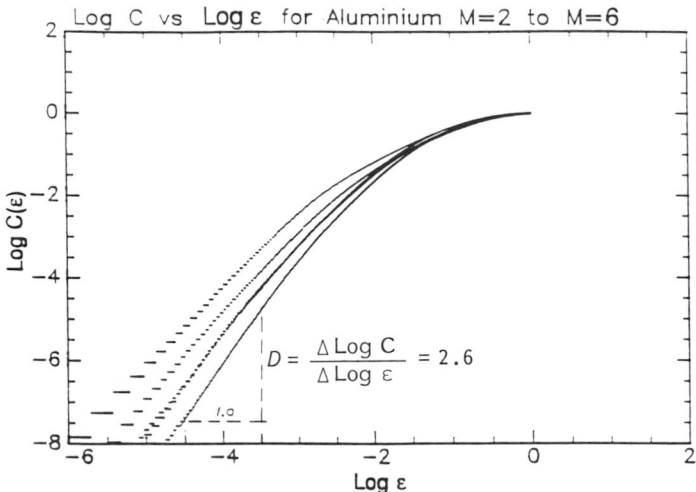

Figure 2.8 Fractal-dimension calculation, $\log C(\varepsilon)$ v. $\log \varepsilon$ for low-speed cutting of aluminium alloy using a low-pass filter at 25 Hz.

2.5 NONLINEAR CUTTING DYNAMICS MODELS

The implications of the measurement of a fractal dimension in the phase space as being less than three is that the tool dynamics can be modelled by a low-order dynamical system. The fact that the motion may be chaotic means that the phase-space dimension must be three or higher. In conventional dynamic modelling in mechanics, this would require a third- or fourth-order system (two degrees of freedom). And in fact the studies by Grabec (1986) used a two-degree-of-freedom model for orthogonal cutting of the form

$$m_1\ddot{x} + g_1\dot{x} + f_1(x) = F,$$
$$m_2\ddot{y} + g_2\dot{y} + f_2(y) = KF, \tag{1}$$

where f_1, f_2 are cutting forces and F, K represent a nonlinear tool friction force and coefficient of friction.

However, as has been noted in the introduction, the cutting forces depend not only on the current motion, $x(t)$, but also on the tool displacement at the earlier revolution of the workpiece, $x(t - \tau)$. The study of time-delay effects on the stability of cutting-tool dynamics has had a long history (Tobias 1965; Tlusty 1985; Stépán 1989). The important point here is that a differential equation with time delay has an infinite-dimensional phase space. Thus we may expect to find chaos in a time-delay equation with only one state variable, as has recently been demonstrated by Ueda et al. (1993) for a phase-locked loop circuit. In our mechanical problem this means that we could expect to see chaos in a single-degree-of-freedom model with both time-delay effects and a strong nonlinearity such as velocity-dependent friction.

The model used in this study is derived from the standard shear-plane and tool friction theory used by previous authors as shown in Figure 2.9. To develop the simplest model we assume a single-degree-of-freedom motion normal to the workpiece with linear

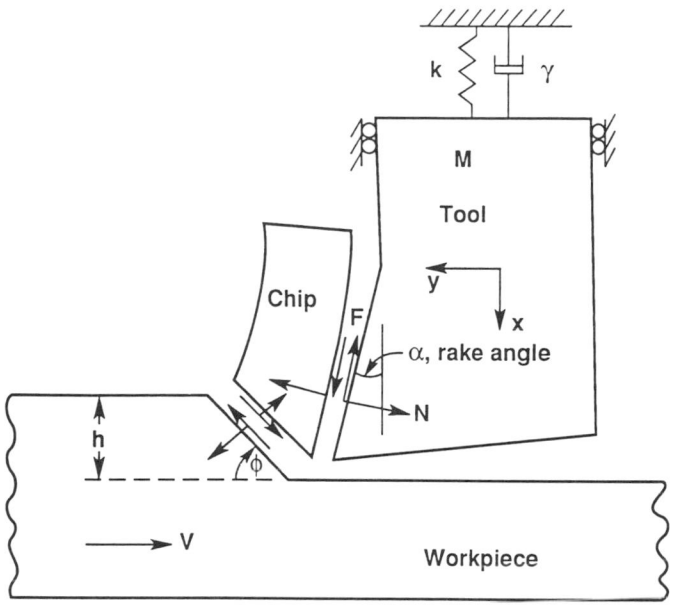

Figure 2.9 Sketch of forces on cutting tool.

machine-tool fixture dynamics. The forces on the tool include a normal force N and a friction force F, which is assumed to be related to the normal force through a coefficient of friction m. The single-degree-of-freedom model then takes the form

$$m\ddot{x} + \gamma\dot{x} + kx = N\sin\alpha - F\cos\alpha,$$

$$F = \mu N, \tag{2}$$

where α is the rake angle of the tool. We further assume, following Tlusty (1985), that the normal force N is proportional to the tool width w and the chip thickness h. In general, for high rake angle $\alpha > 35°$, this relation is nonlinear, i.e.

$$N = w(C_1 h + C_2 h^3). \tag{3}$$

However, for small rake angles there is evidence that the linear term is dominant (Kalpakjian 1988). Time-delay effects enter the problem since the chip thickness depends not only on the tool displacement $x(t)$ but on the displacement at the previous pass, $x(t - \tau)$:

$$h = h_0 + x(t) - x(t - \tau), \tag{4}$$

where $\tau = \pi D/V$ and D is the workpiece diameter and V is the cutting velocity.

Using the expressions for N and h, the equation of motion takes the form

$$\ddot{x} + 2\zeta\dot{x} + \omega_0^2 x = \lambda[h_0 + x - x_\tau][\sin\alpha - \mu(x, x_\tau, \dot{x})\cos\alpha] \tag{5}$$

where

$$x_\tau = x(t - \tau), \quad \text{and} \quad \lambda = \omega_0^2 w C_1/k.$$

The friction coefficient $\mu(x, x_\tau, \dot{x})$ is written as a function of the tool velocity. To be more explicit, μ depends on the relative sliding velocity between the chip and the moving tool; i.e. $V_c + \dot{x}\cos\alpha$. The chip velocity is in turn dependent on the cutting velocity V and the chip thickness h_c: $hV = h_c V_c$. The dependence of h on x and x_τ gives the general dependence of μ on the displacement as well as the velocity. However, in our study we assumed that $h/h_0 \ll 1$ and adapted a simplified model with $\mu(V_c + \dot{x}\cos\alpha)$ and $V_c = rV$, $r < 1$. The friction function chosen is similar to that used in other friction dynamics studies. One could choose μ as a discontinuous Coulomb model, i.e. $\mu = \mu_s \operatorname{sgn}(V_c + \dot{x}\cos\alpha)$ or a modified Coulomb law with both static and dynamic friction. In our study we choose a continuous-friction law used in earlier papers (Feeny and Moon 1989, 1994):

$$\mu = [\mu_k + (1 - \mu_k)\operatorname{sech}(\beta v)]\tanh(\alpha v) \tag{6}$$

shown in Figure 2.10. This law is similar to that used by Grabec (1986) and is similar in character to the velocity dependence of cutting forces measured for steel (Hastings *et al.* 1980).

Using these assumptions and nondimensionalizing the tool displacement x by the average chip thickness h_0, $z = x/h_0$ we obtain a simple material-cutting dynamics model which exhibits some of the complex nonlinear behaviour seen in metal-turning experiments for $\alpha \neq 0$:

$$\ddot{z} + 2\zeta\dot{z} + \omega_0^2 z = \lambda_0[1 + z - z_\tau][1 - \hat{\mu}(\hat{v} + \dot{z})], \tag{7}$$

where $\hat{\mu} = (\cotan\alpha)\mu$ has the functional form shown in Figure 2.10 and $\hat{v} = rVh_0\cos\alpha$. When $\alpha = 0$, as in the Grabec model, an even simpler form is obtained:

$$\ddot{z} + 2\zeta\dot{z} + \omega_0^2 z = -\lambda_1[1 + z - z_\tau]\mu(\hat{v} + \dot{z}). \tag{8}$$

Note that if the Coulomb model for μ were used, a piecewise linear equation would result which might be amenable to semi-analytical methods.

Numerical integration of the model represented by equation (7) shows both self-excited periodic and quasi-periodic-looking dynamics (Figures 2.11, 2.12). This model also predicts a threshold below which the system does not oscillate as in the linear theory.

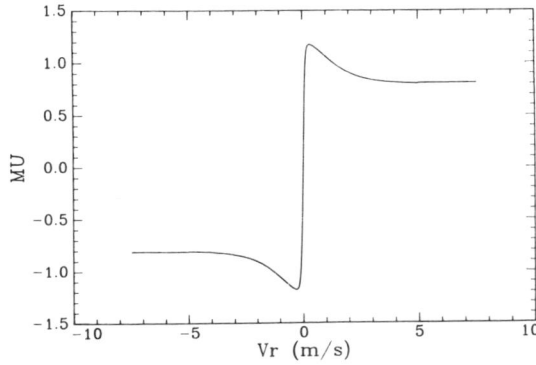

Figure 2.10   Plot of dynamic friction coefficient versus relative speed of tool and chip.

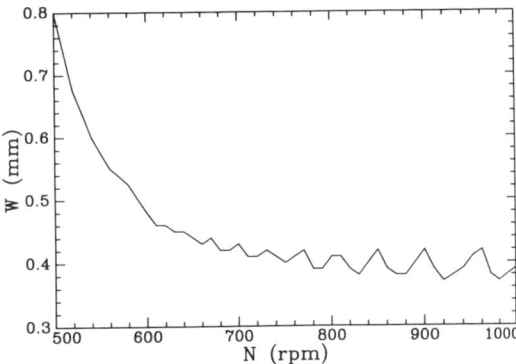

Figure 2.11 Stability boundary for one-degree-of-freedom model of cutting dynamics (model parameters: $\omega_0 = 775$ rad/s, $C = 8000$ N/mm^2, $m = 50$ kg, $h_0 = 1$ mm, $\alpha = 10°$, $\zeta = 38.8$ s^{-1}).

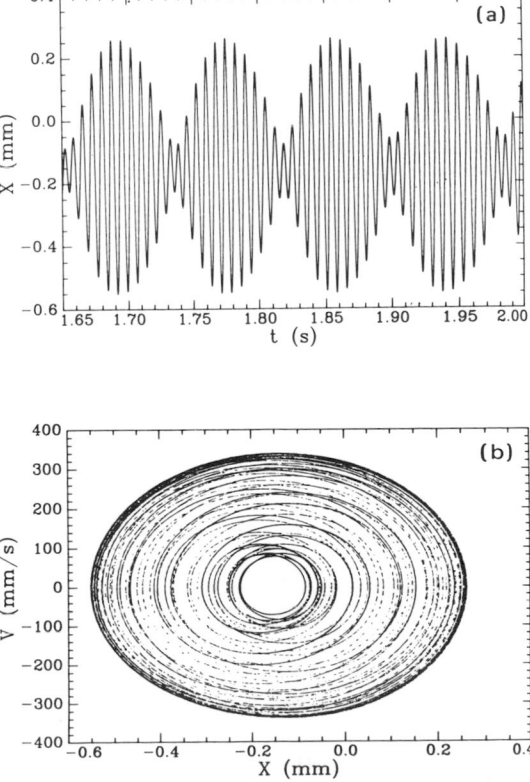

Figure 2.12 Numerical simulation of cutting dynamics: (a) time history; (b) phase-plane plot ($N = 960$) rev/min, $w = 0.6$ mm depth of cut).

These results represent only preliminary studies, and further analysis and experiments will be presented in later work. However, the experimental data seems to support the thesis that chaotic phenomena are common in machine-tool chatter. The numerical model data may indicate that at least two degrees of freedom are necessary as in the Grabec model, to obtain a transition to chaos. Our next step will be to show how fractal properties in the phase space affect the surface quality of the machined workpiece.

ACKNOWLEDGEMENTS

The author wishes to recognize the contributions of W. Holmes and H. van Essen in carrying out the numerical simulation and experiments. Supported in part by the U.S. National Science Foundation.

REFERENCES

Chiriacescu, S. T. (1990) *Stability in the Dynamics of Metal Cutting*, Elsevier Press, Amsterdam.

Feder, J. (1988) *Fractals*, Plenum Press, New York.

Feeny, B. F. and Moon, F. C. (1989) Autocorrelation on symbol dynamics for a chaotic dryfriction oscillator, *Phys. Lett.* **141A**, (8,9), 397–400.

Feeny, B. F. and Moon, F. C. (1994) Chaos in a forced dry-friction oscillator: experiments and numerical modelling, *J. Sound and Vibra*, to appear.

Fineberg, J., Gross, S. P., Marden, M. and Swinney, H. (1991) Instability in dynamic fractures, *Phys. Rev. Lett.*, **67**(4), 457–460.

Gagnepain, J. J. and Roques-Carmes, C. (1986) Fractal approach to two-dimensional and three-dimensional surface roughness, *Wear*, **109**, 119–126.

Grabec, J. (1986) Chaos generated by the cutting process, *Phys. Lett.*, **117A**(8).

Grabec, J. (1988) Chaotic dynamics of the cutting process, *Int. J. of Machine Tools Manufacturing*, **28**(1), 19–32.

Hastings, W. F., Mathew, P. and Oxley, P. L. B. (1980) A machining theory for predicting chip geometry, cutting forces etc. from work material properties and cutting conditions, *Proc. Roy. Soc. Lond.*, **371A**, 569–587.

Kalpakjian, S. (1988) *Manufacturing Processes for Engineering Materials*, 2nd Edn. Chap. 8, Addison-Wesley, Reading, Mass.

Lin, J. S. and Weng, C. I. (1991) Nonlinear dynamics of the cutting processing, *Int. J. Mech. Sic.*, **33**, 645–657.

Majumdar, A. and Bhushan, B. (1991) Fractal model of elastic-plastic contact between rough surfaces, *J. Tribology*, **113**, 1–11.

Majumdar, A. and Tien, C. L. (1990) Fractal characterization and simulation of rough surfaces, *Wear*, **136**, 313–327.

Mandelbrot, B. B., Passoja, D. E. and Paullay, A. J. (1984) Fractal character of fracture surfaces of metals, *Nature*, **308**, 19 April, 721–722.

Merchant, M. E. (1945) *J. Appl. Phys.*, **16**, 267.

Moon, F. C. (1992) *Chaotic and Fractal Dynamics*, J. Wiley & Sons, New York.

Plaut, R. H. and Hsieh, J.-C. (1987) Nonlinear structural vibrations involving a time delay in damping, *J. Sound and Vibr.*, **117**(3), 497–510.

Popp, K. and Stelter, P. (1990) Nonlinear oscillations in structures induced by friction. In: *Nonlinear Dynamics in Engineering Systems*, W. Schiehlen (ed.), Springer-Verlag, Berlin, pp. 233–240.

Postnikov, S. N. (1978) *Electrophysical and Electrochemical Phenomena in Friction, Cutting and Lubrication*, Van Nostrand Reinhold Co., New York.

Scott, P. J. (1989) Nonlinear dynamic systems in surface metrology, *Surface Topography*, **2**, 345–366.

Srinivasan, R. S. and Wood, K. L. (1992) Fractal-based geometric tolerancing for mechanical design. Prepared for the *1992 Design Theory and Methodology Conference*, Phoenix, Ariz September 1992.

Stépán, G. (1989) Retarded dynamical systems: stability and characteristic functions, *Pitman Research Notes in Mathematics Series* **210**, H. Breziz, R. G. Douglas and A. Jeffrey (eds.), Longman Harlow.

Tlusty, J. (1985) Machine dynamics, Chapter 3 of: *Handbook of High Speed Machining Technology*, R. I. King (ed.), Chapman & Hall, New York.

Tlusty, J. and Ismail, F. (1981) Basic non-linearity in machining chatter, *Ann. CIRP*, **30**, 299–304.

Tobias, S. A. (1965) *Machine-Tool Vibration*, Blackie Press, London.

Turcotte, D. (1992) *Fractals and Chaos in Geology and Geophysics*, Cambridge University Press.

Ueda, Y., Ohta, H. and Stewart, H. B. (1993) Bifurcations in a system described by a differential equation with delay, submitted to *Chaos*.

Weck, M. (1978) *Werkzeugmaschinen*. Vol. 4. VDI Verlag, Düsseldorf.

Welbourn, D. B. and Smith, J. D. (1970) *Machine-Tool Dynamics: An Introduction* (Chapters 4 and 5), Cambridge University Press, Cambridge.

F. C. Moon, *204 Upson Hall, Mechanical and Aerospace Engineering, Cornell University, Ithaca, NY 14853, USA*

3 TRANSFER OF ENERGY FROM HIGH-FREQUENCY TO LOW-FREQUENCY MODES

A. H. Nayfeh, S. A. Nayfeh, T. A. Anderson and B. Balachandran

In some recent experimental studies, we observed that energy can be transferred from high-frequency to low-frequency modes in structures with weak nonlinearities. In these experiments, a structure was subjected to a simple-harmonic, low-amplitude excitation. The frequency of the excitation was near one of the high natural frequencies: yet after a very long time the contribution of the first mode to the multi-frequency response was larger than the contribution of the high-frequency mode that was directly excited. Subsequently, we developed analytical models to explain the interactions between widely spaced modes of structures and used them to determine conditions under which energy can be transferred from high-frequency to low-frequency modes, as observed in the experiments.

3.1 INTRODUCTION

Nonlinear modal interactions have been the subject of a great deal of recent research. It has been found that, in weakly nonlinear systems where there exists a special relationship between two or more natural frequencies of the linear modes and an excitation frequency, the long-time responses can contain significant contributions in many modes of vibration (Nayfeh and Mook, 1979; Nayfeh and Balachandran, 1989). The presence of significant responses in more than one mode increases the number of modal equations that must be treated and this generally serves to complicate the dynamics of the system. More importantly, modal interactions can lead to dangerously large responses in modes that are predicted by linear analysis to have insignificant response amplitudes.

Most of the research on modal interactions focuses on autoparametric resonances in systems where the linear natural frequencies ω_i are commensurate or nearly commensurate. The types of possible internal resonances depend on the degree of the nonlinearity. When the nonlinearity is cubic, to the first approximation, internal resonances may occur it $\omega_n \approx \omega_m$, $\omega_n \approx 3\omega_m$,

Nonlinearity and Chaos in Engineering Dynamics
Edited by J. M. T. Thompson and S. R. Bishop,

$\omega_n \approx |\pm 2\omega_m \pm \omega_k|$, or $\omega_n \approx |\pm \omega_m \pm \omega_k \pm \omega_l|$. If quadratic nonlinearities are added, additional resonances may occur if $\omega_n \approx 2\omega_m$ or $\omega_n \approx \omega_m + \omega_k$. These autoparametric resonances have been successfully treated with perturbation methods. There also exists a large body of experimental results which are in good general agreement with the perturbation results. Autoparametric resonances may provide a coupling or an energy exchange between the coupled modes. Consequently, an excitation of a high-frequency mode may produce a large-amplitude response in a low-frequency mode involved with it in an autoparametric resonance.

In externally excited multi-degree-of-freedom systems, combination resonances may occur in response to a single-harmonic external excitation of frequency Ω. The type of combination resonance that can be excited depends on the degree of the nonlinearity, the number of modes involved, and Ω. For a cubic nonlinearity, to the first approximation, combination resonances may occur if $\Omega = |\pm \omega_m \pm \omega_k|/2$, $\Omega \approx |\pm 2\omega_m \pm \omega_k|$, or $\Omega \approx |\pm \omega_m \pm \omega_k \pm \omega_l|$. If quadratic nonlinearities are added, additional combination resonances may occur if $\Omega \approx |\pm \omega_m \pm \omega_k|$. Thus, a high-frequency excitation may produce large-amplitude responses in low-frequency modes that are involved in the combination resonance. Dugundji and Mukhopadhyay (1973) conducted experiments on a cantilever beam subjected to external base excitation at a frequency close to the sum of the natural frequencies of the first torsional and first bending modes, which are approximately in the ratio of 18 to 1. They found that the high-frequency excitation can produce a large-amplitude response in the low-frequency (first bending) mode.

In parametrically excited systems, modal interactions can occur when the excitation frequency is near the sum or difference of two or more linear natural frequencies. These so-called combination resonances have been studied extensively in the literature. Again, these combination resonances can lead to interactions between high- and low-frequency modes.

Often, when the response of a system becomes chaotic, low-frequency modes can be excited. Haddow and Hasan (1988) conducted an experiment by parametrically exciting a cantilever beam near twice the natural frequency of its fourth mode. They found that, as the excitation frequency was decreased, a planar periodic response consisting essentially of the fourth mode lost stability, giving way to a non-planar chaotic motion. They observed that as a result the energy seemed to cascade down through the modes, resulting eventually in a low-frequency steady-state response. Burton and Kolowith (1988) conducted an experiment similar to that of Haddow and Hasan. In certain regions of the parameter space, they observed chaotic motions where the first seven in-plane bending modes as well as the first torsional mode were present in the response. Cusumano and Moon (1989) conducted an experiment with an externally excited cantilever beam. They observed a cascading of energy to low-frequency components in the response associated with chaotic non-planar motions.

Two recent studies suggest that another type of interaction may occur between high-frequency and low-frequency modes. In the first study (Anderson et al. 1992), conducted experiments on, we parametrically excited cantilever. We found that interactions occur between two high-frequency modes and the first mode. The presence of the first mode is accompanied by slow modulation of the amplitudes and phases of the high-frequency modes with the frequency of the modulation being equal to that of the first mode. Our results indicate that the mechanism for the excitation of the first mode is neither a classical internal resonance nor an external or parametric combination resonance involving the first mode. Rather, it seems that slow modulation of the high-frequency modes allows for the energy to be transferred to the first mode.

In the second study (Nayfeh and Nayfeh, 1994), we conducted experiments on an externally excited, circular cross-section, cantilever rod. Because of the axial symmetry, one-to-one autoparametric resonances occur at each natural frequency of the rod and the mode in the plane of the excitation interacts with the out-of-plane mode at the same natural frequency, resulting in non-planar whirling motions. In addition, it is found that when the rod is excited near the natural frequency of its third or any higher mode, a large first-mode response occurs. Moreover, the degree of the coupling between the first mode and the higher modes is qualitatively observed to increase as we drive progressively higher modes. As in the first experiment, the appearance of the first mode is accompanied by modulation of the amplitudes and phases of the high-frequency modes.

The interaction between high- and low-frequency modes observed experimentally is of great practical importance. In many engineering systems, high-frequency excitations can be caused by rotating machinery. Through this mechanism, energy from high-frequency sources can be transferred to low-frequency modes of supporting structures or foundations, resulting in harmful large oscillations. Moreover, some preliminary results indicate that the use of conventional methods for decreasing modal interactions, such as increasing the dissipation or decreasing the forcing amplitude, may have undesirable effects. In the next two sections, we present a summary of our experiments.

3.2   EXPERIMENTS ON A PARAMETRICALLY EXCITED CANTILEVER BEAM

A schematic of the experimental set-up for a base excitation along the axis of the beam is shown in Figure 3.1. The test specimen is a vertically mounted carbon-steel cantilever beam of dimensions 33.56″ × 0.75″ × 0.032″. The beam was clamped to a 250 lb modal shaker with a custom table and suspension to allow base excitation of the beam. We note that the beam is slightly bent in the static configuration. The first four natural frequencies of the beam are 0.65 Hz, 5.65 Hz, 16.19 Hz and 31.91 Hz.

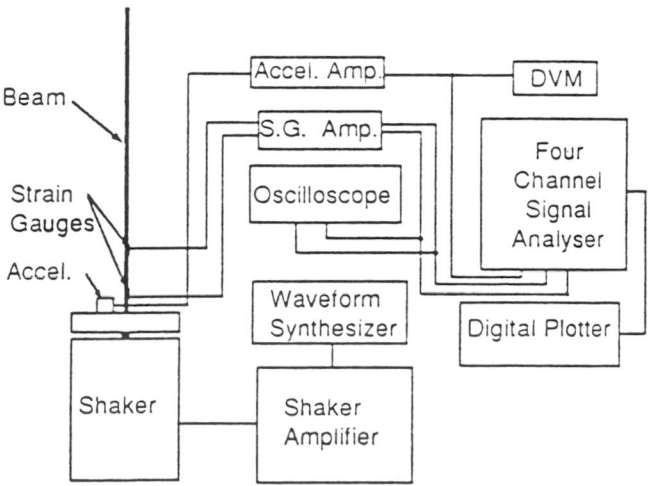

Figure 3.1   Experimental set-up for the parametrically excited cantilever beam.

The base motion was monitored with an accelerometer. A digital voltmeter was used to measure the root-mean-square (r.m.s.) value of the acceleration. A measure of the response was obtained from two strain gauges: one located at $x/L = 0.06$ and the other located at $x/L = 0.25$, where x is the distance along the undeformed beam measured from the base and L is the length of the beam.

The accelerometer and strain-gauge spectra were monitored as the excitation frequency was varied. Also, the autocorrelation function R_{xx} and the pointwise dimension were examined for selected motions. For the spectral analyses, we used 1280 lines of resolution in a 40 Hz baseband. A flat-top window was used during periodic excitations, and a Hanning window with thirty overlap averages was used during random excitations. The two strain-gauge signals were plotted against each other on the digital oscilloscope, thereby producing a pseudo-phase plane. To obtain a Poincaré section, we used the excitation frequency as the clock frequency for the oscilloscope. Due to the manner in which the points are stored in the digital scope, the final Poincaré section effectively corresponds to one obtained at one-half of the clock frequency. We used Fourier spectra, pseudo-phase planes, autocorrelation functions, and dimension calculations to analyse the different motions.

The excitation frequency was chosen to be the control parameter, and the base acceleration was held constant at 0.85g r.m.s. where the symbol g stands for the acceleration due to gravity. Initially when the excitation frequency f_e was at 33.5 Hz, there was a peak in the response spectrum at the excitation frequency. This peak is due to a primary resonance of the fourth mode. As the excitation frequency was gradually decreased to 32.31 Hz, the third mode appeared in the response. It was excited by a principal parametric resonance. The response spectrum, shown in Figure 3.2a, has peaks at f_e and $f_e/2$. The Poincaré section,

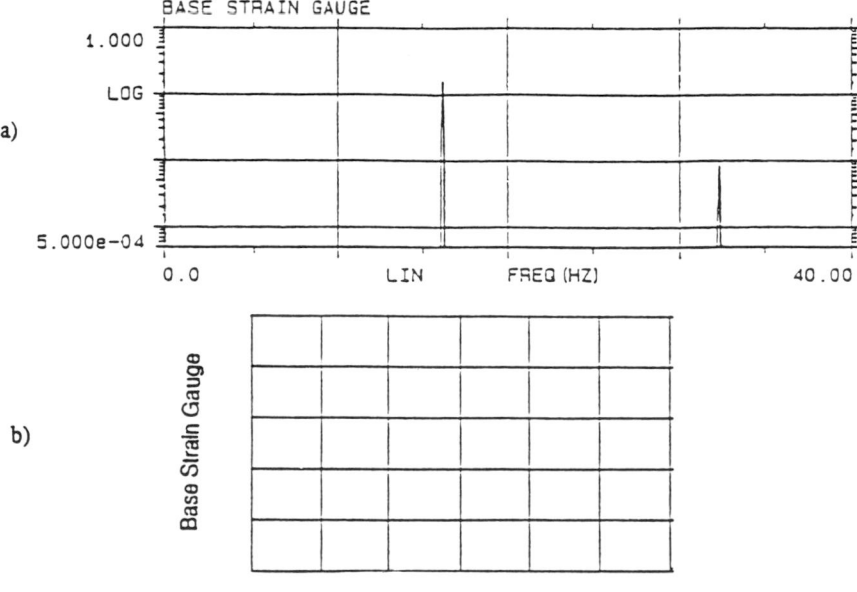

Figure 3.2 Response at $f_e = 32.31$ Hz: (a) Fourier spectrum, (b) Poincaré section.

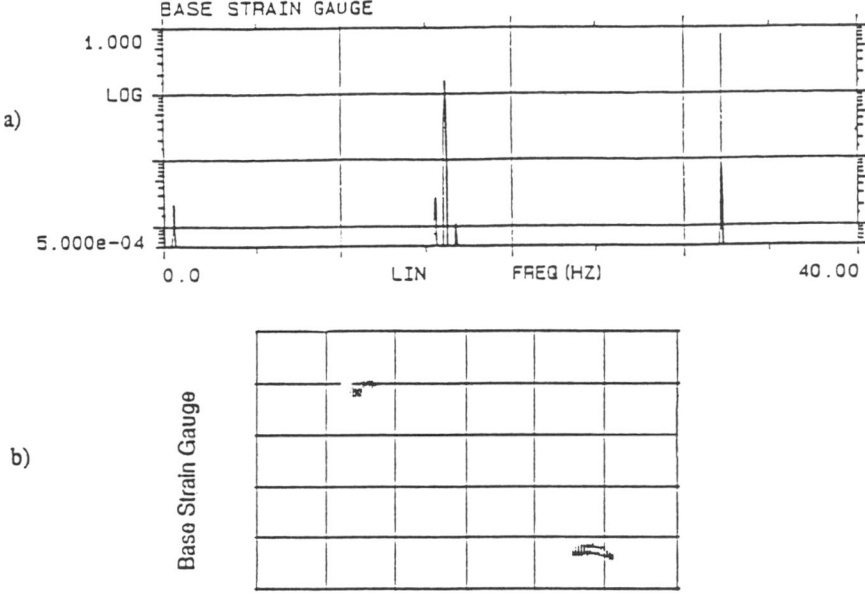

Figure 3.3 Response at $f_e = 32.298$ Hz: (a) Fourier spectrum, (b) Poincaré section.

shown in Figure 3.2b, is characteristic of a periodic motion. The scatter in the points is due to noise. When the excitation frequency was decreased to 32.298 Hz, the periodic response consisting of the third and fourth modes lost stability, resulting in a modulated motion. During this modulated motion, the amplitudes and phases associated with the third and fourth modes varied with time. The spectrum of this response is shown in Figure 3.3a, with the sidebands around the carrier frequency at $f_e/2$ indicating a modulated motion. The sideband spacing f_m is 0.58 Hz, which is close to the first natural frequency of the beam.

Once the modulated motions set in, the contribution of the first mode to the response became large. The modulated response is indicative of an energy transfer from the third and fourth modes to the first mode. During the experiments, the presence of the first mode was very apparent visually. The presence of the first mode in the response leads to a scattering of points along a curve in the corresponding Poincaré section, as shown in Figure 3.3b. This observation indicates that the response is not periodic. When the excitation frequency was further reduced to 32.289 Hz, the motion appeared to become chaotic with a large out-of-plane component. The associated spectrum, shown in Figure 3.4a, has a continuous character in many frequency bandwidths. This characteristic is a signature of chaotic motion (Moon 1987). The Poincaré section, displayed in Figure 3.4b, is typical of non-periodic motions; it does not have any obviously discernible structure.

In Figure 3.5, we show a time record obtained during the transient phase of the motion after f_e was changed from 32.298 Hz to 32.289 Hz. During the initial phase, the third and fourth modes are dominant in the response. Subsequently, there is a transition from the response composed mainly of modulated high-frequency (fast-time scale) motion to one

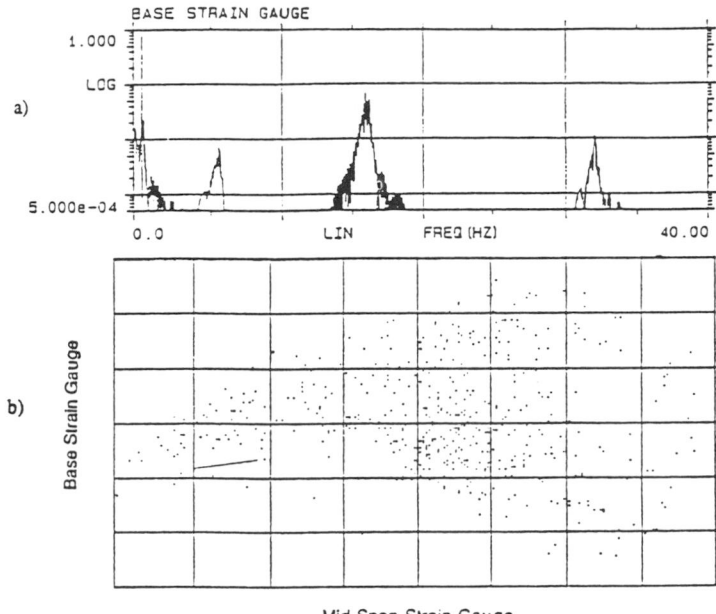

Figure 3.4   Response at $f_e = 32.289$ Hz: (a) Fourier spectrum, (b) Poincaré section.

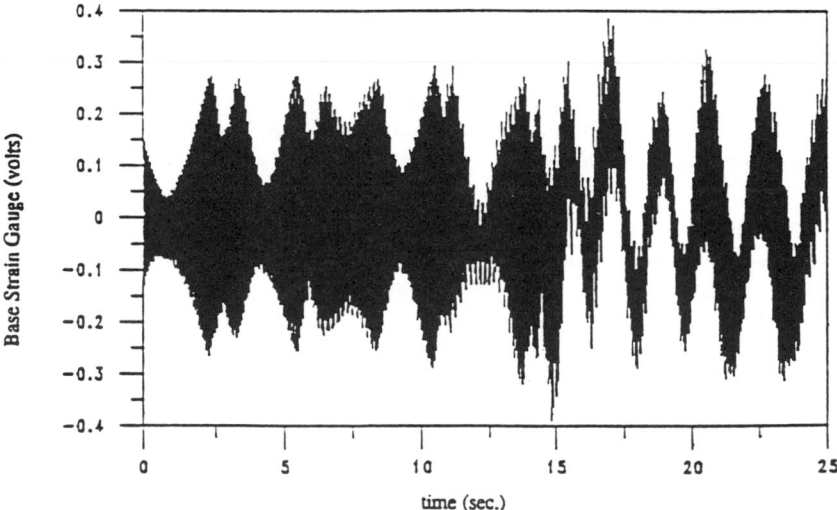

Figure 3.5   A time trace of a transient motion.

dominated by the low-frequency (slow-time scale) first mode. In Figure 3.5, the strain is plotted versus time. The displacements observed during the low-frequency-dominated phase of the motion are much larger than those observed during the high-frequency-dominated phase. The change in the time history is striking.

3.3 EXPERIMENTS ON AN EXTERNALLY EXCITED ROD

In the second study, we conducted experiments on a slender, circular cross-section, steel cantilever rod. The length of the cantilever is 34.5″ and the diameter of its cross-section is 0.0625″. The first five linear natural frequencies of the rod, as determined by examination of the frequency spectra of decaying free oscillations, are shown in Table 3.1.

Figure 3.6 is a schematic diagram of the experimental set-up. A vertical beam is clamped to a 100 lb shaker that supplies a simple-harmonic motion at the base so that an external (i.e. transverse to the axis of the beam) excitation is supplied. The excitation is monitored by means of an accelerometer mounted to the shaker head. The motion of the tip of the beam is measured by two linear-array cameras, one oriented to measure the motion in the plane of the excitation (camera 0) and the other oriented to measure the motion out of the plane of the excitation (camera 1).

Table 3.1 The first five natural frequencies of the test specimen

Mode	Natural Frequency (Hz)
1	1.303 ± 0.005
2	9.049 ± 0.005
3	25.564 ± 0.005
4	50.213 ± 0.007
5	83.105 ± 0.011

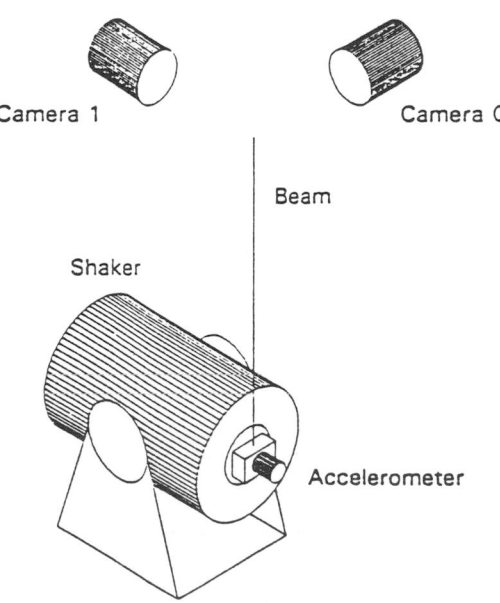

Figure 3.6 Experimental set-up for the externally excited cantilever rod.

The linear-array camera system employs a hardware-implemented peak detector to determine the location of the target in real time. At a specified sampling frequency, it returns two eleven-bit numbers representing the displacement of the tip of the beam in the in-plane and out-of-plane directions. This data is acquired in real time by a personal computer where it is displayed, processed and stored.

We present frequency–response curves for the fifth in-plane and out-of-plane modes of the cantilever beam. The excitation level was held constant at 2.00g r.m.s. and the excitation frequency was varied in the neighbourhood of the fifth natural frequency. Changes in the excitation frequency were made very gradually and transients were allowed to die out before

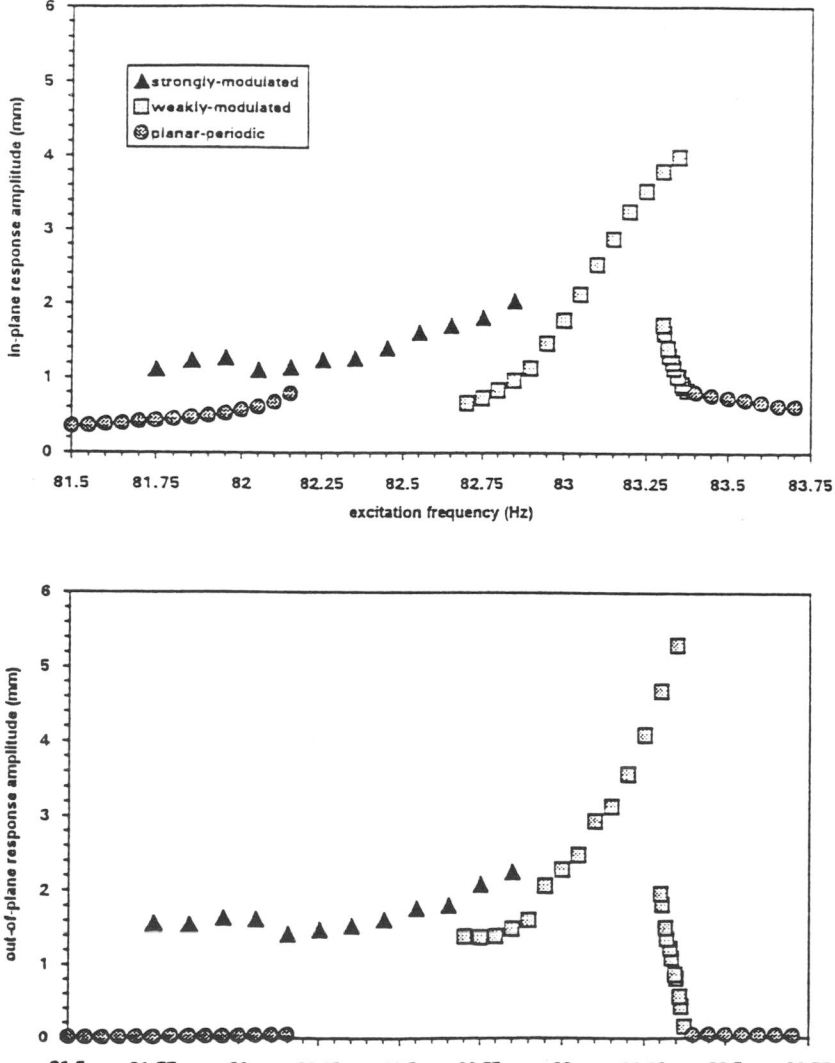

Figure 3.7   Frequency–response curves of the fifth mode for an excitation amplitude of 2.00 g r.m.s.

the amplitude of the response was recorded. The data in the plots is a composite of the responses obtained by performing both forward and backward frequency sweeps. In addition, to ensure that even isolated branches of the frequency–response curves were located, we performed a third sweep where at increments in the excitation frequency, we applied several disturbances to the beam in an effort to find all possible long-time responses.

The results of this procedure are shown in Figure 3.7. Well away from the fifth natural frequency of 83.105 Hz, the only possible response is planar and periodic. The response of the beam is strictly in the plane of the excitation and a visual inspection of the motion indicates that the response is composed almost entirely of the fifth mode. This is confirmed

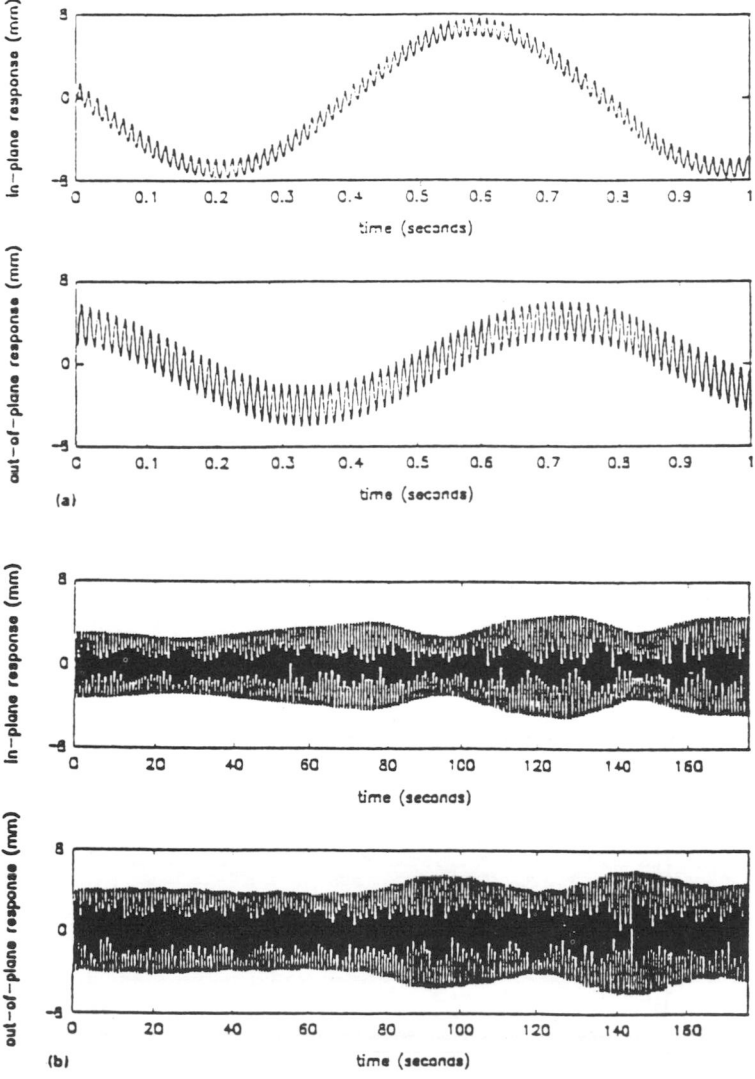

Figure 3.8 Time traces of a typical weakly modulated motion: (a) trace showing 1 s of the response, (b) trace showing 170 s of the response.

by examination of the response spectrum which shows only a single peak at the excitation frequency.

As the frequency of excitation is swept upward from well below the fifth natural frequency, a jump occurs from a planar-periodic to a nonplanar strongly modulated motion. Here, visual inspection of the response clearly detects the modulation of the response of the fifth mode as well as the presence of a low-frequency component in the response.

Increasing the excitation frequency further, we observe a jump to a nonplanar weakly modulated whirling motion. Again, visual inspection of the motion clearly reveals a large low-frequency component in the response. In this case, however, visual inspection does not detect any modulation of the fifth mode. A more detailed discussion of both the weakly and strongly modulated motions follows.

The observed weakly·modulated responses contain a large low-frequency component superimposed on a nearly constant amplitude fifth-mode whirling motion. Typical time traces of in-plane and out-of-plane responses of this type are shown in Figure 3.8a. Visual inspection of these plots does not readily reveal any modulation of either the high- or low-frequency components of the response.

A typical FFT of this type of response is shown in Figure 3.9. The FFT shows two main peaks, one at the frequency of the excitation (near the fifth natural frequency) and the other at the natural frequency of the first mode. Sidebands around the peak corresponding to the fifth mode indicate that the response of the fifth mode is modulated. Moreover, the frequency spacing between the fifth-mode peak and its sidebands is equal to the first natural frequency, confirming that the frequency of modulation of the fifth-mode response is equal to the natural frequency of the first mode.

As indicated by the dense set of sidebands clustered around the peak at the first natural frequency, the response of the first mode is also modulated. Examination of the time-domain data from which this FFT was computed, shown in Figure 3.8b, confirms that the amplitude of the first-mode response is not constant. The time traces in Figure 3.8b contain 170 seconds of data, illustrating the extremely slow variation of the amplitude of the first mode.

The most obvious feature of the strongly modulated motions is the modulation of the fifth mode. A typical time trace of this type of motion is shown in Figure 3.10a. In

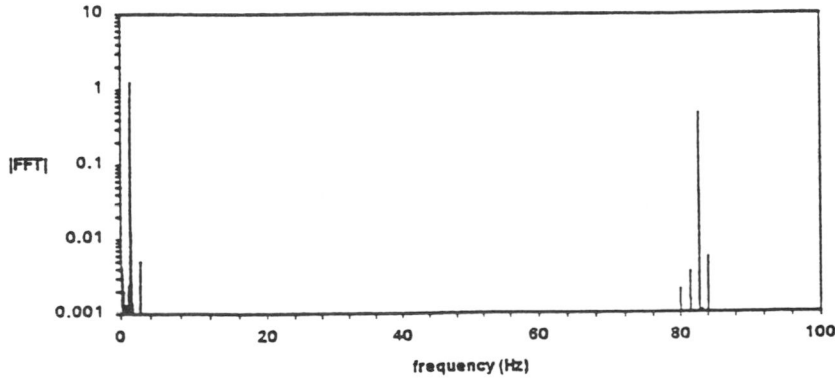

Figure 3.9   Magnitue of the FFT of a typical weakly modulated motion.

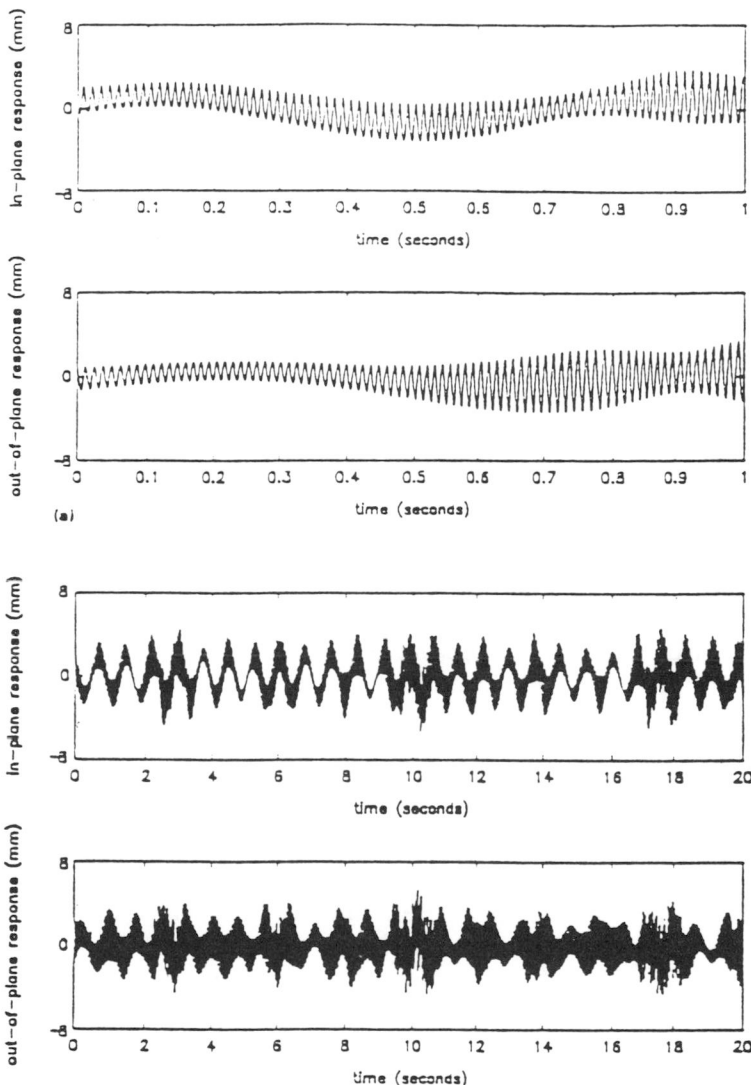

Figure 3.10 Time traces of a typical strongly modulated motion: (a) trace showing 1 s of the response, (b) trace showing 20 s of the response.

contrast to the case of the weakly modulated motions of Figure 3.8, the modulation of the fifth mode is clearly distinguishable without the aid of FFTs. It is also apparent from the asymmetry in the envelope of the traces shown in Figure 3.10a that there is a significant low-frequency component present in the response.

In Figure 3.10b a longer time trace of this motion is presented. Here, the scaling of the time axis is such that both the low-frequency component present in the response and the envelope of the fifth-mode response are clearly discernible. The erratic character of the evolution of the fifth-mode response suggests that the fifth mode is chaotically modulated. This assertion can be further sustained by examination of the FFT of this signal shown in

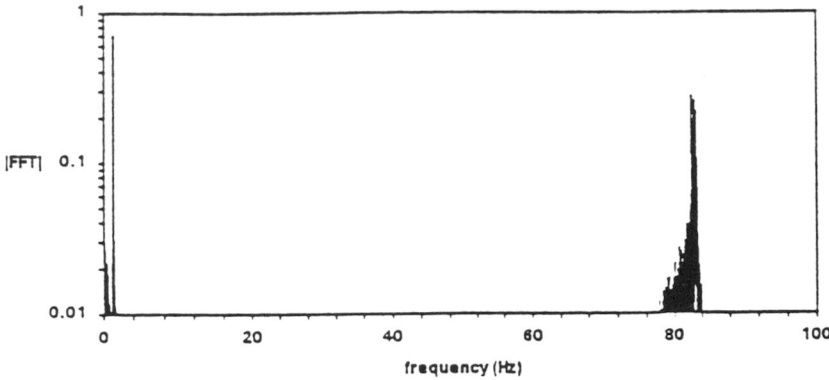

Figure 3.11   Magnitude of the FFT of a typical strongly modulated motion.

Figure 3.11 where the narrow band of response present in the neighbourhood of the fifth natural frequency is characteristic of chaotically modulated motions.

Turning our attention to the low-frequency component present in the response, we find that there is a peak at the first natural frequency of the system and conclude that the low-frequency component in the response is due to the first mode. As in the case of the weakly modulated motions, there appears a dense set of sidebands clustered about the first natural frequency peak, indicating that the first-mode response is also modulated.

3.4   A PARADIGM FOR THE TRANSFER OF ENERGY FROM HIGH-FREQUENCY TO LOW-FREQUENCY MODES

The results presented in Sections 3.2 and 3.3 show modal interactions occurring between high- and low-frequency modes in a flexible structure. The mechanism for the interaction appears to be neither a classical internal resonance nor an external or parametric resonance involving the low-frequency modes. Rather, it seems that these interactions can occur whenever there exist modes whose natural frequencies are much lower than the natural frequencies of the modes being directly driven.

To investigate possible mechanisms for the transfer of energy from high- to low-frequency modes, we studied a representative system made up of two coupled oscillators (Nayfeh and Nayfeh, 1993). These equations are in a form that may be obtained by a two-mode discretization of a continuous system with cubic nonlinearities or, alternatively, they could model a discrete two-degree-of-freedom system. The equations are given by

$$\ddot{u}_1 + 2\varepsilon\mu_1\dot{u}_1 + \varepsilon^2 u_1 = -\varepsilon^2(4\alpha_1 u_1^3 + \alpha_2 u_1 u_2^2), \tag{1}$$

$$\ddot{u}_2 + 2\varepsilon\mu_2\dot{u}_2 + u_2 = \varepsilon(\alpha_3 u_2^3 + \alpha_4 u_1^2 u_2 + f\cos\Omega t), \tag{2}$$

where ε, the ratio of the linear natural frequencies of the system, is positive and small. The high-frequency mode, whose undamped linear natural frequency is nondimensionalized to unity, has coordinate u_2, and the low-frequency mode, whose normalized undamped linear natural frequency is ε, has coordinate u_1. The system has linear viscous damping given by the coefficients μ_1 and μ_2, cubic nonlinearities with the coefficients α_i, and an external forcing function $f \cos \Omega t$ which is applied only to the high-frequency mode of the system. Of principal interest is whether an excitation applied to the high-frequency mode near its linear natural frequency can, as observed in the experiments, generate a large response in the low-frequency mode. To answer this question, we used the method of averaging to construct an approximation of the solutions of equations (1) and (2).

The method of averaging is based on the assumption that small perturbations, such as weak nonlinearities or light damping, cause slow (low-frequency) variations in the response of a system (Nayfeh, 1973). The fast (high-frequency) variations due to the perturbations are assumed to be insignificant. Essentially, the averaging approximation yields a simplified mathematical representation of the dynamics of the system by smoothing away these fast variations. Thus, it is of basic importance that the components which make up the response be correctly classified as either fast or slow.

Neglecting the damping and nonlinearities, one can write the solution to equation (1) as $u_1 = A_0 \cos(\varepsilon t + \phi_0)$. In this solution, it is apparent that whereas u_1 is an $O(1)$ quantity, u_1 is $O(\varepsilon)$, and $\ddot{u}_1$ is order $O(\varepsilon^2)$. This leads us to assume that u_1 itself is slowly varying. Because the natural frequency of u_2 is not small, its motion can be treated in the usual way by assuming that its amplitude and phase are slowly varying as described below.

To explicity show that u_2 is driven near its linear natural frequency, we set $\Omega^2 = 1 + \varepsilon \sigma$, where σ is a measure of the closeness of the excitation frequency to the unperturbed natural frequency of u_2. Next, we apply the variation-of-parameters transformation

$$u_2 = a(t) \cos(\Omega t + \beta(t)), \tag{3}$$

$$\dot{u}_2 = -a(t) \Omega \sin(\Omega t + \beta(t)), \tag{4}$$

to equations (1) and (2) and obtain

$$\ddot{u}_1 + 2 \varepsilon \mu_1 \dot{u}_1 + \varepsilon^2 u_1 = -\varepsilon^2 (4 \alpha_1 u_1^3 + \alpha_2 u_1 a^2 \cos^2(\Omega t + \beta)), \tag{5}$$

$$\dot{a} \Omega = -\varepsilon g \sin(\Omega t + \beta), \tag{6}$$

$$a \dot{\beta} \Omega = -\varepsilon g \cos(\Omega t + \beta), \tag{7}$$

where

$$g = \sigma a \cos(\Omega t + \beta) + \alpha_3 a^3 \cos^3(\Omega t + \beta) + \alpha_4 u_1^2 a \cos(\Omega t + \beta)$$
$$+ 2 \mu_2 a \Omega \sin(\Omega t + \beta) + f \cos \Omega t. \tag{8}$$

Keeping only the slowly varying terms on the right-hand sides of equations (5)–(7), we obtain the averaged or modulation equations

$$\ddot{u}_1 + 2 \varepsilon \mu_1 \dot{u}_1 + \varepsilon^2 u_1 = -\varepsilon^2 (4 \alpha_1 u_1^3 + \tfrac{1}{2} \alpha_2 u_1 a^2) \tag{9}$$

and

$$\dot{a} = -\varepsilon (\mu_2 a + \tfrac{1}{2} f \sin \beta), \tag{10}$$

$$\dot{\beta} = -\varepsilon \left(\frac{1}{2} \sigma + \frac{1}{2} \alpha_4 u_1^2 + \frac{3}{8} \alpha_3 a^2 + \frac{f}{2a} \cos \beta \right), \tag{11}$$

where we have set $\Omega \approx 1$. Equation (9) can be rewritten as a pair of first-order equations:

$$\dot{u}_1 = \varepsilon v_1, \tag{12}$$

$$\dot{v}_1 = -\varepsilon (u_1 + 2 \mu_1 v_1 + 4 \alpha_1 u_1^3 + \tfrac{1}{2} \alpha_2 u_1 a^2). \tag{13}$$

The fixed-point solutions of the averaged equations represent constant amplitude and phase motions of the high-frequency mode accompanied by static (DC) responses of the low-frequency mode. Setting the time derivatives in equations (10)–(13) equal to zero and solving for σ and u_1 in terms of a, we obtain

$$u_1 = 0 \tag{14}$$

or

$$u_1 = \pm \sqrt{-\frac{1 + \frac{1}{2}\alpha_2 a^2}{4\alpha_1}} \tag{15}$$

and

$$\sigma = -\frac{3}{4}\alpha_3 a^2 - \alpha_4 u_1^2 \pm \sqrt{\frac{f^2}{a^2} - 4\mu_2^2}. \tag{16}$$

The stability of a fixed-point solution is studied by examination of the eigenvalues of the Jacobian matrix of equations (10)–(13) evaluated at the fixed point of interest. If all of the eigenvalues have negative real parts, the fixed point is asymptotically stable and any motion in the neighbourhood of this fixed point is expected to be attracted to it. These solutions are called *stable nodes* and are denoted by solid lines in the frequency–response curves of Figure 3.12. If a real eigenvalue becomes positive, the fixed point loses stability and the motion is expected to diverge from it. These unstable solutions are called *saddles* and are denoted by dotted lines in Figure 3.12.

If, instead, a Hopf bifurcation occurs (a complex conjugate pair of eigenvalues crosses transversely from the left half of the complex plane into the right half of the complex plane), the fixed point loses stability, but in this case the motion is expected to oscillate about the fixed point. These unstable fixed points (called *unstable foci* and denoted by dashed lines in Figure 3.12) are of great interest because, in their neighbourhood, we expect to find motions where u_1 oscillates and u_2 is modulated at the frequency of oscillation of u_1 as observed in the experiments.

In Figure 3.12, we present frequency–response curves for a case in which non-trivial solutions for u_1 occur. It should be noted that, although we show only the fixed-point solutions corresponding to positive values of u_1 in Figure 3.12, there exists a second set of solutions corresponding to negative values of u_1. The trivial solutions are unstable with a positive real eigenvalue in the central region of the plot. In this region, a non-trivial solution for u_1 exists. The upper branch of this solution consists of two regions of stable nodes joined by a region of unstable foci. Where the stable nodes exist, the motion will consist of periodic oscillations in u_2 and either a positive or negative non-zero static deflection in u_1. Where the unstable foci exist, oscillatory u_1 motions accompanied by modulated u_2 responses will occur.

In Figure 3.13, we present Hopf bifurcation sets for the values of the α_i used in Figure 3.12 and various values of the damping coefficients. Below these curves, oscillations in u_1 decay of a constant value and above them oscillations in u_1 are sustained. From the curves in Figure 3.13a, it is apparent that at any particular excitation frequency, increasing the damping coefficient μ_2 of the high-frequency mode increases the critical forcing amplitude required to generate oscillations in u_1. From Figure 3.13b, however, we see that increasing the damping coefficient μ_1 of the low-frequency mode does not always increase the value

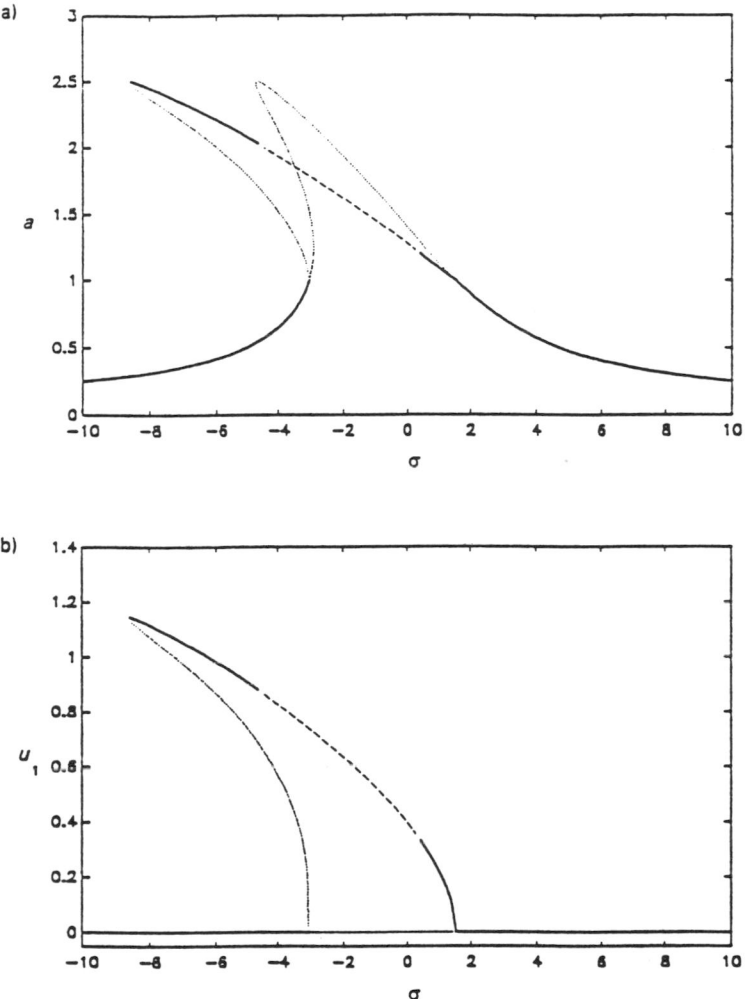

Figure 3.12 Frequency–response curves for $\alpha_1 = \alpha_3 = 1$, $\alpha_2 = -2$, $\alpha_4 = 3$, $\mu_1 = 0.25$, $\mu_2 = 0.5$ and $f = 2.5$. Solid lines denote stable solutions, dotted lines denote unstable solutions with a positive real eigenvalue, and dashed lines denote unstable solutions with a complex-conjugate pair of eigenvalues in the right half-plane.

of the critical forcing amplitude. At some excitation amplitudes and frequencies, increasing μ_1 actually destabilizes the system.

To study the dynamics of the system in the neighbourhood of unstable foci, we employed a fourth-order Runge–Kutta–Fehlberg algorithm and integrated the averaged equations using the same parameter values as used in Figure 3.12. As predicted by the stability analysis, oscillatory responses of u_1 are found to occur here. The dynamics of the system are very complicated in these regimes and various nonlinear phenomena, such as period-doubling bifurcations culminating in chaos, symmetry-breaking bifurcations, the existence of multiple attractors, and the merging of attractors are found.

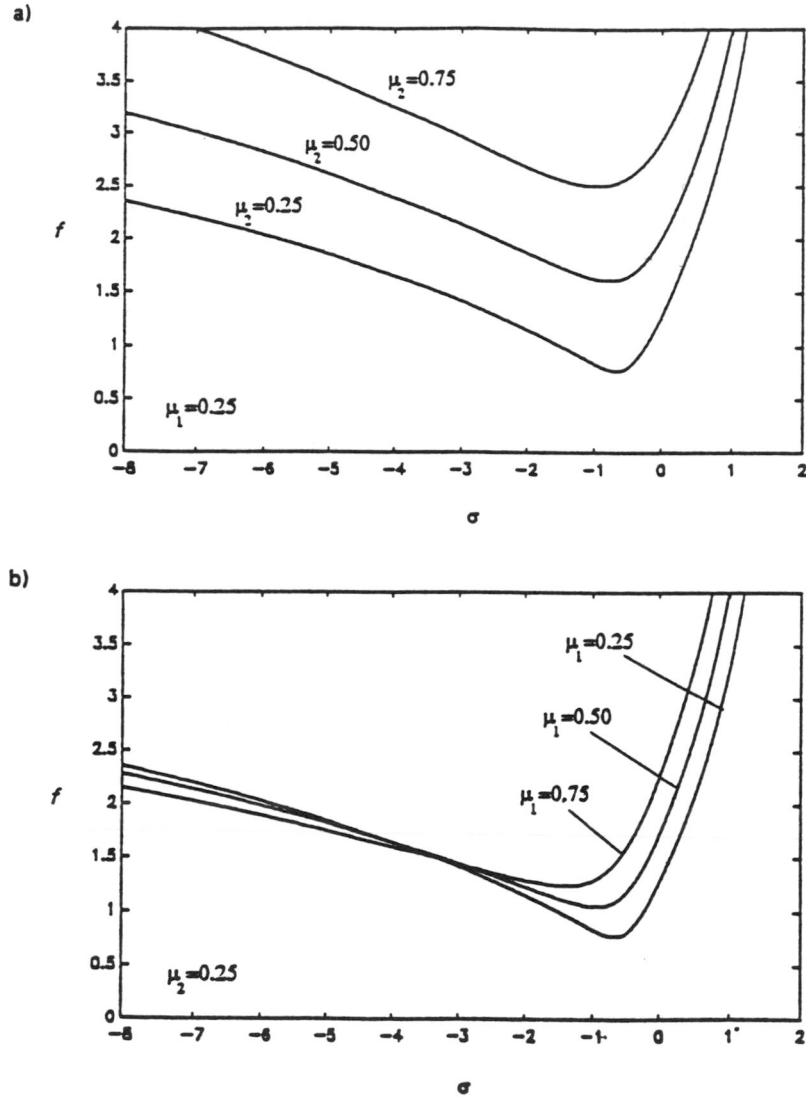

Figure 3.13 Boundaries between constant and oscillatory motions of u_1 for $\alpha_1 = \alpha_3 = 1$, $\alpha_2 = -2$, $\alpha_4 = 3$, and various damping values.

In Figure 3.14, we present a sequence of responses obtained for the parameter values used in Figure 3.12, $f = 2.5$, and various values of σ. As shown in Figure 3.12, as σ is decreased through $\sigma = 0.349$, a Hopf bifurcation occurs. In Figure 3.14a, we plot the motion in the $a–u_1$ plane just before the supercritical Hopf bifurcation occurs. As expected, the long-time response consists of only the stable fixed point. It should be noted that there exists a second fixed-point solution corresponding to negative values of u_1 which is not plotted here.

In Figure 3.14b, we show the motion just after the bifurcation. As predicted, the response changes from the point in the plane shown in Figure 3.14a to the limit cycle shown in

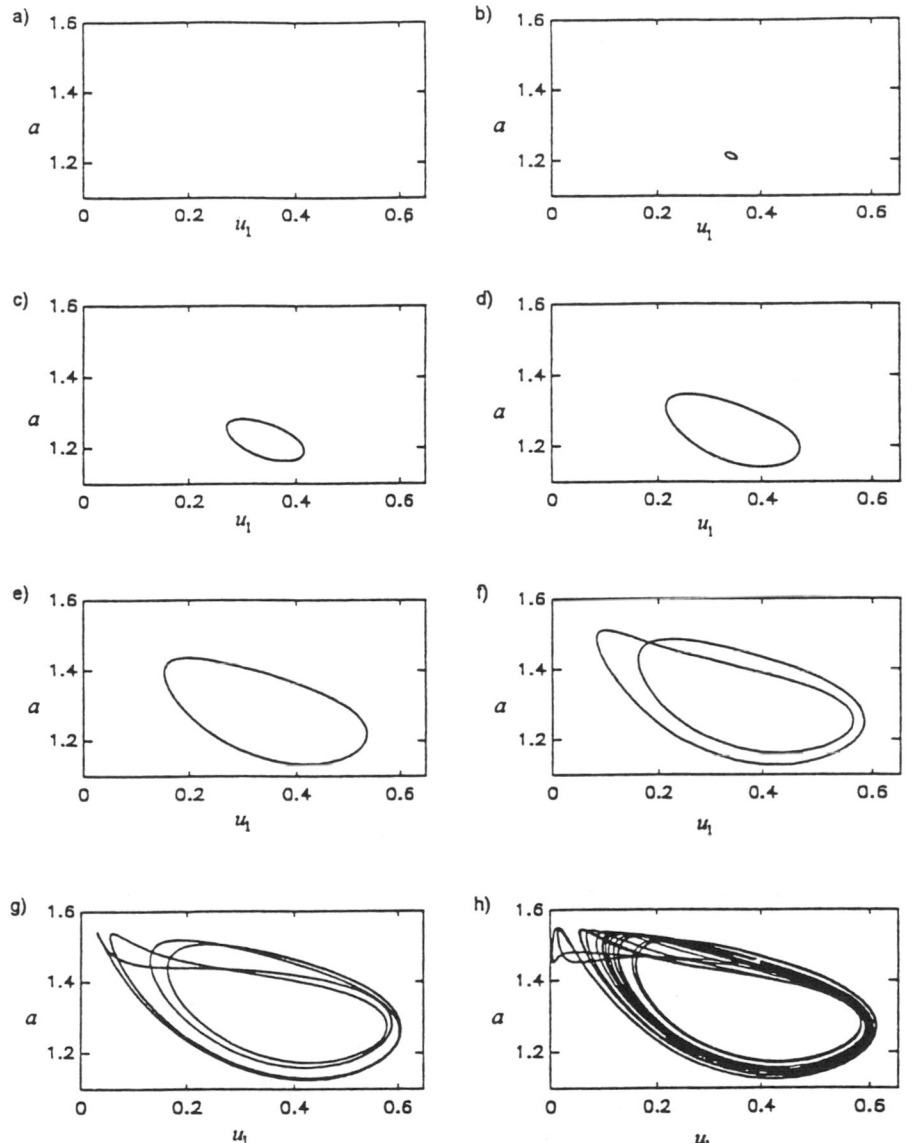

Figure 3.14 Numerical simulation of the averaged equations for $\alpha_1 = \alpha_3 = 1$, $\alpha_2 = -2$, $\alpha_4 = 3$, $\mu_1 = 0.25$, $\mu_2 = 0.5$, $f = 2.5$, and $\sigma = $ (a) 0.350, (b) 0.348, (c) 0.300, (d) 0.200, (e) 0.000, (f) -0.170, (g) -0.243, (h) -0.260.

Figures 3.14b. As σ is further decreased, the size of the limit cycle increases as shown in Figures 3.14b–e. Decreasing σ further, we obtain the period-doubling bifurcation sequence of Figure 3.14f–g which culminates in the creation of the chaotic attractor shown in Figure 3.14h. It should be noted that only a short sample of the chaotic attractor is shown. As the motion continues, the trajectory would fill the area outlined roughly by the portion of the trajectory shown.

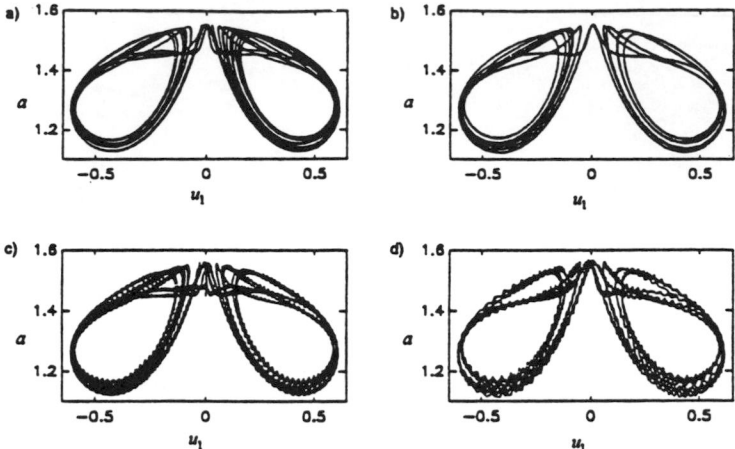

Figure 3.15   Numerical simulation of (a) the averaged equations and (b)–(d) the exact equations for $\alpha_1 = \alpha_2 = 1$, $\alpha_2 = -2$, $\alpha_4 = 3$, $\mu_1 = 0.25$, $\mu_2 = 0.5$, $f = 2.5$, $\sigma = -0.27$, and $\varepsilon =$ (b) 0.01, (c) 0.03, (d) 0.05.

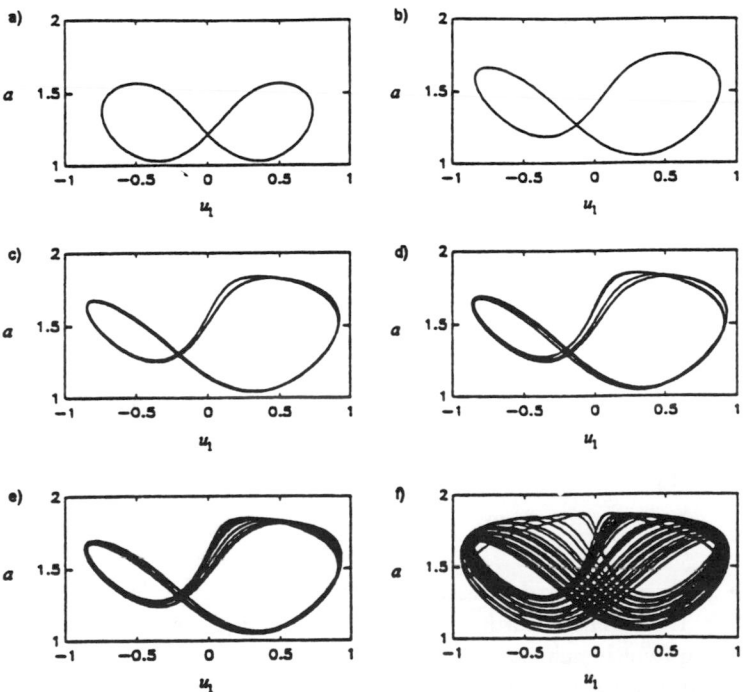

Figure 3.16   Numerical simulation of the averaged equations for $\alpha_1 = \alpha_3 = 1$, $\alpha_2 = -2$, $\alpha_4 = 3$, $\mu_1 = 0.25$, $\mu_2 = 0.5$, $f = 2.5$, and $\sigma =$ (a) -0.41, (b) -1.10, (c) -1.28, (d) -1.31, (e) -1.32, (f) -1.36.

For all of the responses shown in Figure 3.14, there exists a mirror image with opposite signs of u_1 and v_1 in the left half of the plane. Decreasing σ further, the chaotic attractors in the left and right halves of the $a-u_1$ plane merge into a single attractor. That is, the motion does not remain in either the left or the right half of the plane but rather jumps erratically from one to the other. This response is shown in Figure 3.15a. As σ is further decreased through roughly -0.41, this attractor loses stability. The motion is no longer attracted to it but rather diverges from it after some time and jumps to another attractor (depicted in Figure 3.16a) which, for values of σ less than roughly -0.2, coexists with the attractors discussed thus far.

As shown in Figure 3.16a this attractor is periodic and symmetric at $\sigma = -0.41$. As σ is decreased, a symmetry-breaking bifurcation occurs; one of the two resulting non-symmetric attractors is shown in Figure 3.16b. As σ is further decreased, these attractors undergo a period-doubling bifurcation sequence leading to chaos as shown in Figures 3.16c–f. The chaotic attractor in Figure 3.16e is non-symmetric and an attractor with its mirror image also exists. Another decrease in σ causes these attractors to merge, resulting in the symmetric attractor shown in Figure 3.16f. As σ is further decreased, a great variety of nonlinear dynamical phenomena are observed until a reverse Hopf bifurcation occurs at the end of the unstable branch leading to stable fixed-point solutions.

The analysis of this two-degree-of-freedom system shows interactions between high- and low-frequency modes through which an excitation applied to the high-frequency mode results in large-amplitude responses in the low-frequency mode. The response of the system is similar to that reported in the two experiments.

3.5   CONCLUDING REMARKS

The interaction between high- and low-frequency modes observed experimentally and demonstrated theoretically is of great practical importance. In many engineering systems, high-frequency excitations can be caused by rotating machinery. Through the mechanism discussed in this chapter, energy from high-frequency sources can be transferred to low-frequency modes of supporting structures or foundations, resulting in harmful large oscillations. Moreover, the results obtained in this research indicate that the use of conventional methods for decreasing modal interactions, such as increasing the dissipation or decreasing the forcing amplitude, may have undesirable effects.

ACKNOWLEDGEMENT

This work was supported by the United States Air Force Office of Scientific Research under Grant No. F49620-92-J-0197.

REFERENCES

Anderson, T. J., Balachandran, B., and Nayfeh, A. H. (1992) Observations of modal interactions among modes with widely spaced frequencies, *ASME Winter Annual Meeting, Nonlinear Vibration Symposium*, *November 8–13, 1992.*

Burton, T. D. and Kolowith, M. (1988) Nonlinear resonances and chaotic motion in a flexible parametrically excited beam, *Proceedings of the Second Conference on Non-Linear Vibrations, Stability, and
  Dynamics of Structures and Mechanics, Blacksburg, VA, June 1–3, 1988.*

Cusumano, J. P. and Moon, F. C. (1989) Low dimensional behaviour in chaotic nonplanar motions of
  a forced elastic rod: experiment and theory, *Nonlinear Dynamics in Engineering Systems, IUTAM
  Symposium, Germany, 1989.*

Dugundji, J. and Mukhopadhyay, V. (1973) Lateral bending-torsion vibrations of a thin beam under
  parametric excitation, *Journal of Applied Mechanics,* **40**, 693–698.

Haddow, A. G. and Hasan, S. M. (1988) Nonlinear oscillations of a flexible cantilever: experimental
  results, *Proceedings of the Second Conference on Non-Linear Vibrations, Stability, and Dynamics of
  Structures and Mechanics, Blacksburg, VA, June 1–3, 1988.*

Moon, F. C. (1987) *Chaotic Vibrations.* New York, Wiley-Interscience.

Nayfeh, A. H. (1973) *Perturbation Methods,* New York, Wiley-Interscience.

Nayfeh, A. H. and Balachandran, B. (1989) Modal interactions in dynamical and structural systems,
  ASME Applied Mechanics Reviews, **42**, 175–202.

Nayfeh, A. H. and Mook, D. T. (1979) *Nonlinear Oscillations,* New York, Wiley-Interscience.

Nayfeh, S. A. and Nayfeh, A. H. (1994) Energy transfer from high- to low-frequency modes in flexible
  structures via modulation, *Journal of Vibration and Acoustics,* **116**, 203–207.

Nayfeh, S. A. and Nayfeh, A. H. (1993) Nonlinear interactions between widely spaced modes, *International Journal of Bifurcation and Chaos,* **3**, 417–427.

A. H. Nayfeh, S. A. Nayfeh, T. A. Anderson and B. Balachandran, *Department of Engineering
Science and Mechanics, Virginia Polytechnic Institute and State University, Blacksburg, VA 24061–0219,
USA*

4 STABILITY MEASUREMENTS IN NONLINEAR MECHANICAL EXPERIMENTS GUIDED BY DYNAMICAL SYSTEMS THEORY

P. V. Bayly, L. N. Virgin, J. A. Gottwald and E. H. Dowell

Physical experiments remain an important aspect of research in nonlinear dynamics. This chapter presents some recent results from a series of experiments conducted to illuminate nonlinear behaviour in mechanical oscillators. The emphasis is on the measurement of stability in real devices, using local linear approximations to Poincaré maps. For stable periodic orbits, the local behaviour of the map is inferred from transients after imposed perturbations. For chaotic motion the maps can be estimated by fitting the behaviour of nearly recurrent points. Three experiments are discussed: a physical realization of Duffing's equation using a cart rolling in a potential-energy track; an impacting pendulum; and a two-degree-of-freedom spring pendulum. Characteristic multipliers are estimated for stable periodic motion in the impact oscillator and the spring pendulum, and for unstable periodic orbits embedded in chaotic attractors of the impact oscillator and the Duffing cart.

4.1 INTRODUCTION

The stability of nonlinear dynamical systems is an important issue. Analytical and numerical techniques to describe the stability of mathematical models have been available for some time and are well documented (Bolotin 1964; Hayashi 1985; Guckenheimer and Holmes 1983). In contrast to the output of analyses and simulations, data from experiments are affected by noise and coupling with the laboratory environment. Measurements of state variables are often inaccessible or uncertain. Quantitative techniques developed within the framework of dynamical systems theory are much more valuable if they are robust enough to apply to experimental data. Moon has included a discussion of experimental methods in nonlinear dynamics in his recent book (Moon 1992).

Nonlinearity and Chaos in Engineering Dynamics
Edited by J. M. T. Thompson and S. R. Bishop,   © 1994 John Wiley & Sons Ltd

Characteristic multipliers (CMs), or Floquet multipliers, represent the rate of attraction (or repulsion) of motion towards (or away from) a periodic orbit (Thompson and Stewart 1986). The CMs of a periodic orbit are the eigenvalues of the associated Poincaré map, linearized about its corresponding fixed point. Eigenvalues of magnitude greater than one belong to unstable orbits; their eigenvectors are the principal directions of repulsion.

This chapter describes the estimation of CMs purely from experimental data. Perturbations were applied to stable periodic motion of the impact oscillator and the spring pendulum. Using data from Poincaré sections during the resulting transients, CMs of the orbits were tracked as a forcing parameter was varied. To eliminate spurious eigenvalue estimates in the two-degree-of-freedom (2-DOF) spring pendulum, the Karhunen–Loeve decomposition was used to reduce the local linear map to its least stable eigenspace. Finally, CMs of unstable periodic orbits were extracted from Poincaré sections of chaotic motion in the Duffing cart and the impact oscillator.

A related study by Murphy *et al.* (1994) discusses the application of these methods to a jump instability in a geometrically nonlinear 1-DOF oscillator. Other behaviour under investigation includes the spinning motion of a simple pendulum (Bayly and Virgin 1992) and the escape of a periodically forced mass from a potential well.

4.2   OVERVIEW OF EXPERIMENTS

In this section we describe three experiments which exhibit bifurcations and/or chaos: a cart rolling in a potential well, designed to mimic Duffing's equation; an impacting pendulum; and an elastic pendulum.

Each experiment consists of a forced, dissipative, physical system which undergoes at least one bifurcation as a parameter is varied. The Duffing cart and the impacting pendulum both respond to periodic forcing with subharmonic and chaotic orbits. The spring pendulum furnishes an example of a higher-order system which exhibits internal resonance at certain parameter values. Schematic diagrams of the three experiments are shown in Figure 4.1.

4.2.1   The Duffing oscillator

Duffing's equation describes oscillators with cubic stiffness terms. It also roughly describes the motion of a cart rolling on a quartic potential-energy track. The experimental track was constructed from plexiglass; the cart's displacement along the track was recorded by a potentiometer geared to a chain–sprocket pair. Periodic forcing was incorporated by imposing a sinusoidal lateral displacement to the track.

Duffing's equation with damping and harmonic forcing takes the form

$$X'' + 2\zeta X' - \frac{X}{2}(1 - X^2) = A\eta^2 \cos(\eta\tau), \tag{1}$$

where the displacement X is analogous to the horizontal projection of the cart's position and η is the ratio of forcing frequency to the natural frequency of small oscillations. This equation includes only viscous (ζ) damping and neglects some effects of the imposed track motion. A more accurate mathematical model and further details of the experimental system

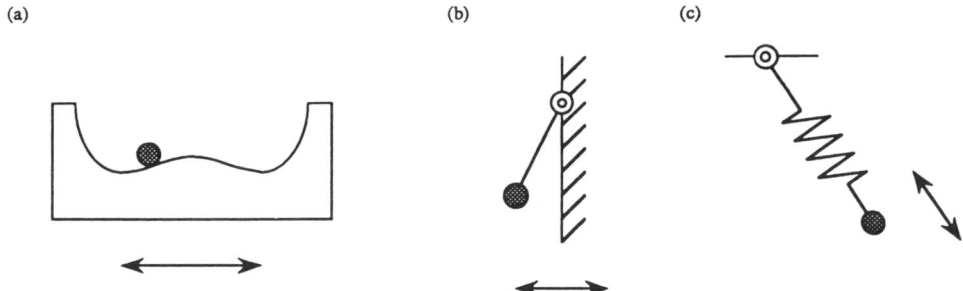

Figure 4.1 Schematic diagrams of three experiments: (a) the Duffing cart in a double potential well, driven by lateral displacement of the track; (b) the impacting pendulum, also excited by lateral support motion; (c) the radially forced spring pendulum.

can be found in Gottwald *et al.* (1992). In the current study the emphasis is on the quantitative description of experimental data, not on the construction of the most faithful numerical model.

4.2.2 The impacting pendulum

The impacting pendulum was used to examine chaotic motion and bifurcations to subharmonic behaviour. The device consisted of a rubber ball mounted on a freely pivoting aluminium arm; motion of the ball was limited by a rigid stop at $\theta = 0$. The experiment was excited by a harmonic lateral displacement of its base, and angular displacement was measured with a potentiometer mounted to the pendulum pivot.

The equation of motion for the system is

$$\theta'' + \frac{2\zeta\Omega}{\eta}\theta' + \frac{\Omega^2 \sin\theta}{4\eta^2} = \frac{A}{L}\cos\theta \sin\tau, \tag{2}$$

where θ is the angle measured from the plane of the pivot axis. The parameter $\Omega = 1$ in the free-play region ($\theta > 0$); the impact condition is incorporated in equation (2) by letting Ω become very large ($\Omega \gg 1$) when $\theta \leqslant 0$. Chaos is present in this system for certain parameter values, and subharmonic orbits are particularly common (Thompson and Ghaffari 1983; Shaw and Holmes 1983).

4.2.3 The spring pendulum

The 2-DOF spring pendulum exhibits combination resonance due to nonlinear coupling when its primary mass is excited in the radial direction. The experimental system was made of a cylindrical brass bob which slid on linear bearings along a hardened-steel pendulum arm. The bob was attached via slender cables to a pair of springs. Radial excitation was accomplished by applying a harmonic displacement to the end of the springs. Angular displacement was measured by a potentiometer at the pendulum pivot, and radial displacement was measured by a potentiometer which recorded the rotation of one of the cable pulleys.

The governing equations of motion of the spring pendulum are

$$R'' + 2\frac{\zeta_1}{\eta_1}R' + \frac{1}{\eta_1^2}R - (1+R)(\theta')^2 + \frac{1}{\eta_2^2}(1 - \cos\theta) = F\sin\tau \tag{3}$$

and

$$\theta'' + 2\frac{\zeta_2}{\eta_2}\theta' + \frac{2}{1+R}R'\theta' + \frac{1}{\eta_2^2}\frac{\sin\theta}{(1+R)} = 0. \tag{4}$$

The ratios of forcing frequency to the natural frequencies of radial and angular oscillation are η_1 and η_2 respectively. The ratio of natural frequencies of radial and swinging motion was fixed at 2.76 (i.e. $\eta_2 = 2.76\ \eta_1$ at all times). The spring pendulum provides a physical testbed on which to try to extend algorithms hitherto applied to 1-DOF systems.

4.2.4   Forcing and data acquisition

All three of the experiments were forced by the same mechanism: a harmonically oscillating aluminium table driven by a Scotch yoke. Unmodelled characteristics of the forcing table included the following: a small amount of backlash existed in the Scotch yoke; the timing belts between the motor and the yoke had finite stiffness; and a small amount of low-frequency drift' in motor speed was usually present. Analogue voltage signals from the potentiometers were converted to 12-bit digital data and acquired by a Mac II microcomputer using Lab View software and a National Instruments NB-MIO-16 A–D board. Poincaré sections were taken by sampling only when a marker on the Scotch yoke triggered a photodetector.

4.3   CHARACTERISTIC MULTIPLIERS OF STABLE ORBITS NEAR BIFURCATIONS

A periodic orbit of a continuous system is associated with a fixed point of the corresponding Poincaré map. The stability of a stable periodic orbit near a bifurcation may be explored by examining the linearized Poincaré map after an imposed perturbation.

A local linear approximation to the Poincaré map takes the form

$$\boldsymbol{x}_{n+1} = A\boldsymbol{x}_n + \boldsymbol{b}, \tag{5}$$

where $\boldsymbol{x}_n$ (an N-dimensional column vector) is the nth iterate of the map and A is the local Jacobian matrix ($N \times N$). Constructing the matrices:

$$X_0 = [\boldsymbol{x}_{n_1}, \boldsymbol{x}_{n_2}, \dots, \boldsymbol{x}_{n_M}], \quad (N \times M)$$
$$X_1 = [\boldsymbol{x}_{n_1+1}, \boldsymbol{x}_{n_2+1}, \dots, \boldsymbol{x}_{n_M+1}], \quad (N \times M)$$
$$B = [\boldsymbol{b}, \boldsymbol{b}, \dots, \boldsymbol{b}], \quad (N \times M)$$

the local linear equation can be written for a number of points simultaneously:

$$X_1 = AX_0 + B. \tag{6}$$

Here M is the number of Poincaré points in matrices X_0 and X_1. If $M = N + 1$ for an N-dimensional Poincaré map, equation (6) can be solved for the unknown elements of A and B. In practice, many more Poincaré points should be taken and the equation should be solved in a least-squares sense (Murphy *et al.* 1994). The eigenvalues of A are the CMs of the associated periodic orbit. The magnitude of the CMs describe the rate of attraction of transients to the limit cycle; CMs with magnitude greater than 1 belong to unstable cycles.

4.3.1 Period-doubling bifurcation of the impacting pendulum

The amplitude of angular displacement of the impacting pendulum is plotted in Figure 4.2a as a function of the frequency ratio, η. Both simulation and experiment experience a period-doubling bifurcation over this range of frequencies. The phase portraits of experimental and simulated period-2 motion display good agreement. They also illustrate the use of time-delay embedding to represent state-space orbits.

In the experiment the perturbation consisted of the brief application of a brake to the pendulum pivot. Four displacement samples were taken every forcing period, separated by a quarter-cycle delay. This allowed Poincaré sections to be reconstructed from lagged values of the displacement:

$$x_n = [\theta(\tau_n), \theta(\tau_n + \pi/2)]^T, \quad \tau_n = 2\pi n + \phi. \tag{7}$$

For each of a number of parameter values, periodic orbits were perturbed, several series of Poincaré points were obtained. These points were substituted into equation (6) (with $N = 2$) and estimates of the Jacobian, A, and its eigenvalues were extracted. Plots of the paths of the CMs in the complex plane are shown in Figures 4.3a and 4.3b. For small-amplitude

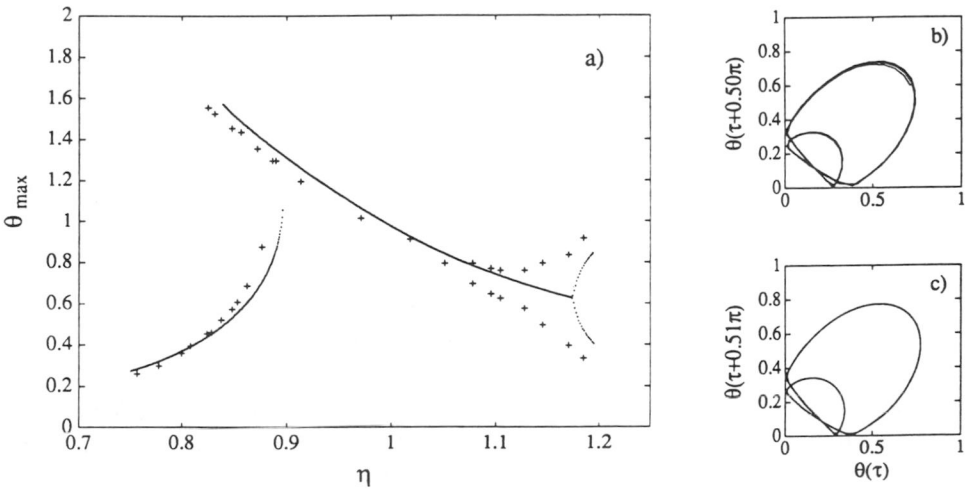

Figure 4.2 (a) The amplitude of swinging of the impacting pendulum in response to harmonic lateral support excitation. The frequency ratio η was varied from 0.75 to 1.20 and the forcing amplitude A/L was fixed at 0.14. A viscous damping factor $\zeta = 0.038$ was used in the numerical model. Experiment: (+) signs; simulation: dots. (b) An experimental phase portrait with $\eta = 1.23$ and $A/L = 0.13$. (c) A phase portrait from a simulation with $\eta = 1.23$, $A/L = 0.13$.

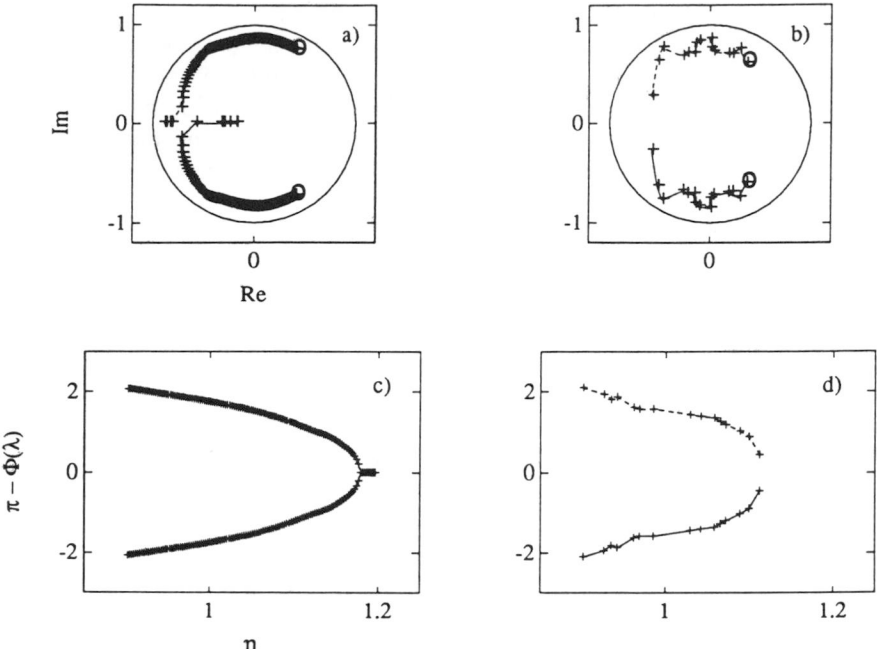

Figure 4.3 The loci of characteristic multipliers (CMs) of period-1 motion in the impacting pendulum as the forcing frequency was varied. (a) Estimates from numerical simulations as the frequency ratio η was increased from 0.90 to 1.20. (b) The corresponding trajectories for the experiment as η was increased from 0.90 to 1.12. (c) The difference between the phase angle, $\Phi = \arg(\text{Im}, \text{Re})$ and π, plotted as a function of η for numerical simulations. This representation shows the approach of the CMs to the negative real axis. (d) Same as (c) but for the experimental impacting pendulum.

motion, the estimates of the CMs have been shown (Murphy *et al.* 1994) to approach the eigenvalues of the exact Poincaré map obtained by Shaw and Holmes for the piecewise linear impact oscillator (Shaw and Holmes 1983). The rotation of the eigenvalues towards the real axis, illustrated in Figures 4.3c and 4.3d, has been proposed as a predictor of incipient instability (Thompson and Virgin 1986).

4.3.2 Parametric instability of the spring pendulum in four-dimensional (4D) state space

Radial excitation of the spring pendulum leads to stable oscillations with coupled radial and angular motion, if the forcing frequency is close to twice the angular natural frequency. A plot of radial amplitude against frequency ratio is shown in Figure 4.4, along with two steady-state trajectories in (θ, R)-space. In Figure 4.4a the deviation from the linear frequency response marks the regime of coupled swinging motion. The stability of these orbits was investigated experimentally.

Perturbations were imposed by abruptly changing the forcing frequency, and samples of radial and angular displacement were obtained during the ensuing transients. Poincaré points

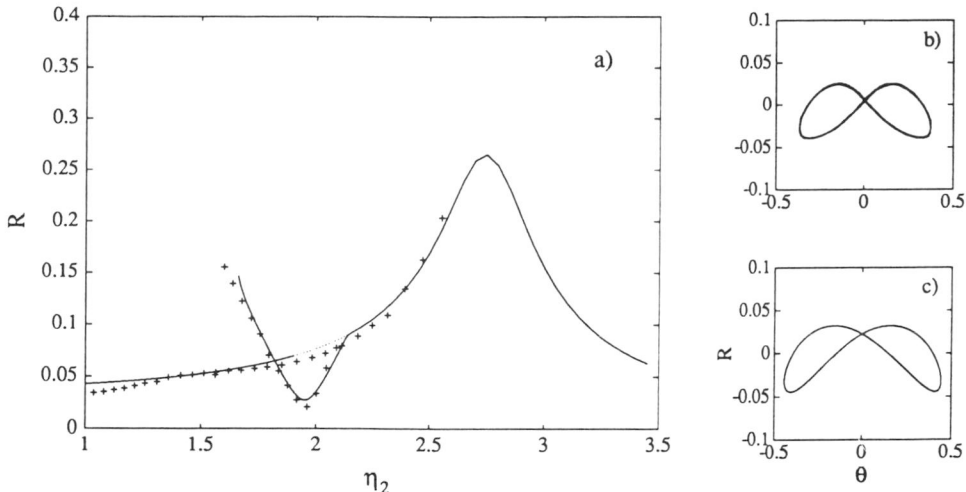

Figure 4.4 (a) The amplitude of radial oscillations of the spring pendulum as a function of the frequency of harmonic radial forcing. The forcing amplitude F was fixed at 0.28. Viscous damping coefficients $\zeta_1 = 0.07$ and $\zeta_2 = 0.02$ modelled damping in the simulations. Experimental values are shown by the (+) signs. Simulation results are indicated by the solid curve (dotted in the unstable regime near $\eta_2 = 2$). (b) An experimental phase projection in the (θ, R) plane with $\eta_2 = 2.01$. (c) A phase portrait from a simulation with $\eta_2 = 2.01$.

were constructed using time-delay embedding. In this experiment the points were embedded in a 4D state space to take into account the two degrees-of-freedom in the device:

$$x_n = [R(\tau_n), \theta(\tau_n), R(\tau_n + \pi/2), \theta(\tau_n + \pi/2)]^T, \quad \tau_n = 4\pi n + \phi. \tag{8}$$

The Karhunen–Loeve decomposition was applied to the ensemble of Poincaré points to locate the least stable 2D subspace. (Bayly and Virgin 1993a). This method identifies the

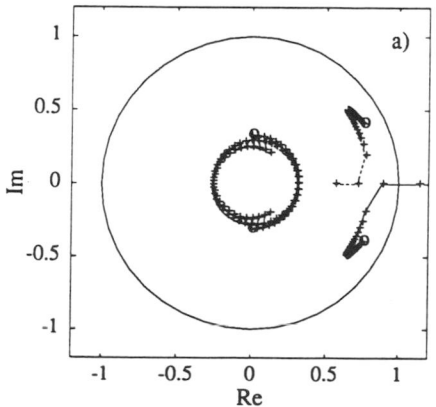
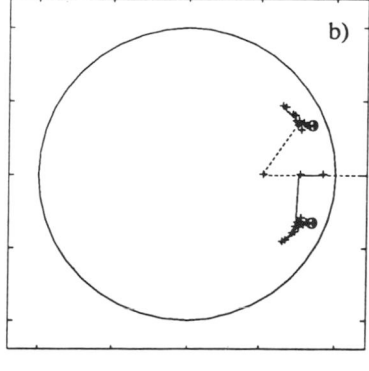

Figure 4.5 The loci of characteristic multipliers (CMs) of period-2 coupled motion of the spring pendulum as the frequency ratio η_2 was decreased from 2.05 to 1.60. (a) Analytical estimates from Floquet theory (circles indicate the start of each locus, estimated at $\eta_2 = 2.05$). (b) The trajectories of two CMs estimated from the projection of experimental data onto the plane spanned by the two dominant Karhunen–Loeve eigenvectors. These should correspond to the largest-magnitude eigenvalues of the full system.

linear subspace which contains the most variance of a data set. The most weakly attracting eigenspace should generally contain the most variance, since strongly attracting dynamics tend to limit excursions. Estimates of CMs were obtained by projecting the original data onto the least stable 2D eigenspace and fitting an $N = 2$ version of equation (6).

Plots of CMs obtained from the variational equations of an analytical solution (Bayly and Virgin 1993a) are shown, together with experimental estimates of the two least stable CMs, in Figure 4.5. Estimates of CMs from numerical simulations were also found to agree very well with analytical predictions (Bayly and Virgin 1993a). The loci of the CMs were obtained as the frequency ratio η_2 was varied from approximately 2.05 to 1.6. The small-magnitude eigenvalues in the analytical results correspond to strongly attracting directions in state space.

4.4   CHARACTERISTIC MULTIPLIERS OF UNSTABLE ORBITS IN CHAOTIC ATTRACTORS

Similar methods to those described above can be applied to characterize unstable periodic orbits, when such cycles can be found. One place to find unstable orbits is in a chaotic attractor (Grebogi *et al.* 1988). Following Lathrop and Kostelich (1989), we identify points in the Poincaré section that are nearly recurrent. Points whose future iterates return to their location in state space, within a small normalized distance ε, after, say, p cycles are called (p, ε)-recurrent points. Using data near these recurrent points, one can build a local linear approximation to the Poincaré map. The methods of Lathrop and Kostelich were modified slightly and applied to the Poincaré sections of chaotic motion in the Duffing cart and the impacting pendulum.

4.4.1   An unstable period-1 cycle of the Duffing cart

Poincaré sections of chaotic motion taken from the experiment are very similar to those obtained from simulations of Duffing's equation, both qualitatively and in terms of measures such as the correlation dimension and Lyapunov exponents. A comparison is shown in Figure 4.6. Lyapunov exponents were estimated from the experimental data using the program *LCE_EXP*, written by Kruel and Eiswirth (Kruel *et al.* 1993). In such algorithms, which rely on successive linear approximations to the local flow, the largest exponent is expected to be the most reliable. In simulations, Lyapunov experiments were found by integrating the exact locally linearized equations as described by Eckman and Ruelle (1985).

An autocorrelation function of chaotic motion of the cart is shown in Figure 4.7a. The local maxima reflect a degree of underlying period-5 and period-1 recurrence. The location of an unstable period-1 orbit in the chaotic attractor was suggested by a cluster of $(1, 0.05)$-recurrent points near $x^* = [-1.33, 0.55]^T$. All points x_m within a normalized radius $\delta = 0.03$ of x^* were found and stored with their pre-images x_{m-1} and their images x_{m+1}. Karhunen–Loeve decomposition indicated that both sets lay on approximately 1-dimensional linear spaces, and identified the directions of these manifolds. Projecting the original data onto these manifolds and using a simple 1D linear fit (the $N = 1$ version of equation (6)), estimates of CMs were obtained.

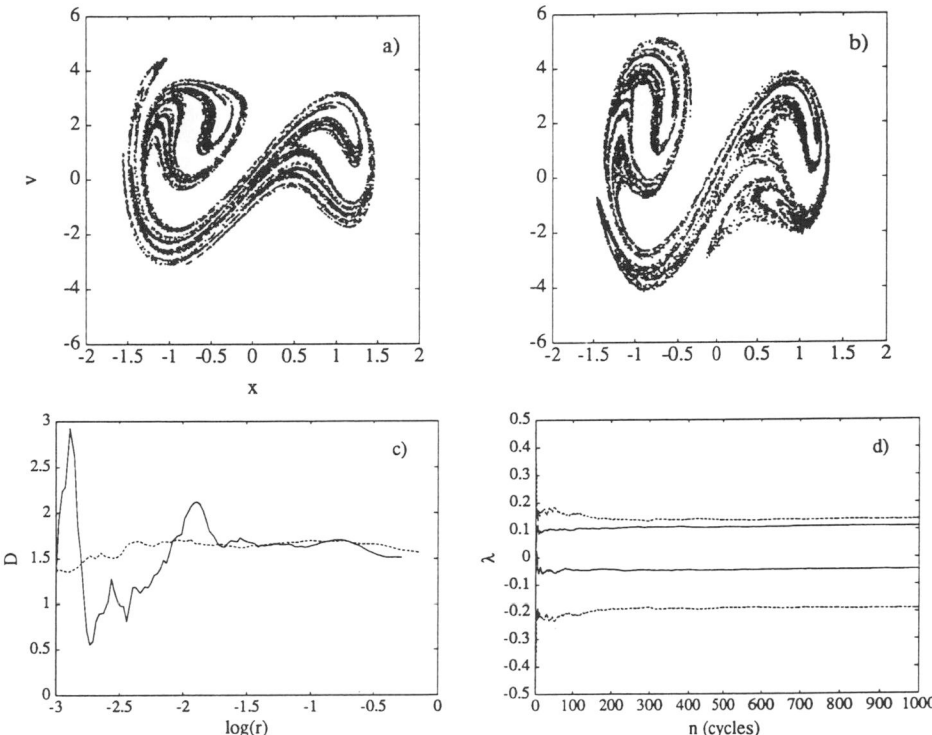

Figure 4.6 A comparison of chaos in the forced, damped Duffing's equation, and in the experimental Duffing cart. (a) Poincaré section obtained by sampling the numerical simulation at forcing phase $\phi = 0$. Parameters were $F = 0.188$, $\Omega = 0.84$, $\zeta = 0.025$. (b) The corresponding Poincaré section from the experiment. (c) The curves of correlation dimension, $D = d[\log C(r)]/d[\log r]$ versus $\log r$ showing a well-defined plateau near $D = 1.7 \pm 0.1$. Experiment: solid curve; simulation: dashed curve. (d) Estimates of Lyapunov exponents plotted as a function of the number of forcing cycles used in the average divergence calculation. Experimental estimates: $\lambda_1 = 0.11$, $\lambda_2 = -0.04$ (solid curves); simulation estimates: $\lambda_1 = 0.14$, $\lambda_2 = -0.19$ (dashed curves).

A magnified view of the chaotic attractor showing the near-recurrent points and their images under the map, along with the principal directions of attraction and repulsion, is shown in Figure 4.7b. The CMs were approximately 2.9 and -0.2. The eigenvectors are estimates of the tangent vectors to the stable and unstable manifolds.

4.4.2 An unstable period-3 cycle in the impacting pendulum

Figure 4.8 illustrates the very good qualitative agreement between a chaotic attractor from the impacting pendulum experiment and one from a numerical simulation with the same parameters (Bayly and Virgin 1993b).

An unstable period-3 orbit of the experimental impacting pendulum (see Figure 4.9a) was located by finding a group of (3, 0.05)-recurrent points in the experimental Poincaré section. As before, points within $\delta = 0.03$ were used to generate a local linear approximation to the

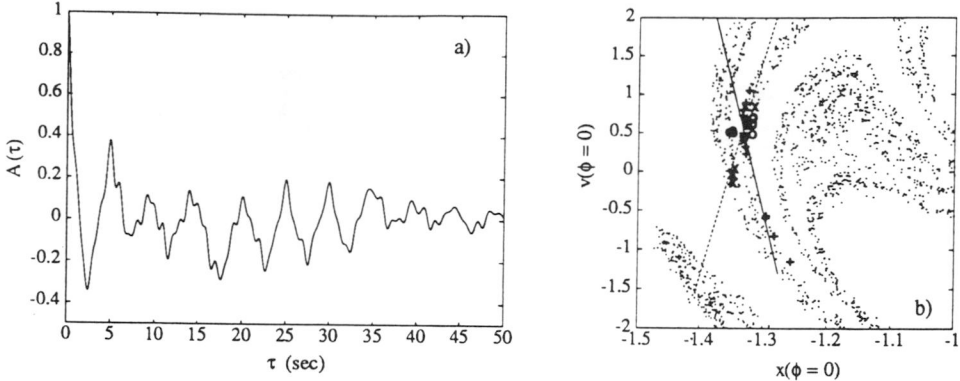

Figure 4.7 (a) The autocorrelation function of the displacement of the experimental Duffing cart. Parameters are the same as in Figure 4.6, corresponding to an actual forcing cycle length of 1.0 second. Local maxima in the autocorrelation function reflect a degree of recurrent behaviour. (b) Close-up view of the experimental Poincaré section, showing 14 points close to the (1, 0.05)-recurrent point $[-1.33, 0.55]^{T}$ ($\circ$). Also shown are their pre-images ($\times$) and their images ($+$). The unstable eigenvector (solid) and the stable eigenvector (dashed) of the saddle point are also shown.

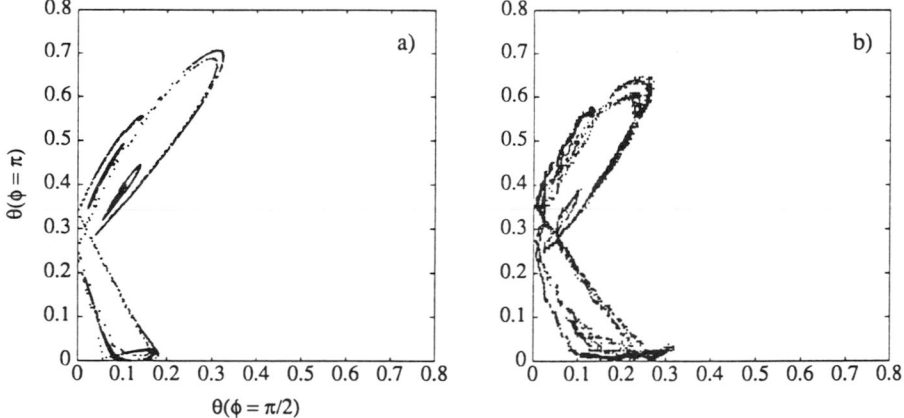

Figure 4.8 A qualitative comparison between numerical and experimental Poincaré seections of chaos in the impacting pendulum. (a) The time-delay embedded Poincaré section obtained by sampling the displacement at forcing phase $\phi = \pi/2$ and at $\phi = \pi$ during each forcing cycle of a simulation. Parameters were $A/L = 0.13$, $\eta = 1.29$, $\zeta = 0.038$. (b) The corresponding Poincaré section from the experiment.

Poincaré map, and from these to find its characteristic multipliers. The magnified Poincaré section showing the local dynamics is shown in Figure 4.9b; the CMs were estimated to be -4.8 and 0.3.

4.5 DISCUSSION

Methods of characterizing the stability of nonlinear systems were applied to three physical experiments. The basic techniques were drawn, from classical dynamical systems theory. It

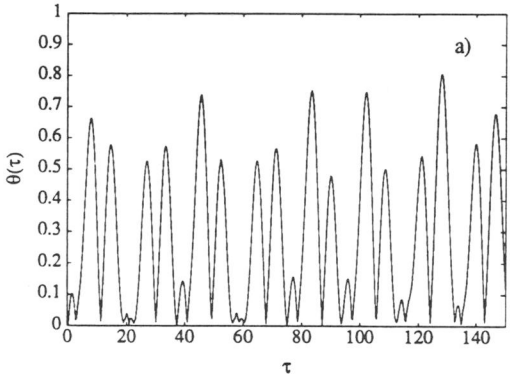 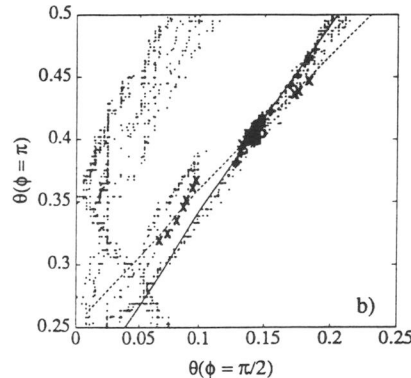

Figure 4.9 (a) A time series of chaotic motion in the impacting pendulum. Two cycles of a nearly recurrent period-3 orbit can be seen between $\tau \approx 75$ and $\tau \approx 115$. (b) Close-up view of the experimental Poincaré section, showing 21 points close to the (3, 0.05)-recurrent point $[0.14, 0.40]^{\mathrm{T}}$ (∘). Also shown are their pre-images (×). The unstable eigenvector (solid) and the stable eigenvector (dashed) of the saddle point are also shown.

was necessary, however, to use statistical 'tricks', like least-squares solution and the Karhunen−Loeve decomposition, to eliminate spurious results due to noise and imprecision. With this pragmatic approach we have been able to obtain quantitative stability measurements from physical systems; these results may be used in applications like the prediction of instability (Thompson and Virgin 1986) or the control of chaos (Ott *et al.* 1990).

ACKNOWLEDGEMENT

This research was supported by grants from the National Science Foundation and the US Army Research Office.

REFERENCES

Bayly, P. V. and Virgin, L. N. (1992) Evidence of diffusive dynamics and random walking in a deterministic dynamical system: the shaken pendulum, *International Journal Of Bifurcations and Chaos*, **2** (4), 983−988.

Bayly, P. V. and Virgin, L. N. (1993a) An empirical study of the stability of periodic motion in the forced spring-pendulum, *Proc. R. Soc.*, **A443**, 391−408.

Bayly, P. V. and Virgin, L. N. (1993b) An experimental study into an impacting pendulum, *Journal of Sound and Vibration*, **164**, 364−374.

Bolotin, V. V. (1964) *Dynamic Stability of Elastic Systems*, San Francisco, Holden-Day.

Eckman, J. P. and Ruelle, D. (1985) Ergodic theory of chaos and strange attractors, *Reviews of Modern Physics*, **57** (3), 617−656.

Gottwald, J. A., Virgin, L. N. and Dowell, E. H. (1992) Experimental mimicry of Duffing's equation, *Journal of Sound and Vibration*, **158** (3), 447−467.

Grebogi, C., Ott, E. and Yorke, J. (1988) Unstable periodic orbits and the dimensions of multifractal chaotic attractors, *Physical Review A*, **37**, 1711−1724.

Guckenheimer, J. and Holmes, P. J. (1983) *Nonlinear Oscillations, Dynamical Systems, and Bifurcations of Vector Fields*, New York, Springer-Verlag.

Hayashi, C. (1985) *Nonlinear Oscillations in Physical Systems*, Princeton University Press, Princeton, NJ.

Kruel, Th.-M., Eiswirth, M. and Schneider, F. W. (1993) Computation of Lyapunov spectra: effect of interactive noise and application to a chemical oscillator, *Physica*, **63D**, 117–137.

Lathrop, D. E. and Kostelich, E. J. (1989) Characterization of an experimental strange attractor by periodic orbits, *Physical Review A*, **40**, 4028–4031.

Moon, F. C. (1992) *Chaotic and Fractal Dynamics*, New York, Wiley.

Murphy, K. D., Bayly, P. V., Virgin, L. N. and Gottwald, J. A. (1994) Measuring the stability of periodic attractors using perturbation induced transients: application to two nonlinear oscillators, *Journal of Sound and Vibration*, **172**, 85–102.

Ott, E., Grebogi, C. and Yorke, J. (1990) Controlling chaos, *Physical Review Letters*, **64** (11), 1196–1199.

Shaw, S. W. and Holmes, P. J. (1983) A periodically forced impact oscillator, *Journal of Sound and Vibration*, **90** (1), 129–155.

Thompson, J. M. T. and Ghaffari, R. (1983) Chaotic dynamics of an impact oscillator, *Physical Review A*, **27** (3), 1741–1743.

Thompson J. M. T. and Stewart, H. B. (1986) *Nonlinear Dynamics and Chaos*, John Wiley & Sons, Chichester.

Thompson, J. M. T. and Virgin, L. N. (1986) Predicting a jump to resonance using transient maps and beats, *International Journal of Nonlinear Mechanics*, **21** (3), 205–216.

P. V. Bayly, L. N. Virgin, J. A. Gottwald and E. H. Dowell, *School of Engineering, Duke University, Durham, NC 27708-0300 USA*

5 EXPERIMENTAL OBSERVATION OF BASINS OF ATTRACTION AND HOMOCLINIC BIFURCATION IN A MAGNETO-MECHANICAL OSCILLATOR

J. P. Cusumano and B. W. Kimble

A new experimental technique for observing global phase-space structures is presented. The method is based on the idea of stochastic interrogation: an ensemble of initial conditions is generated by switching between random and deterministic excitation. Basins of attraction can be found by following the evolution of points in this ensemble. The method is applied to the study of local and global bifurcations in a driven two-well magneto-mechanical oscillator. We observe the evolution of basins of attraction in the nonlinear oscillator as the forcing amplitude is increased, and find evidence for homoclinic bifurcation long before the onset of chaos. Since the entire transient is collected for each initial condition, the same data can be used to obtain pictures of the flow of phase space under the action of the Poincaré map. Using Liouville's theorem, we obtain global damping estimates by calculating the contraction of areas under the action of the map, and show that they are in good agreement with the results of more conventional damping estimation methods.

5.1 INTRODUCTION

The structure and evolution of basins of attraction is now understood to be a central issue in the study of instability phenomena and chaotic transitions, but virtually nothing has been done in this area experimentally, particularly for mechanical systems. In this chapter, we present an experimental technique that can be used to obtain images of global phase-space structures, such as basins of attraction, for a wide range of nonlinear oscillators. The method uses what can be thought of as stochastic interrogation: an ensemble of initial conditions, which are needed to construct basin images, is generated by switching between random and deterministic excitation. Using our method, we are able to observe basin evolution and find evidence for homoclinic bifurcation long before the onset of chaos.

Nonlinearity and Chaos in Engineering Dynamics
Edited by J. M. T. Thompson and S. R. Bishop,   © 1994 John Wiley & Sons Ltd

We apply the technique to a periodically driven oscillator similar to the well-known two-well magneto-elastic oscillator of Moon and Holmes (1979) and Moon (1980). After a detailed picture of the local bifurcation structure of the system over a specific path in the parameter space is obtained using an automated bifurcation data-acquisition system, the evolution of basins of attraction is studied for forcing amplitudes well below the first period-doubling. The transition from simple to complex basin boundaries as the forcing amplitude is increased is observed.

Extensive research has been conducted on numerically produced basins of attraction, and it is well understood that the global stability of an attractor is determined by the size and shape of its basin of attraction, as well as the nature of its basin boundaries. The appearance of fractal basin boundaries can greatly change the transient behaviour of a system: long chaotic transients can occur after the transition to fractal boundaries (Grebogi et al. 1983b). These transients can severely hamper the effectiveness of systems which need short settling times, such as phase-locked loop circuits, in which 'pull-in' times needed to create synchroniz-ation become extremely lengthy (Endo and Chua 1990). The appearance of fractal boundaries greatly increases the final-state uncertainty, even for very small initial-condition uncertainty (McDonald et al. 1980; Grebogi et al. 1983a). Finally, fractal boundaries can lead to rapid erosion in the size of basins of attraction, severely reducing global stability even when local stability is unchanged (Thompson and Soliman 1990; Cusumano et al. 1992).

The theory of homoclinic bifurcation furnishes conditions for the existence of complicated invariant sets in the phase space of a dynamical system: in specific applications, necessary conditions for the onset of chaos are most often obtained by application of the Holmes–Melnikov perturbation method (Guckenheimer and Holmes 1983). The intricate phase-space structures resulting from homoclinic bifurcation do not necessarily give rise to strange attractors. However, homoclinic bifurcation is associated with the phenomena of fractal basin boundaries, fractal–fractal transitions, chaotic transients, and final-state uncertainty, even when the possible attractors are all non-chaotic (Moon and Li 1985; Yamaguchi and Mishima 1985; Grebogi et al. 1986, 1987). Thus, it has become one of the major goals of researchers to be able to predict the transition from smooth to fractal boundaries. Using the experimental method described here, the same data used to obtain basins of attraction furnishes pictures of the *flow* of phase space under the action of the Poincaré map. Dissipation rapidly contracts the phase space, making it possible to obtain an image of the complicated unstable manifolds associated with chaotic transients, *even when the steady-state solutions are merely periodic*. Thus, our method makes it possible to experimentally observe homoclinic bifurcations much earlier than has been possible heretofore.

As a demonstration of how physical information can be extracted from the basin data for practical applications, we obtain damping estimates by calculating the contraction of areas under the action of the map. Using this 'volume decrement' data together with Liouville's theorem for a dissipative system, the damping coefficient is calculated in each well of the oscillator, and shown to be in excellent agreement with coefficients obtained using more conventional methods. This estimate also serves as further validation of the basin data.

In Section 5.2, the experimental two-well oscillator and the automated data-acquisition system are described. In Section 5.3, after a brief presentation of the local bifurcation structure of the system, basins of attraction are found for a region of parameter space in which almost all initial conditions are attracted to two period-1 orbits. In Section 5.4, experimental evidence for homoclinic bifurcation is obtained. The evidence consists of a simple-to-complex transition in the basin structure, an increase in the transient time, and the development of a persistant image of a complicated unstable manifold. In Section 5.5, we show how the basin data can be used to obtain damping estimates for the system using Liouville's theorem. A final discussion and conclusions are presented in Section 5.6.

5.2 DESCRIPTION OF THE EXPERIMENTAL SYSTEM

The mechanical system studied here is a stiffened beam buckled by two magnets. This system is similar to the magneto-elastic oscillator of Moon and Holmes (1979). Moon and Holmes showed that the steady-state behaviour of this system is in many respects well modelled by the two-well Duffing equation, and they used the experimental system to demonstrate the physical existence of strange attractors. However, whereas the Moon–Holmes system used a flexible beam, we have added extra stiffness in the form of steel bars (each 19.21 cm × 0.52 cm × 1.28 cm) epoxied and bolted along the length the thin steel beam (20.96 cm × 0.07 cm × 1.28 cm), away from the clamped end (see Figure 5.1). The additional constraint is necessary in our case to force the system to have one degree of freedom even during transients: our method of stochastic interrogation would excited higher modes in a flexible beam, making the study of basins of attraction more difficult. A 1.75 cm portion of the beam is left uncovered near the clamped end to act as a hinge from which we are

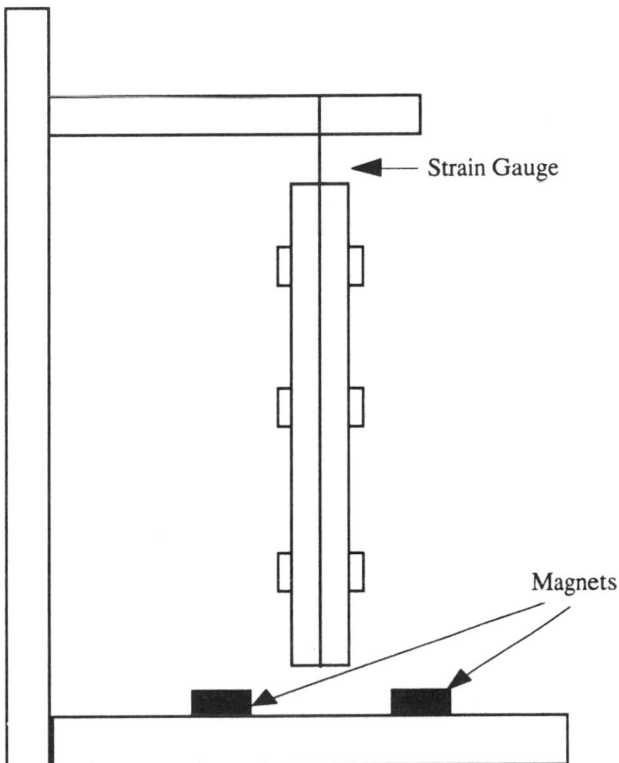

Figure 5.1 Schematic of the two-well magneto-mechanical oscillator (see text for dimensions). The constrained beam–magnet system is driven by an electromechanical shaker attached to the rigid frame.

able to determine position using strain gauges. Two rare-earth permanent magnets (1.27 cm diameter × 0.32 cm) are placed on the base of the frame holding the beam to create the two-well potential. The vertical distance between the undeformed beam tip and the magnets is 0.20 cm.

The magneto-mechanical oscillator is fixed to a rigid mount which is attached to an electromagnetic shaker. The shaker is controlled by the output of a digital-to-analogue (D/A) converter, the output of which passes through a low-pass filter set at 20 Hz before being sent to the shaker amplifier. The strain at the clamp and its time derivative (obtained by analogue differentiation) measure, respectively, the position x and velocity dx/dt of the beam. The strain and its time derivative are sent through a low-pass elliptical filter set at 50 Hz. Data was acquired using 12-bit data-acquisition (A/D) boards. The noise signal output from the filters with no forcing was found to be on the order of 1 mV r.m.s. Since this is below the resolution of the A/D converters as scaled for the experiments (2.44 mV for the least-significant bit), the noise floor in our measurements was as low as possible with 12-bit digitization.

The linear frequency response about the static equilibria of the left and right wells of Figure 5.1 was obtained using low-level random excitation to verify that the system did indeed have only one degree of freedom. For both equilibria, only one resonant peak was found below 100 Hz: the natural frequencies were found to be 9.14 ± 0.08 Hz for the right well and 9.62 ± 0.08 Hz for the left well. Since the next highest peaks in both transfer functions were about 30 dB down at approximately 225 Hz, we conclude that the system is acting like a single-degree-of-freedom oscillator in the bandwidth of relevance to this study (< 50 Hz).

Data acquisition was carried out with a workstation-based system, and control programs for bifurcation and basin diagrams were written using a library of real-time Fortran subroutines. Experimental bifurcation diagrams are obtained in a straight-forward manner, although current methods limit data to stable solution branches. The control program outputs a sinusoidal forcing signal with amplitude V, waits a predetermined number of forcing periods for transients to dissipate, and then collects data, again for a predetermined number of periods. The bifurcation parameter is then incremented and the cycle is repeated until the parameter path of interest is traversed.

Bifurcation and basin diagrams are conceptually complementary: bifurcation diagrams use steady-state data to explore essentially local phase-space phenomena; by contrast, basin diagrams use *transient* data to visualize *global* phase-space structures. This complementarity is reflected in the required control strategy, and necessarily makes collecting basin data a bit more complex, since an *ensemble* of transients is required, as opposed to a single steady state.

The state of our system is given by $(x, \theta) \in \mathbb{R}^2 \times S^1$ where $x = (x, \dot{x})$ and the driving phase $\theta = \omega t \pmod{2\pi}$ for a given driving frequency ω. As with the bifurcation diagrams, we take the Poincaré section Σ^{θ_0} to be

$$\Sigma^{\theta_0} = \{(x, \dot{x}, \theta) | \theta = \theta_0\}. \tag{1}$$

The dynamics is then completely determined by the action of the period-1 Poincaré map $\phi_T : \Sigma^{\theta_0} \to \Sigma^{\theta_0}$, where $T = 2/\pi\omega$.

Typically, basins of attraction are represented by two-dimensional images in which the initial conditions in each basin are coded with an identifying colour. When numerically generating a basin image (or 'initial-condition map'), one discretizes the region of interest into a finely spaced grid of initial conditions. Then the trajectory of each initial condition is simulated until it has converged to an attractor. In most physical experiments, however, it is very difficult to specify initial data in this manner. Our control program overcomes this problem by using an interval of stochastic excitation to generate a random initial condition x_0 before switching to deterministic forcing. The transient data (i.e. the image of x_0 under repeated application of ϕ_T) needed to discern the asymptotic behaviour is collected as soon as the output of the deterministic forcing begins. By repeating this cycle a large number of times, one is able to fill out a portion of the initial-condition space near the attractors. Postprocessing of the ensemble of orbits is then done to correlate the initial conditions with the appropriate attractor.

5.3   EXPERIMENTAL BASINS OF ATTRACTION

The bifurcation structure of the two-well oscillator is shown in Figure 5.2 for a fixed forcing frequency of 11.009 Hz. This frequency was chosen since, at it, the system had a simple basin structure for low forcing amplitudes. Only a brief description of the diagram will be given here: a more detailed discussion of the various solutions represented in the diagram, including dimension estimates as well as additional global results and experimental details, can be found in Cusumano and Kimble (1994). Bifurcation data was collected at each fixed forcing amplitude by first discarding 400 periods of data as transients: a subsequent 200 samples of the position x as measured by the strain gauge were collected in the zero-phase stroboscopic Poincaré

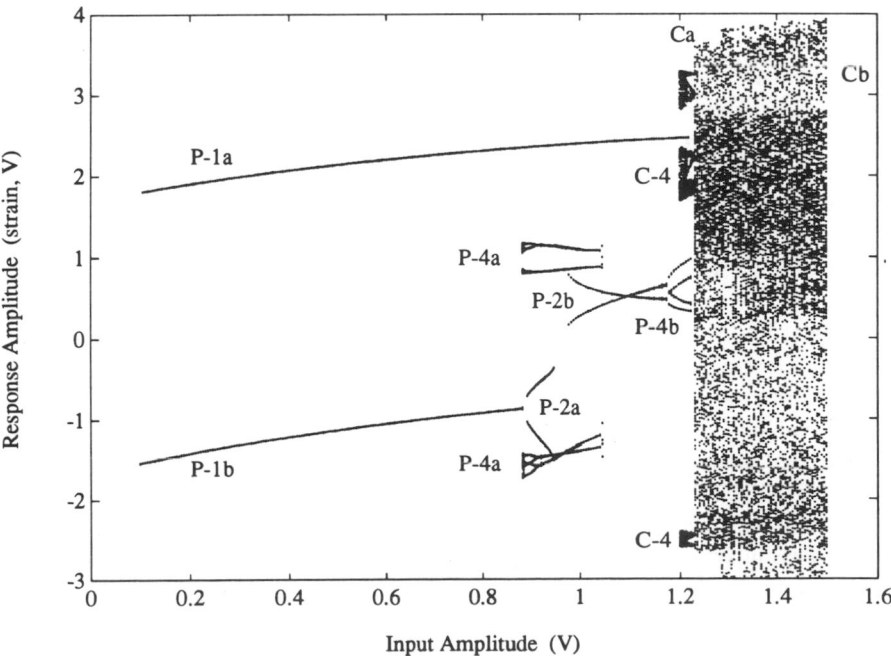

Figure 5.2 Experimental bifurcation structure for the system with driving frequency $\omega = 11.009$ Hz. Solution branches are marked with 'P-n' for period-n orbits, and 'C' for chaotic orbits.

section Σ^0. Typical runs took about 8–16 hours (depending on voltage step size) to scan up and down the entire parameter range.

Two initial branches of solutions were obtained by starting at low forcing levels with one initial condition on the left (negative) attractor and one started on the right (positive) attractor, corresponding to small-amplitude periodic orbits in each well (branches P-1a and P-1b). The amplitude V of the sinusoidal voltage input to the shaker ranged from 0.1 to 1.5 V and back down to 0.1 V, with an initial step of 4.88 mV (2 least-significant bits). In addition to the period-1 orbits (which are the main interest of the remainder of this chapter), period-2 (P-2), period-4 (P-4), chaotically modulated period-4 (C4), and chaotic (C) orbits are visible in the diagram. Several of the branches (P-4a, P-4b and C4) required backtracking in order to trace them out completely. No other solution branches were found with this procedure, although other disconnected branches with small basins of attraction may exist. (As described below, analysis of the basin data reveals that a period-3 solution with a very small basin of attraction exists near $V = 0.498$ V).

Possibly the most notable feature of the bifurcation diagram is the large number of jumps between branches and the fact that, though period-doubling is observed, the classic period-doubling route to chaos does not occur. (Our experience to date with various electro-mechanical systems suggests that this is the norm.) Experimental noise and discretization error can cause branch-jumping before a solution branch has lost stability if the basin of attraction is in some sense too small. To minimize the effect of possible spurious jumps, we scanned forward and backward over the parameter range of interest a number of times in each run. In addition, lowering the voltage step size to the minimum possible with our set-up (2.44 mV, or 1 least-significant bit) did not change the information in the diagram, although in one case it changed the way in which jumping occurred between the different branches. Thus, experimental data-acquisition parameters are not believed to be the primary cause of the jumps which remain in Figure 5.2. In general, one would require global information concerning the arrangement in phase space of coexisting solution branches in order to understand the jump phenomena, sudden branch annihilations, and chaotic–chaotic transitions seen in the bifurcation diagram. A reasonable hypothesis is that boundary crises (Grebogi *et al.* 1983c), in which an attractor collides with its basin boundaries, are responsible for at least some of these sharp transitions. To sort this out one would have to correlate the bifurcation results with experimental basin measurements: we leave such an analysis for future work; however, a method which could be used to do such studies experimentally is described below.

Experimental basins of attraction for the two-well oscillator are presented for four runs carried out at a forcing frequency of 11.009 Hz with forcing amplitudes ranging from 0.098 V to 0.498 V. This corresponds to a portion of the simplest part of the bifurcation diagram, in which only two period-1 orbits are found. The basin images were each generated in runs consisting of 10^4 stochastic/deterministic interrogation cycles (i.e. with 10^4 different initial conditions). Each cycle started with the time equivalent of 40 periods of stochastic excitation by digitally generated, uniformly distributed pseudo-random noise which was passed through a 20 Hz low-pass filter before being sent to the shaker amplifier. Each basin image took about 35 hours to complete. The locations of the two period-1 attractors were identified from the data sets. The orbit for each initial condition was then examined to determine its limiting behaviour. Convergence was defined as having the last point in the 50-iterate orbit fall within 10 least-significant bits (i.e. 24.4 mV) of one of the attractors: points not falling within these neighbourhoods were labelled 'non-convergent'. In this way, each initial condition and its

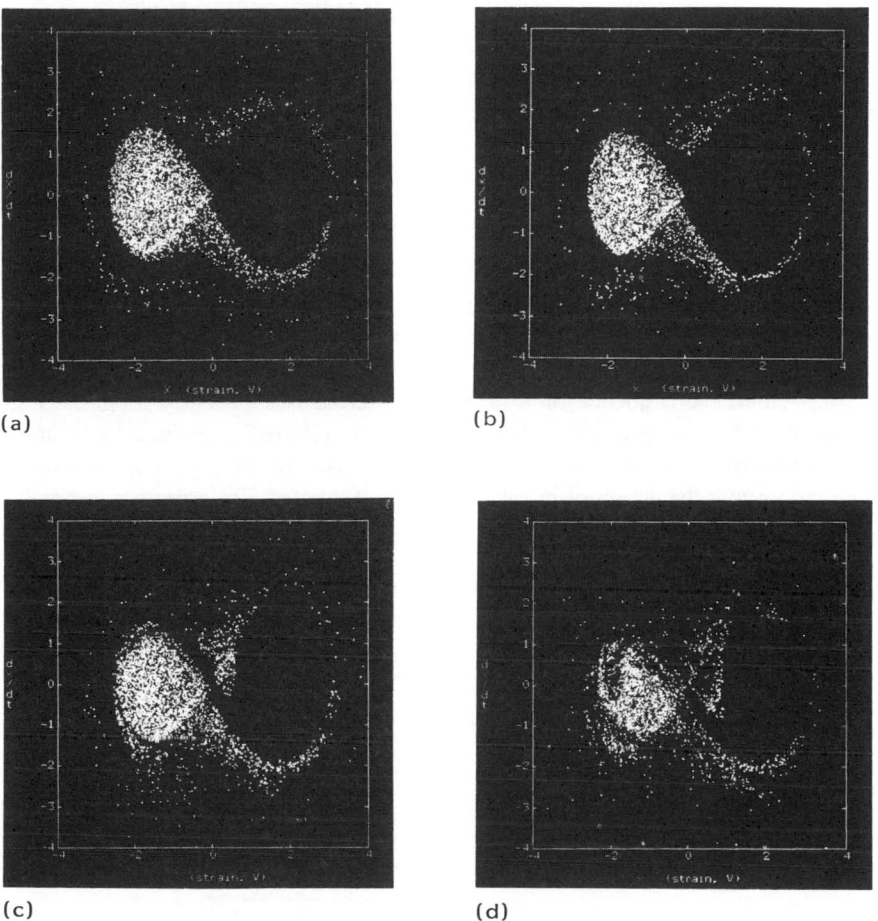

(a)   (b)

(c)   (d)

Figure 5.3 Experimental basins of attraction with $\omega = 11.009$ Hz: (a) $V = 0.098$ V; (b) $V = 0.198$ V; (c) $V = 0.298$; (d) $V = 0.498$ V. Each image consists of 10^4 initial conditions. For all cases, red (respectively white) marks initial conditions which asymptotically approach the right (respectively left) period-1 attractor.

subsequent trajectory were identified as belonging to one of the period-1 basins of attraction, or to the set of non-converging points.

In Figure 5.3, two-colour basin plots for the four runs are shown (only initial-condition data is shown in the plots). The basin structure clearly undergoes a metamorphosis from simple to complex as V increases, as evidenced by the change from smooth to irregular boundaries and the increased mixing of coloured regions. This suggests a transition to fractal basin boundaries, but the resolution of the experiment is not sufficient to unequivocally say this. However, the known theoretical correlation of smooth-to-fractal basin transitions with homoclinic bifurcation can be used to obtain experimental evidence supporting this hypothesis, as we now show. (In principle, one should be able to use the data to estimate the fractal dimension of the basin boundaries, following Grebogi *et al.* (1983a), however, this calculation has not yet been implemented.)

5.4   EVIDENCE OF HOMOCLINIC BIFURCATION

Conclusive evidence of homoclinic bifurcation can be obtained by exploiting the fact that the basin data allows one to construct images showing the flow of points in the phase space. The basin boundary consists of the stable manifold to a saddle-type orbit: however, sequences of points on the Poincaré section tend to be repelled from this stable manifold. Instead, dissipation in the map tends to push orbits towards the unstable manifold of the saddle-type orbit. A good example of this is in Figure 5.3a: a thin ridge of increased density is visible in the centre of the figure at a nominal $45°$ angle with the horizontal, suggesting the location of an unstable manifold emanating from an unstable periodic orbit near $(x, \dot{x}) = (0, 0)$. During the deterministic excitation phase of the interrogation cycle, one can hope to see the approximate location of the unstable manifold because of the increased density of points along it, provided the residence time near the manifold is sufficiently long. This situation can be expected to arise precisely when a homoclinic bifurcation occurs. In Figure 5.4, the entire ensemble of initial data for each of the four cases of Figure 5.3 is presented, along with its image under the action of ϕ_T after 3, 6 and 9 iterates. The variation in point density

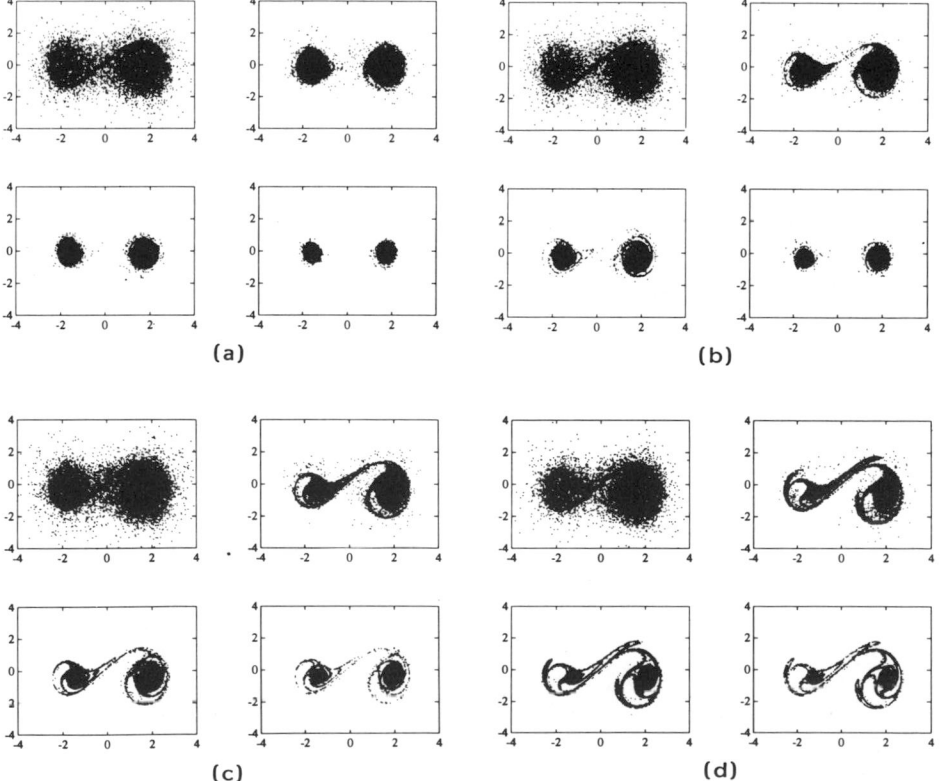

Figure 5.4   The flow of solutions on the Poincaré section: (a) $V = 0.098$ V; (b) $V = 0.198$ V; (c) $V = 0.298$ V; (d) $V = 0.498$ V. Let S be the ensemble of initial conditions. For each value of the forcing, the image of S under the nth iterate of the Poincaré map, $\phi_T^n(S) = \phi_{nT}(S)$, is shown for $n = 0$, 3, 6 and 9 (read left to right, top to bottom). As in Figure 5.3, $\omega = 11.009$ Hz.

in the initial conditions (the 0th iterate in each figure) reflects the invariant probability distribution achieved at the end of the random excitation phase of the interrogation cycle, and hence in each case it is approximately the same. For the lowest forcing level, $V = 0.098$ V (Figure 5.4a), the phase space is contracted smoothly into roughly circular neighbourhoods which collapse down onto the attractors as the number of iterates increases. In this case, the residence time for all orbits starting near the unstable manifold is short, and points are rapidly swept towards the period-1 attractors: hence, the unstable manifold is essentially invisible.

At higher forcing levels, however, the flow collapses onto a complicated structure in the phase space: we believe that this structure contains the highly folded unstable manifold that one expects to arise near a homoclinic bifurcation. A sufficient number of orbits are slowed down near this structure that it becomes visible (Figures 5.4b–d). At the highest forcing levels (Figures 5.4c and 5.4d), the orbits linger long enough that phase-space contraction makes the structure look locally very much like a 1-dimensional manifold for several forcing periods, thus revealing an approximate image of the unstable manifold (at least far enough away from the attractors). The unstable manifold can be seen in the flow precisely when the basins begin to lose their simple boundaries (i.e. at $V = 0.198$ V, corresponding to Figures 5.3b and 5.4b). This is consistent with the hypothesis that Figure 5.3 shows a fractal transition. While all (except, as explained shortly, for a very small number) of the points eventually approach the period-1 attractors, points near the folded unstable manifold take a long time to escape. In Figure 5.5, we plot the number of points that remain

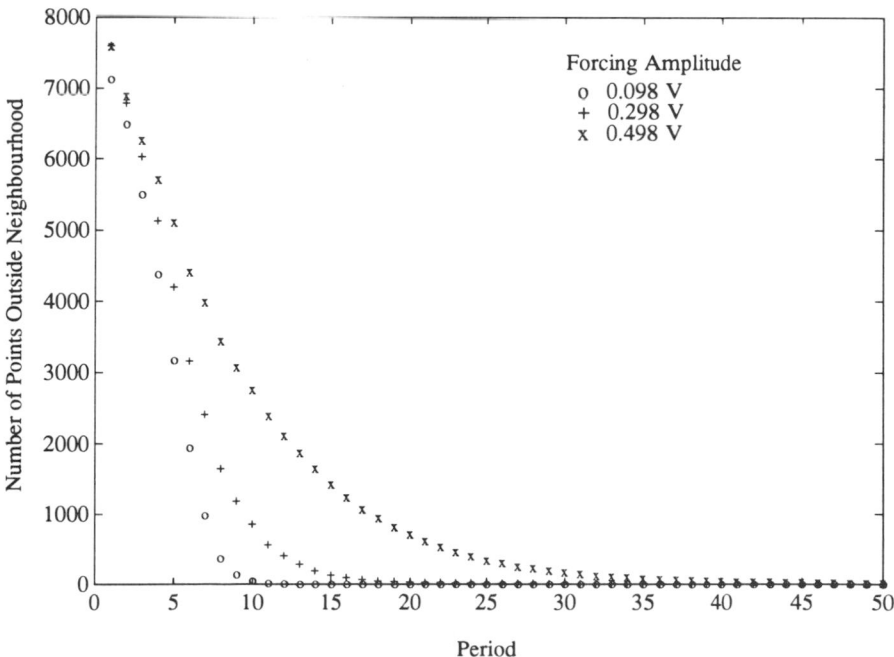

Figure 5.5 The number of points remaining outside of a neighbourhood of the attractors as a function of the number of iterates (or forcing periods). The neighbourhood used for the plots had a radius of 150 least significant bits, or 0.366 V.

outside of a specific neighbourhood of the attractors versus the number of iterates (forcing periods). All points rapidly enter the neighbourhood within about ten iterates in the smooth-basin case, but after the basin boundaries become complex the number of points staying outside of the neighbourhood at any fixed period rapidly increases.

The numbers of points classified as non-convergent after 50 forcing periods were 0, 0, 49 and 593 for V equal to 0.098, 0.198, 0.298 and 0.498, respectively. Examination of the non-convergent points in the last case revealed a period-3 orbit with a very small basin: only about 74 points, or less than 1% of the initial data, fell within the period-3 basin. The initial conditions in this period-3 basin were very close to the unstable manifold of Figure 5.4d, a result consistent with the observations of Grebogi *et al.* (1987) in their analysis of accessible boundary orbits.

All of this data together leads to the conclusion that a homoclinic bifurcation occurs near $V = 0.198$ V. Chaotic transients result from the trapping of orbits near the unstable manifold after bifurcation; however, as a practical matter, it is difficult to recognize such behaviour by looking at time series unless the transients are unusually long. Even in the worst case, over 94% of the initial conditions converge to the attractor within 50 forcing periods: in the 4–5 seconds typically available, an observer could easily fail to recognize the system as possessing chaotic transients. Thus, the stochastic interrogation approach to collecting ensembles of data allows this behaviour to be seen much closer to the homoclinic tangency.

5.5   GLOBAL DAMPING ESTIMATES USING LIOUVILLE'S THEOREM

It is obvious that the data collected for the basin images possesses a large amount of dynamical information. As a simple example of how this information can be extracted to measure physical parameters, we apply Liouville's theorem to the flow of Figure 5.4a and obtain damping estimates for the system.

Liouville's theorem (Arnold 1978), which relates the divergence of the vector field of a system to the rate of contraction of volume elements, is given by

$$\frac{dv}{dt} = \int_v \nabla \cdot f \, d\mu, \qquad (2)$$

where v is the volume of a region of phase space and f is the governing vector field. We assume, as would be the case for the two-well Duffing oscillator with linear damping, that the system equations have the form:

$$\dot{x} = y,$$
$$\dot{y} = -2\zeta y - g(x) + F \cos \Omega t, \qquad (3)$$

where ζ is the critical damping ratio, and the relative driving frequency $\Omega = \omega/\omega_0$ for a given natural frequency ω_0 (time has been rescaled by ω_0 for convenience). The divergence of the time-dependant vector field f (the right-hand side of equation (3)) is then easily found to be

$$\nabla \cdot f = -2\zeta, \qquad (4)$$

which is a constant. Thus, equation (2) becomes

$$\frac{dv}{dt} = -2\zeta v, \qquad (5)$$

which yields

$$\ln\frac{v}{v_0} = -2\zeta(t - t_0). \qquad (6)$$

Since the basin data evolves under the action of ϕ_T, it is natural to take $t - t_0 = T$ and define the *volume logarithmic decrement* δ_v as

$$\delta_v \equiv \ln\frac{v}{v_0} = -2\zeta T. \qquad (7)$$

Rearranging equation (7) gives the following relationship between the critical damping ratio ζ and δ_v:

$$\zeta = \frac{-4\pi\delta_v}{\Omega}. \qquad (8)$$

In elementary vibration theory, the equivalent formula relating the *amplitude logarithmic decrement* δ_x (see, for example, Meirovitch 1986) to the critical damping ratio is

$$\zeta = \frac{-\delta_x}{2\pi}, \qquad (9)$$

where the change in amplitude is defined as taking place over one period of *free* vibration.

Equations (8) and (9) are both used to estimate the damping coefficient. Figure 5.6a shows a semi-log plot of basin 'volume' (area) versus forcing period of each cycle for the right (positive) attractor of the system with parameters as in Figures 5.3a and 5.4a. The area of the region occupied by the ensemble of points in the right basin of Figure 5.3a was estimated at each period by multiplying the standard deviations in the x and dx/dt directions. This is not a very good area estimate initially, but becomes quite good as the points cluster around the attractor, and yields a nice scaling region until the points are all contracted into the noise of the measurements (after about 40–45 periods). The slope of the fitted line is an estimate of δ_v. Similarly, Figure 5.6b shows a semi-log plot of vibration amplitude versus number of periods for a free vibration about the right potential well. The slope of the fitted line in this case is the amplitude logarithmic decrement δ_x. The resulting damping coefficient estimates for this and other trials are shown in Table 5.1. We remark that study of the amplitude-logarithmic-decrement data produced from the free-vibration time series showed that the damping for the system is approximately piecewise linear, depending on the size of oscillation. Therefore, to compare damping estimates from the two methods, only free oscillations comparable in amplitude to those used to find δ_v from the basin data were used to estimate δ_x. With this in mind, the damping estimates based on Liouville's theorem and the volume decrement are in very close agreement with those obtained using the more

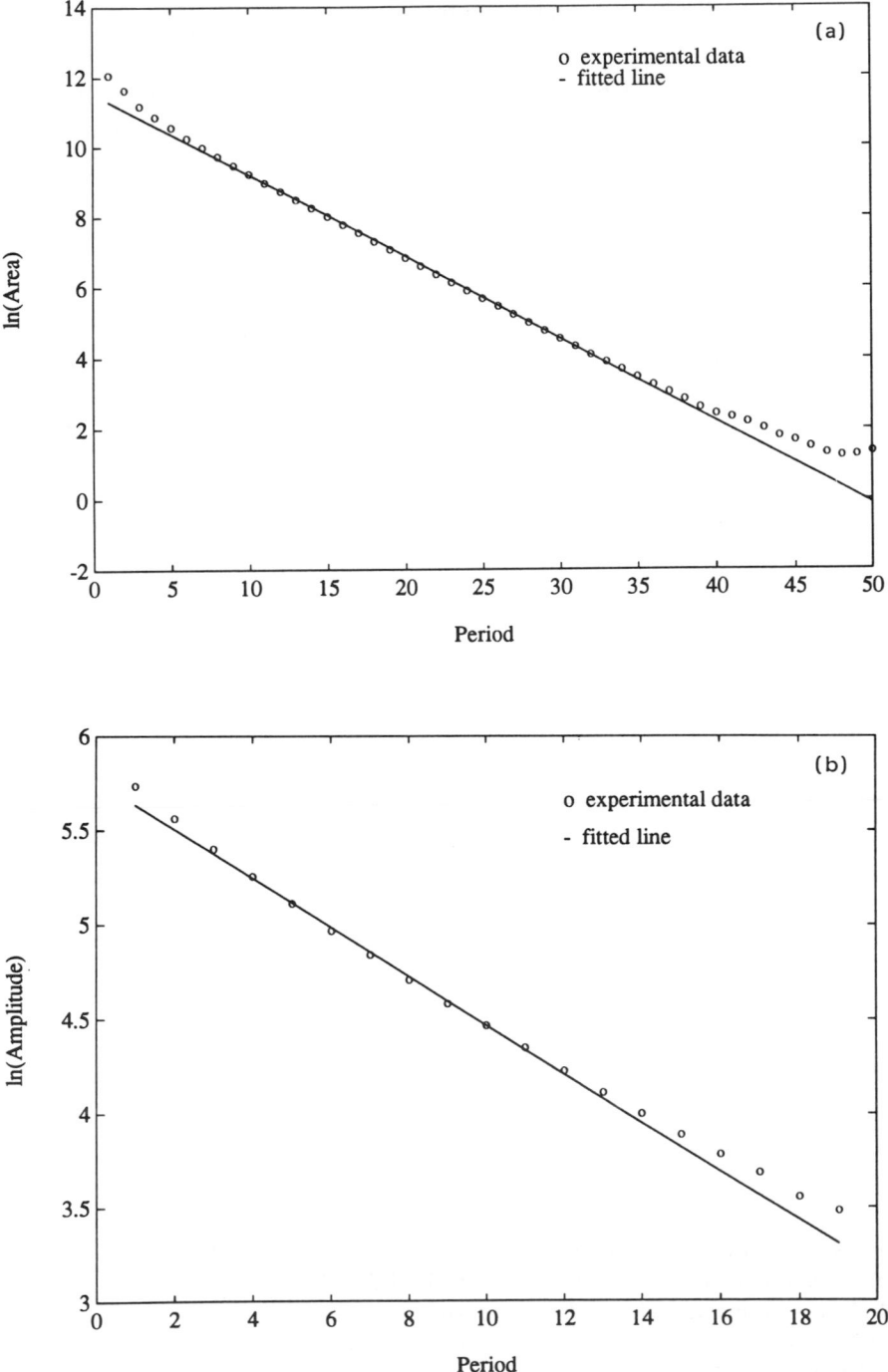

Figure 5.6 Comparison of (a) volume logarithmic decrement data for the forced system to (b) amplitude logarithmic decrement data for free vibrations. Both data sets are for the right attractor with $\omega = 11.009\,\text{Hz}$ and $V = 0.098\,\text{V}$. The slope of the fitted line gives δ_v for plot (a) and δ_x for plot (b). See Table 5.1.

Table 5.1 Critical damping ratios estimated using the volume-logarithmic-decrement method, compared to a standard amplitude logarithmic estimate using free-vibration data

Forcing frequency (Hz)	Forcing amplitude (V)	ζ	
		Right well	Left well
11.009	0.098	0.022 ± 0.003	0.020 ± 0.004
11.009	0.198	0.022 ± 0.003	0.021 ± 0.004
9.105	0.098	0.024 ± 0.005	0.02 ± 0.01
9.404	0.098	0.024 ± 0.003	0.022 ± 0.004
	Free vibration	0.023 ± 0.001	0.021 ± 0.001

conventional method. However, the method based on Liouville's theorem is actually a global technique, since there is no assumption of 'small vibrations' in its definition.

5.6   DISCUSSION AND CONCLUSIONS

The stochastic interrogation method described here generates data sets which allow for the experimental visualization and analysis of global features in the phase space of nonlinear oscillators. We have demonstrated the usefulness of this approach by applying it to the study of a driven, two-well nonlinear magneto-mechanical oscillator. The resulting data yielded images of basins of attraction for the system, visualization of the flow of initial states on the Poincaré section, and an estimate of the linear damping coefficient using a novel approach based on Liouville's theorem.

By combining the basins of attraction with the flow images, we obtained experimental evidence of homoclinic bifurcation well before the onset of chaos. As part of this evidence, a simple-to-complex ('fractal') transition in the basin boundaries was shown to coincide with the appearance of a complicated structure believed to contain the unstable manifold of a saddle-type orbit. This is a significant improvement in experimental technique, since, previously, theoretical criteria for homoclinic bifurcation (such as supplied by the Holmes–Melnikov method) could only be checked by locating regions in the parameter space where chaotic attractors exist. Since homoclinic bifurcation is only a necessary condition for the existence of a chaotic attractor, such studies are typically only able to put loose, one-sided bounds on the theoretical prediction (see, for example, Moon *et al.* 1987). In principle, our method can check the prediction directly and precisely. Admittedly, for low-bandwidth systems, checking a Melnikov curve at more than a few points will be difficult, given the time-consuming nature of the data-acquisition task.

Without question, data collected by stochastic interrogation contains a wealth of dynamical information that the results presented here have barely begun to tap. The estimation of the damping coefficients using Liouville's theorem and the volume logarithmic decrement, though simple, promises to be a useful technique because it should be capable of yielding results far away from the equilibria. The main limitation as applied here was the crude volume estimate employed to generate Figure 5.6. Beyond the determination of basic physical parameters, the stochastic interrogation data contains the information required to construct nonlinear models of the system. In work to be presented elsewhere, we have used the data to construct transition probability matrices for finite partitions of the phase space. These matrices allow the evolution of probability densities to be studied (see Shaw 1984 and Lasota and Mackey 1985). In particular, quantities such as the entropy of the Poincaré map and the invariant distribution for the system can be obtained. Future work will aim at constructing noisy deterministic models using the interrogation data.

ACKNOWLEDGEMENTS

The work of the first author was supported by the Air Force Office of Scientific Research, and that of the second author by a National Science Foundation Graduate Research Fellowship. The authors would like to thank Matt Davies for helping them get started with real-time programming. The assistance of Brian Moquin and Tim Kohler in preparing the colour figures (at the last minute, of course!) was greatly appreciated.

REFERENCES

Arnold, V. I. (1978) *Mathematical Methods of Classical Mechanics*, Springer-Verlag, New York.

Cusumano, J. P. and Kimble, B. W. (1994) A stochastic interrogation method for experimental measurements of global dynamics and basin evolution: application to a two-well oscillator, *Nonlinear Dynamics*, in press.

Cusumano, J. P., Lin, D., Morooney, K and Pepe, L. (1992) Sensitivity analysis of nonlinear systems using animated basins of attraction, *Journal of Computers in Physics*, **6** (6), 647–655.

Endo, T. and Chua, L. O. (1990) Bifurcation diagrams and fractal basin boundaries of phase-locked loop circuits, *IEEE Transactions on Circuits and Systems*, **37** (4), 534–540.

Grebogi, C., McDonald, S., Ott, E. and Yorke, J. A. (1983a) Final state sensitivity: an obstruction to predictability, *Physics Letters*, **99A** (9), 416–418.

Grebogi, C., Ott, E. and Yorke, J. A. (1983b) Fractal basin boundaries, long-lived chaotic transients and unstable-unstable pair bifurcation, *Physical Review Letters*, **50** (13) 935–938.

Grebogi, C., Ott, E. and Yorke, J. A. (1983c) Crises, sudden changes in chaotic attractors, and chaotic transients, *Physica*, **7D**, 181.

Grebogi, C., Ott, E. and Yorke, J. A. (1986) Metamorphoses of basin boundaries in nonlinear dynamical systems, *Physical Review Letters*, **56** (10), 1011–1014.

Grebogi, C., Ott, E. and Yorke, J. A. (1987) Basin boundary metamorphoses: changes in accessible boundary orbits, *Physica*, **24D**, 243–262.

Guckenheimer, J. and Holmes, P. (1983) *Nonlinear Oscillators, Dynamical Systems and Bifurcations of Vector Fields*, New York, Springer-Verlag.

Lasota, A. and Mackey, M. (1985) *Probabilistic Properties of Deterministic Systems*, Cambridge University Press, Cambridge.

McDonald, S. W., Grebogi, C., Ott, E. and Yorke, J. A. (1985) Fractal basin boundaries, *Physica*, **17D**, 125–153.

Meirovitch, L. (1986) *Elements of Vibration Analysis*, New York, McGraw-Hill .

Moon, F. C. (1980) Experiments on chaotic motion of a forced nonlinear oscillator: strange attractors, *ASME Journal of Applied Mechanics*, **47**, 638–644.

Moon, F. C. and Holmes, P. (1979) A magnetoelastic strange attractor, *Journal of Sound and Vibration*, **65** (2), 275–296.

Moon, F. C. and Li, G-X. (1985) Fractal basin boundaries and homoclinic orbits for periodic motion in a two-well potential, *Physical Review Letters*, **55** (14), 1439–1442.

Moon, F. C., Cusumano, J.P. and Holmes, P.J. (1987) Evidence for homoclinic orbits as a precursor to chaos in a magnetic pendulum, *Physica*, **24D**, 383–390.

Shaw, R. (1984) *The Dripping Faucet as a Model Chaotic System*, Santa Cruz, Aerial Press.

Thompson, J. M. T. and Soliman, M.S. (1990) Fractal control boundaries of driven oscillators   and their relevance to safe engineering design, *Proceedings of the Royal Society of London A*, **428**, 1–13.

Yamaguchi, Y. and Mishima, N. (1985) Fractal basin boundary of a two-dimensional cubic map, *Physical Review Letters*, **109A** (5), 196–200.

J. P. Cusumano and B. W. Kimble, *Department of Engineering Science and Mechanics, Penn State University, University Park, PA 16802, USA.*

PART II

Impact and Friction

Discontinuous systems are an important class of non-smooth dynamical systems incorporating impact oscillators, stick–slip, and piecewise linear systems. Many physical situations of practical engineering relevance undergo instantaneous changes of state and it is necessary to fully assess their dynamic characteristics to predict extreme motions or reduce wear. Such systems exhibit many of the typical characteristics of smooth nonlinear systems such as generic bifurcations, multiple solutions and chaos and thus, for the most part, the topological concepts of smooth dynamical systems theory can still be applied. In particular the idea of viewing a dynamical system in geometrical terms remains a useful concept. Significantly, though, discontinuous systems may also display new phenomena, the importance of which still needs to be established.

IMPACT OSCILLATORS

Impact oscillator is the term used here to represent a system which is driven in some way and which also undergoes intermittent or a continuous sequence of contacts with motion-limiting constraints. Even if the governing system is linear, the constraints introduce a severe nonlinearity into the overall system, bringing discontinuity into the mathematics. Many physical systems of practical importance in engineering undergo impacts at motion-limiting stops, for example in marine engineering, where ships collide with fenders (see for example Thompson and Stewart 1986, chapter 15); in mechanical engineering, where gears rattle; and many other examples.

Impacting systems are not new; indeed engineers have been faced with problems of wear and fatigue caused by repeated impacts for many years. To analyse the effects of impacts statistical methods have been developed and in some cases novel devices incorporated to avoid inherent dangers.

One basic premise of many of the studies to date has been that the time spent during impact is small compared to the dynamic time. Furthermore, if no deformation of the two impacting surfaces occurs then a good approximation to the impacting process is one which involves an instantaneous reversal of velocities. A common technique in nonlinear dynamics is the study of the associated map of the system formed by either stroboscopically sampling the full trajectory or, specifically for systems with impacts, via an examination of the system at each impact. It is the instantaneous change in velocity that produces discontinuities in the system (both in the map itself and, importantly, in its gradient) which can be considered to be a severe form of nonlinearity.

For smooth systems close to a bifurcation point the use of centre-manifold theory (Carr 1981) allows the dynamics to be reduced to a low-(possibly one-)dimensional manifold. On this centre manifold, coordinate transformations yield a normal form of the bifurcation which allows its classification. The same cannot be carried out in a straightforward manner for non-smooth systems with impact. It is now known that a bifurcation occurs for low-velocity impacts, termed a *grazing bifurcation*, which is not one of the simple bifurcations classified for smooth systems (Nordmark 1991; Foale and Bishop to appear). At these bifurcations the centre mani-fold does not exist but many of the other tools of nonlinear dynamical systems can still be applied, provided that special attention is paid to those areas in which discontinuities occur. These points are addressed in the chapter by *Foale and Bishop* which compares experimentally observed impact bifurcations with theoretical predictions.

SYSTEMS WITH DRY FRICTION

When a force is applied to a body that is in contact with another the friction force acts in two ways. The stick phase is characterized when the friction opposes any impending motion while during the slip phase the friction acts as a resistance against the induced motion. Vibrations caused by self-sustained oscillations induced by dry friction are common in engineering systems (Rayleigh 1945; Stoker 1950; Nayfeh and Mook 1979; Hagedorn 1982). Some early studies of the stick–slip phenomenon were directed towards an understanding of the behaviour of a bowed string with particular reference to a violin. Here the friction induced by the stroking of the bow induces an oscillation which in turn produces a sound often thought to be pleasurable. In other cases self-sustained oscillations produce noise and subsequent wear that is most undesirable (for instance the squealing of brakes). We now know that for an accurate description the bowed string needs to be modelled as a continuum but simplified discrete-mass analogues are extremely useful in illustrating behaviour and form a paradigm for stick–slip behaviour (Popp and Stelter 1990).

In an attempt to explain the apparent differences between static and dynamic friction forces recent work (Tworzydlo *et al.* 1992) has highlighted the importance of more accurate modelling of constitutive relations for the normal and tangential displacements of the interface, even when contact is maintained at all times. The field of dry friction has been reviewed in a number of papers within the book by Ibrahim and Soom (1993). In the present volume, the chapter by *Glocker and Pfeiffer* illustrates the common use of friction to provide damping by detailing the application of stick–slip models to turbine-blade dampers.

BIFURCATION AND STABILITY ANALYSIS

The two previous sections have indicated the mathematical complexities of discontinuous systems when trying to understand the dynamical response. These have shown a clear need for a development of the theory to overcome these difficulties. In the following chapters we shall see examples of discontinuous systems and, in the chapter by *Kleczka and Kreuzer*, a method of analysis based on the use of generalized coordinates with superimposed constraints. In the chapter by *Fey et al.* we see how discontinuities can be built into standard finite-element packages and the subsequent system analysed to identify bifurcations and plot response curves. This last concept allows the geometrical ideas of nonlinear dynamics and chaos to become part of engineering design.

REFERENCES

Carr, J. (1981) *Applications of Center Manifold Theory*, Springer-Verlag, New York.

Foale, S. and Bishop, S.R. Bifurcations in impact oscillators. To appear in *Nonlinear Dynamics*.

Hagedorn, P. (1982) *Nonlinear Oscillations*, London, Oxford Science Publications.

Ibrahim, R. A. and Soom, A. (1992) *Friction-Induced Vibration, Chatter, Squeal and Chaos*, ASME DE Vol. 49, New York.

Nayfeh, A. H. and Mook, D. T. (1979) *Nonlinear Oscillations*, New York, Wiley.

Nordmark, A. B. (1991) Non-periodic motion caused by grazing incidence in an impact oscillator, *J. Sound and Vibration*, **145** (2), 279–297.

Popp, K. and Stelter, P. (1990) Stick–slip vibrations and chaos, *Phil. Trans. R. Soc. Lond. A*, **332**, 89–105.

Rayleigh, J. W. S. (1945) *The Theory of Sound*, New York, Dover Publications.

Stoker, J.J. (1950) *Nonlinear Vibrations*, New York, Interscience Publishers.

Thompson, J. M. T. and Stewart, H. B. (1986) *Nonlinear Dynamics and Chaos*, Chichester, Wiley.

Tworzydlo, Becker, E. B. and Oden, J. T. (1992) Numerical modelling of friction-induced vibrations and dynamic instabilities. In: *Friction-Induced Vibration, Chatter, Squeal and Chaos*, ASME DE Vol. 49., R. A. Ibrahim and A. Soom (eds.), pp. 13–32.

6 BIFURCATIONS IN IMPACT OSCILLATORS: THEORY AND EXPERIMENTS

S. Foale and S. R. Bishop

Forced oscillators with impacts at rigid stops have been widely studied as examples of simple nonlinear systems. Such systems are of interest because a large number of physical systems display behaviour which can be classified as impacting, where it is important to use a dynamical analysis to identify and thus avoid the noise or wear caused by repeated unacceptably large impacts. These include examples as diverse as gear rattle, heat-exchanger tube wear in nuclear power stations, and ships colliding against fenders. In this chapter we study the bifurcations which arise in impact oscillators both theoretically and with a simple laboratory experiment.

6.1 INTRODUCTION

Many studies have highlighted the practical physical and engineering problems which involve rigid impacts between two components. Mechanical engineering provides many examples of systems with impacts such as rattling gears (Karagiannis and Pfeiffer 1991; Kahraman and Singh 1990), vibration absorbers (Sharif-Bakhtiar and Shaw 1988), car suspensions (Stennson et al. 1992), and impact print hammers (Tung and Shaw 1988). In these mechanical-engineering examples of impact oscillators, the primary problems caused by the successive impacts are noise and wear. Another rich source of impact oscillator problems is the offshore engineering environment. Work by Thompson and coworkers on the problem of a ship moored to an articulated mooring tower, essentially an inverted pendulum with buoyancy, undergoing wave-driven oscillations was extensively studied (Thompson 1983; Thompson et al. 1984). A similar problem, that of a ship moored against a fender, was studied by Lean (1971), and more recently by Sterndorff et al. (1992). Other offshore impacting problems arise in the installation of a structure over a guiding 'indexing' system, discussed in more detail in Foale (1993). Indexing systems can comprise bumper piles, which guide the struture into position (Nelson et al. 1983; Stahl et al. 1993) or pile/sleeve arrangements (Robinson and Ramzan 1989). The effect of earthquakes on various structures has motivated other studies of dynamical

Nonlinearity and Chaos in Engineering Dynamics
Edited by J. M. T. Thompson and S. R. Bishop,   © 1994 John Wiley & Sons Ltd

impact-type problems, for example, the responses of a slender block which rocks under external excitation (Hogan 1989; Tso and Wong 1989). The 'pounding' (i.e. collision) of nearby buildings under earthquake excitation is another example (Jing and Young 1990). The electricity-generating industry has also produced impact oscillator problems, such as the cross-flow-induced impacting of heat-exchanger tubes (Paidoussis and Li 1992).

6.2   THE IMPACT OSCILLATOR MODEL

There are several different ways in which the impact can be modelled in an impact oscillator. Probably the simplest is the coefficient of restitution rule, $\dot{x} \rightarrow -r\dot{x}$, which is applied when a stop is reached. This rule provides an instantaneous reversal of velocity and, if the coefficient of restitution is $r > 0$, a loss of energy at the impact. We call the coefficient of restitution model the COR model. Other approaches to modelling the impact have been to use a stiffness function which rapidly increases after impact, such as a piecewise linear stiffness (Shaw and Holmes 1983) or the Hertz impact law (Jing and Young 1990; Foale and Bishop in press). Away from the stop, we assume that the equation governing the motion is the simple forced linear oscillator given by

$$\ddot{x} + d\dot{x} + x = \alpha \cos(\omega t), \quad x < a, \tag{1}$$

which is valid when the displacement is $x < a$, where $x = a$ is the position of the stop. In equation (1) an overdot represents differentiation with respect to the time t, d is the linear damping coefficient, α the amplitude of the forcing function, and ω the forcing frequency. In almost all cases, one more system parameter can be eliminated from the above by rescaling the displacement, x, so that the stop is at $x = 1$. However, this rescaling rules out the special case of $x = 0$, so we will use the above form. In equation (1) we assume that time and displacement have been rescaled in order to scale to one any mass or stiffness terms. We also note that variation of the forcing amplitude α is directly equivalent (after rescaling) to adjusting the position of the stop. A further rescaling of x by putting $x \rightarrow \alpha x$ in equation (1) gives exactly the same linear oscillator with the stop at a/α, i.e. increasing the amplitude of the forcing by some factor is equivalent to moving the stop closer to the equilibrium position by the same factor. In addition to equation (1), when the displacement x reaches the position of the stop $x = a$ the rule

$$\dot{x} \rightarrow -r\dot{x}, \quad x = a \tag{2}$$

is applied, where the coefficient of restitution, r, lies in the range $0 < r \leqslant 1$. This coefficient, r, is determined empirically for the impact between two surfaces of different material properties. After the rule, equation (2), has been applied then the linear oscillator, equation (1), takes over again.

6.3   DISCONTINUITIES IN GRADIENT FROM THE COR MODEL

Numerical observations (Foale and Bishop 1992) indicate that some sort of bifurcation occurs in a COR impact oscillator when part of an orbit just touches a stop with zero velocity.

This bifurcational event occurs when a stable fixed point of the map crosses a line of discontinuity in gradient (in fact, a square-root singularity in the derivative of the map). It is possible to show how this square-root singularity in the derivative arises by expanding for a small time backwards and forwards in time from a low-velocity impact. Let us take a dynamical system which is governed by a smooth second-order ordinary differential equation with periodic forcing away from the impact, with the coefficient of restitution rule $\dot{x} \rightarrow -r\dot{x}$ applied at impact. Suppose that a low-velocity impact occurs with velocity $\dot{x} = v \ll 1$ at time t_0. We will try to obtain the mapping from the plane defined by $t = t_1$ to the plane $t = t_2$ where $t_1 < t_0 < t_2$, $\Delta t_1 = t_0 - t_1 \ll 1$, $\Delta t_2 = t_2 - t_0 \ll 1$, by expanding in the small variables defined above. The mapping takes the point (ξ_1, η_1, t_1) to (ξ_2, η_2, t_2), undergoing a low-velocity impact at (a, v, t_0) in the process. Expanding backwards from the impact we have

$$\xi_1 = a - v\Delta t_1 + A_1 \Delta t_1^2 + \ldots, \tag{3}$$

$$\eta_1 = v - \Delta t_1 A_1 + \ldots \tag{4}$$

and expanding forwards from the impact, after the application of the impact rule we have

$$\xi_2 = a - rv\Delta t_2 + A_2 \Delta t_2^2 + \ldots, \tag{5}$$

$$\eta_2 = -rv + A_2 \Delta t_2 + \ldots, \tag{6}$$

where A_1 and A_2 are the accelerations at the impact with positive and negative velocities respectively. Combining these two equations, the lowest-order terms of the mapping from the constant-time planes t_1 to t_2 via a low-velocity impact are given by

$$\xi_2 = (a - \xi_1)\left(2r + \frac{A_1}{A_2}\right) + a - r\eta_1, \tag{7}$$

$$\eta_2 = -\sqrt{\eta_1^2 - 2A_1(\xi_1 - a)}\left(\frac{A_2}{A_1} + r\right) + \eta_1 \frac{A_2}{A_1} + A_2 \Delta t. \tag{8}$$

We note that

$$v = \sqrt{\eta_1^2 - 2A_1(\xi_1 - a)} + \ldots \tag{9}$$

so that, as the velocity of the impact tends to zero the expression under the square root tends to zero. The mapping from t_1 to t_2 given by equations (7) and (8) is only valid where the term under the square root is greater than zero. If this term is less than zero there is no impact between t_1 and t_2 and so a simple linear mapping which is such that the whole mapping is continuous takes a point (ξ_1, η_1, t_1) to (ξ_2, η_2, t_2) if Δt is small enough. The further mapping from t_2 back to $t_1 + T$ (assuming that the oscillator is periodically forced with period T this forms the stroboscopic Poincaré map sampled at phase t_1), if it has no further low-velocity impacts, will aslo be a simple linear mapping locally. The total map, composed of the mapping around the low-velocity impact and the further mapping back onto the plane $t = t_1$ has two distinct regions separated by the line $v = 0$. When the term under the square root is less than zero both parts of the total mapping behave in a simple locally linear manner, and so the resulting total mapping will also behave in a simple locally linear manner. When $v^2 > 0$ there is a square-root term in the mapping around the low-velocity impact, and thus in the total mapping. As this term goes to zero from above this square-root term goes to zero, and the first differential of this mapping with respect to ξ_1 and η_1 will have terms involving $1/v$. There will therefore be a square-root singularity in the derivative of the mapping on the impact side of the line $v = 0$, although continuity will still be preserved. When a stable fixed point of a Poincaré map crosses this line of discontinuity of gradient under a change of parameter then there will be a change in the qualitative behaviour of the system which we call a grazing bifurcation. The above demonstration of the existence of a square-root singularity in gradient in an impact oscillator is similar to that given in Nordmark (1991). These grazing bifurcations which arise in impact oscillators are of the type called 'border collision bifurcations' by Nusse and Yorke (1992). Border-collision bifurcations occur when a fixed point of a map crosses a line of discontinuity in gradient, and the grazing bifurcation is a special case of this.

6.4   GRAZING BIFURCATIONS

The strategy for the location of particular steady-state periodic solutions of COR impact oscillators which behave linearly between impacts has been widely used (for example, Shaw and Holmes 1983). The general idea is that the solution of the linear oscillator away from impacts is known, and so conditions for simple steady-state solutions with low period and low numbers of impacts can often be written down and solved analytically. A point on a possible period-one, one-impact steady-state orbit is found by matching the time and velocity (t_i and y_i) at a stop with the time and velocity one period later. Solving the equations for displacement and velocity in these two unknowns we obtian a quadratic equation for y_i and a trigonometric expression for t_i. In Foale and Bishop (in press) it is shown that, by rearranging the quadratic equation for the velocity at impact, we obtain an expression for the forcing amplitude α:

$$\alpha^2 = \alpha_c^2 + 2\alpha_c ry_i z(\alpha, \omega, d, r) + r^2 y_i^2 f(\alpha, \omega, d, r), \tag{10}$$

where

$$\gamma = \sqrt{(d\omega)^2 + (\omega^2 - 1)^2}, \quad \beta = \frac{d}{2}, \quad \Omega = \sqrt{1 - \beta^2}, \quad v = e^{-2\pi\beta/\omega}. \tag{11}$$

Here, both z and f are known functions of the system parameters and α_c is the value of the forcing amplitudé α at which the non-impacting limit cycle has maximum a, i.e. the non-impacting stable orbit first grazes the stop. It is easy to see from the general solution of equation (1) that $\alpha_c = a\gamma$.

From equation (10) we can see that if a period-one orbit has one low-velocity impact, i.e. $0 < y_i \ll 1$ then $\alpha \to \alpha_c$ from above or below depending on the sign of the coefficient of the linear term of this quadratic in y_i. Since $r, y_i, \alpha_c > 0$ we have that the sign of z, where

$$z = \frac{\gamma v \left(1 + \dfrac{1}{r}\right) \sin\left(\dfrac{2\pi\Omega}{\omega}\right)}{\Omega\left(1 + v^2 - 2v \cos\left(\dfrac{2\pi\Omega}{\omega}\right)\right)}, \tag{12}$$

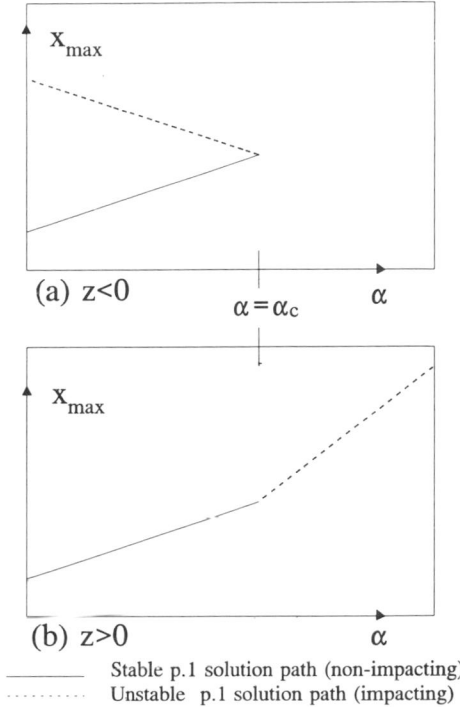

Figure 6.1 The two types of first grazing bifurcation for (a) $z < 0$, (b) $z > 0$.

controls the type of grazing bifurcation. There are two distinct type of grazing bifurcation depending on whether z is positive or negative: the two cases are sketched in Figure 6.1, where the maximum amplitude of the period-one solutions are plotted against the forcing amplitude α with all other parameters fixed. In both cases when $\alpha < \alpha_c$ there is a stable non-impacting period-one solution whose amplitude grows linearly with α (the solid lines). The stability characteristics of this non-impacting solution do not change as α changes, since this is just the solution to the simple forced linear oscillator, equation (1). In Figure 6.1a, with $z < 0$, as the velocity of the period-one, one-impact solution $y_i \to 0$, $\alpha \to \alpha_c$ from below, with $\alpha = \alpha_c$ when $y_i = 0$. In Figure 6.1b, with $z > 0$, $\alpha \to \alpha_c$ from above as $y_i \to 0$. It can be seen from equation (12) that the sign of z is governed by that of $\sin(2\pi\Omega/\omega)$, and this changes sign whenever $\omega = 2\Omega/n$, $n = 1, 2, 3, \ldots$. We call the two grazing bifurcations shown in Figures 6.1a and 6.1b saddle–node-type and flip-type respectively.

Local codimension-one bifurcations (flips and saddle–nodes) of the period-one, one-impact orbits are also easy to locate analytically since expressions can be found for their stability, see for example Shaw and Holmes (1983). Figure 6.2 shows the loci of saddle–node and flip bifurcations in the two-dimensional parameter space, α against ω, as well as the locus of grazing bifurcations. The fixed parameters here are $r = 0.7$, $d = 0.1$, $a = 1.0$ in equation (1). We can see that in this case z changes sign at $\omega \approx 2$, $\omega \approx 1$, $\omega \approx 2/3$, $\omega \approx 1/2, \ldots$. Along the solid line representing the locus of first-grazing bifurcations, the type of grazing bifurcation alternates as z changes sign.

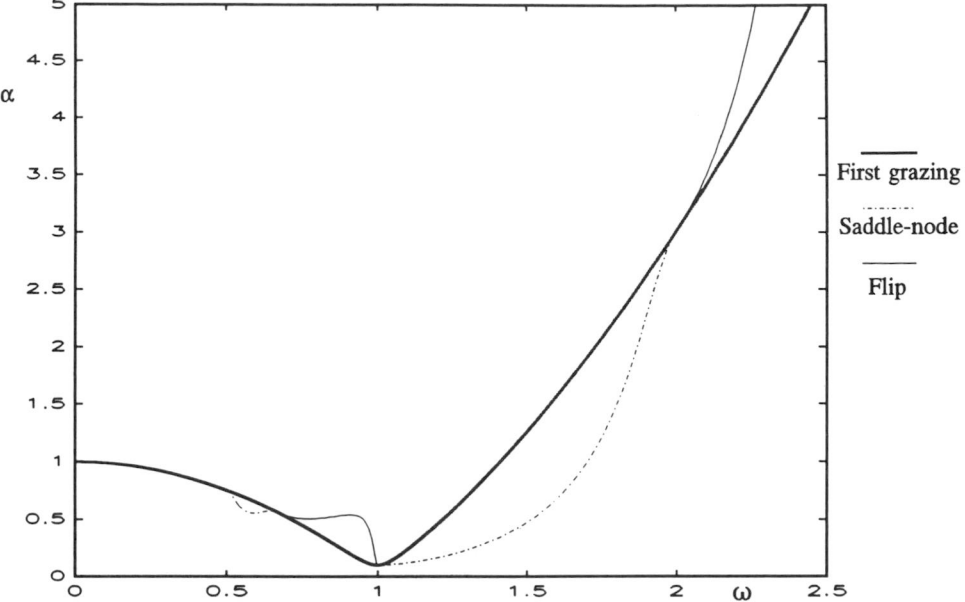

Figure 6.2   Bifurcation loci for the one-sided COR impact oscillator.

6.5   EXPERIMENTAL INVESTIGATIONS

The grazing bifurcations which are found to occur in the COR model when an orbit undergoes an additional low-velocity impact seem to be very important to the overall bifurcational behaviour of an impact oscillator. In Foale and Bishop (in press) it is shown that, even when the discontinuous-coefficient-of-restitution-impact rule is replaced by a continuous and differentiable stiffness function which rises rapidly at impact to oppose the motion (the Hertz law model), the qualitative bifurcational behaviour appears to be the same. In this chapter an experiment is described which is devised in order to test whether the qualitative bifurcational behaviour of the COR model and an experiment with one-sided impact are the same.

Several experimental studies of impact oscillators have been carried out. For example, Moon and Shaw (1983) investigated the chaotic vibrations of a beam with impacts at a stop, and compared the experimental results with a piecewise linear model obtained from a Galerkin approximation using only the first mode. Poincaré maps and Fourier transforms of chaotic motions from the experiment and from the theoretical model were compared, and found to agree reasonably well. Stennson and Nordmark (1992) used a vibrating spring/mass system with impacts at a stop and compared this experimental set-up with a simple coefficient-of-restitution theoretical model. Bifurcation diagrams and chaotic attractors in Poincaré maps obtained from numerical simulations of the model and from the experiment were compared. Some of these results showed a remarkable agreement. The main concern of the present work has not, however, been addressed in previous experimental studies. The grazing bifurcations which are found to occur in the COR model when an orbit undergoes an additional low-

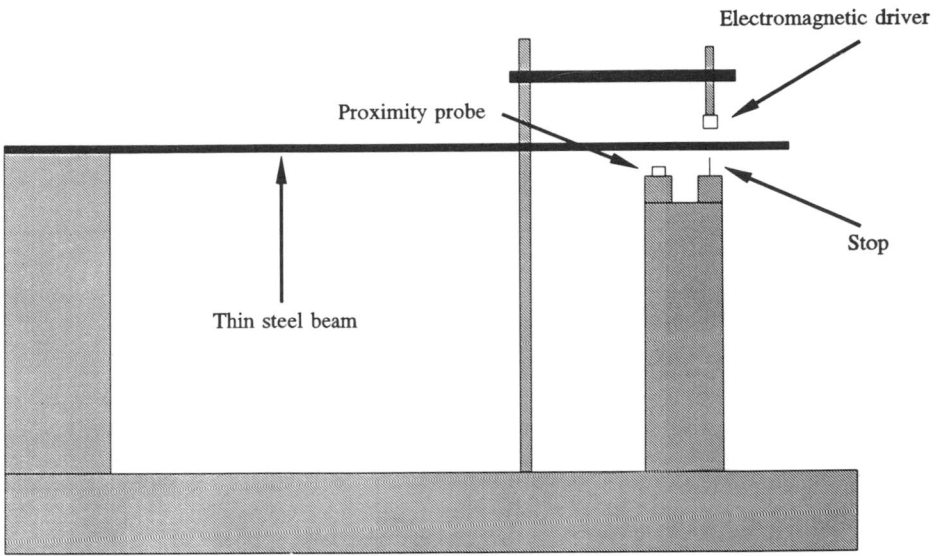

velocity impact seem to be very important to the overall bifurcational behaviour of an impact oscillator.

Figure 6.3 shows a diagram of the experimental apparatus used. A long, slender steel beam is attached to a metal base at one end. At the other, free, end of the beam there is an electromagnet which is used to force the beam, a proximity probe to measure the displacement of the beam, and a steel pin which acts as a stop. The pin is attached to a micrometer and so can be moved towards the beam or away from it, and the distance from the stop to the beam can be measured. A microcomputer with an analogue-to-digital/digital-to-analogue converter (ADC/DAC) card is used both to drive the electromagnet with some waveform and to read the voltage from the proximity probe. A continuous sine-wave function of a given frequency and amplitude is sent to the output channel. A range of frequencies is chosen and the software steps through this range of frequencies, allowing 30 seconds at each step for any transient motions to decay. The ADC is then used to acquire 10 000 readings of the voltage at the proximity probe at a rate of 10 000 Hz, i.e. one second of data. The maximum absolute voltage in this one-second acquisition period is recorded and the timebase for waveform generation is then altered to start the DAC generating the waveform at the next frequency. After a further 30 seconds another reading is taken, and so on. When the end of this range of frequencies is reached the process is reversed so that a frequency sweep is performed in the opposite direction.

6.6 FREQUENCY SWEEPS WITH NO STOP

The beam was forced at a frequency near to the natural frequency of the first mode. The amplitude of the forcing is also chosen so as to be small enough that the beam is not forced into a region with nonlinear response characteristics. With the stop well away

from the beam, three frequency sweeps were carried out. In this way it is confirmed that the
frequency–response curves of the beam alone are very close to linear. Preliminary studies
had indicated a natural frequency around 28 Hz, so the frequency range chosen was 25.9 Hz
to 30.3 Hz and in each case a sweep up in 56 equal steps and down in 55 equal steps was
performed. The parameters of the beam can then be estimated from this experimentally
obtained frequency–response curve. We are assuming that only the first linear mode is being
excited, and further assume that there is linear damping of the beam. The idealized equation
describing the motion of a point along the beam (when no stop is present) is then

$$\ddot{x} + \Delta \dot{x} + \omega_c^2 x = A \cos(\omega t), \tag{13}$$

where an overdot represents differentiation with respect to time t, Δ is the linear damping
coefficient, ω_c the natural frequency, A the forcing amplitude after the mass has been scaled
out, and x is the displacement of the beam from the stop in millimetres. The asymptotic
steady-state solution of this equation is $x = (A/\gamma)\cos(\omega t)$, where $\gamma = ((\Delta \omega)^2 + (\omega^2 - \omega_c^2)^2)^{1/2}$.
The maximum absolute value of the displacement is therefore A/γ. By performing a nonlinear
minimization on the function

$$s = \sum_{i=1}^{i=N} \left(r_i - \frac{A}{\gamma(\omega_i)} \right)^2 \tag{14}$$

with respect to the three parameters Δ, A and ω_c we can estimate the best fit to the
model, equation (13), for the experimental data set (r_i, ω_i), $i = 1, 2, \ldots, N$ where N is the total
number of data points and r_i is the maximum displacement of the beam (measured in volts
from the proximity probe) at the frequency ω_i. In this way we estimated the parameters of
the equivalent linear system as

$$\begin{aligned}
&\text{Forcing amplitude } A\text{:} &&1.420 \text{ mm s}^{-2}, \\
&\text{Natural frequency } \omega_c\text{:} &&176.19 \text{ rad s}^{-1}, \\
&\text{Damping coefficient } \Delta\text{:} &&0.0776 \text{ s}^{-1}.
\end{aligned}$$

6.7 FREQUENCY SWEEPS WITH IMPACT

Since the parameters A, ω_c and Δ have been fitted to the linear model given by equation
(13), the free parameter which is altered in the experiment is the position of the stop. Three

Table 6.1

Run number	Measured displacement of stop (volts)	Measured displacement at stop (mm)	Frequency of first grazing (rad s^{-1})	Frequency of second grazing (rad s^{-1})	Frequency of saddle node (rad s^{-1})
1	0.09	4.65	171.53	179.07	186.42
2	0.17	4.60	173.98	177.81	184.79
3	0.26	4.55	174.99	177.37	182.65
4	0.40	4.50	175.43	176.56	188.43
5	0.49	4.45	175.99	177.06	180.52
6	0.61	4.40	175.87	176.49	180.14

frequency sweeps up and down for each position of the stop were then performed, with the maximum displacement at each frequency being recorded as described for the case with no stop. The three runs for each position of the stop were averaged. From these experimental averaged frequency–response curves the positions of three bifurcational events were estimated for each position of the stop. These bifurcations are: first grazing where the amplitude of response first reaches the displacement of the stop as frequency is increased and the response curve flattens out; second grazing where the amplitude of response first reaches the displacement of the stop as frequency is decreased; saddle–node bifurcation where the impacting solution loses its stability as the frequency is increased. The results of the experimental runs with the stop are summarized in Table 6.1.

6.8   COMPARISON OF EXPERIMENTAL AND THEORETICAL RESULTS

Figure 6.4 shows a typical response curve for run 4 along with the numerically calculated theoretical response curve assuming a coefficient-of-restitution model with the parameters calculated from the linear response curves (the runs with no stop). There is still one parameter which remains to be estimated: the coefficient of restitution, r. In Figure 6.4 we have chosen to have $r = 0.2$ which is quite a low value (Goldsmith 1960 gives the coefficient of restitution between steel and steel as somewhere between 0.7 and 0.8). Despite the good fit between the runs with no stop and a linear response curve, confirming that predominantly only one mode is being excited, once a stop has been imposed there is a possibility of other modes

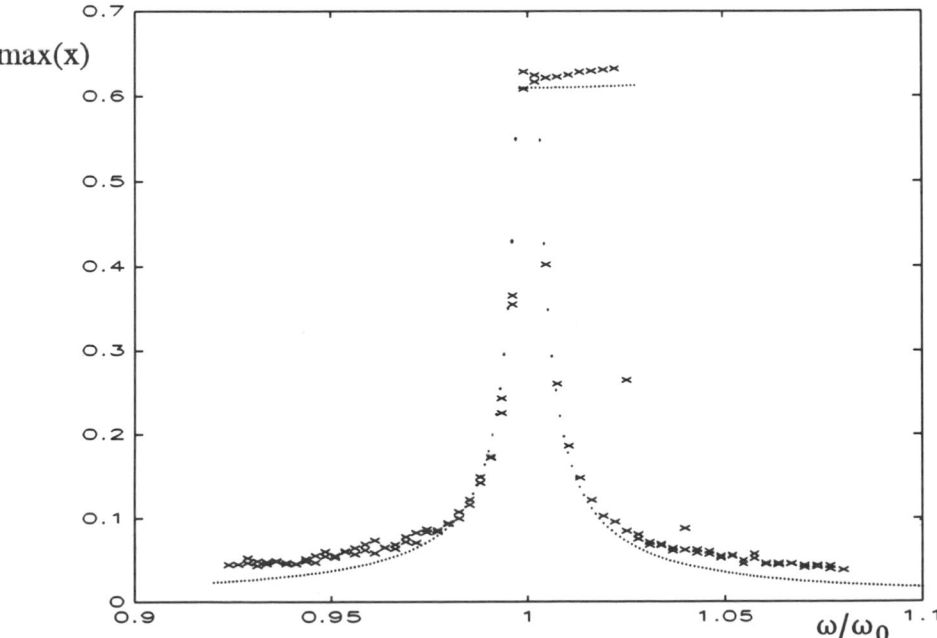

Figure 6.4   Response curves for experimental run 4, experimentally measured (x) and numerically determined (.).

being excited. The coefficient of restitution must be made small in order to account for the energy which is being transferred to these higher modes at each impact. There is a good qualitative agreement between the frequency–response curves obtained experimentally and the theoretical response curves. For each position of the stop the response curve flattens out after the first grazing as the frequency is increased from below the ω_c, the natural frequency of the first linear mode.

A more detailed comparison between the bifurcational behaviour of the theoretical and experimental models is made in Figure 6.5 where the positions in parameter space (position of stop v. forcing frequency or d v. ω) of the various bifurcations listed in the table are shown. For a given forcing frequency, a grazing bifurcation will occur if the maximum amplitude of the response with no impacts is equal to the position of the stop, $x = d$. The locus of grazing bifurcations in d v. ω parameter space then is given by

$$d = \frac{A}{\sqrt{(\Delta\omega)^2 + (\omega^2 - \omega_c^2)^2}}. \tag{15}$$

The location of the saddle–node bifurcations in this parameter space for the theoretical model with the fitted parameters can be calculated for a given coefficient of restitution using the explicit expressions for the eigenvalues of the first differential matrix of the impact map given in Foale and Bishop (in press) (with suitable rescaling). The loci of saddle–node bifurcations for a range of values of the coefficient of restitution r is plotted in Figure 6.5. Clearly a very low coefficient of restitution is required for the theoretical curve to lie near to the experimentally observed one.

The overall qualitative behaviour of a one-sided impact oscillator modelled using a simple

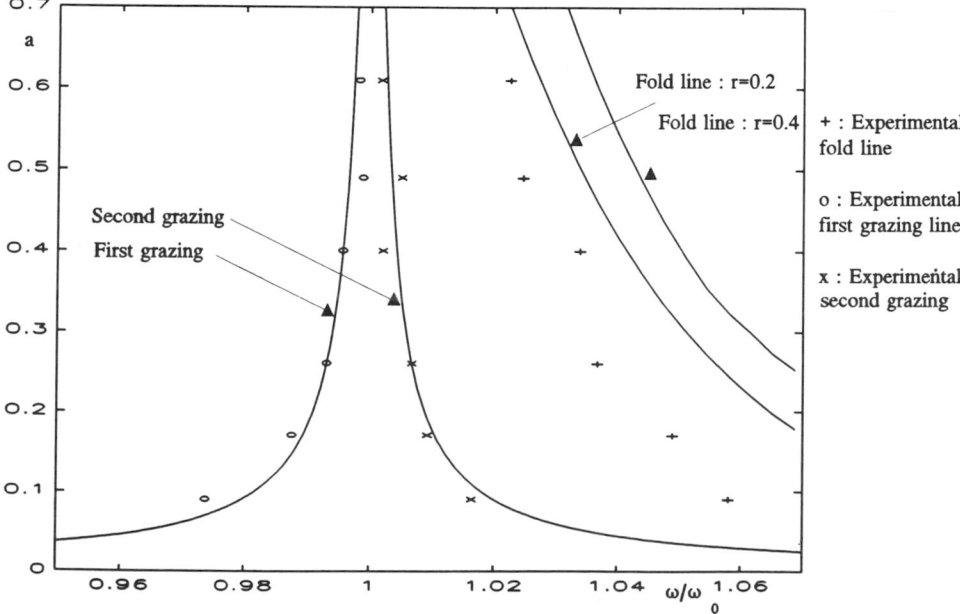

Figure 6.5   Experimental and theoretical bifurcation locus diagram.

coefficient-of-restitution rule to model the impact appears to agree well with the results of the experiments using an electromagnetically excited beam described above. In particular, the bifurcations which are predicted by the theory (including both the conventional smooth bifurcations and the 'new' grazing bifurcations), and which determine the overall shape of the frequency—response curve, are found experimentally. It is clear that the grazing bifurcations play an important part in the overall dynamical response of an impact oscillator.

REFERENCES

Foale, S. (1993) Bifurcations in impact oscillators: theoretical and experimental studies. *Doctoral thesis*, University of London.

Foale, S. and Bishop, S. R. (1992) Dynamical complexities of forced impacting systems, *Philosophical Transactions of the Royal Society of London A*, **338**, 547–556.

Foale, S. and Bishop, S. R. (in press) Bifurcations in impact oscillators, *Nonlinear Dynamics*.

Goldsmith, W. (1960) *Impact*, London, Edward Arnold.

Hogan, S. J. (1989) On the dynamics of rigid-block motion under harmonic forcing, *Proceedings of the Royal Society of London A*, **425**, 441–476.

Jing, H. S. and Young, M. (1990) Random response of a single degree of freedom vibro impact system with clearance, *Earthquake Engineering and Structural Dynamics*, **19**, 789–798.

Kahraman, A. and Singh, R. (1990) Nonlinear dynamics of a spur gear pair, *Journal of Sound and Vibration*, **142** (1), 49–75.

Karagiannis, K. and Pfeiffer, F. (1991) Theoretical and experimental investigations of gear rattling, *Nonlinear Dynamics*, **2**, 367–387.

Lean, G. H. (1971) Subharmonic motions of moored ships subjected to wave action, *Transactions of the Royal Institute of Naval Architects*, **113**, 387–399

Moon, F. C. and Shaw, S. W. (1983) Chaotic vibrations of a beam with nonlinear boundary conditions, *International Journal of Nonlinear Mechanics*, **18** (6), 465–477.

Nelson, W. E., Benton, S. M. and Bernhard, S. (1983) Bumper pile design for mating platform and subsea drilling template, *Offshore Technology Conference, 1983, Houston*.

Nordmark, A. B. (1991) Non-periodic motion caused by grazing incidence in an impact oscillator, *Journal of Sound and Vibration*, **145** (2), 279–297.

Nusse, H. E. and Yorke, J. A. (1992) Border collision bifurcations including 'period two to period three' for piecewise smooth systems, *Physica*, **57D**, 39–57.

Paidoussis, M. P. and Li, G. X. (1992) Cross-flow-induced chaotic vibrations of heat exchanger tubes impacting on loose supports, *Journal of Sound and Vibration*, **152** (2), 305–326.

Robinson, R. W. and Ramzan, F. A. (1989) Prediction of jacket to template docking forces during installation, Preprint.

Sharif-Bakhtiar, M. and Shaw, S. W. (1988) The dynamic response of a centrifugal pendulum vibration absorber with motion limiting stops, *Journal of Sound and Vibration*, **126** (2), 221–235.

Shaw, S. W. and Holmes, P. J. (1983) A periodically forced piecewise linear oscillator, *Journal of Sound and Vibration*, **90** (1), 122–155

Stahl, B., Nelson, W. E. and Baur, M. P. (1983) Motion monitoring of a moored floating platform during installation over a subsea template, *Journal of Petroleum Technology*, **35**, 1239–1248.

Stennson, A. and Nordmark, A. B. (1992) Chaotic vibrations of a spring/mass system with a unilateral displacement limitation – theory and experiments, Preprint, Lulea University of Technology, Sweden.

Stennson, A., Asplund, C. and Karlsson, L. (1992) The nonlinear behaviour of a MacPherson strut wheel suspension, Preprint, Lulea University of Technology, Sweden.

Sterndorff, M. J., Waegter, J. and Eilersen, C. (1992) Design of fixed offshore platforms to dynamic ship impact loads, *Journal of Offshore Mechanics and Arctic Engineering*, **114**, 146–153.

Thompson, J. M. T. (1983) Complex dynamics of compliant offshore structures, *Proceedings of the Royal Society of London*, **A 387** (1793), 407–427.

Thompson, J. M. T., Bokaian, A. R. and Ghaffari, R. (1984) Subharmonic and chaotic motions of compliant offshore structures and articulated mooring towers, *Journal of Energy Resources Technology*, **106**, 191–198.

Tso, W. K. and Wong, C. M. (1989) Steady state rocking response of rigid blocks part 1: analysis, *Earthquake Engineering and Structural Dynamics*, **18**, 89–106.

Tung, P. C. and Shaw, S. W. (1988) The dynamics of an impact print hammer, *Journal of Vibration, Acoustics, Stress and Reliability in Design*, **110**, 193–200.

S. Foale and S. R. Bishop, *Centre for Nonlinear Dynamics, University College London, Gower Street, London WC1E 6BT, UK*

7 STICK–SLIP PHENOMENA AND APPLICATIONS

Ch. Glocker and F. Pfeiffer

This chapter deals with the dynamics of mechanical systems with topology variations due to stick–slip phenomena. These systems show a changing number of degrees of freedom because the superimposed constraints become active or passive with the changing system dynamics. This leads to unsteady changes in the contact forces and hence in the accelerations. Due to the coupled nature of the problem, a state transition in one of the contacts can cause state transitions in any of the other contact points. For solving the equations of motion using numerical integration a system description is selected where the equations of motion are written in terms of generalized coordinates of the unbounded system. Each of the additional constraints is taken into account by including a Lagrange multiplier in the equations of motion. A sufficient condition for all potentially active constraints to remain active or become passive is provided by the solution of a linear complementarity problem. Thus the numerical solution procedure of the system can be regarded as a sequence of initial-value problems of ordinary differential equation systems with varying dimensions, which are selected by the solution of complementarity problems.

7.1 INTRODUCTION

In practical applications nonlinear dynamics often arise from impacts and stick–slip phenomena. Typical examples are walking machines, frictional dampers between turbine blades, assembly processes of manipulators and impact dynamics in gearboxes, all of which are discussed in Pfeiffer (1991). One important property of such systems is that their number of degrees of freedom varies with time, which results from different states of contacts between the rigid bodies of the mechanical system. In detail each of the possible contacts may show sliding, sticking, or separation, and these contact situations may occur in any combination. The dynamic system achieves its maximal number of degrees of freedom f under the assumption that all contacts show separation. This state is used to describe the system by a set of generalized coordinates $q \in \mathbb{R}^f$. Each of the possible contact constraints is controlled by kinematic indicators like relative distance g_N and relative velocity in normal and tangential directions $\dot{g}_N$ and $\dot{g}_T$. A normal or tangential contact constraint is said to be potentially

---

Nonlinearity and Chaos in Engineering Dynamics
Edited by J. M. T. Thompson and S. R. Bishop,   © 1994 John Wiley & Sons Ltd

active if the necessary conditions for contact, $g_N = \dot{g}_N = 0$, or sticking, $g_N = \dot{g}_N = \dot{g}_T = 0$, are fulfilled. A normal or tangential constraint is said to be active if in addition the relative acceleration is equal to zero, $\ddot{g}_N = 0$ or $\ddot{g}_T = 0$. Each active constraint reduces the number of degrees of freedom by one. Thus the constraints on the acceleration level are taken into account as algebraic secondary conditions in the equations of motion and are included as additional contact forces λ_N or λ_T. A representation is thus achieved where an index-1 system can be integrated if the set of active constraints is known. Thereofore the last step is the determination of the active set. Under the assumption of certain contact laws, such as a unilateral constraint in the normal and a Coulomb friction constraint in the tangential direction, it is possible to determine whether a potentially active constraint is active or not. Common models of impacts and Coulomb friction introduce unsteady changes in the contact forces. Thus a state transition in one of the contacts, for example, a transition from sliding to sticking, can cause transitions in any of the other active contact constraints. To avoid the combinatorial problem of testing all contact combinations for the feasible solution, special algorithms from optimization theory are applied. They are used for similar problems, for example in Lötstedt (1981), Moreau (1988), Panagiotopoulos (1977) and Klarbring and Björkman (1988). The main point of this chapter consists of the formulation of the coupled contact problem as a linear complementarity problem in standard form, whose solution provides the active set of constraints. A brief summary of the contact kinematics and the dynamical equations of rigid-body systems is given. A frictional damper which is used to reduce the oscillations of turbine blades is used as an example.

7.2   KINEMATICS

From the evaluation of the kinematics of any contact problem a representation of the constraints on the displacement, velocity and acceleration level must be available which will be used later to describe the necessary and sufficient conditions for contact and sticking. As an example for the contact problem, bodies with smooth convex contours Σ are investigated in Glocker and Pfeiffer (1992), see Figure 7.1. The results are summarized briefly in the following.

It is always possible to orient the normal and tangential vectors $(\mathbf{n}, \mathbf{t})_{1,2}$ of the bodies antiparallel to each other and to define a distance vector $\mathbf{r}_D$ that is perpendicular to both tangents. The points which are connected by $\mathbf{r}_D$ are called the contact points of the problem.

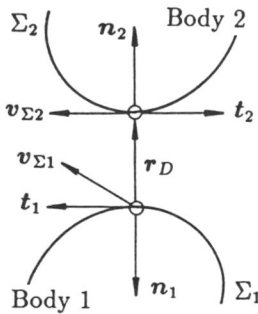

Figure 7.1   Contact kinematics.

The distance g_N of the contact points is then given by

$$g_N = r_D^T n_2 = - r_D^T n_1. \tag{1}$$

The terms $v_{\Sigma 1,2}$ denote the absolute velocities of the contact points. A projection of $(v_{\Sigma 1} - v_{\Sigma 2})$ onto the normal and tangential vector of body 1 leads to the relative velocities in normal and tangential directions, $\dot{g}_N$ and $\dot{g}_T$, which can be stated in terms of the generalized velocities $\dot{q}$ of the system,

$$\dot{g}_N = w_N^T \dot{q} + \tilde{w}_N, \quad \dot{g}_T = w_T^T \dot{q} + \tilde{w}_T. \tag{2}$$

The relative accelerations of the contact points are derived by differentiation of equations (2) with respect to time:

$$\ddot{g}_N = w_N^T \ddot{q} + \bar{w}_N, \quad \ddot{g}_T = w_T^T \ddot{q} + \bar{w}_T, \tag{3}$$

and the structure of the constraint vectors w_N and w_T in equations (2), (3) is given by

$$w_N = J_{\Sigma 1}^T n_1 + J_{\Sigma 2}^T n_2, \quad w_T = J_{\Sigma 1}^T t_1 + J_{\Sigma 2}^T t_2, \tag{4}$$

where $J_{\Sigma 1,2}$ are the Jacobians $(\partial v_{\Sigma 1,2})/(\partial \dot{q})$ of the contact points.

7.3 CONSTRAINED MOTION

The dynamics of a multi-body system with f degrees of freedom is usually described by using a set of generalized coordinates $q \in \mathbb{R}^f$. This means here that the system is uniquely determined by such a q if all contacts show separation. Then the dynamical equations are of the form

$$\sum_{i=1}^n \{ J_T^T (\dot{p} - F) + J_R^T (\dot{L} - M) \}_i = 0, \tag{5}$$

where equation (5) consists of the momentum terms p_i, the moments of momentum L_i, the sum of all active forces F_i, and active moments M_i on body i. The projection into configuration space $\mathbb{R}^f$ is done by premultiplying the terms $(\dot{p} - F)_i \in \mathbb{R}^3$ and $(\dot{L} - M)_i \in \mathbb{R}^3$ using the Jacobians $J_{Ti} = (\partial v_i)/(\partial \dot{q})$ and $J_{Ri} = (\partial \Omega_i)/(\partial \dot{q})$. The resulting equation,

$$M(q, t)\ddot{q} - h(q, \dot{q}, t) = 0 \in \mathbb{R}^f, \tag{6}$$

however, does not yet take into account the contact forces $F_{N1,2}$ and $F_{T1,2}$ in the normal and tangential directions. Thus a complete description of the system under the influence of these forces is achieved by adding one additional term for each of the bodies to equation (5),

$$... - J_{\Sigma 1}^T (F_{N1} + F_{T1}) - J_{\Sigma 2}^T (F_{N2} + F_{T2}) = 0. \tag{7}$$

Together with the properties of the contact forces, see Figure 7.2,

$$F_{N1} = n_1 \lambda_N, \quad F_{N2} = n_2 \lambda_N, \quad F_{T1} = t_1 \lambda_T, \quad F_{T2} = t_2 \lambda_T, \tag{8}$$

we can reformulate (7) by using (4) to get a representation of a system (6) which contains

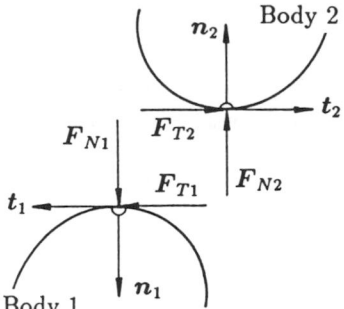

Body 1

Figure 7.2 Contact forces.

one additional contact,

$$M\ddot{q} - h - w_N \lambda_N - w_T \lambda_T = 0. \tag{9}$$

The arising pairs of normal and tangential contact forces are proportional to the multipliers λ_N and λ_T which are determined by certain contact laws. Next, we assume that n_N of the normal and $n_T \leqslant n_N$ of the tangential constraints are potentially active. Together with the relative accelerations (3) and under the assumption of Coulomb friction in the $(n_N - n_T)$ sliding contacts, the available set of equations is given by

$$M\ddot{q} - h - \sum_{i=1}^{n_N} (w_N \lambda_N + w_T \lambda_T)_i = 0,$$

$$\ddot{g}_{Ni} = w_{Ni}^T \ddot{q} + \bar{w}_{Ni}, \qquad i = 1, \ldots, n_N, \tag{10}$$

$$\ddot{g}_{Ti} = w_{Ti}^T \ddot{q} + \bar{w}_{Ti}, \qquad i = 1, \ldots, n_T,$$

$$\lambda_{Ti} = -\mu_i \operatorname{sign}(\dot{g}_{Ti}) \lambda_{Ni}, \qquad i = (n_T + 1), \ldots, n_N,$$

where μ_i is the coefficient of friction of the ith sliding contact and the sign function takes into account the opposite directions of relative velocity and friction force. Equation (11) is a representation of (10) in matrix notation,

$$M\ddot{q} - h - (W_N + H_R | W_T) \begin{pmatrix} \lambda_N \\ \lambda_T \end{pmatrix} = 0,$$

$$\begin{pmatrix} \ddot{g}_N \\ \ddot{g}_T \end{pmatrix} = \begin{pmatrix} W_N^T \\ W_T^T \end{pmatrix} \ddot{q} + \begin{pmatrix} \bar{w}_N \\ \bar{w}_T \end{pmatrix}, \tag{11}$$

where the $(n_N - n_T)$ tangential forces λ_{Ti} of the sliding contacts are expressed by the corresponding normal forces with respect to the last equation in (10). Thus the last $(n_N - n_T)$ columns of H_R consist of the terms $- w_{Ti} \mu_i \operatorname{sign}(\dot{g}_{Ti})$.

7.4   CONTACT LAWS

For establishing the equations of the contact laws we consider each of the tangential constraints as a pair of simultaneously acting constraints such that the resulting tangential forces λ_T are expressed by the difference

$$\lambda_T = \lambda_T^{(+)} - \lambda_T^{(-)}. \tag{12}$$

Thus the basic set of equations (11) is now

$$M\ddot{q} - h - (W_N + H_R| + W_T| - W_T) \begin{pmatrix} \lambda_N \\ \lambda_T^{(+)} \\ \lambda_T^{(-)} \end{pmatrix} = 0,$$

$$\begin{pmatrix} \ddot{g}_N \\ +g_T \\ -\ddot{g}_T \end{pmatrix} = \begin{pmatrix} W_N^T \\ +W_T^T \\ -W_T^T \end{pmatrix} \ddot{q} + \begin{pmatrix} \bar{w}_N \\ +\bar{w}_T \\ -\bar{w}_T \end{pmatrix}. \tag{13}$$

This formulation is necessary to achieve a representation of the problem where even dependent constraints can be taken into consideration. From the definition of the relative accelerations (3) and the normal forces (8) it is obvious that the contact problem in the normal direction is determined by the n_N complementarity conditions

$$\ddot{g}_N \geqslant 0, \quad \lambda_N \geqslant 0, \quad \ddot{g}_N^T \lambda_N = 0. \tag{14}$$

The structure of the tangential subproblem is much more complicated. In Appendix B similar conditions are derived from Coulomb's law of friction, see also Seyfferth (1993),

$$-\ddot{g}_T = -\ddot{g}_T^+ + z^-, \quad \ddot{g}_T^+ \geqslant 0, \quad \lambda_T^{(+)} \geqslant 0, \quad \ddot{g}_T^{+T} \lambda_T^{(+)} = 0;$$

$$\lambda_{T0}^{(-)} = H_0 \lambda_N - \lambda_T^{(+)}, \quad z^- \geqslant 0, \quad \lambda_{T0}^{(-)} \geqslant 0, \quad z^{-T} \lambda_{T0}^{(-)} = 0;$$

$$+\ddot{g}_T = -\ddot{g}_T^- + z^+, \quad \ddot{g}_T^- \geqslant 0, \quad \lambda_T^{(-)} \geqslant 0, \quad \ddot{g}_T^{-T} \lambda_T^{(-)} = 0;$$

$$\lambda_{T0}^{(+)} = H_0 \lambda_N - \lambda_T^{(-)}, \quad z^+ \geqslant 0, \quad \lambda_{T0}^{(+)} \geqslant 0, \quad z^{+T} \lambda_{T0}^{(+)} = 0. \tag{15}$$

The $4n_T$ additional equations in the left-hand column of (15) consist of the tangential accelerations $\ddot{g}_T$, which are split into their positive and negative parts, and the definition of the friction saturations $\lambda_T^{(\pm)}$ as the difference of maximal transferable forces $(H_0 \lambda_N)$ and actual tangential forces $\lambda_T^{(+)}$. The matrix H_0 contains the friction coefficients μ_{0i} for vanishing sliding velocity and is explained in Appendix B. Using the abbreviations

$$\ddot{g}_M := \begin{pmatrix} \ddot{g}_N \\ +\ddot{g}_T \\ -\ddot{g}_T \end{pmatrix}, \quad G := \begin{pmatrix} W_N^T \\ +W_T^T \\ -W_T^T \end{pmatrix} M^{-1} \begin{pmatrix} W_N^T \\ +W_T^T \\ -W_T^T \end{pmatrix}^T,$$

$$\lambda_{T0} := \begin{pmatrix} \lambda_{T0}^{(-)} \\ \lambda_{T0}^{(+)} \end{pmatrix}, \quad z := \begin{pmatrix} z^- \\ z^+ \end{pmatrix},$$

$$\ddot{g} := \begin{pmatrix} \ddot{g}_N \\ \ddot{g}_T^+ \\ \ddot{g}_T^- \end{pmatrix}, \quad N_R := \begin{pmatrix} W_N^T \\ +W_T^T \\ -W_T^T \end{pmatrix} M^{-1} \begin{pmatrix} H_R^T \\ 0 \\ 0 \end{pmatrix}^T, \quad I := \begin{pmatrix} 0 & E & 0 \\ 0 & 0 & E \end{pmatrix},$$

$$\lambda := \begin{pmatrix} \lambda_N \\ \lambda_T^{(+)} \\ \lambda_T^{(-)} \end{pmatrix}, \quad \bar{g} := \begin{pmatrix} W_N^T \\ +W_T^T \\ -W_T^T \end{pmatrix} M^{-1} h + \begin{pmatrix} \bar{w}_N \\ +\bar{w}_T \\ -\bar{w}_T \end{pmatrix}, \quad N_0 := \begin{pmatrix} H_0 & 0 & 0 \\ H_0 & 0 & 0 \end{pmatrix},$$

$$\tag{16}$$

the conditions (13), (14), (15) are now rewritten as

$$\ddot{g}_M = (G + N_R)\lambda + \bar{g}, \quad \ddot{g}_M = \ddot{g} - I^T z, \quad \lambda_{T0} = (N_0 - I)\lambda,$$

$$\ddot{g} \geqslant 0, \quad \lambda \geqslant 0, \quad \ddot{g}^T \lambda = 0, \quad \lambda_{T0} \geqslant 0, \quad z \geqslant 0, \quad \lambda_{T0}^T z = 0. \qquad (17)$$

The first equation in (17) results from (13) after elimination of the generalized accelerations $\ddot{q}$. The second and third equations consist of the split accelerations and the friction saturations from (15). In matrix notation (17) is stated as

$$\begin{pmatrix} \ddot{g} \\ \lambda_{T0} \end{pmatrix} = \begin{pmatrix} G + N_R & I^T \\ N_0 - I & 0 \end{pmatrix} \begin{pmatrix} \lambda \\ z \end{pmatrix} + \begin{pmatrix} \bar{g} \\ 0 \end{pmatrix},$$

$$\begin{pmatrix} \ddot{g} \\ \lambda_{T0} \end{pmatrix} \geqslant 0, \quad \begin{pmatrix} \lambda \\ z \end{pmatrix} \geqslant 0, \quad \begin{pmatrix} \ddot{g} \\ \lambda_{T0} \end{pmatrix}^T \begin{pmatrix} \lambda \\ z \end{pmatrix} = 0. \qquad (18)$$

This structure is well known as a linear complementarity problem in standard form,

$$y = Ax + b, \quad y \geqslant 0, \quad x \geqslant 0, \quad y^T x = 0, \qquad (19)$$

whose solution $y \in \mathbb{R}^{n_N + 4n_T}$, $x \in \mathbb{R}^{n_N + 4n_T}$ is provided by a modified simplex algorithm, see Murty (1988), and contains all unknown contact forces and accelerations. In the frictionless case, $N_0 = N_R = 0$, the matrix A is positive semidefinite, which leads to a unique solution of $\ddot{q}$ in (13). Friction destroys this structure, and uniqueness with respect to $\ddot{q}$ can only be expected for sufficientely small coefficients of friction.

7.5   TURBINE BLADE DAMPER

Parabolic friction elements which are mounted between the platforms of neighbouring turbine blades give an application of the theory presented, see e.g. Glocker and Pfeiffer (1992). The dampers are designed to achieve a friction-induced reduction of the blade oscillations which result from an excitation by periodically changing gas forces. The mechanical model shown in Figure 7.3 consists of two platforms (masses m_1, m_3) with turbine blades (masses m_2, m_4) which are connected by a friction damper (mass m_5, moment of inertia J_5). The displacements of the platforms and blades are determined by the translational coordinates (q_1, q_2, q_3, q_4). When we consider the unbounded system the damper has two translational degrees of freedom (q_6, q_7), and one rotational degree of freedom (q_5). A constant force F_Z, representing the centrifugal force, affects the damper and presses it against the oblique planes of the platforms. The excitation of the turbine blades is taken into account by the forces $F_i(t) = A \sin(\omega t + \varphi_i)$, $i = 2, 4$. Altogether the resulting configuration contains a coupled two-point contact problem, where the normal and tangential forces of the two contact points influence each other and depend on the system dynamics.

The aim of the design of a friction damper is to obtain the largest possible dissipation of energy by friction. Thus the damper has to exhibit sufficiently long phases of sliding and to show a low tendency to lock up. Using an original data set from industry, Figure 7.4 shows the computed accelerations $\ddot{q}_5$, $\ddot{q}_7$ as a function of their velocities for a static damper equilibrium $q_{50} \neq 0$. The unsteady changes in the accelerations arise from stick–slip transitions and indicate that the damper is working.

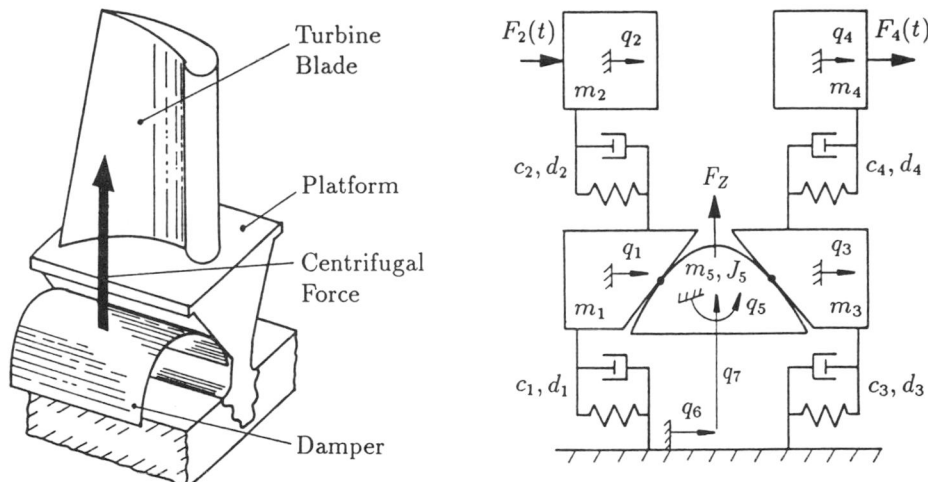

Figure 7.3 Turbine blade damper and mechanical model.

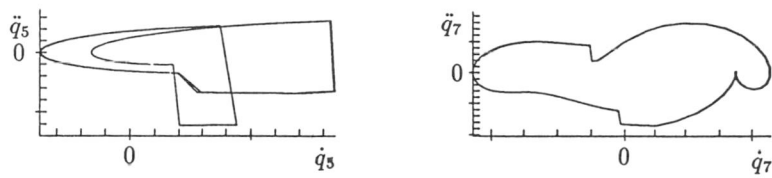

Figure 7.4 Dynamical behaviour of the damper.

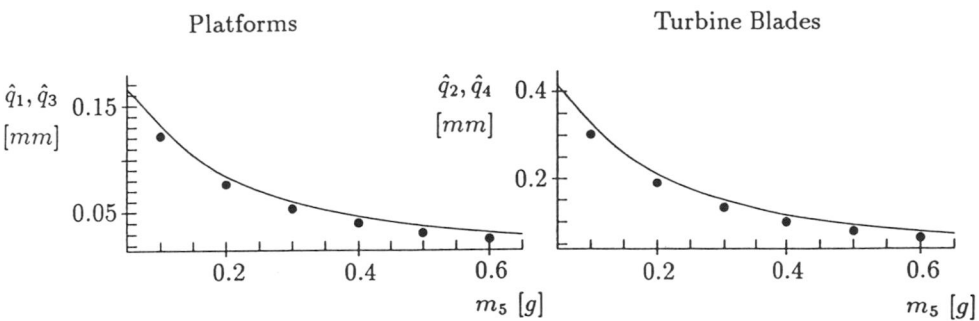

Figure 7.5 Comparison of linear and nonlinear simulations.

Another task is to examine how much the blade and platform amplitudes can be reduced by installing different dampers. Figure 7.5 shows the platform and blade amplitudes ($\hat{q}_{1,3}$) and ($\hat{q}_{2,4}$) depending on the damper mass m_5. The dots correspond to the amplitudes computed by using the model of Figure 7.3. The solid line results from a simulation of a linear system which is achieved by the assumption that the damper is not working. Thus, the damper together with the two platforms is regarded as one rigid body and the model of Figure 7.3

becomes linear. The difference between the dots and the solid line then shows exactly the reduction of the amplitudes by friction.

APPENDIX A

This appendix contains a summary of some definitions and theorems of convex analysis which have been used to derive the complementarity conditions of the frictional contacts in Appendix B and can be found in detail in Rockafellar (1972).

A subset C of $\mathbb{R}^n$ is said to be *convex* if $(1 - \lambda)x + \lambda y \in C$ whenever $x \in C$, $y \in C$ and $0 < \lambda < 1$.

A vector y is said to be *normal* to a convex set C at a point $x \in C$ if y does not make an acute angle with any line segment in C with x as the endpoint. The set of all such vectors y is called the *normal cone* to C at x:

$$N_C(x) = \{y \,|\, y^T(x^* - x) \leqslant 0; \quad x \in C; \quad \forall x^* \in C\}. \tag{20}$$

If x is, for example, an element of the relative interior of C, then it readily follows that the normal cone consists of the single element 0 only.

A function $f: S \subset \mathbb{R}^n \to \mathbb{R}$ is said to be *convex* if the set $\{(x, \mu) \,|\, x \in S; \, \mu \in \mathbb{R}; \, \mu \geqslant f(x)\}$ is a convex subset of $\mathbb{R}^{n+1}$. One very important convex function in the indicator function of a convex set C:

$$\Psi_C(x) = \begin{cases} 0 & \text{if } x \in C, \\ +\infty & \text{if } x \notin C. \end{cases} \tag{21}$$

Let $f_1, \ldots, f_m$ be convex functions on $\mathbb{R}^n$, and let

$$f(x) = \inf_{x_i} \{f_1(x_1) + \cdots + f_m(x_m) \,|\, x_i \in \mathbb{R}^n; \quad x_1 + \cdots + x_m = x\}$$

$$:= (f_1 \,\square\, \cdots \,\square\, f_m)(x). \tag{22}$$

Then f is a convex function on $\mathbb{R}^n$, and the operation $\square$ is called *infimal convolution*.

A vector y is said to be a *subgradient* of a convex function f at a point x if $f(x^*) \geqslant f(x) + y^T(x^* - x)$, $\forall x^*$. The set of all subgradients of f at x is called the *subdifferential* of f at x and is denoted by $\partial f(x)$, where

$$\partial f(x) = \{y \,|\, f(x^*) \geqslant f(x) + y^T(x^* - x); \forall x^*\}. \tag{23}$$

The subdifferential is a convex set. If it consists of just one vector, then the function f is differentiable at x, and the subdifferential is the same as the gradient of f. For example, the subdifferential of the indicator function $\Psi_C(x)$ is the normal cone to C at x, which follows immediately from the definitions,

$$\partial \Psi_C(x) = N_C(x). \tag{24}$$

Let C be a non-empty convex set which is defined by

$$C = \{x \,|\, f(x) \leqslant 0\}. \tag{25}$$

A vector y is then normal to C at a point x if and only if there exists a $\lambda \geqslant 0$ such that $y \in \lambda \partial f(x)$. This means together with the definition of the normal cone that

$$y \in N_C(x) \Leftrightarrow y \in \lambda \partial f(x), \quad f(x) \leqslant 0, \quad \lambda \geqslant 0, \quad \lambda f(x) = 0. \tag{26}$$

Let f be a convex function on $\mathbb{R}^n$. The function f^* defined by

$$f^*(x^*) = \sup_x \{x^T x^* - f(x)\} \tag{27}$$

is called the *conjugate* of f. It is a convex function on $\mathbb{R}^n$. The conjugacy operation $f \rightarrow f^*$ is closely related to the classical Legendre transformation in the case of differentiable convex functions. There exist various operations between functions and their conjugates: for example, the correspondence between addition and infimal convolution,

$$f(x) = (f_1 \square \cdots \square f_m)(x) \leftrightarrow f^*(x^*) = f_1^*(x^*) + \cdots + f_m^*(x^*). \tag{28}$$

From the definition of the subgradients and the conjugates it can be shown that the two following conditions are equivalent to each other:

$$x^T x^* = f(x) + f^*(x^*) \Leftrightarrow x^* \in \partial f(x). \tag{29}$$

APPENDIX B

In this appendix the differential inclusions for the tangential subproblem under the assumption of one-dimensional Coulomb friction are derived. For convenience the inequality notation $a \leqslant b$ is used to express the condition $\{a_i\} \leqslant \{b_i\}$ for each of the components of a and b. The upper indices $+$ and $-$ denote the positive and negative parts of a vector, $a^+ = \frac{1}{2}\{|a_i| + a_i\}$, $a^- = \frac{1}{2}\{|a_i| - a_i\}$, $a = a^+ - a^-$. In contrast, the upper indices $(+)$ and $(-)$ are only chosen for reasons of distinction and do not have the above meaning.

The frictional contact law of Coulomb for a single contact and a normal force $\lambda_N \geqslant 0$ can be state as

$$|\lambda_T| \leqslant \mu_0 \lambda_N, \begin{cases} |\lambda_T| < \mu_0 \lambda_N \Rightarrow \ddot{g}_T = 0, \\ \lambda_T = +\mu_0 \lambda_N \Rightarrow \ddot{g}_T \leqslant 0, \\ \lambda_T = -\mu_0 \lambda_N \Rightarrow \ddot{g}_T \geqslant 0. \end{cases} \tag{30}$$

Here μ_0 is the coefficient of friction for vanishing relative velocity $\dot{g}_T$. For a collection of n_T contacts the admissible values of the tangential forces are then given by the convex set

$$C_T = \{\lambda_T^* | -H_0 \lambda_N \leqslant \lambda_T^* \leqslant H_0 \lambda_N\}, \tag{31}$$

where $H_0 \in \mathbb{R}^{n_T, n_N}$ and $\{H_{0ii}\} = \mu_{0i}$, $\{H_{0ij}\} = 0$. It is well known that (30) can be expressed in the form of a variational inequality,

$$-\ddot{g}_T^T(\lambda_T^* - \lambda_T) \leqslant 0, \quad \lambda_T \in C_T, \quad \forall \lambda_T^* \in C_T, \tag{32}$$

which is by (20) and (24) equivalent to the differential inclusion

$$-\ddot{g}_T \in N_{C_T}(\lambda_T) = \partial \Psi_{C_T}(\lambda_T), \tag{33}$$

where Ψ_{C_T} is the indicator function of C_T and N_{C_T} is the corresponding normal cone. For each $\lambda_T \in C_T$ the indicator Ψ_{C_T} is equal to zero and hence with the help of (29) we can write instead of (33):

$$- \ddot{g}_T^T \lambda_T = \Psi_{C_T}^*(- \ddot{g}_T). \tag{34}$$

Here $\Psi_{C_T}^*$ is the conjugate (27) of Ψ_{C_T}. Now let us introduce the convex sets

$$C_T^{(+)} = \{\lambda_T^{(+)*} | - \lambda_T^{(+)*} \leqslant 0, \quad \lambda_T^{(+)*} - H_0 \lambda_N \leqslant 0\},$$

$$C_T^{(-)} = \{\lambda_T^{(-)*} | - \lambda_T^{(-)*} \leqslant 0, \quad \lambda_T^{(-)*} - H_0 \lambda_N \leqslant 0\}. \tag{35}$$

The relation between (35) and (31) is given by the difference

$$C_T = C_T^{(+)} - C_T^{(-)}, \tag{36}$$

which means that for each $\lambda_T^{(+)} \in C_T^{(+)}$, $\lambda_T^{(-)} \in C_T^{(-)}$ the vector λ_T defined by

$$\lambda_T = \lambda_T^{(+)} - \lambda_T^{(-)} \tag{37}$$

is an element of C_T. With the property (36) the indicator function of (31) and its conjugate can be expressed according to (22) and (28) by

$$\Psi_{C_T}(\lambda_T) = \Psi_{C_T^{(+)}}(\lambda_T^{(+)}) \square \Psi_{-C_T^{(-)}}(- \lambda_T^{(-)}),$$

$$\Psi_{C_T}^*(- \ddot{g}_T) = \Psi_{C_T^{(+)}}^*(- \ddot{g}_T) + \Psi_{-C_T^{(-)}}^*(- \ddot{g}_T). \tag{38}$$

When we put (37) and the second equation of (38) into (34) we get

$$- \ddot{g}_T^T(\lambda_T^{(+)} - \lambda_T^{(-)}) = \Psi_{C_T^{(+)}}^*(- \ddot{g}_T) + \Psi_{-C_T^{(-)}}^*(- \ddot{g}_T). \tag{39}$$

Equation (39) is fulfilled, for example, if

$$- \ddot{g}_T^T \lambda_T^{(+)} = \Psi_{C_T^{(+)}}^*(- \ddot{g}_T), \quad + \ddot{g}_T^T \lambda_T^{(-)} = \Psi_{-C_T^{(-)}}^*(- \ddot{g}_T), \tag{40}$$

which due to (29) and $\Psi_{-C_T^{(-)}}(- \lambda_T^{(-)}) = \Psi_{-C_T^{(-)}}(\lambda_T^{(-)})$ is equivalent to the differential inclusions

$$- \ddot{g}_T \in \partial \Psi_{C_T^{(+)}}(\lambda_T^{(+)}) = N_{C_T^{(+)}}(\lambda_T^{(+)}), \quad + \ddot{g}_T \in \partial \Psi_{C_T^{(-)}}(\lambda_T^{(-)}) = N_{C_T^{(-)}}(\lambda_T^{(-)}). \tag{41}$$

Thus we conclude that for each $\ddot{g}_T$ and $\lambda_T^{(+)}$, $\lambda_T^{(-)}$ from (41) the frictional laws (32) are satisfied by $\lambda_T = \lambda_T^{(+)} - \lambda_T^{(-)}$. An evaluation of the differential inclusions (41) by using the normal cone conditions (25), (26) leads directly to the conditions (15).

REFERENCES

Glocker, Ch. and Pfeiffer, F. (1992) An LCP-approach for multibody systems with planar friction, *Proc. Contact Mechanics Int. Symp., Lausanne*, 13–30.

Klarbring, A. and Björkman, G. (1988) A mathematical programming approach to contact problems with friction and varying contact surface, *Comp. and Struct.*, **30** (5), 1185–1198.

Lötstedt, P. (1981) Coulomb friction in two-dimensional rigid body systems, *Z. Ang. Math. Mech.*, **61**, 605–615.

Moreau, J. J. (1988) Unilateral contact and dry friction in finite freedom dynamics, *Non-Smooth Mech. and Appl.*, CISM Courses and Lectures, **302**, Vienna, Springer Verlag.

Murty, K. G. (1988) *Linear Complementarity, Linear and Nonlinear Programming*, Sigma Series in Applied Mathematics, **3**, Berlin, Heldermann Verlag.

Panagiotopoulos, P.D. (1977) *Ungleichungsprobleme in der Mechanik*, Aachen, Habilitation.

Pfeiffer, F. (1991) Dynamical systems with time-varying or unsteady structure, *Z. Ang. Math. Mech.*, **71** (4), T6–T22.

Rockafellar, R. T. (1972) *Convex Analysis*, Princeton NJ, Princeton University Press.

Seyfferth, W. (1993) *Modellierung unstetiger Montageprozesse mit Robotern. Fortschrittbeichte VDI*, Reihe 11, Nr. 199, VDI-Verlag, Dusseldorf.

Ch. Glocker and F. Pfeiffer, *Lehrstuhl B für Mechanik, TU München, Postfach 202420, D-80290 München, Germany.*

8 ON APPROXIMATIONS OF NON-SMOOTH FUNCTIONS IN BIFURCATION ANALYSIS

W. Kleczka and E. Kreuzer

The governing equations of technical systems often contain non-smooth functions due to clearances, frictional effects, impacts, or other phenomena, which can only be defined in a piecewise manner. This class of nonlinearities if of major technical relevance (e.g. gear rattling, stick–slip effects), but it is difficult to analyse (Hsu *et al.* 1990). This is mainly due to the lack of a local power-series representation of the non-smooth function, which constitute the basis for many perturbation techniques in nonlinear dynamics. Specifically, partial derivatives up to the third order are needed for a local symbolic closed-form approximation of the Poincaré map in the neighbourhood of a periodic solution, which is the starting point for the local bifurcation analysis (Kleczka 1991).

A generally applicable method for the treatment of mechanical systems with non-smooth characteristics is outlined which allows a nonlinear stability analysis of periodic solutions on the basis of a discrete map. Thereby, the problems typically encountered for smooth approximations of non-smooth functions are avoided, namely the lengthy expressions needed for the calculation of the partial derivatives of the model equations and numerical problems due to the inclusion of numerical parameters defining the approximation.

8.1 APPROXIMATION EQUATIONS

Many discontinuities $d(x)$ in engineering systems can be expressed in the following general form:

$$d(x) = \begin{cases} h(x) & \text{for} \quad s(x) \geqslant 0, \\ -h(x) & \text{for} \quad s(x) < 0. \end{cases} \tag{1}$$

The scalar function $h(x)$ describes the non-smooth quantity while the scalar function $s(x)$ stands for the switching criterion. Both functions can depend on the state vector denoted by x.

Nonlinearity and Chaos in Engineering Dynamics
Edited by J. M. T. Thompson and S. R. Bishop,   © 1994 John Wiley & Sons Ltd

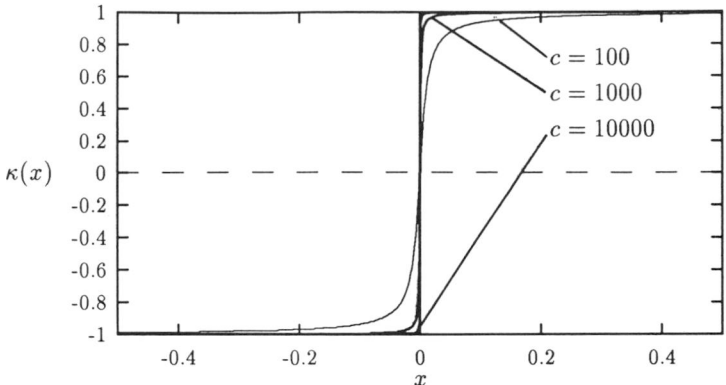

Figure 8.1   Normalized inverse tangent function for different values of c

From this formulation two problems arise:

1. The governing equations are not given in a closed symbolic form so that it is very difficult to apply analytical algorithms.
2. The need for partial derivatives in the nonlinear stability analysis of the dynamic equations produces problems.

Both drawbacks can be overcome by introducing a smooth approximation $k(x)$ of the function $d(x)$ in terms of a suitable inverse tangent function:

$$d(x) \approx k(x) = h(x)\frac{2}{\pi} \lim_{c \to \infty} \arctan cs(x). \tag{2}$$

In Figure 8.1 the normalized inverse tangent function

$$\kappa(x) = \frac{2}{\pi} \arctan cx \tag{3}$$

is plotted for different values of the numerical parameter c. Clearly it can be seen that this function provides the required sign change at the origin. The accuracy increases with increasing c.

8.2   CALCULATION OF PARTIAL DERIVATIVES

The first derivative of $k(x)$ with respect to x_j becomes

$$\frac{\partial k(x)}{\partial x_i} = \frac{\partial h(x)}{\partial x_i} \frac{2}{\pi} \lim_{c \to \infty} \arctan cs(x) + \mathscr{R}_1 \tag{4}$$

with

$$\mathscr{R}_1 = h(x)\frac{\partial s(x)}{\partial x_i} \lim_{c \to \infty} \frac{2c}{\pi(1 + c^2 s(x)^2)}. \tag{5}$$

The second derivative of $k(\boldsymbol{x})$ with respect of x_i and x_j is given by

$$\frac{\partial^2 k(\boldsymbol{x})}{\partial x_i \partial x_j} = \frac{\partial^2 h(\boldsymbol{x})}{\partial x_i \partial x_j} \frac{2}{\pi} \lim_{c \to \infty} \arctan cs(\boldsymbol{x}) + \mathcal{R}_2 \tag{6}$$

with

$$\mathcal{R}_2 = \left(\frac{\partial h(\boldsymbol{x})}{\partial x_i} \frac{\partial s(\boldsymbol{x})}{\partial x_j} + \frac{\partial h(\boldsymbol{x})}{\partial x_j} \frac{\partial s(\boldsymbol{x})}{\partial x_i} + h(\boldsymbol{x}) \frac{\partial^2 s(\boldsymbol{x})}{\partial x_i \partial x_j} \right) \lim_{c \to \infty} \frac{2c}{\pi(1 + c^2 s(\boldsymbol{x})^2)}$$

$$- h(\boldsymbol{x}) s(\boldsymbol{x}) \frac{\partial s(\boldsymbol{x})}{\partial x_i} \frac{\partial s(\boldsymbol{x})}{\partial x_j} \lim_{c \to \infty} \frac{4c^3}{\pi(1 + c^2 s(\boldsymbol{x})^2)^2}. \tag{7}$$

Analogously, the third derivative of $k(\boldsymbol{x})$ with respect to x_i, x_j and x_k has the same structure:

$$\frac{\partial^3 k(\boldsymbol{x})}{\partial x_i \partial x_j \partial x_k} = \frac{\partial^3 h(\boldsymbol{x})}{\partial x_i \partial x_j \partial x_k} \frac{2}{\pi} \lim_{c \to \infty} \arctan cs(\boldsymbol{x}) + \mathcal{R}_3. \tag{8}$$

The explicit expression for $\mathcal{R}_3$ is rather involved and is omitted for brevity.

It can be seen that the analytic calculation of the partial derivatives of the smooth approximation (2) becomes more and more complex the higher the derivative. Another problem is the appearance of powers of the numerical parameter c in the remainders $\mathcal{R}_i$ leading to numerical problems during the evaluation of the expressions because the parameter c has to be chosen as large as possible in order to guarantee a good approximation.

These observations lead to the concept of passive and active state variables as explained in the next section.

8.2.1 Passive and active state variables

The main point is that – if no jumps occur – the quantities $\mathcal{R}_i$ are of the order $\mathcal{O}[c^{-1}]$ and therefore approach zero as $c \to \infty$. This means that in this case they can be neglected for the calculation of the partial derivatives.

For the efficient calculation of the partial derivatives of (2) it is thus advantageous to distinguish between active and passive state variables:

$$d(\boldsymbol{x}) \approx k(\boldsymbol{x}, \tilde{\boldsymbol{x}}) = h(\boldsymbol{x}) \frac{2}{\pi} \lim_{c \to \infty} \arctan cs(\tilde{\boldsymbol{x}}). \tag{9}$$

Passive variables ($\tilde{\boldsymbol{x}}$)

- always have the same value as the corresponding active variables ($\boldsymbol{x}$),
- are treated as constants for the calculation of the derivatives,
- preserve smoothness of the partial derivatives.

This means, that for the calculation of the partial derivatives the inverse tangent function is treated as a non-constant coefficient which preserves all smoothness properties but does not increase the complexity of the already rather involved expressions.

The partial derivatives with respect to the state variables are thus given in the simplified form

$$\frac{\partial^n k(x, \tilde{x})}{\partial x_i \dots} = \frac{\partial^n h(x)}{\partial x_i \dots} \frac{2}{\pi} \lim_{c \to \infty} \arctan cs(\tilde{x}). \tag{10}$$

This concept is well suited for a computer algebra implementation.

8.2.2 Making use of structural properties of mechanical systems

For an efficient calculation of partial derivatives it is also of major importance to utilize generic properties of the equations of motion of multi-body systems.

The symbolic vector differential equation of motion of a multi-body system is given in the form

$$M(y, p, t)\ddot{y} + k(y, \dot{y}, p, t) = q(y, \dot{y}, p, t). \tag{11}$$

Here, t denotes the time, M is the $(f \times f)$ generalized mass matrix, y is the $(f \times 1)$ vector of generalized coordinates, p is the $(r \times 1)$ vector of control parameters, k is the $(f \times 1)$ vector of gyroscopic and centrifugal forces, and q is the $(f \times 1)$ vector of generalized applied forces, where f is the number of degrees of freedom and r is the number of control parameters (Kreuzer and Schiehlen 1990; Kreuzer and Leister 1991).

The first-order state equation has the following general structure:

$$\frac{d}{dt} \underbrace{\begin{bmatrix} y \\ \dot{y} \\ p \end{bmatrix}}_{\dot{x}} = \underbrace{\begin{bmatrix} \dot{y} \\ M^{-1}(q - k) \\ 0 \end{bmatrix}}_{f(x)}, \tag{12}$$

where M^{-1} is given by

$$M^{-1} = \frac{1}{\det(M)} \operatorname{adj}(M). \tag{13}$$

The only non-trivial equation for the acceleration is given by

$$\ddot{y} = \frac{1}{s(y, p, t)} A(y, p, t) \, v(y, \dot{y}, p, t) \equiv fa(y, \dot{y}, p, t), \tag{14}$$

with a scalar $s = \det M$, an $(f \times f)$ matrix $A = \operatorname{adj} M$, and an $(f \times 1)$ vector $v = q - k$. The calculation of the partial derivatives of the state equation $f(y, \dot{y}, p, t)$ with respect to the state variables y and $\dot{y}$ and the control parameters p is performed via generally applicable computer algebra tools in two steps:

1. All non-trivial partial derivatives of the explicitly evaluated problem-dependent quantities s, v and A are precalculated and separately stored.
2. All partial derivatives of equation (14) with its implicitly given (and hence problem-

independent) dependencies are formally calculated by successively applying the chain rule up to the desired order. Trivial relations calculated in Step 1 are explicitly inserted.

For the numerical evaluation of the partial derivatives, the stored symbolic relations of the partial derivatives calculated in Step 1 are evaluated once and then used repetitively for forming the entire expressions.

With these methods partial derivatives may also be efficiently provided for systems with non-smooth characteristics. Now it is possible to apply a general scheme for the local bifurcation analysis of periodic motions as discussed in the next section.

8.3 LOCAL BIFURCATION ANALYSIS

The logical structure of the analysis is outlined in Figure 8.2. The first step is to map the considered real-world physical system onto a simplified mechanical system and to determine relevant parameter ranges. The derivation of a mathematical model is the next step. The detection of periodic solutions by numerical routines follows.

For harmonically driven oscillators, it is advantageous to replace the ordinary differential system equations by a discretized map $\xi_{k+1} = g(\xi_k)$ in terms of the vector of deviation variables $\xi = x - \bar{x}$, where $\bar{x}$ denotes the periodic solution under examination. Its stability is determined by that of the respective fixed point of the map. In order to apply analytical procedures to the map, a closed symbolic formulation is necessary. Therefore, a local approximation for $g(\xi)$ is computed in terms of the coefficients of a Taylor series expansion about the periodic solution $\bar{x}$:

$$g(\xi) = \frac{\partial g(x)}{\partial x}\bigg|_{\bar{x}} \xi + \frac{1}{2}\frac{\partial^2 g(x)}{\partial x^2}\bigg|_{\bar{x}} \{\xi, \xi\} + \frac{1}{6}\frac{\partial^3 g(x)}{\partial x^3}\bigg|_{\bar{x}} \{\xi, \xi, \xi\} + \mathcal{O}(\|\xi\|^4), \tag{15}$$

where $\{\xi, \ldots, \xi\}$ denotes the outer product. It can be shown that the linear coefficients $g_{i,j}$, the quadratic coefficients $g_{i,jk}$, and the cubic coefficients $g_{i,jkl}$ for the ith equation are given by the solution of

$$\frac{d}{dt}\begin{bmatrix} x_i \\ g_{i,j} \\ g_{i,jk} \\ g_{i,jkl} \end{bmatrix} = \begin{bmatrix} f_i \\ f_{i,a}g_{a,j} \\ f_{i,a}g_{a,jk} + f_{i,ab}g_{a,j}g_{b,k} \\ f_{i,a}g_{a,jkl} + f_{i,ab}g_{a,jk}g_{b,l} + f_{i,ab}g_{a,j}g_{b,kl} \\ + f_{i,ab}g_{a,jl}g_{b,k} + f_{i,abc}g_{a,j}g_{b,k}g_{c,l} \end{bmatrix} \tag{16}$$

at $t = T$ (Kleczka *et al.* 1990). Here, T is the discretization time, namely the period of the solution to be studied, and a, b, c are summation indices*. The only non-zero initial conditions are given by $x_i(t = 0) = \bar{x}_i$, $g_{i,j}(t = 0) = \delta_{ij}$, where the Kronecker δ is introduced.

---

*Einstein's summation convention holds for the entire index formulation.

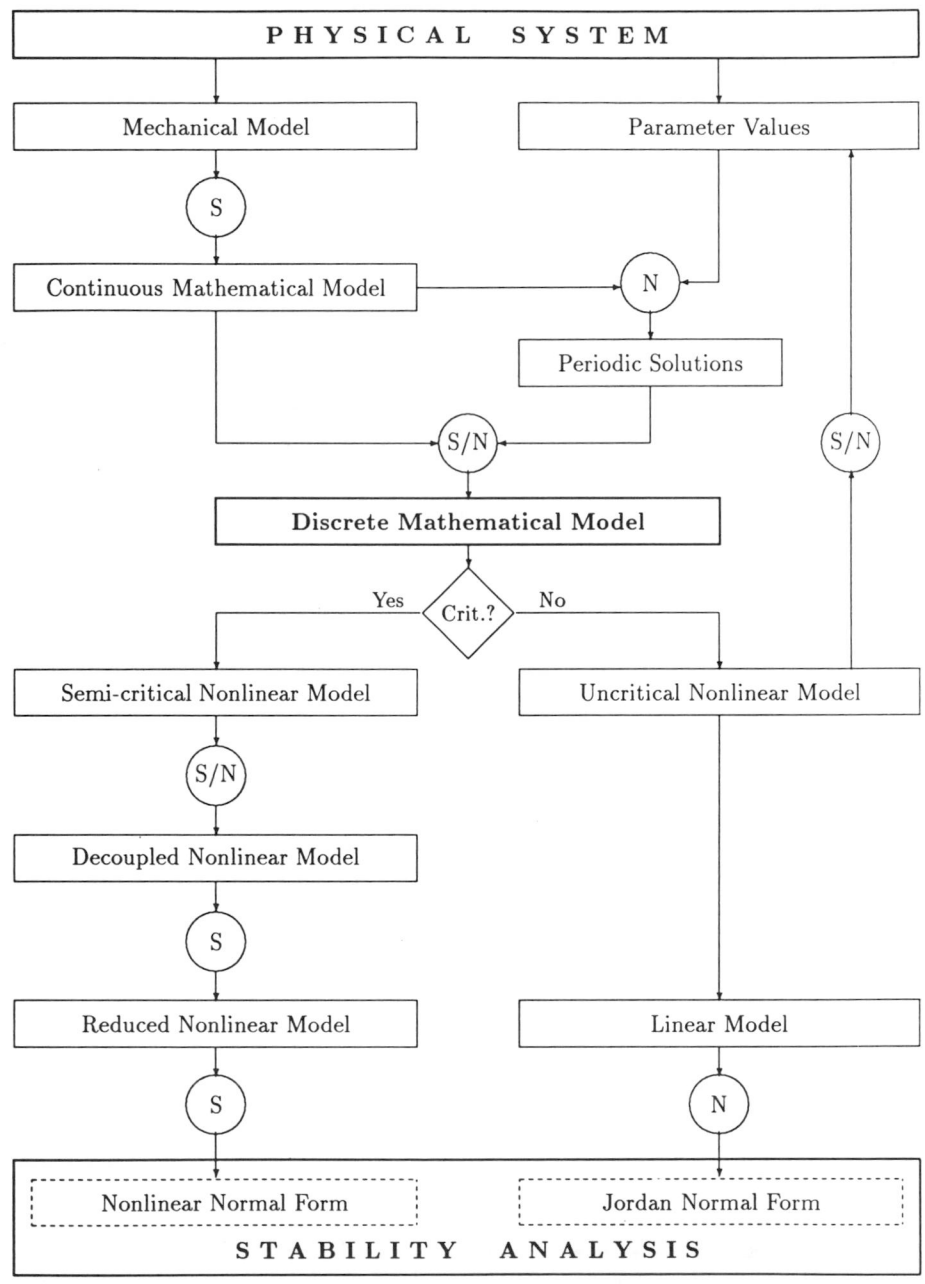

S: Symbolic Computations   N: Numeric Computations

Figure 8.2 Logical structure of the analysis.

In order to study the influence of variations of the parameter vector p on the map, the set of differential equations has to be extended by the trivial equation $\dot{p} = 0$.

For the nonlinear stability analysis at least a third-order approximation of the underlying Poincaré map is required to avoid meaningless results. The entire set of ODEs is symbolically generated by MAPLE as Fortran code. The solution, however, is obtained by standard numerical integration routines.

By means of an iterative procedure, an approximation for the bifurcation parameter value can be calculated on the basis of the linear map with nonlinear parameter dependence. Linear stability analysis, however, fails for nonlinear systems with critical eigenvalues, i.e. eigenvalues for which a bifurcation takes place. It can be shown that for this case the system can be reduced to a lower-dimensional critical subsystem without losing information about the stability properties (Carr 1981). Afterwards, the reduced system is transformed to a nonlinear normal form, from which the stability can be determined (Guckenheimer and Holmes 1986; Wiggins 1990).

8.4 EXAMPLE: SUBMERGED DOUBLE PENDULUM

A simple mechanical model of the planar submerged inverted double pendulum is given in Figure 8.3. Applied forces acting at the bars are buoyancy forces, gravity forces, and hydrodynamic forces. The excitation is characterized by the angular amplitude a and frequency ω and is transmitted through a torsional spring to the lower bar (Kleczka and Kreuzer 1994).

The upright position of this two-degrees-of-freedom (2-DOF) rigid-body system is a stable equilibrium position due to the buoyancy forces acting at the bars. For the mathematical description, the absolute angles α_1 and α_2 and the absolute angular velocities $\dot{\alpha}_1$ and $\dot{\alpha}_2$ are chosen as state variables which are summarized in the state vector $x = [\alpha_1 \, \alpha_2 \, \dot{\alpha}_1 \, \dot{\alpha}_2]^T$.

According to Morison's formula, the hydrodynamic damping-force component dF acting at an infinitesimal slice dl of the bar takes the form

$$dF(x) \sim \begin{cases} u(x)^2 & \text{for} \quad u(x) \geqslant 0, \\ -u(x)^2 & \text{for} \quad u(x) < 0, \end{cases} \tag{17}$$

where $u(x)$ is the normal or tangential velocity component of the surrounding current relative to the bar.

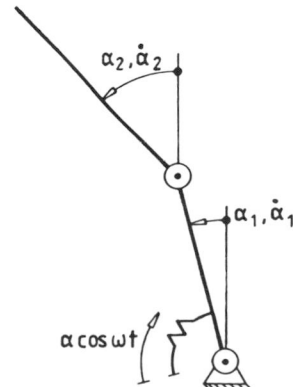

Figure 8.3 Mechanical model of the system.

Table 8.1 Iteration sequence for pitchfork bifurcation parameter value

Iteration	a^*	λ^*
0	0.1000000	0.48912898 + 0.32400075 j
1	0.1407018	0.84036745
2	0.1605386	0.99358588
3	0.1617132	1.00161352
4	0.1614095	0.99954580
5	0.1614944	1.00012418
10	0.1614761	0.99999996

This formulation is analogous to equation (1) with $h(x) = u(x)^2$ and $s(x) = u(x)$. Therefore, a continuous approximation of the damping force in terms of the inverse tangent function is possible:

$$dF(x) \approx u(x)^2 \frac{2}{\pi} \lim_{c \to \infty} \arctan cu(x). \tag{18}$$

By making use of the concept of passive and active variables and by utilizing structural properties of mechanical systems as discussed before, the calculation of the partial derivatives of the model equations resulted in 5012 Fortran lines instead of 187,204 Fortran lines without any optimization.

Now, the nonlinear stability analysis for the double pendulum can be performed according to the steps shown in Figure 8.2.

A local third-order Poincaré map of the double pendulum with one control parameter – the excitation amplitude a requires the generation and solution of 780 ODEs.

Applying the procedure for a stable one-periodic solution of the double pendulum, a symmetry-breaking pitchfork bifurcation with critical eigenvalue $\lambda^* = +1$ and a period-doubling flip bifurcation with critical eigenvalue $\lambda^* = -1$ could be detected. On the basis of the symbolically given map, a very accurate estimation of the bifurcation parameter values a^* of the excitation amplitude a could be obtained. An example of the iteration sequence for the pitchfork bifurcation is given in Table 8.1. It is interesting to note that the algorithm converges even though the periodic solution is unstable for $a > a^*$.

Centre manifold theory applied to the symbolically given discrete map clearly indicates a stable supercritical pitchfork bifurcation and an unstable subcritical flip bifurcation.

Numerical simulations of the steady-state behaviour of the double pendulum confirm the obtained analytical results (Kleczka and Kreuzer 1994).

8.5   CONCLUSION

The computer costs for the calculation of partial derivatives especially in the case of smooth approximations of non-smooth characteristics can be reduced considerably, and so a further step is performed towards a general treatment of a class of systems whose general analysis usually involves great difficulties.

The entire procedure applied for the bifurcation analysis of a submerged double pendulum clearly reveals the advantages of the methods presented — namely the concept of passive and active variables and the utilization of structural properties of mechanical systems. The computational burden of obtaining the partial derivatives of the model equation for this system is reduced by approximately 97% in terms of Fortran lines.

The partial derivatives are used to compute a local approximation of the Poincaré map which is the basis for the subsequent bifurcation analysis; the results are confirmed by numerical simulations.

It is of major importance that all presented tools are not specific to a particular problem (Kleczka and Kreuzer 1991), and that the application of software for computer algebra is a prerequisite for the approach because the analytical operations and expressions become rather involved.

REFERENCES

Carr, J. (1981) *Applications of Centre Manifold Theory*, New York, Springer-Verlag.

Guckenheimer, J. and Holmes, P. (1986) *Nonlinear Oscillations, Dynamical Systems, and Bifurcations of Vector Fields*, New York, Springer-Verlag.

Hsu, C. S., Kreuzer, E. and Kim, M. C. (1990) Bifurcation characteristics of piecewise linear mappings and their implications, *J. Dynamics and Stability of Systems*, **5** (4), 227–254.

Kleczka, M. (1991) *Methoden zur Verzweigungsanalyse mit Anwendung auf einen Spielschwinger*, Fort-schrittberichte VDI, Series 11, No. 153, Düsseldorf, VDI-Verlag.

Kleczka, W. and Kreuzer, E (1991) Systematic computer-aided analysis of dynamic systems, *Proc. of the International Symposium on Symbolic and Algebraic Computation (Bonn, F.R.G., 1991)*, S. M. Watt (ed.), New York, ACM Press, pp. 429–430.

Kleczka, W. and Kreuzer, E. (1994) On the systematic analytic-numeric bifurcation analysis, *J. Nonlinear Dynamics*, to appear.

Kleczka, M., Kleczka, W. and Kreuzer, E. (1990) Bifurcation analysis: a combined numerical and analytical approach, *Proc. of the Workshop on Cont. and Bif. (Leuven, Belgium, 1989)*, D. Roose, B. DeDier and A. Spence (eds.) Kluwer Acad. Publ. Dordrecht (NATO ASI series), pp. 123–137.

Kreuzer, E. and Leister, G. (1991) *Programmpaket NEWEUL'90*, Universität Stuttgart, Inst. B. für Mechanik, Anleitung AN-23.

Kreuzer, E. and Schiehlen, W. (1990) NEWEUL – Software for the generation of symbolical equations of motion. In: *Multibody Systems Handbook*, W. Schiehlen (ed.), Berlin, Springer-Verlag, pp. 181–202.

Wiggins, S. (1990) *Introduction to Applied Nonlinear Dynamical Systems and Chaos*, New York, Springer-Verlag.

W. Kleczka and E. Kreuzer, *Meerestechnik II-Strukturmechanik, Technische Universität Hamburg-Harburg, Eissendorfer Strasse 42, D-21071 Hamburg, Germany*

9 CHAOS AND BIFURCATIONS IN A MULTI-DOF BEAM SYSTEM WITH NONLINEAR SUPPORT

R. H. B. Fey, E. L. B. van de Vorst, D. H. van Campen, A. de Kraker, G. J. Meijer and F. H. Assinck

This chapter deals with the long-term behaviour of periodically excited mechanical systems consisting of linear components with many degrees of freedom and local nonlinearities. The system which is investigated is a two-dimensional beam supported in the middle by a one-sided spring and excited by a periodic force. The linear part of this system is modelled using the finite-element method and reduced using a component mode synthesis method. Models with one and four degrees of freedom are investigated . Branches of periodic solutions and branches of bifurcation points are calculated. Also stable and unstable manifolds are calculated. The system shows (sub)harmonic, quasi-periodic and chaotic behaviour. An intermittency route to chaos is investigated in detail. Experimental results for the beam system are compared with theoretical results.

9.1   INTRODUCTION

This chapter deals with the long-term behaviour of periodically excited mechanical systems, which consist of linear components with many degrees of freedom (DOFs) and local nonlinearities. As a practical example of such a system, one may think of a mooring buoy which is connected by a cable to a ship and is excited by the waves of the sea (see Thompson and Stewart 1986). In this system the local nonlinearity is the cable (one-sided spring).

The linear components, which are supposed to be slightly or proportionally damped, are reduced using a component mode synthesis method based on free-interface eigenmodes and residual flexibility modes (Craig 1985). Only eigenmodes up to a cut-off frequency are kept in the reduced component. The residual flexibility modes guarantee unaffected (quasi-)static load behaviour of the reduced components. The highest kept eigenfrequency gives an indication of the highest frequency for which the reduced model is valid. The nonlinear components and the reduced linear components are coupled in order to approximate the unreduced system. The highest excitation frequency must be much lower than the cut-off

Nonlinearity and Chaos in Engineering Dynamics
Edited by J. M. T. Thompson and S. R. Bishop,   © 1994 John Wiley & Sons Ltd

frequency because nonlinear system response contains higher frequencies than the excitation frequency. The choice of this cut-off frequency is not obvious. However, using this reduction method, it still remains possible to investigate efficiently the long-term behaviour of systems with local nonlinearities, which are modelled using the finite-element method (see Fey *et al.* 1992).

In Section 9.2 a periodically excited beam system with nonlinear support is introduced. The long-term behaviour of two reduced models (1 and 4 DOFs) of this system is analysed. These models are derived using the reduction method mentioned above. In Section 9.3 the 1-DOF model is investigated. Periodic solutions are calculated by solving a two-point boundary-value problem with the finite difference method (Fey 1992). Branches of periodic solutions are followed using a path following technique. The local stability of these periodic solutions is investigated using Floquet theory. Basins of attraction of the 1-DOF model are calculated using a method for calculating manifolds of unstable periodic solutions (Parker and Chua 1989). In Section 9.4 the 4-DOF model is investigated. Branches of bifurcation points are calculated using the shooting method in combination with path following (Meijaard 1991). Unstable manifolds are calculated for the 4-DOF model. An experimental set-up of the beam system has been constructed and experimental results are compared with numerical results in Section 9.5. All calculations presented in this paper were carried out using a development release of the finite-element package DIANA (1993).

9.2   BEAM SYSTEM SUPPORTED BY ONE-SIDED SPRING

Figure 9.1 shows the beam system which is analysed both numerically and experimentally. The beam is supported at both ends by leaf springs, and supported in the middle by a one-sided leaf spring. The periodic force is realized by a rotating mass.

The beam and leaf springs are modelled using the finite-element method and reduced using the component mode synthesis method mentioned above. The one-sided leaf spring is modelled by a one-sided spring, so mass influences are neglected. Because of symmetry only half the beam is modelled.

9.3   SINGLE-DEGREE-OF-FREEDOM MODEL

The equation of motion for the 1-DOF model is:

$$m\ddot{y} + b\dot{y} + k(\alpha_p + 1)y = m_e r_e \omega^2 \cos(\omega t), \tag{1}$$

where

$$\alpha_p = \begin{Bmatrix} \alpha, & y < 0, \\ 0, & y \geqslant 0, \end{Bmatrix}$$

and $\alpha = 6$, $\omega = 2\pi f_e$, $b = 2\xi\sqrt{mk}$, $\xi = 0.03$, $m = (\varphi_1^T M \varphi_1)/\varphi_{1y}^2 = 2.37\,\mathrm{kg}$, $k = (\varphi_1^T K \varphi_1)/\varphi_{1y}^2 = 18\,960\,\mathrm{N/m}$, $m_e = 10.75\,\mathrm{g}$, $r_e = 0.014\,\mathrm{m}$. Here M and K are the mass and the stiffness matrix of the linear part of the system, i.e. of the beam without the one-sided spring. φ_1 is the first eigenmode of the linear system with an eigen-frequency of 14.2 Hz. φ_{1y} is the element of φ_1 corresponding to the displacement of the middle of the beam y. Branches of periodic solutions of this system are calculated for varying excitation frequency f_e. The

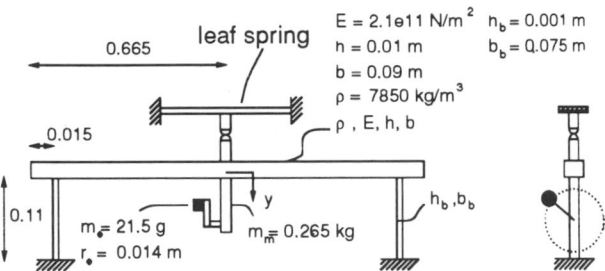

Figure 9.1 Beam system.

maximum displacements of these periodic solutions are given in Figure 9.2. As well as the harmonic peak near 20.8 Hz, we also find 1/2, 1/3, 1/5 and 1/7 subharmonic resonances.

The stable and unstable manifolds (W^s_3 and W^u_3) of the unstable 1/3 subharmonic which exists at $f_e = 70$ Hz, are shown in Figure 9.3. This figure shows that one half of W^u_3 goes to the stable 1/3 subharmonic attractor and the other half of W^u_3 goes to the stable harmonic attractor. W^s_3 represent the boundaries of the basins of attraction of these attractors.

Figure 9.4 shows the unstable and stable manifolds of the unstable harmonic and unstable 1/7 subharmonic at $f_e = 50.5$ Hz. Homoclinic and heteroclinic intersections of the stable and unstable manifolds exist and cause a chaotic structure of the manifolds. In this area of the Poincaré section no chaotic attractor could be found: only one 1/2 subharmonic and one 1/7 subharmonic attractor exist. Because of the homoclinic and heteroclinic intersections the boundaries of the basins of attraction of these attractors are fractal. The dimension of the boundaries of the basins of attraction is calculated to be 1.85 (Grebogi *et al.* 1983).

Note that the distance between the stable manifolds and the 1/7 subharmonic attractor is very small. This attractor is a so-called noisy attractor; a small disturbance on the system vibrating in this attractor can easily result in a jump to the 1/2 subharmonic attractor (Soliman and Thompson 1990).

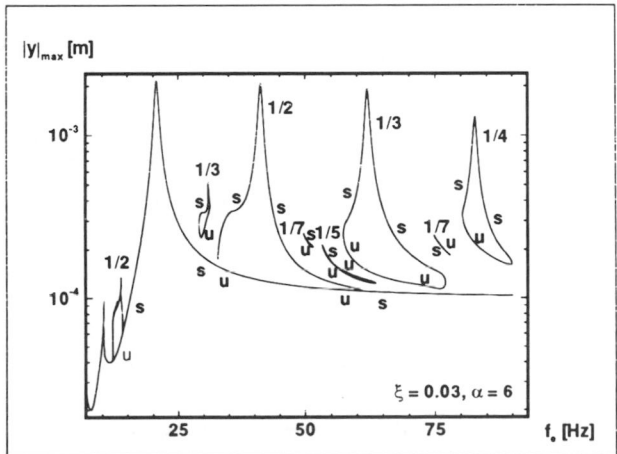

Figure 9.2 Maximum displacements of periodic solutions of 1-DOF: (s) stable, (u) unstable.

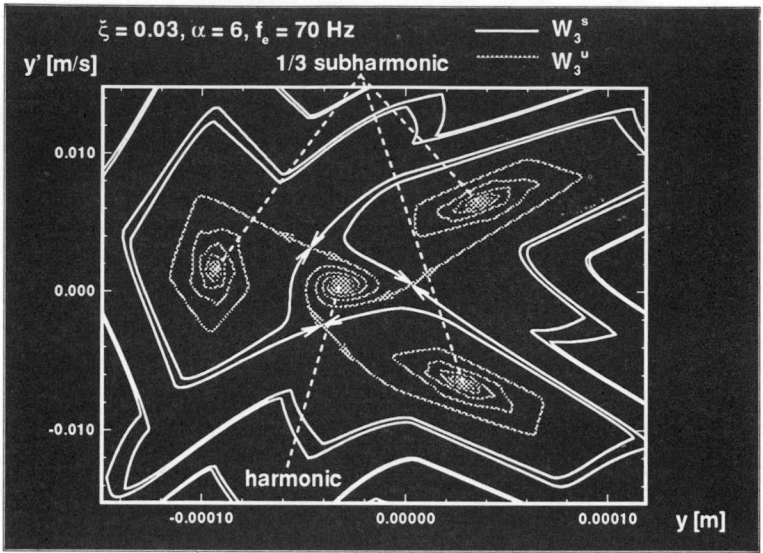

Figure 9.3   Poincaré section with stable and unstable manifolds (W_3^s and W_3^s) of 1-DOF system at $f_e = 70$ Hz.

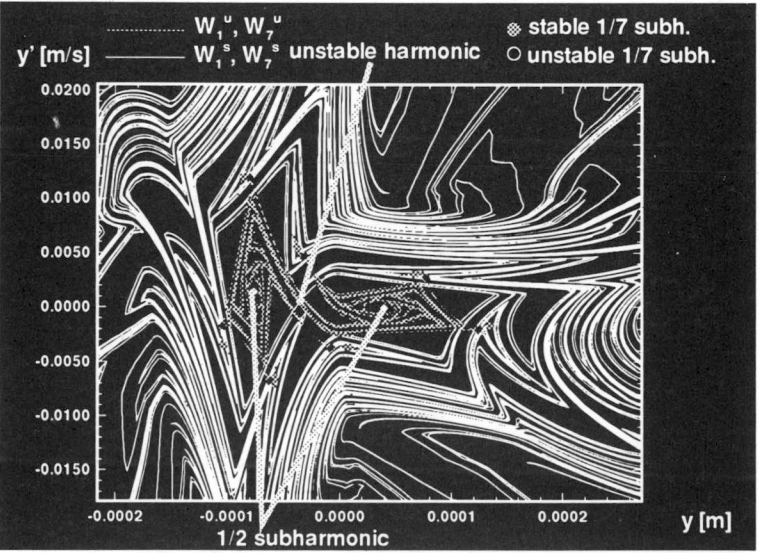

Figure 9.4   Poincaré section with stable and unstable manifolds of 1-DOF system at $f_e = 50.5$ Hz.

9.4   FOUR-DEGREE-OF-FREEDOM MODEL

If the linear system is reduced with the component mode synthesis method to four degrees of freedom (three eigenmodes with eigenfrequencies $f_1 = 14.2$ Hz, $f_2 = 127.3$ Hz, $f_3 = 352.8$ Hz and one residual flexibility mode), the linear model is valid up to a frequency of 350 Hz. Earlier investigation of a similar system showed that the 4-DOF model including

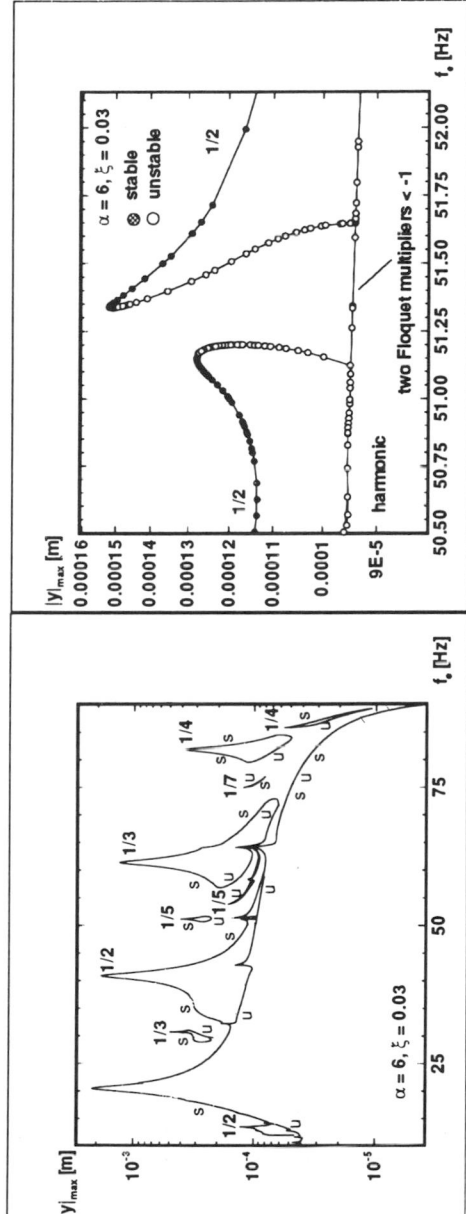

Figure 9.5 Maximum displacements of periodic solutions of the 4-DOF model. y is the displacement of the middle of the beam. (s) stable, (u) unstable.

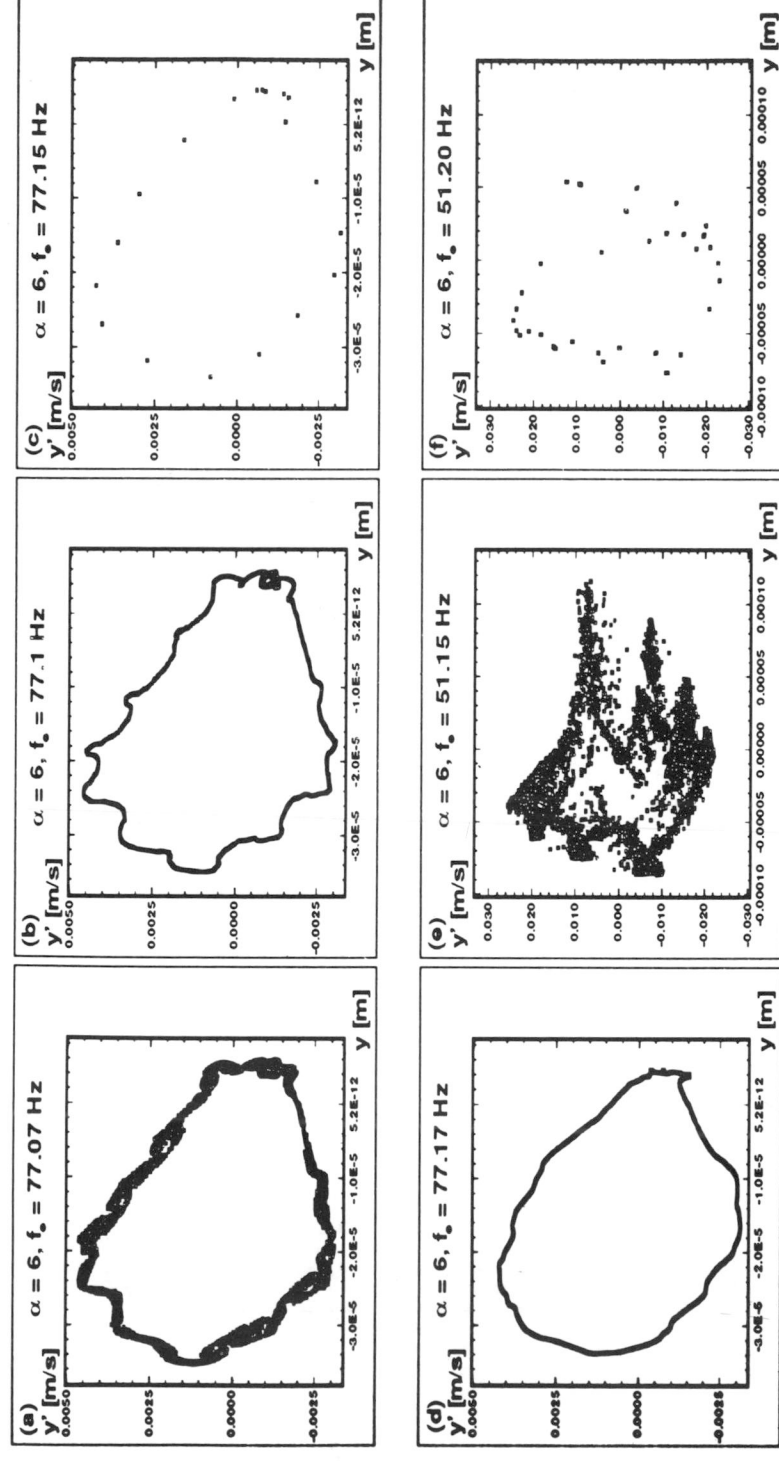

131

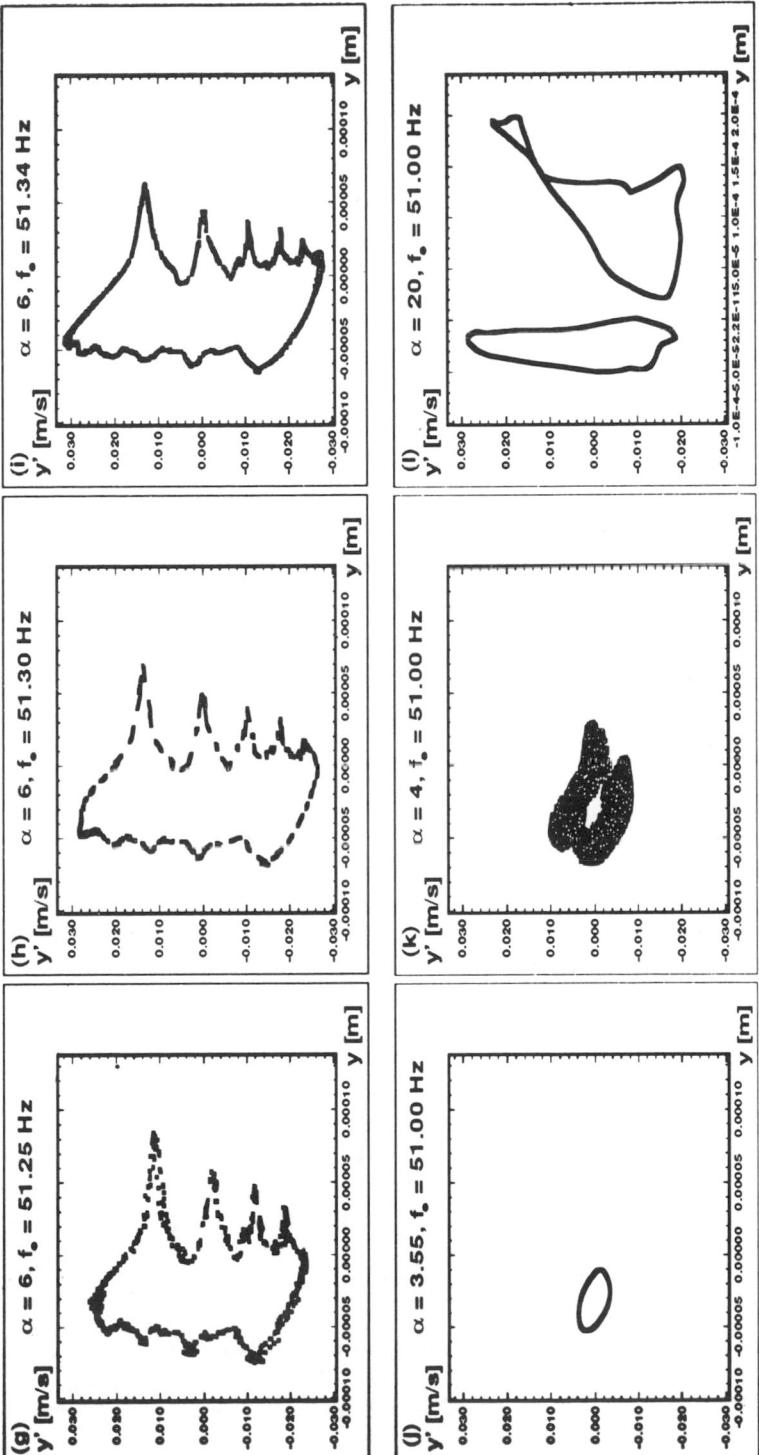

Figure 9.6 Poincaré sections with steady-state behaviour of the 4-DOF model.

the one-sided spring with $\alpha = 6$ is a good approximation of the unreduced system for excitation frequencies f_e lower than 80 Hz (Fey 1992). Again branches of periodic solutions are calculated for this 4-DOF model for varying excitation frequency f_e. The maximum displacements of the periodic solutions for the middle of the beam are given in Figure 9.5.

Figure 9.5 shows that the response of the 4-DOF model is very similar to the response of the 1-DOF model (Figure 9.2). At the right-hand side of the 1/3 subharmonic branch the harmonic branch is now unstable and marked by two secondary Hopf bifurcations. In this frequency interval quasi-periodic behaviour and mode-locking leading to chaos are found (Newhouse *et al.* 1978) (see Figures 9.6a–d).

Note that the 1/2 subharmonic branch is now divided into two parts in the frequency range near 50 Hz. Between the two 1/2 subharmonic branches only an unstable harmonic branch exists, which has *two* real Floquet multipliers lower than -1. The steady-state behaviour of the 4-DOF model in this frequency interval is shown in Figures 9.6e–i and 9.7. Figure 9.7 shows that if the excitation frequency f_e is near 51.34 Hz the steady-state behaviour seems to be periodic over a long time but then suddenly shows a small burst, after which the almost-periodic behaviour recovers. This kind of chaotic behaviour is called intermittency (see Pomeau and Manneville 1980). If the excitation frequency is decreased the periodic-behaviour intervals become smaller until the behaviour is entirely chaotic.

At $f_e = 51.3425$ Hz besides the unstable harmonic with *two* Floquet multipliers lower than -1, there also exist a stable and an unstable 1/2 subharmonic and a stable and an unstable 1/5 subharmonic (Figure 9.5). The unstable manifolds of the unstable 1/2 subharmonic W_2^u are shown in Figure 9.8a. In this figure, the cross-sections of the manifolds are not homoclinic points because these are manifolds of a multi-DOF model (unstable manifolds do not intersect one another of course). The figure shows that one half of W_2^u goes to the 1/2 subharmonic attractor. The other half of W_2^u almost returns to the unstable 1/2 subharmonic and almost creates a homoclinic orbit. A homoclinic orbit exists when the unstable manifold returns to its unstable periodic solution: it is simultaneously the stable manifold of that unstable periodic solution. This is not the case here, because W_2^u has a fractal structure in the neighbourhood of the unstable 1/2 subharmonic which means that W_2^u crosses the stable manifolds of the unstable 1/2 subharmonic. In Figure 9.8b a transient is shown, which has the same structure as W_2^u. Because W_2^u finally approaches the unstable 1/2 subharmonic, this transient finally ends in the stable 1/2 subharmonic. The unstable manifolds of the harmonic W_2^u solutions are planes in the Poincaré section, spanned by the two eigenmodes corresponding to the two Floquet multipliers which are lower than -1. These manifolds finally approach a curve which has the same structure as W_2^u. One half of the unstable manifolds of the unstable 1/5 subharmonic also approache a curve with the same structure as W_2^u, where the other half end in the stable 1/5 subharmonic. If f_e is decreased the stable 1/2 subharmonic solution disappears via a cyclic fold bifurcation and the steady-state behaviour becomes chaotic and has almost the same structure as W_2^u at $f_e = 54.3425$ Hz; this behaviour is called intermittency (Figure 9.6i).

The bifurcation points (one real Floquet multiplier lower than -1 and one Floquet multiplier equal to -1), which mark out the intermittency interval are followed. Here, two system parameters must be varied which are chosen to be the excitation frequency f_e and the relative stiffness of the one-sided spring α. The branches of bifurcation points are given in Figure 9.9. In Figure 9.10 the response is given in the frequency range near 50 Hz for different values of the relative stiffness of the one-sided spring α.

Figures 9.9 and 9.10a show that for $\alpha = 3.2$ only one 1/2 subharmonic branch exists, next

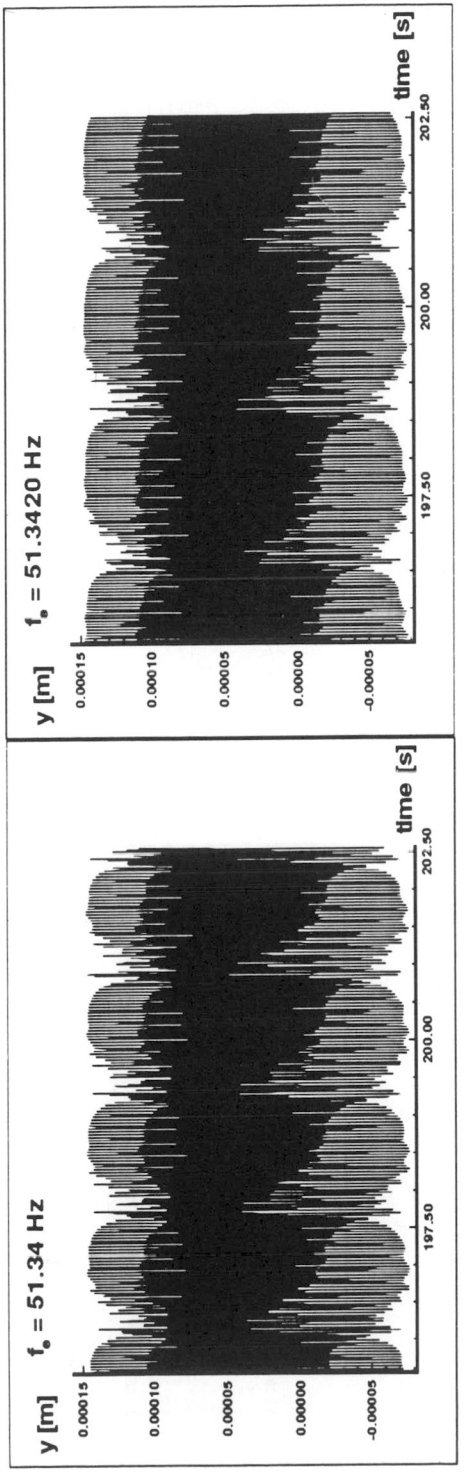

Figure 9.7 Steady-state behaviour of the 4-DOF model: intermittency.

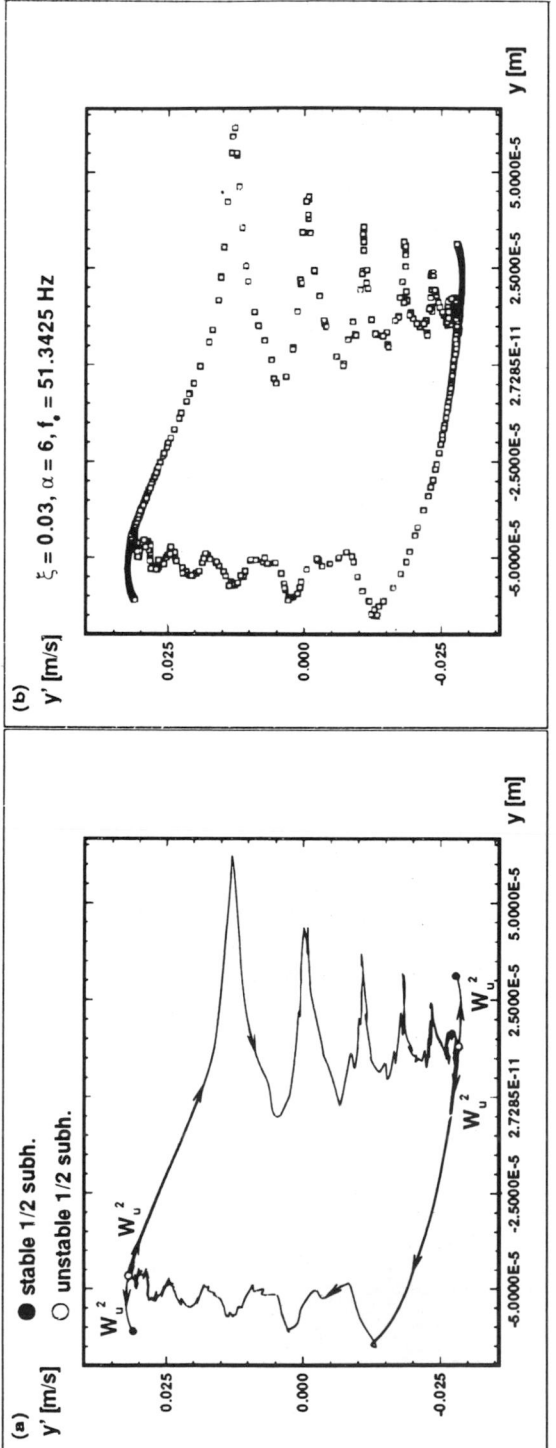

Figure 9.8 (a) Unstable manifolds of unstable 1/2 subharmonic. (b) Transient of 4-DOF model.

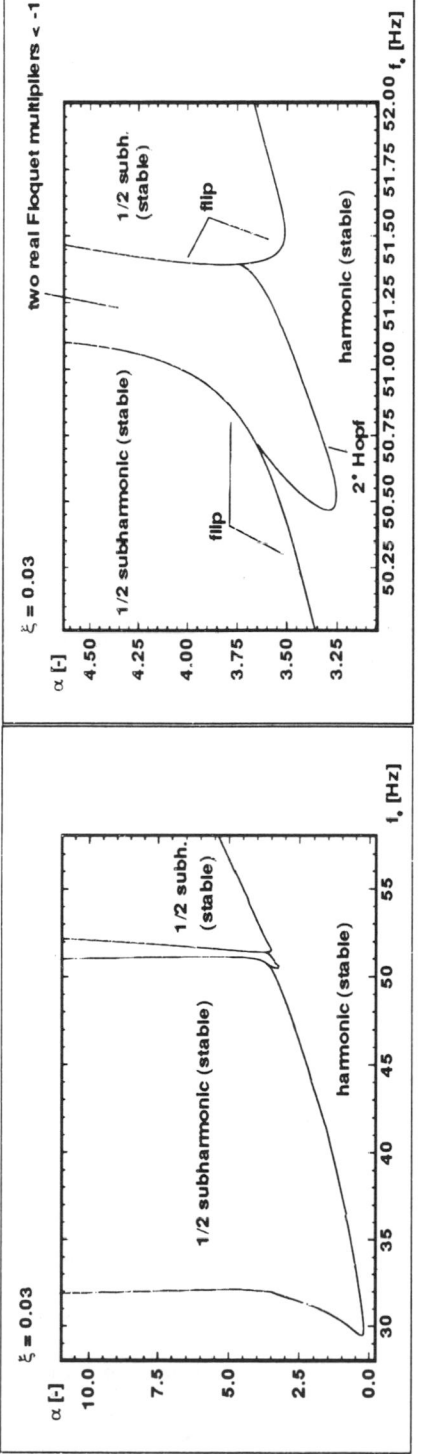

Figure 9.9 Bifurcation curves of 4-DOF model.

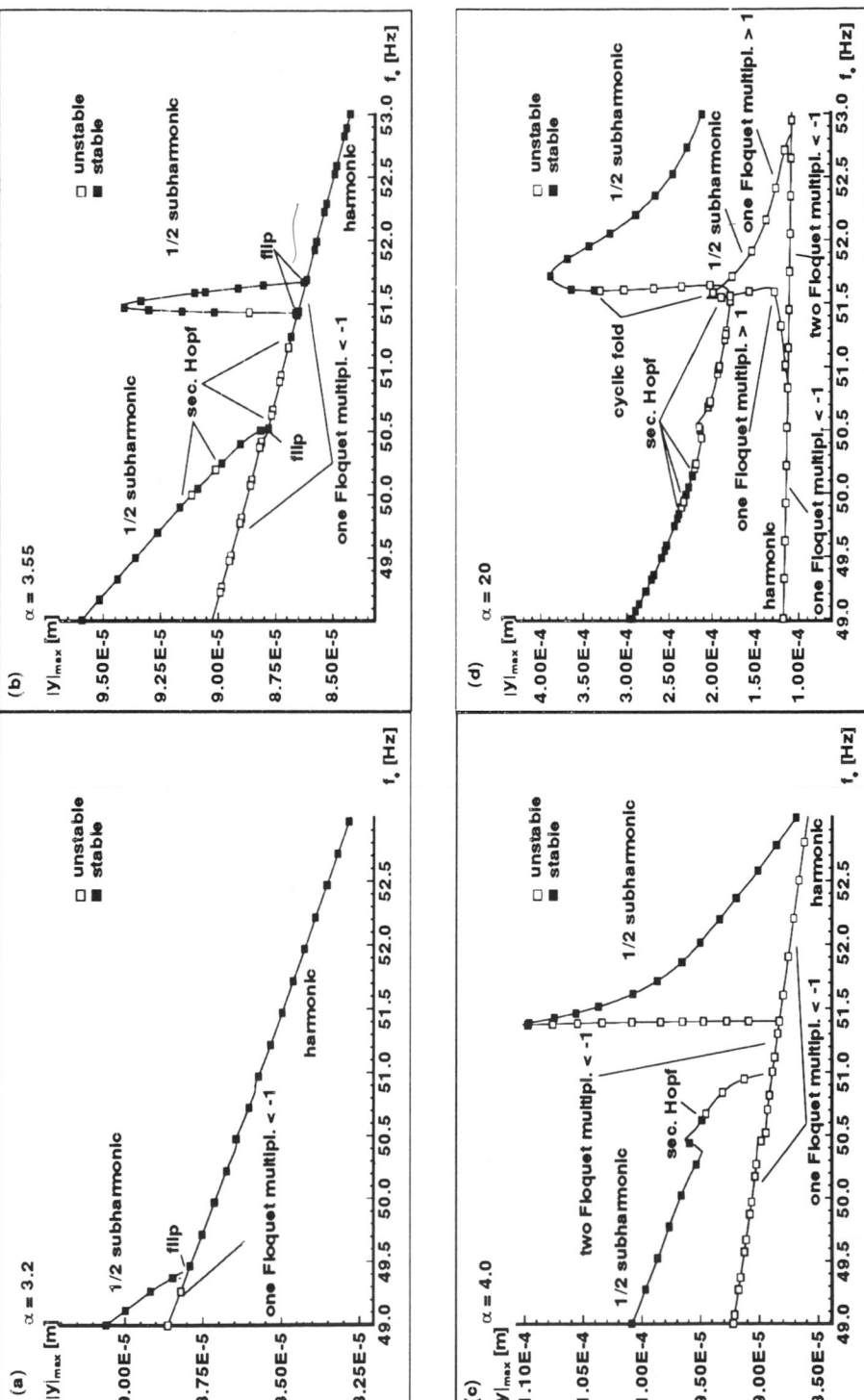

Figure 9.10 Maximum displacement of periodic solutions of 4-DOF model.

to the harmonic branch, which becomes unstable via a flip bifurcation. If α is increased a small part of the stable harmonic branch becomes unstable via two secondary Hopf bifurcations at the right-hand side of the 1/2 subharmonic branch. Quasi-periodic behaviour was found in this frequency interval (Figure 9.6j). Further increasing α to 3.55 (Figure 9.10b) results in a second 1/2 subharmonic branch. If α is increased once again the flip bifurcation, which marks the left-hand 1/2 subharmonic branch, approaches the left-hand secondary Hopf bifurcation. At the point where they merge, $\alpha = \alpha_{bl}$, a codimension-2 bifurcation exists. Here, three Floquet multipliers are equal to -1. If α is increased again to α_{br}, another codimention-2 bifurcation is created, when the right-hand secondary Hopf bifurcation touches the left-hand flip bifurcation of the right-hand 1/2 subharmonic branch. The middle part of the harmonic branch is unstable with two Floquet multipliers lower than -1 (Figure 9.10c) Chaotic behaviour was found in this frequency interval (Figure 9.6k).

Note that the two codimension-2 bifurcations do not exist for the same value of α. If α is between α_{bl} and α_{br} the middle unstable part of the harmonic branch is marked at the left-hand side by one Floquet multiplier equal to -1 and one real Floquet multiplier lower than -1, while at the right-hand side a secondary Hopf bifurcation exists. At the left-hand side of the middle unstable part of the harmonic branch *two* real Floquet multipliers are lower than -1 and at the right-hand side two complex Floquet multipliers have a modulus larger than 1. In Figure 9.11a the Floquet multipliers are shown as a function of

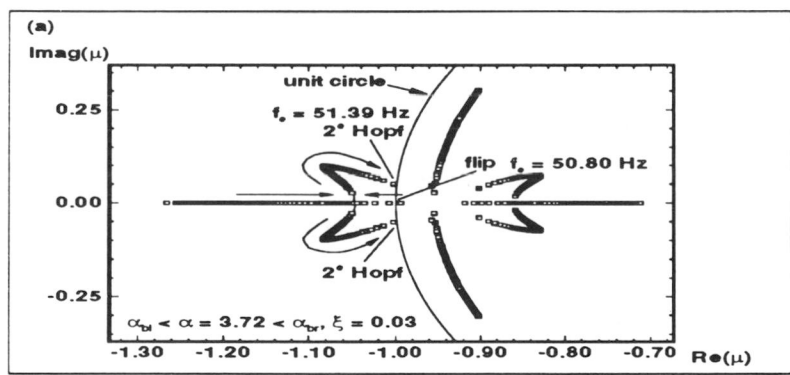

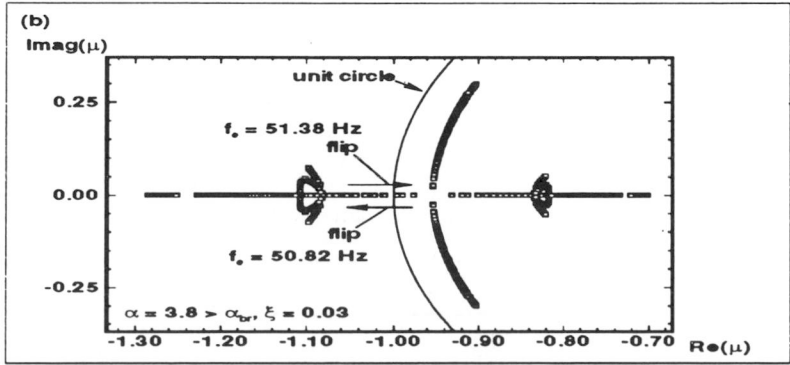

Figure 9.11 Floquet multipliers μ of 4-DOF model.

the excitation frequency f_e. The two real Floquet multipliers lower than -1 become equal at some point and subsequently they become conjugate complex; at the secondary Hopf bifurcation their modulus has been decreased to 1. Figure 9.11b shows that if α is larger than α_{br} the Floquet multipliers still become complex in this frequency range but not longer pass the unit circle.

Figure 9.10d shows the response of the system in the frequency area of 50 Hz for $\alpha = 20$. The left-hand and right-hand 1/2 subharmonic branches have been connected, resulting in an upper and a lower 1/2 subharmonic branch. An unstable part exists at the upper 1/2 subharmonic branch marked by two secondary Hopf bifurcations. Here, stable quasi-periodic behaviour exists with a base frequency which is half the excitation frequency. In the Poincaré section this is represented by two closed curve (Figure 9.6l).

9.5   EXPERIMENTS

An experimental set-up of the beam system presented in Figure 9.1 has been constructed with $m_e = 46.86$ g, $\alpha = 6.49$ and $\xi = 0.015$. In Figure 9.12 numerical and experimental results are compared. The experimental results correspond very well with the numerical results (4-DOF model) represented by the solid lines; almost all subharmonics are observed in the experiments.

No periodic behaviour was found in the experiment in the frequency range near 50 Hz (T symbols). The numerical results predict two unstable 1/5 subharmonics and one stable 1/2 subharmonic. In Section 9.3 fractal boundaries of basins of attraction were found for the 1-DOF system in this frequency range. Earlier investigation of a similar system showed that the 1/2 subharmonic attractor is a noisy attractor in this frequency range (van de Vorst *et al.* 1992). This explains why in the experiment no periodic behaviour is found in this frequency range because system parameters will not be time-invariant exactly; this is particularly true for the excitation frequency f_e.

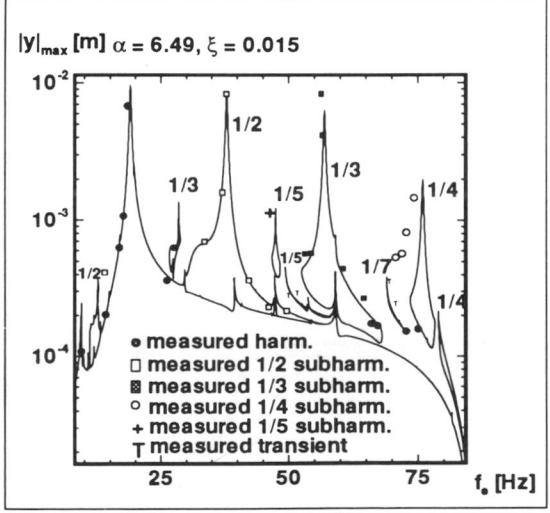

Figure 9.12   Experimental results.

9.6   CONCLUSIONS

In this chapter a nonlinear supported beam system was investigated. The beam was modelled using the finite-element method. This model was reduced using a component mode synthesis method. In this system the first few eigenfrequencies cover a broad frequency range. This is the reason why the 1-DOF model shows almost the same frequency response as the 4-DOF model in the low-frequency range. The experimental results confirm this. Sometimes it is insufficient to investigate only local stability using Floquet theory. This theory does not detect fractal boundaries of basins of attraction which may cause noisy attractors. In the experiment no periodic behaviour could be found in frequency intervals where the numerical results predicted noisy periodic attractors. The general conclusion is that it is possible to investigate complicated systems with local nonlinearities both efficiently and accurately by using the reduction method.

REFERENCES

Craig Jr. R. R. (1985) A review of time-domain and frequency-domain component mode synthesis methods. In: *Combined Experimental/Analytical Modelling of Dynamic Structural Systems Using Substructure Synthesis*, D.R. Martinez and A.K. Miller (eds), pp. 1–31.

DIANA (1993) *DIANA User's Manual*, 6.0 edn., TNO Building and Construction Research Delft.

Fey, R. H. B. (1992) Steady-state behaviour of reduced dynamic systems with local nonlinearities. *PhD thesis*, Eindhoven University of Technology.

Fey, R. H. B., Campen, D.H. van and Kraker, A. de (1992) Long term structural dynamics of mechancial systems with local nonlinearities, *Nonlinear Vibrations*, DE-50, AMD-144, WAM of the ASME, Anaheim, Calif., pp. 159–166.

Grebogi, C., McDonald, S. W., Ott, E. and Yorke, Y. A. (1983) Final state sensitivity: an obstruction to predictability, *Physics Letters*, **99A**, 415–418.

Meijaard, J. P. (1991) Dynamics of mechanical systems, algorithms for a numerical investigation of the behaviour of non-linear discrete models, *PhD thesis*, Delft University of Technology.

Newhouse, S., Ruelle, D. and Takens, F. (1978) Occurrence of strange axiom-a attractors near quasi-periodic flow on T^m, $m \leqslant 3$, *Commun. Math Phys.*, **64**, 35–40.

Parker, T. S. and Chua, L.O. (1989) *Practical Numerical Algorithms for Chaotic Systems*, Berlin, Springer.

Pomeau, Y. and Manneville, P. (1980) Intermittent transition to turbulence in dissipative dynamical systems, *Commun. Math. Phys.*, **74**, 189–197.

Soliman, M. S. and Thompson, J. M. T. (1990) Stochastic penetration of smooth and fractal basin boundaries under noise excitation, *Dynamics and Stability Systems*, **5**, 281–298.

Thompson, J. M. T. and Stewart, H. B. (1986) *Nonlinear Dynamics and Chaos*, John Wiley, Chichester.

Vorst, E. L. B. van de, Fey, R. H. B., Campen, D. H. van and Kraker, A. de (1992) Manifolds of nonlinear dynamic single-DOF systems, In: *Topics in Applied Mechanics, Integration of Theory and Applications in Applied Mechanics*, J. F. Dÿksman and F. T. M. Nieuwstadt (eds), pp. 293–304.

R. H. B. Fey and G. J. Meijer, *TNO Building and Construction Research, Centre for Mechanical Engineering, PO Box 29, 2600 AA Delft, The Netherlands*

E. L. B. van de Vorst, D. H. van Campen, A. de Kraker and F. H. Assinck, *Department of Mechanical Engineering, Eindhoven University of Technology, PO Box 513, 5600 MB Eindhoven, The Netherlands*

PART III
Control

In many engineering systems chaos is seen as an undesirable state leading to unpredictable behaviour. As a consequence, in many environments much effort is spent on the avoidance of chaotic regimes. However, in practice as systems evolve it may be impossible to avoid chaotic motions, which changes the emphasis towards the control of chaos rather than its avoidance. This new and developing area is currently perhaps the most exciting research in the dynamical systems field.

The first researchers to take an interest in the control of chaos was the group at Maryland. The main idea, formulated in the paper by Ott *et al.* (1990) and extended in Romeiras *et al.* (1992), is to make use of one of the infinite number of unstable periodic orbits that are embedded within the chaotic attractor. These unstable solutions are first located and examined so that one with suitably improved system performance can be chosen. The next stage is to stabilize the selected orbit (or steady state) by the use of small, carefully chosen, temporal perturbations to an accessible parameter of the system. Use is made of the ergodic nature of the chaotic trajectory so that when it falls in the neighbourhood of the selected unstable orbit it may be controlled and stabilized onto the desired system response. In this way solutions of almost any periodicity can be selected from within the chaotic attractor providing greater flexibility so that chaos may be seen as useful rather than something to be avoided. Examples of applications this method and a brief review of the concepts are given in a later chapter by *Grebogi and Lai* in Part VII.

This method of controlling chaotic systems (termed the OGY method after its originators Ott, Grebogi and Yorke) has drawn growing attention and has now been employed in a variety of scientific disciplines. A further extension of the ideas is discussed in the chapter by *Kovács et al.* in which the control process is carried out while the governing system is operating on a fractal basin boundary. Trajectories initiated close to a fractal basin boundary exhibit a chaotic transient motion before settling down onto one of the coexisting attractors; this approach therefore provides insight into the control of transient chaos.

A different question of control is covered in the chapter by *Stépán* in which it is assumed that the desired response is intrinsically unstable and control is constantly required in order to maintain the system in this state. Both human and digital processors have an in-built delay in any control system and invariably what results is a small-amplitude, essentially stochastic, vibration that is called micro-chaos (μ-chaos). The inverted pendulum is chosen to theoretically and experimentally test ideas of capturing μ-chaos with a deterministic model rather than a white-noise term. This deterministic approach allows amplitudes and parameter zones to be predicted so that design can be improved.

If a chaotic system is thought to be an undesirable characteristic then perhaps the simplest solution is to identify those zones in parameters space where chaos occurs and make sure that the system does not take on these values. However, if we are not able to control the system parameters then the alternative solution is to adapt the system so as to shift the undesirable behaviour away from the parameter zones of interest. Similar ideas are used in vibration absorbers where small modifications to the system can be implemented to achieve the desired response. A method for carrying out this approach to the suppression of chaos is summarized in the chapter by *Brindley and Kapitaniak.*

REFERENCES

Ott, E., Grebogi, C. and Yorke, J. (1990) Controlling chaos, *Phys. Rev. Lett.*, **64**, 1196–1199.
Romeiras, J., Grebogi, C., Ott, E. and Dayawansa, W. P. (1992) Controlling chaotic dynamical systems, *Physica*, **58D**, 165–192.

10 μ-CHAOS IN DIGITALLY CONTROLLED MECHANICAL SYSTEMS

G. Stépán

The human and/or computer control of a machine is always needed when the desired stationary motion of the machine is otherwise unstable. A careful choice of the control strategy is required because of the presence of time delays. However, experiments clearly show that the desired motions or equilibria are never stable and small-amplitude stochastic vibrations appear even in the case of the best controllers. A part of these oscillations is due to the finite sensitivity of the human operator and to the digital effects in the computer control which can be described by deterministic models. The resulting micro-chaos (μ-chaos) is also analysed in the presence of dry friction, and the life expectancy of transient or temporary chaos is estimated by means of one-dimensional mappings.

10.1 INTRODUCTION

The prescribed equilibrium positions or stationary motions of machines are often unstable. In these cases, a human operator is required to control the machine in a way that results in the desired operation of the machine. Nowadays, computer control may replace the human operator and more and more machine elements are controlled by microprocessors.

There is a common feature of both types of controllers, namely the time delay which is introduced into the system. In the case of a human operator, the delay is caused by the reflex time (of the order of 100 ms), while a digital processor has a sampling delay (usually less than 10 ms). Delay tends to destabilize any dynamical system, which means that stabilizing an otherwise unstable mechanical system is not a trivial task in the presence of after-effects. Car- and trailer-driving, crane-handling and master–slave systems are typical examples where a human operator is involved (see e.g. Scheidl *et al.* 1985 or Stépán 1987). Robots, computer-controlled bearings, valves and car suspensions are systems in which the digital processor appears as controller (see e.g. Stépán *et al.* 1990 or Palkovics *et al.* 1993).

Nonlinearity and Chaos in Engineering Dynamics
Edited by J. M. T. Thompson and S. R. Bishop,   © 1994 John Wiley & Sons Ltd

When dealing with basic research on the engineering problems mentioned above, researchers often choose the challenging task of stabilizing the inverted pendulum. Since this is a clear basic example, many publications concerning it have appeared over the last forty years (see e.g. Higdon and Cannon 1963; Mori *et al.* 1976; Stépán 1984; Henders and Sondack 1992; Kawazoe 1992). These cover either the experimental or the theoretical aspects.

This chapter also considers the balancing of an inverted pendulum. A brief summary of the model is followed by stability analysis and comparison with experimental results. The experimentally detected small-amplitude stochastic vibration is central to the chapter. The characterization of the μ-chaos (micro-chaos) is important from mechanical engineering and design viewpoints. The careful design of the mechanical parameters as well as control parameters including sampling time and finite-digit discretization helps to obtain a tolerable chaotic motion around the desired equilibrium. In the presence of Coulomb friction, the μ-chaos may be temporary only and the system settles down. In these cases, the estimation of the life expectancy of the chaos provides useful information about the system, as the damping factor does in elementary vibration theory. The expected escape-time calculation concludes the chapter.

10.2   INVERTED PENDULUM

Consider the simplest planar mechanical model of the inverted pendulum shown in Figure 10.1. Its lowest point slides smoothly along the horizontal line. The system has two degrees of freedom (DOF) described by the general coordinates φ and q. The angle φ is detected together with its derivatives and the horizontal control force Q is determined by them in such a way that the upper $\varphi = 0$ position should be asymptotically stable.

The nonlinear equations of motion assume the form

$$
\begin{pmatrix} \frac{1}{3}ml^2 & \frac{1}{2}ml\cos\varphi \\ \frac{1}{2}ml\cos\varphi & m \end{pmatrix}
\begin{pmatrix} \ddot{\varphi} \\ \ddot{q} \end{pmatrix}
- \begin{pmatrix} \frac{1}{2}mlg\sin\varphi \\ \frac{1}{2}ml\dot{\varphi}^2\sin\varphi \end{pmatrix}
= \begin{pmatrix} 0 \\ Q \end{pmatrix}
$$

from which the 'cyclic' coordinate q can easily be eliminated to leave the single second-order equation

$$
(4 - 3\cos^2\varphi)\ddot{\varphi} + \frac{3}{2}\dot{\varphi}^2\sin(2\varphi) - \frac{6g}{l}\sin\varphi = -\frac{6}{ml}Q\cos\varphi. \tag{1}
$$

In these equations, g stands for the gravitational acceleration.

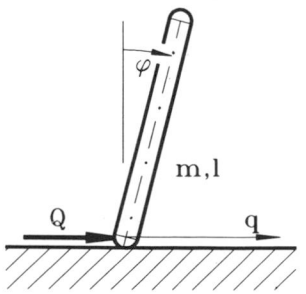

Figure 10.1   Mechanical model.

The control force Q is considered in the simplest form of a *PD* controller with constant parameters P and D chosen appropriately:

$$Q(t) = D\dot{\varphi}(t - r(t)) + P\varphi(t - r(t)) + \text{higher-order terms.} \qquad (2)$$

In this formula, the dependence on the time t is also emphasized because of the presence of the time delay or retardation $r:[t_0, \infty) \to \mathbb{R}_+$. This delay describes the past effects which always appear in the control. The time dependence of this delay provides a unified handling of the two types of control we consider here. The human analogue control can be modelled by a constant reflex delay τ, that is,

$$r_a(t) \equiv \tau, \quad \tau \in \mathbb{R}_+, \qquad (3)$$

while the sampling effect of the digital computer control is presented by the periodic function

$$r_d(t) = \tau + (t - \tau \operatorname{int}(t/\tau)), \quad \tau \in \mathbb{R}_+, \qquad (4)$$

where τ now stands for the sampling time. The numerical values of τ are usually quite different in the two cases as mentioned in Section 10.1. Nevertheless, the right-hand side of Figure 10.2 shows the two delay functions for the same value of τ which is considered as the measure of the length of the past effect in both cases.

The nonlinear higher-order terms (h.o.t.) in equation (2) of the control force have, of course, an important role, in the nonlinear analysis. In the case of computer control, however, there is another 'strong' nonlinearity in the control force Q which is related to the finite number of digits used in the A/D and D/A converters of the system. If h denotes the value of one digit converted into control force, the formula

$$Q(t) = h \operatorname{int} \frac{D\dot{\varphi}(t - r(t)) + P\varphi(t - r(t))}{h} \qquad (5)$$

gives a simple way of describing this type of nonlinearity which is quite similar to the description of sampling delay in equation (4).

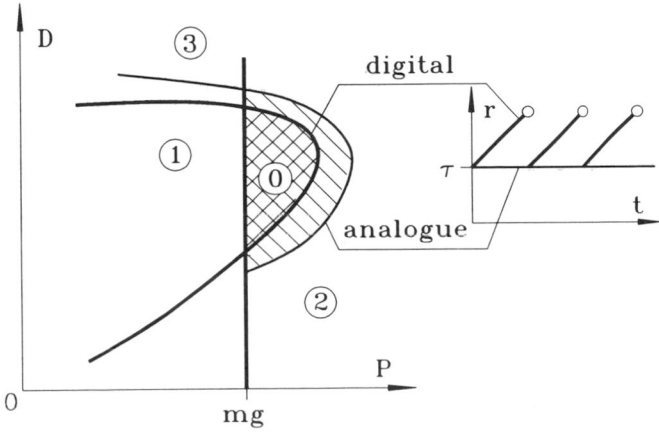

Figure 10.2 Stability charts.

Later on, the Coulomb friction is also to be used as another 'strong' nonlinearity in the mechanical system. Although this is not presented in the equation of motion (1) and will not be modelled precisely, the existence of a zone around the equilibrium $\varphi = 0$ will, however, be considered where the system stops moving when it takes a zero angular velocity $\dot{\varphi}$ there.

10.3 STABILITY AND NONLINEAR VIBRATIONS

A stability analysis of the $\varphi = 0$ position is required to find suitable control parameters P and D in equation (2). The variational system of the motion equations (1) and (2) at the trivial solution assumes the form of a linear retarded functional differential equation (RFDE) with a periodic coefficient in the delay:

$$\ddot{\varphi}(t) - \frac{6g}{l}\varphi(t) = -\frac{6}{ml}(D\dot{\varphi}[t - r(t)] + P\varphi[t - r(t)]). \tag{6}$$

The application of the Floquet theory (see Hale 1977) usually does not provide closed-form analytical results in these infinite-dimensional problems. However, the above-mentioned cases of analogue and digital control can be investigated analytically as shown below.

It there is no delay in the system, i.e. $r(t) \equiv 0$, then the Routh–Hurwitz criterion gives the simple necessary and sufficient conditions of asymptotic stability in the form

$$P > mg \quad \text{and} \quad D > 0.$$

Experimental work clearly shows that the gain parameters P and D cannot be increased above a certain limit as the time delay is always present in the system. The following two theorems clearly express this fact.

Theorem 1. *Let the time delay r in equation (6) be constant according to equation (3) (the case of analogue control). The trivial solution of the RFDE equation (6) is asymptotically stable if and only if*

$$mg < P < \left(mg + \frac{ml}{6\tau^2}\omega^2\right)\cos\omega, \tag{7}$$

where ω is the only value satisfying

$$D\omega = P\tau \tan\omega$$

in the interval $(0, \pi/2)$.

Since the delay is the constant τ, the application of Floquet theory leads to the analysis of the corresponding transcendental characteristic function, and the proof of the theorem can be carried out with the methods presented by Stépán (1989).

Theorem 2. *Let the time delay r in equation (6) be periodic according to equation (4) (the case of digital control). The trivial solution of the RFDE (6) is asymptotically stable if and only if*

$$mg < P$$

and

$$\frac{6\tau}{ml}\left(\frac{\cosh\lambda-1}{\lambda^2}\tau P - \frac{\sinh\lambda}{\lambda}D\right)^2 + \frac{(\cosh\lambda-1)(1+2\cosh\lambda)}{\lambda^2}\tau P$$

$$+ \frac{\sinh\lambda(1-2\cosh\lambda)}{\lambda}D < 0, \tag{8}$$

where

$$\lambda = \tau\sqrt{6g/l}.$$

The proof of this theorem can be based on standard methods if it is realized that the monodromy operator used in Floquet theory has a finite spectrum in this special case. Using the dimensionless time $T = t/\tau$, the linearized equation of motion (6) with the delay (4) can be rewritten as

$$\varphi''(T) - \frac{6g}{l}\tau^2\varphi(T) = u_k, \quad T \in [k, k+1),$$

$$u_k = -\frac{6\tau}{ml}(D\varphi'(k-1) + P\tau\varphi(k-1)), \quad k = 1, 2, \ldots$$

where prime signifies differentiation with respect to T, and $t_k = k\tau (k = 1, 2, \ldots)$ represents the kth sampling moment, so $T_k = k$. Since the right-hand side of this linear equation is piecewise constant, it is easy to solve it analytically for each sampling interval having length 1 in the dimensionless time. With this solution, a three-dimensional linear discrete mapping can be constructed for

$$\mathbf{x}_k = \begin{pmatrix} x_{1k} \\ x_{2k} \\ x_{3k} \end{pmatrix} = \begin{pmatrix} \varphi(k) \\ \varphi'(k) \\ u_k \end{pmatrix}$$

in the form

$$\mathbf{x}_{k+1} = \mathbf{A}\mathbf{x}_k, \quad \mathbf{A} = \begin{pmatrix} \cosh\lambda & \dfrac{\sinh\lambda}{\lambda} & \dfrac{\cosh\lambda-1}{\lambda^2} \\ \lambda\sinh\lambda & \cosh\lambda & \dfrac{\sinh\lambda}{\lambda} \\ -\dfrac{6\tau^2}{ml}P & -\dfrac{6\tau}{ml}D & 0 \end{pmatrix}, \quad \lambda = \tau\sqrt{\dfrac{6g}{l}}. \tag{9}$$

The application of Jury's criterion (Kuo 1977) for the third-degree characteristic polynomial of the coefficient matrix $\mathbf{A}$ completes the proof of the theorem: the fact that the spectrum is located in the open unit disc of the complex plane is equivalent to the conditions (8).

Figure 10.2 shows qualitatively the two stability charts determined by the theorems above in the plane of the control parameters P and D. The lower stability limit for P is the same in both cases, but the shaded stability domain is somewhat smaller in the digital case in accordance with the fact that the average time delay is greater than that of the analogue case.

The numbers of characteristic roots causing instability are also given with encircled numbers in each domain of the (P, D) plane. It is easy to prove that Hopf bifurcation occurs as the proportional gain P is increased through its upper critical value at constant D. Considering the full nonlinear equations (1)–(3) in the analogue case, we can also prove the supercriticality of this Hopf bifurcation if the nonlinearity in the control force Q is symmetric having h.o.t. $= - P_3 \varphi^3 (t - \tau)$, $P_3 \in \mathbb{R}_+$ in equation (2). This means the existence of an orbitally asymptotically stable limit cycle around the equilibrium which is unstable due to the application of too great a proportional gain P.

In both cases, the shaded stability domains shrink as the measure τ of the delay increases, and at a certain critical value they disappear. This critical time delay and sampling time are calculated in the following theorem.

Theorem 3. *There always exist parameters P, D such that the trivial solution of equation (6) is asymptotically stable if $\tau < \tau_{cr}$, and the trivial solution is always unstable if $\tau > \tau_{cr}$, where the critical values are*

$$\tau_{cr,a} = \sqrt{\frac{l}{3g}} \quad and \quad \tau_{cr,d} = \sqrt{\frac{l}{6g}} \ln \frac{3 + \sqrt{5}}{2} \tag{10}$$

in the analogue case of equation (3) and in the digital case of equation (4), respectively.

10.4   EXPERIMENTAL OBSERVATIONS

The above analysis of the analogue model of the human-controlled inverted pendulum is very approximate since only some of the main elements of the human operator's behaviour are captured by the PD control and the reflex delay. However, the results are quite reliable even quantitatively. For example, $\tau_{cr,a}$ in equation (10) gives about 0.1 s critical delay in the case of a 0.3 m long stick. It can be checked by anybody that the balancing of a stick shorter than 30 cm is impossible for most of us since our reflex delay is of the order of 100 ms.

The recent experiments of Kawazoe (1992) show also a strong periodic component in the angle signals produced by untrained operators. This is due to the application of too great a proportional gain which is compensated at great angles by a degressive (saturation-like) characteristic of the control force. This refers to the case of supercritical Hopf bifurcation as described by means of the parameter P_3 in the previous section.

An obvious, but important observation is that an asymptotically stable upper position cannot be produced even by a trained operator. The signal is always stochastic in the vicinity of the equilibrium. This is partly explained by the stochasticity of the human parameters (like the reflex delay), but deterministic equations modelling the finite sensitivity of the operator's 'detectors' may also describe small chaotic motions.

The critical sampling time $\tau_{cr,d}$ in equation (10) for computer control is somewhat smaller than $\tau_{cr,a}$. In case of the 0.3 m long stick it gives 69 ms. However, for a simple task like this, computer control may run far below this value, with 1–10 ms sampling time, so the performance of the computer control is usually much better than that of the human operator.

Figure 10.3 shows the time history and spectrum of the angle φ we obtained in our experiments with an inverted pendulum attached to a cart. The cart was driven by an IBM

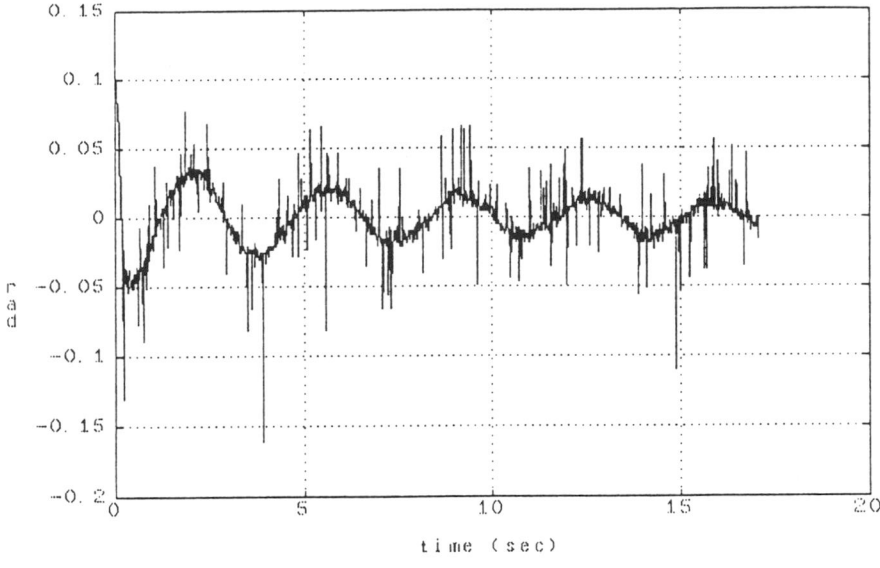

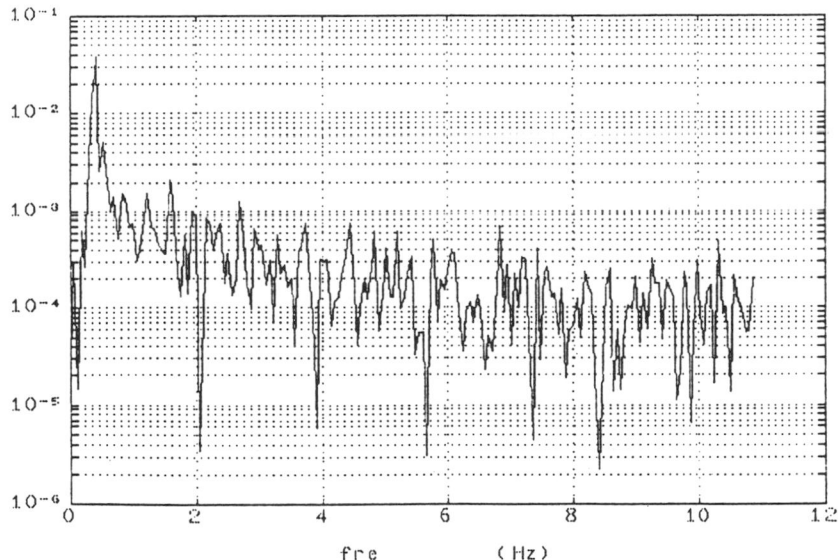

Figure 10.3 Measured signal and its spectrum.

PC-controlled DC motor. The sampling time was about 1.2 ms. The damped oscillation experienced in the transient motion, and which caused the peak in the spectrum at about 0.35 Hz, was clearly identified by means of the characteristic roots of A in the linear mapping (9) when the parameters of the experimental set-up were substituted there. The stochastic vibrations which have a low overall amplitude, but exist also at higher frequencies, can be explained by some nonlinear effects only, since external noise and stochastic parameters are negligiable in this case. The nonlinear control force (5) related to quantization will serve as a basis for the qualitative explanation of the experienced chaotic vibrations in the subsequent section.

Another important observation should also be mentioned here. The pendulum was attached to the cart by means of a ball bearing. With some elementary experiments, we estimated the value of the moment of the Coulomb friction there as about 6×10^{-4} Nm. This means that the pendulum could take an equilibrium position if $|\varphi| < 10^{-3}$ radians and $\dot{\varphi} = 0$. In our experiments, after 8–15 minutes, the inverted pendulum stopped with zero control force; that is, the chaotic motion found that tiny region provided by the slight Coulomb friction after a certain time. The length of this time seemed to vary stochastically. We have not been able to carry out enough experiments for a reliable statistical analysis in this respect.

10.5   μ-CHAOS

Following the above ideas about the cause of stochastic oscillations in the experiments, we consider the linearized equation of motion (1) with the sampling delay (4) and nonlinear control force (5). As the linear mapping (9) was constructed in the proof of Theorem 2, a nonlinear mapping is now obtained in the form

$$\mathbf{x}_{k+1} = \mathbf{B}\mathbf{x}_k + \mathbf{g}(\mathbf{x}_k), \qquad \mathbf{B} = \begin{pmatrix} \cosh \lambda & \dfrac{\sinh \lambda}{\lambda} & \dfrac{\cosh \lambda - 1}{\lambda^2} \\ \lambda \sinh \lambda & \cosh \lambda & \dfrac{\sinh \lambda}{\lambda} \\ 0 & 0 & 0 \end{pmatrix},$$

$$\lambda = \tau \sqrt{\dfrac{6g}{l}}, \qquad \mathbf{g}(\mathbf{x}_k) = \begin{pmatrix} 0 \\ 0 \\ -\dfrac{6\tau^2}{ml} h \operatorname{int}\left(\dfrac{P}{h}x_{1k} + \dfrac{D}{\tau h}x_{2k}\right) \end{pmatrix}. \qquad (11)$$

Since the characteristic roots of $\mathbf{B}$ are $e^{\pm \lambda}$ and 0, the trivial solution is always unstable. It is also clear, however, that for sufficiently small h, and for gains P and D satisfying conditions (8) for system (9), there is a 'small' attractive and invariant set around the origin. In order to get a view of the solutions within this attractive set, let us drop two dimensions in equation (11). This will result a one-dimensional 'cartoon' of the real system in the form

$$x_{k+1} = bx_k - c \operatorname{int}(x_k), \qquad (12)$$

where the scalar x has also been rescaled for convenience. Now, b can be thought of as the spectral radius $e^{\lambda} (> 1)$ of $\mathbf{B}$, and $c (> 0)$ stands for a control gain.

Figure 10.4 shows an iteration with parameters $b > 2$ and $b - 1 < c < b$ in equation (12). On a large scale, the upper part of the figure shows an attractive origin since the iteration seems to run along $x_{k+1} = (b - c)x_k$ as if there were no quantization effect. But the small-scale figure shows an invariant set C_μ which attracts the solutions, while the origin itself is unstable, of course. This set can easily be determined in this one-dimensional case as a function of the parameters b and c; the task is more difficult for the real system (11),

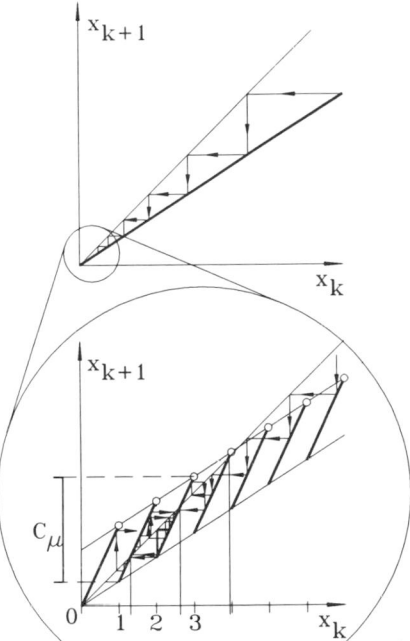

Figure 10.4 One-dimensional mapping.

however. As Figure 10.4 clearly shows, there may be several fixed points and periodic solutions of equation (12) in C_μ but they are all unstable and we may well expect the existence of a chaotic hyperbolic set that occupies most of the attractor C_μ. This is what we call μ-chaos.

Delchamps (1990) gave a proposition for the existence of the above positively invariant attractive set and also compared the statistical features of the chaos to white noise. Haller and Stépán (1994) proves the existence of a hyperbolic strange attractor and investigates the geometric properties, entropy and Haussdorf dimension of μ-chaos. These results along with the relatively simple determination of the attractor C_μ give the possibility of designing the stochastic motion of the digitally controlled machine around the desired position.

10.6   LIFE EXPECTANCY OF μ-CHAOS

The last paragraph of Section 10.4 described the experienced transient chaotic motion of the inverted pendulum. The mapping

$$x_{k+1} = \begin{cases} x_k & \text{if} \quad x_k \in [0, C_0) \\ b x_k - c \, \text{int}(x_k) & \text{if} \quad x_k \in [C_0, \infty) \end{cases} \tag{13}$$

can now be considered as a 'cartoon' of the real system (11) to which Coulomb friction is added. The interval $[0, C_0)$ represents the set where the system stops moving and remains in equilibrium by means of the Coulomb friction.

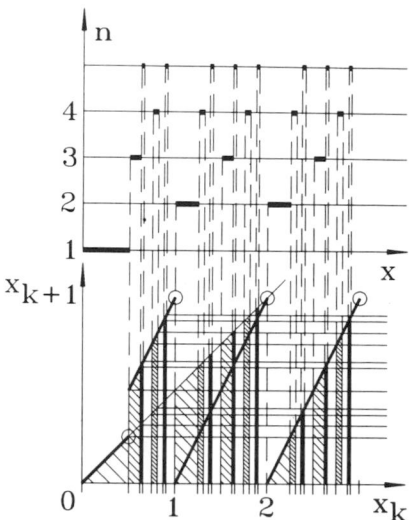

Figure 10.5   Calculation of life expectancy of transient μ-chaos.

The lower part of Figure 10.5 shows this mapping in the special case when $b = c$, i.e. when the 'large-scale' mapping defined by $x_{k+1} = (b - c)x_k$ has a zero characteristic root. Note that the real structure with an extended PID controller (e.g. when $a_{33} \neq 0$ in equation (9)) may also be tuned to have zero characteristic roots only. Furthermore, setting the parameters $b = c = 2$ and $C_0 = 1/2$ enables us to carry out exact calculations conveniently and to represent the calculation of the expected value of the length of the transient μ-chaotic behaviour.

An iteration starting from $[2, \infty)$ reaches the set $C_\mu = [0, 2)$ in one step. From this set, it may reach the set $C_0 = [0, 1/2)$ only after a great number of steps. Depending on the open interval from which a solution starts, the upper part of Figure 10.5 shows the number n of iterations the actual solution needs to reach C_0. The length s_n of an interval from which n iterations lead to C_0 is given by

$$s_n = \frac{f_n}{2^n}, \qquad f_1 = 0, \ f_2 = 1, \qquad f_n = f_{n-1} + f_{n-2},$$

where $\{f_n\}$ is the so-called Fibonacci series. Since $\sum_{n=1}^{\infty} s_n = 1$, the solutions will reach the set C_0 with probability 1. However, a measure-zero Cantor set also exists containing several unstable solutions including strange ones. Starting from this Cantor set, the solutions will never reach an equilibrium in C_0. The average number of the iteration steps n gives the expected lifetime of a μ-chaotic motion in C_μ. Using integration in the Lebesgue sense, we calculate this expected value E in the following way:

$$E = \lim_{N \to \infty} \frac{\int_0^N n \, dx}{\int_0^N dx} = \sum_{n=1}^{\infty} n s_n = 6.$$

Of course, we are not able to present any quantitative connection between the expected number of iterations $E = 6$ in the specific 'cartoon' model and the measured 8–15 minutes of transient μ-chaotic motion of the real structure. However, the above method is to be developed in this direction since the knowledge of the life expectancy of transient μ-chaos at design stage could be as important as the damping factor is.

10.7 CONCLUSIONS

In robotics, the introduction of a new stability definition called 'stability in a practical sense' shows that digitally controlled machines rarely reach the desired equilibrium position and small-amplitude stochastic vibrations appear. The digital effects arising at A/D converters can explain these micro-chaotic motions. Instead of the time-honoured consideration of an external white noise in the system, a deterministic mathematical model can be given. This model may well serve as a basis for the estimation of the amplitude and frequency ranges of μ-chaos as well as for that of its life expectancy. This might be essential for design of advanced computer-controlled machines.

ACKNOWLEDGEMENTS

This research was supported by the Hungarian Scientific Research Foundation under grant number OTKA 5-328. The author wishes to thank G. Haller (CalTech) for discussions and Prof. S. Thompson (Queen's University of Belfast), and E. Enikov and T. Müller (TU Budapest) for their help in the experimental work.

REFERENCES

Delchamps, D. F. (1990) Stabilizing a linear system with quantized state feedback, *IEEE Transactions on Automatic Control*, **35**, 916–924.

Hale, J. K. (1977) *Theory of Functional Differential Equations*, New York, Springer.

Haller, G. and Stépán, G. (1994) Notes on μchaos.

Henders, M. G. and Sondack, A. C. (1992) 'In-the-large' behaviours of an inverted pendulum with linear stabilization, *Int. J. of Nonlinear Mechanics*, **27**, 129–138.

Higdon, D. T. and Cannon, R. H. (1963) On the control of unstable multiple-output mechanical systems, *ASME Publications*, **63-WA-48**, 1–12.

Kawazoe, Y. (1992) Manual control and computer control of an inverted pendulum on a cart, *Proc. 1st Int. Conf. on Motion and Vibration Control* (*Yokohama, 1992*), pp. 930–935.

Kuo, B. C. (1977) *Digital Control Systems*, Champaign, Ill., SRL Publishing Company.

Mori, S., Nishihara, H. and Furuta, K. (1976) Control of an unstable mechanical system, *Int. J. Control*, **23**, 673–692.

Palkovics, L., Stépán, G. and Michelberger, P. (1993) Chaotic behaviour of the nonlinear wheel suspension system, *Machine Vibration*, **2**, 47–53.

Scheidl, R., Stribersky, A., Troger, H. and Zeman, K. (1985) Nonlinear stability behaviour of a tractor-semitrailer in downhill motion, *Proc. 9th IAVSD-Symposium* (Linköping, 1985), pp. 509–522.

Stépán, G. (1984) A model of balancing, *Periodica Polytechnica*, **28**, 195–199.

Stépán, G. (1987) The role of delay in robot dynamics, *Proc. 6th Symposium on Theory and Practice of Robots and Manipulators* (*Cracow, 1986*), Paris, Hermes, pp. 177–183.

Stépán, G. (1989) *Retarded Dynamical Systems*, Harlow, Longman.

Stépán, G., Steven, A. and Maunder, L. (1990), Design principles of digitally controlled robots, *Mechanism and Machine Theory*, **25**, 515–527.

G. Stépán, *Department of Applied Mechanics, Technical University of Budapest, H-1521 Budapest, Hungary*

11 CONTROLLING CHAOS ON FRACTAL BASIN BOUNDARIES

Z. Kovács, K. G. Szabó and T. Tél

The OGY method of controlling chaos (Ott, Grebogi and Yorke 1990) has been applied to stabilize periodic orbits on fractal basin boundaries. As illustrative examples the fractal boundaries in an invertible two-dimensional map and in the driven damped pendulum are considered. The scaling behaviour in the number of trajectories controlled is discussed.

11.1 INTRODUCTION

Fractal basin boundaries (Moon and Li 1985; Thompson et al. 1987) are typical attributes of nonlinear systems. If two or more attractors coexist in a system, their basins of attraction are separated by basin boundaries. These boundaries might be *fractals* with the consequence that trajectories started in their vicinity exhibit very complicated and unpredictable *transient chaotic* motion (Tél 1990) before settling down onto one of the possible simple attractors. All kinds of basin boundaries are *stable manifolds* along which one or more *saddles*, situated between the attractors, can be approached (Grebogi et al. 1988). In the case of fractal boundaries at least one of the saddles is *chaotic*. In fact, many examples of transient chaos, including experiments (see e.g. Iansiti et al. 1985; Arecchi et al. 1982; Bergé and Dubois 1983; Arecchi and Lisi 1983; Kowalik et al. 1988; Papoff et al. 1988; Dangoisse et al. 1986), are due to trajectories starting close to a fractal boundary separating multiple attractors.

It might occur that one of the attractors is connected with an undesirable outcome, such as the capsizing of a ship (Thompson et al. 1990). Because of intertwined basin boundaries, it is then difficult to direct the trajectory towards a desirable attractor. Thus, it can be of practical importance to stop the trajectory *on the basin boundary*, or more precisely, on a periodic orbit of the chaotic saddle. This means that an orbit hesitating to go to any of the attractors (like Balaam's ass) is converted into an attracting orbit. In this chapter we show how this can be achieved by applying results from the theory of chaos control.

Controlling chaos by means of the OGY method (Ott et al. 1990; Romeiras et al. 1992; Auerbach et al. 1992; Lai et al. 1992) has attracted interest of late among both theoreticians and experimentalists. The approach can be summarized as follows. One selects

Nonlinearity and Chaos in Engineering Dynamics
Edited by J. M. T. Thompson and S. R. Bishop,   © 1994 John Wiley & Sons Ltd

a target region around a predetermined hyperbolic periodic orbit on the chaotic set. Then an ensemble of trajectories is started in some region of the phase space and one waits until trajectories enter the target region where the actual controlling algorithm is applied. The controlling perturbation is adjusted so that the predetermined periodic orbit is stabilized. Only small local perturbations are allowed, smaller in size than some value δ which is proportional to the linear extension of the target region.

The method was originally worked out for systems the dynamics of which is governed by an underlying chaotic attractor in phase space. All trajectories are then controlled since any region on the attractor is visited sooner or later. The time needed to wait at a given maximum perturbation δ, i.e. the expected time $\tau(\delta)$ to achieve control, however, diverges: $\tau(\delta) \sim \delta^{-\gamma}$ as the maximum perturbation approaches zero.

The method has been extended to controlling transient chaos (Tél 1991, 1993; Lai *et al.* 1993) asociated with non-attracting chaotic sets: chaotic repellers or saddles. These are unstable objects in the sense that trajectories starting from their neighbourhoods escape with an average rate κ, called the *escape rate*. One can select again any of the infinitely many periodic orbits on the chaotic set and apply the same procedure as for permanent chaos. The qualitatively new feature compared with the control of permanent chaos is that an orbit is stabilized now which *has nothing to do with the asymptotic behaviour* (attractor) of the system. The control of chaos on fractal basin boundaries is thus a special case of controlling transient chaos.

11.2   CONTROL OF A SIMPLE MAP

First, we investigate a simple two-dimensional invertible mapping introduced in McDonald (1985a, b) to illustrate the existence of a fractal boundary between two fixed point attractors. The map is defined in polar coordinates x and θ as

$$\theta_{n+1} = \theta_n + a \sin(2\theta_n) - b \sin(4\theta_n) - x_n \sin\theta_n, \tag{1}$$

$$x_{n+1} = -J_0 \cos\theta_n.$$

Because of an invariance under the transformation $\theta \to 2\pi - \theta$, it is ʼsufficient to consider the range $0 \leqslant \theta \leqslant \pi$ only.

At parameter values $a = 1.32$, $b = 0.9$, $J_0 = 0.3$ the only attractors are the points $A_1 = (0, -J_0)$ and $A_2 = (\pi, J_0)$. Three further fixed points H_0 and $H_{1,2}$ lie along the line $x = -J_0 \cos\theta$ and are hyperbolic. Their θ values are $\theta_0 = \pi/2$, θ_1, and $\theta_2 = \pi - \theta_1$, respectively, with $\theta_1 = \arccos$ $[(2a + J_0)/4b]/2$. All of these fixed points are on the chaotic saddle and belong, therefore, to the basin boundary (see Figure 11.1). Note that the position of the fixed point H_0 does not depend on the system parameters, and therefore it is more difficult to control than $H_{1,2}$. Only the control of the latter will be discussed here.

We shall demonstrate the stabilization of H_1 below by adding a weak time-dependent perturbation p_n to the parameter a, i.e. by replacing a in equation (1) by $a_n = a + p_n$. In order to achieve this, one has to use a large *ensemble* of points starting from some region of phase space including the chaotic saddle and concentrate on long-lived chaotic transients. Next, define a target region, a disc of radius r, around H_1, and denote by ξ_n the distance of the nth point of the trajectory from the desired fixed point.

Some of the trajectories will stay around the chaotic saddle over many time steps

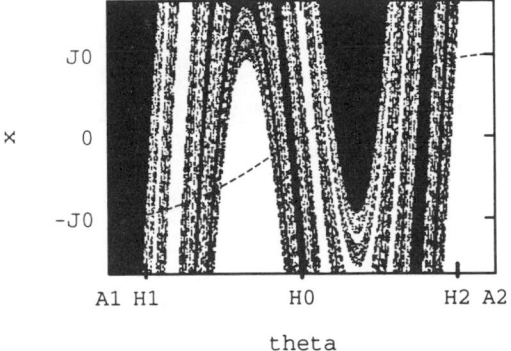

Figure 11.1 Basin boundary for map (1) with parameters $a = 1.32$, $b = 0.9$, $J_0 = 0.3$ plotted in the range $|x| \leqslant 0.5$, $0 \leqslant \theta \leqslant \pi/2$. The fixed points lie on the curve $x = -J_0 \cos \theta$ (dashed line). The θ values of the attractors $A_{1,2}$ as well as of the hyperbolic points H_0 and $H_{1,2}$ are indicated on the horizontal axis.

and might fall near H_1. Therefore, wait until ξ_n of any trajectory enters the target region around H_1 and then change the actual value of p_n to be different from zero. Pick p_n so that the next iterate ξ_{n+1} falls on the stable manifold of H_1 of the *uncontrolled* map. This mechanism is completely the same as for chaotic attractors; therefore, the result for the appropriate choice of p_n can be taken over from OGY. The computation based on the *linearized* dynamics around a fixed point of a two-dimensional map says (Ott *et al.* 1990) that

$$p_n = \frac{\lambda_u}{\lambda_u - 1} \frac{\xi_n f_u}{g f_u}. \tag{2}$$

Here λ_u and f_u are the unstable eigenvalue of the fixed point in the uncontrolled map ($p = 0$) and the corresponding left eigenvector, respectively. The quantity gp yields the shift of the fixed point's position when changing the perturbation parameter by a small amount of p. It is assumed that the parameter p can be varied in a small range $|p| < \delta$ only. Thus, if $|p_n|$ happens to be greater than the maximum perturbation δ, then we set $p_n = 0$. This last condition also specifies the size of the target region where control is activated.

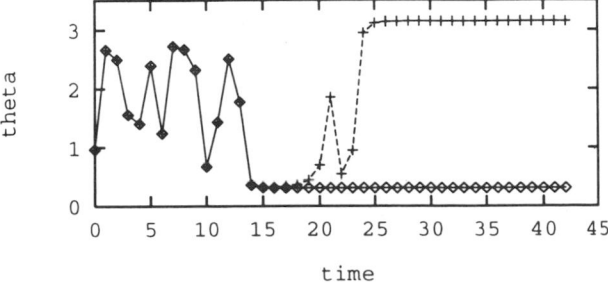

Figure 11.2 The θ component of a trajectory of map (1) starting at the initial point $x_0 = 0.154,839$, $\theta_0 = 0.962\,746$. Crosses and diamonds represents the uncontrolled and controlled trajectories, respectively. Control sets in at step 16 where the trajectory enters the control region, a disc of radius $r = 0.01$, around the hyperbolic point H_1.

Figure 11.2 shows one coordinate of a trajectory that moves around on the basin boundary until, at the 16th step, it enters the target region where control is applied, and then the trajectory becomes captured at H_1. Without control, this trajectory goes away from this fixed point and eventually runs into the attractor A_2.

11.3   THE DRIVEN DAMPED PENDULUM

Our second example, the forced driven pendulum, is a system with a continuous time dynamics. The equation of motion for the angle Φ of the pendulum in appropriate dimensionless variables is (Gwinn and Westervelt 1985, 1986a, b):

$$\ddot{\Phi} + \beta\dot{\Phi} + \sin\Phi = \rho\cos\omega t, \tag{3}$$

where β and ρ measure the damping and driving strengths, respectively, and dots denote time derivatives. Here we assume a pure sinusoidal torque drive of frequency ω without an additional constant torque. In the following, we fix $\beta = 0.2$ and $\omega = 1$, and consider ρ as our control parameter.

It is known (Gwinn and Westervelt 1985, 1986a, b) that for $\rho = 2$ (and other ρ values in a certain region around this point) the system possesses two periodic attractors with fractal basin boundaries. The attractors correspond to a motion in which the pendulum keeps winding around but each rotation is interrupted by a complete swinging cycle. If we introduce a two-dimensional Poincaré section in the system by measuring Φ and $\dot{\Phi}$ when the driving

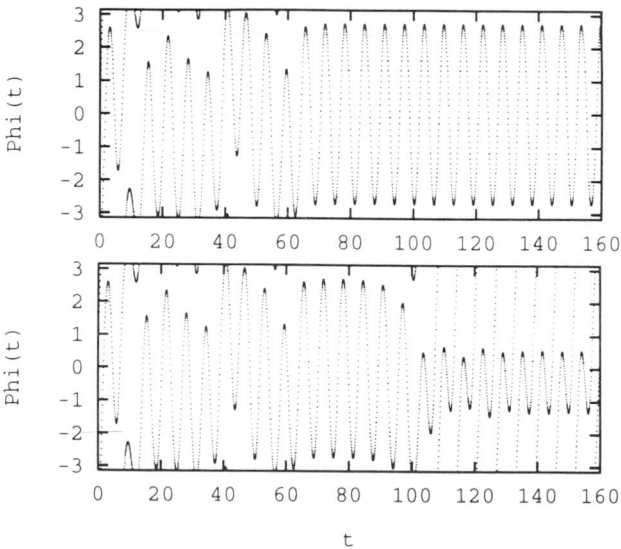

Figure 11.3   Time dependence of the angle Φ (modulo π) started from $\Phi_0 = 0.580\,685\,8$ and $\Phi_0 = 2.752\,419\,0$ at $t = 0$. The trajectory enters the target region, a disc of radius $r = 0.05$ around the unstable fixed point $(-2.631\,308\,5,\ 0.651\,685\,6)$ on the Poincaré surface, after 10 driving cycles where control sets in causing the pendulum to keep swinging without rotation (upper box). Without control (lower box), the same trajectory is attracted to the stable winding motion.

torque is maximal (and positive), the attractors are represented as fixed points of the two-dimensional map induced in the section. The perpetual swinging ('pendulum-like') behaviour never making a full rotation appears as an unstable fixed point of the map lying on the basin boundary.

This swinging motion can be stabilized by applying the control method described earlier. The parameter ρ is used for this purpose: we take $\rho_n = 2 + p_n$ as the driving torque amplitude. p_n is kept fixed between two returns to the Poincaré surface when its value is refreshed according to the control rules. We started a large number of orbits from a region containing a part of the saddle, and looked for those that stay long enough on the boundary to be able to come close to the unstable fixed point. Figure 12.3 presents the continuous time dynamics of an orbit that wanders on the basin boundary until it gets close to the swinging motion. Then control is turned on, pulling the orbit into the desired fixed point. This behaviour can be compared to the motion, also shown in Figure 12.3, started from the same initial condition but never subject to control: it ends up following the attractor characterized by winding in a counterclockwise sense.

11.4 SCALING PROPERTIES

Quantitatively, another novel feature distinct from the permanent case is that many trajectories escape before falling into the target region. Consequently, the number $N(\delta)$ of trajectories controlled at a given maximum perturbation δ behaves as

$$N(\delta) \sim \delta^{\gamma(\kappa)} \tag{4}$$

where $\gamma(\kappa)$ is an exponent depending on the escape rate κ. Its explicit form has been found to be (Tél 1991):

$$\gamma(\kappa) = 1 + \frac{\ln|\lambda_u| - \kappa}{\ln 1/|\lambda_s|} \tag{5}$$

where λ_s denotes the stable eigenvalue of the controlled fixed point. κ is the escape rate from the chaotic saddle the stable manifold of which forms the basin boundary. In other words, $1/\kappa$ is the average chaotic lifetime on the basin boundary.

Since the time to achieve control cannot be longer than the chaotic life time, we obtain that $\tau(\delta)$ is independent of the maximum allowed perturbation δ and is limited from above by $1/\kappa$:

$$\tau(\delta) = \text{const.} \leqslant 1/\kappa. \tag{6}$$

We mention in passing that the escape rate κ which appeared in the formulae above can be estimated in the knowledge of the *uncertainty exponent* α (Grebogi *et al.* 1983a). This exponent plays a central role in characterizing fractal boundaries where uncertainty in initial conditions leads to enhanced uncertainty in the final state. It is easy to see (Tél 1990) that

$$\alpha = \frac{\kappa}{\lambda} + D_1^{(1)} - D_0^{(1)} \leqslant \frac{\kappa}{\lambda} \tag{7}$$

where λ, $D_0^{(1)}$ and $D_1^{(1)}$ stand for the average Lyapunov exponent, and for the partial fractal and information dimensions of the chaotic saddle along its unstable direction, respectively. The inequality is expected to be close to saturation since the difference between $D_0^{(1)}$ and $D_1^{(1)}$ is typically less than a few percentage points. By assuming that the local Lyapunov exponent of the controlled trajectory is close to the average one, the approximate relation

$$\kappa \approx \alpha \ln \lambda_u \tag{8}$$

can also be used.

11.5   CLOSING REMARKS

We have demonstrated that chaotic trajectories moving in the vicinity of a fractal basin boudary can be controlled, i.e. directed towards a periodic orbit on the chaotic saddle organizing the boundary. Such trajectories never reach the attractors of the system and represent a qualitatively new asymptotic behaviour. We saw that controlling chaos on fractal boundaries is more difficult than the control of permanent chaos as one has to use ensembles of trajectories, but it is also simpler since the time needed for control does not grow with decreasing perturbation. By applying recent techniques like targeting (Shinbrot *et al.* 1990) or control via synchronization (Lai and Grebogi 1993) one can hope that even the average number of trajectories needed to achieve control can be drastically reduced.

ACKNOWLEDGEMENT

This work has been supported by the Hungarian National Science Foundation under the grant numbers OTKA 2090, T4439, and F4286.

REFERENCES

Arecchi, F. T. and Lisi, F. (1983) Hopping mechanism generating $1/f$ noise in non-linear systems response. *Phys. Rev. Lett.*, **50**, 1330.

Arecchi, F. T., Meucci, R., Puccioni, G. and Tredicce, J. (1982) Experimental evidence of subharmonic bifurcations, multistability and turbulence in a Q-switched gas laser. *Phys. Rev. Lett.*, **49**, 1217.

Arecchi, F. T., Badii, R. and Politi, A. (1985) Generalized multistability and noise-induced jumps in a nonlinear dynamical system. *Phys. Rev. A*, **32**, 402.

Auerbach, D., Grebogi, C., Ott, E. and Yorke, J. (1992) Controlling chaos in high dimensional systems. *Phys. Rev. Lett.*, **69**, 3479.

Bergé, P. and Dubois, M. (1983) Transient reemergent order in convective spatial chaos. *Phys. Lett.*, **93A**, 365.

Bleher, S., Grebogi, C., Ott, E. and Brown, R. (1988) Fractal boundaries for exit in Hamiltonian dynamics. *Phys. Rev. A*, **38**, 930.

Dangoisse, D., Glorieux, P. and Hennequin, D. (1986) Laser chaotic attractor in crisis. *Phys. Rev. Lett.*, **57**, 2657.

Grebogi, C., McDonald, S. W., Ott, E. and Yorke, J. A. (1983a) Final state sensitivity: An obstruction to predictability. *Phys. Lett.*, **99A**, 415.

Grebogi, C., Ott, E. and Yorke, J. A. (1983b) Fractal basin boundaries, long-lived chaotic transients, and unstable–unstable pair bifurcation. *Phys. Rev. Lett.*, **50**, 935.

Grebogi, C., Ott, E. and Yorke, J. A. (1986) Metamorphoses of basin boundaries in nonlinear dynamical systems. *Phys. Rev. Lett.*, **56**, 1011.

Grebogi, C., Nusse, H., Ott, E. and Yorke, J. (1988) Basic sets: Sets that determine the dimension of basin boundaries. *Lecture Notes in Math.*, **1342**, 22.

Gwinn, E. G. and Westervelt, R. M. (1985) Intermittent chaos and low-frequency noise in the driven damped pendulum. *Phys. Rev. Lett.*, **54**, 1613.

Gwinn, E. G. and Westervelt, R. M. (1986a) Fractal basin boundaries and intermittency in the driven damped pendulum. *Phys. Rev. A*, **33**, 4143.

Gwinn, E. G. and Westervelt, R. M. (1986b) Horse-shoes in the driven, damped pendulum. *Physica*, **23D**, 396.

Holt, R. G. and Schwartz, I. B. (1984) Newton's method as a dynamical system: Global convergence and predictability. *Phys. Lett.*, **105A**, 327.

Iansiti, M., Hu, Q., Westervelt, R. M. and Tinkham, M. (1985) Noise and chaos in a fractal basin boundary regime of a Josephson junction. *Phys. Rev. Lett.*, **55**, 746.

Isomäki, H. M., von Boehm, J. and Räty, R. (1988) Fractal basin boundaries of an impacting particle. *Phys. Lett.*, **126A**, 484.

Kocarev, L. (1987a) The basin boundaries of one-dimensional maps. *Phys. Lett.*, **121A**, 274.

Kocarev, L. (1987b) Quasifractal metamorphoses of 1-d maps. *Phys. Lett.*, **125A**, 389.

Kowalik, Z. J., Franaszek, M. and Pieranski, P. (1988) Self-reanimating chaos in the bouncingball system. *Phys. Rev. A*, **37**, 4016.

Lai, Y.-C. and Grebogi, C. (1993) Synchronization of chaotic trajectories using control. *Phys. Rev. E*, **47**, 2357.

Lai, Y.-C., Ding, M. and Grebogi, C. (1992) Controlling Hamiltonian chaos. *Phys. Rev. E*, **47**, 86.

Lai, Y.-C., Tel, T. and Grebogi, C. (1993) Stabilizing chaotic-scattering trajectories using control. *Phys. Rev. E*, **48**, 709.

McDonald, S. W., Grebogi, C., Ott, E. and Yorke, J. A. (1985a) Fractal basin boundaries. *Physica* **17D**, 125.

McDonald, S. W., Grebogi, C., Ott, E. and Yorke, J. A. (1985b) Structure and crises of fractal basin boundaries. *Phys. Lett.*, **107A**, 51.

Moon, F. C. and Li, G. X. (1985) Fractal basin boundaries and homoclinic orbits for periodic motion in a two-well potential. *Phys. Rev. Lett.*, **55**, 1439.

Ott, E., Grebogi, C. and Yorke, J. (1990a) Controlling chaos. *Phys. Rev. Lett.*, **64**, 1196.

Ott, E., Grebogi, C. and Yorke, J. (1990b) Controlling chaotic dynamical systems. In: *Chaos*, D. K. Campbell (ed.), New York, AIP, pp. 153–172.

Papoff, F., Dangoisse, D., Poite-Hanoteau, E. and Glorieux, P. (1988) Chaotic transients in a CO_2 laser with modulated parameters: Critical slowing-down and crisis-induced intermittency. *Optics Comm.*, **67**, 358.

Romeiras, F. J., Grebogi, C., Ott, E. and Dayawansa, W. P. (1992) Controlling chaotic dynamical systems. *Physica*, **58D**, 165.

Shinbrot, T., Ott, E., Grebogi, C. and Yorke, J. (1990) Using chaos to direct trajectories to targets. *Phys. Rev. Lett.*, **65**, 3215.

Takesue, S. and Kaneko, K. (1984) Fractal basin structure. *Prog. Theor. Phys.*, **71**, 35.

Tél, T. (1990) Transient chaos. In: *Directions in Choaos, Vol. 3*, Bao-lin Hao (ed.), World Scientific, Singapore, pp. 149–221.

Tél, T. (1991) Controlling transient chaos. *J. Phys. A* **24**, L1359.

Tél, T. (1993) Crossover between the control of permanent and transient chaos. *Int. J. Bifurcation and Chaos*, **3**, 757.

Thompson, J. M. T., Bishop, S. R. and Leung, L. M. (1987) Fractal basins and chaotic bifurcations prior to escape from a potential well. *Phys. Lett.*, **121A**, 116.

Thompson, J. M. T., Rainey, R. C. T. and Soliman, M. S. (1990) Ship stability criteria based on chaotic transients from incursive fractals. *Phil. Trans. Roy. Soc. London A*, **332**, 149.

Thompson, J. M. T. and Soliman, M. S. (1990) Fractal control boundaries of driven oscillators and their relevance to safe engineering design. *Proc. Roy. Soc. London A*, **428**, 1.

Varghese, M. and Thorp, J. S. (1988) Truncated-fractal basin boundaries in forced pendulum systems. *Phys. Rev. Lett.*, **60**, 665.

Z. Kovács, K.G. Szabó and T. Tél, *Institute for Theoretical Physics, Eötvös University, Puskin u. 5-7, H-1088 Budapest, Hungary*

12 THE CONTROL OF CHAOS BY A DYNAMICAL ABSORBER

J. Brindley and T. Kapitaniak

We describe an effective method of controlling chaos by joining to a main chaotic system another small system. The idea of this method is similar to that of the socalled dynamic vibration absorber in linear systems. A key feature of the method is its avoidance of the need for feedback mechanisms, which can, in mechanical systems, be complex and costly. It does, however, depend on a knowledge of the dynamics of the main system.

12.1 INTRODUCTION

The presence of chaos both in nature and in man-made devices is very common and has been extensively demonstrated in the last decade. Quite frequently chaos is a beneficial feature, as in some chemical or heat and mass-transport problems. However, in many other situations, chaos may be an undesirable phenomenon, leading to unwanted oscillations, irregular operations etc. Chaotic behaviour may also be detrimental to the operation of various devices because it cannot be predicted in detail.

The problem of controlling chaos, that is, of converting the chaotic behaviour found in a physical system to a periodic time dependence or aperiodicity which is predictable, has attracted recent interest Ott et al. (1990) showed that in a chaotic system we can always stabilize one of many originally unstable periodic orbits embedded in the chaotic attractor. In non-feedback methods (Bandhyopadhyay et al. 1992; Kapitaniak 1992) the controlled dynamical system is slightly modified in such a way that stable and predictable solutions replace chaotic behaviour. Appropriate modification is done by the addition of a small periodic or even random driving force, or modulation of one of the system parameters. Application of a particular method depends on the properties of the controlled system and our knowledge of its dynamics. Generally non-feedback methods are restricted to systems of which the dynamics is known, and to such systems they can be applied in a much simpler way than feedback methods.

In this chapter we describe a method of controlling chaos in mechanical systems in which a control effect is obtained by coupling the chaotic system with another system, usually

Nonlinearity and Chaos in Engineering Dynamics
Edited by J. M. T. Thompson and S. R. Bishop, © 1994 John Wiley & Sons Ltd

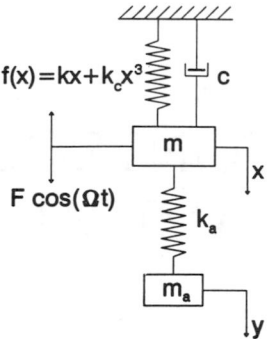

Figure 12.1 Schematic diagram of the main system and the dynamical absorber.

linear. The idea of this method is similar to that of the so-called dynamical vibration absorber, long known in linear systems. A dynamical vibration absorber is a one-degree-of-freedom system, usually a mass on a spring (sometimes viscous damping is also added), which is connected to the main system as shown in Figure 12.1. The additional degree of freedom introduced shifts resonance zones, and in some cases can eliminate oscillations of the main mass. Although such a dynamical absorber can change the overall dynamics substantially it need usually only be physically small in comparison with the main system, and does not require an increase of excitation force. It can be easily added to the existing system without major changes of design or construction. This contrasts with devices based on feedback control, which can be large and costly.

An exactly similar idea of coupling a linear oscillator to a nonlinear system is valuable also in electrical engineering, an example being that of the so-called Chua's circuit coupled to a simple linear circuit. One of us has reported on this configuration elsewhere (Kapitaniak *et al.* 1993), and we comment on the phenomenon more fully in Section 13.3.

12.2   CONTROLLING DUFFING'S OSCILLATOR

To explain the role of dynamical absorbers in controlling chaotic behaviour let us consider the Duffing oscillator, coupled with an additional linear system:

$$\ddot{x} + a\dot{x} + bx + cx^3 + d(x - y) = B_0 + B_1 \cos \Omega t, \tag{1a}$$

$$\ddot{y} + e(y - x) = 0, \tag{1b}$$

where a, b, c, d, e, B and Ω are constant. Here d and e are the characteristic parameters for the absorber, and we take e as control parameter. The parameters of equations (1) are related to those of Figure 13.1 in the following way: $a = c/m\,\Omega$, $b = k/m\,\Omega^2$, $c = k_c/m\,\Omega^2$, $d = k_a/m\,\Omega^2$, $e = k_a/m_a\,\Omega2$, $B_0 = F_0/m\,\Omega$ and $B_1 = F_1/m\,\Omega$. It should be noted here that parameters d and e are related to each other through the absorber stiffness k_a. For simplicity in the rest of this chapter we assume that d is constant and consider e as control parameter, i.e. we take constant stiffness k_a and allow the absorber mass, m_a, to vary.

It is well known that the uncoupled equation (1a) (i.e. without the dynamical absorber) shows chaotic behaviour for certain parameter regions (Kapitaniak 1991a, b). In many cases the route to chaos proceeds via a sequence of period-doubling bifurcations (Kapitaniak 1991b), and in such cases our method provides an easy way of switching between chaotic and periodic behaviour.

To analyse the system with the absorber (d present, $e \neq 0$), we first assume that all parameters of equation (1), except the forcing frequency Ω, are constant, and estimate the Ω-domain where chaos exists. The application of the harmonic balance method enables us to determine the stability domain of approprriate $2\pi/\Omega = T$ periodic solutions, i.e.

$$
\begin{aligned}
x &= C_0 + C_1 \cos(\Omega t + \psi), \\
y &= D_0 + D_1 \cos(\Omega t + \gamma),
\end{aligned}
\tag{2}
$$

and $2T$ periodic solutions

$$
\begin{aligned}
x &= A_0 + A_{1/2} \cos[(\Omega/2)t + \rho] + A_1 \cos \Omega t, \\
y &= E_0 + E_{1/2} \cos[(\Omega/2)t + \beta] + E_1 \cos \Omega t,
\end{aligned}
\tag{3}
$$

where $C_0, C_1, D_0, D_1, A_0, A_{1/2}, A_1, E_0, E_{1/2}, E_1, \psi, \gamma, \rho$ and β are constants which are determined by substituting equations (2) or (3) into equations (1). Approximate boundaries of stability as functions of forcing frequency Ω for each solution can be estimated by adding small perturbations dx and dy to x and y, and considering an appropriate Hill's equation. The whole procedure is fully described in Kapitaniak (1991a, b), so we omit details here. Knowing the period-doubling bifurcation values Ω_1' and Ω_2' at which we have bifurcation from $T \rightarrow 2T$ periodic solutions, and Ω_1'' and Ω_2'' at which we have bifurcation from $2T \rightarrow 4T$ periodic solutions, we can obtain approximate values for the accumulation points Ω_1'' and Ω_2'' as

$$
\begin{aligned}
\Omega_1'' &= \Omega_1' + \Delta \Omega_1/(1 - 1/\delta), \\
\Omega_2'' &= \Omega_2' + \Delta \Omega_2/(1 - 1/\delta),
\end{aligned}
\tag{4}
$$

where $\Delta \Omega_1 = \Omega_1'' - \Omega_1'$, $\Delta \Omega_2 = \Omega_2' - \Omega_2''$ and $\delta = 4.669\ldots$ is a Feigenbaum constant. In Kapitaniak (1991a, b) it has been shown that the interval $[\Omega_1'', \Omega_2'']$ can be considered as an approximation of the Ω frequency domain for which chaos exists.

The above procedure can be easily performed using any symbolic algebra system (we used Mathematica) and by following it for different values of e we are able to obtain a map of behaviour of equations (1) as a function of two parameters: the frequency Ω and the dynamical absorber control parameter e, as shown for example in Figure 12.2 (solid lines). The other parameters of equations (1) have been fixed at the values $a = 0.077$, $b = 0$, $c = 1.0$, $B_0 = 0.045$ and $B_1 = 0.16$. This plot is in good agreement with numerically obtained behaviour domains as shown in Figure 12.2 (broken lines). Numerical results were obtained using a fourth-order Runge–Kutta method with a time step $\pi/200\Omega$, and to determine chaotic behaviour the Lyapunov exponents were calculated using the algorithm of Wolf *et al.* (1985).

From Figure 12.2 it is clear that, for fixed Ω, we can obtain different types of periodic behaviour by making slight changes in e. As an example, consider a system with $\Omega = 0.98$. For $e < 0.09$, the system is chaotic, but by changing e from 0.01 to 0.16 it is easily possible to obtain T, $2T$, $4T$, $8T$ periodic orbits. Theoretically orbits of higher periods are also possible, but their narrow range of existence make them difficult to find either experimentally or numerically. What is of vital significance is that values of the parameter $e \in [0.01, 0.16]$ can

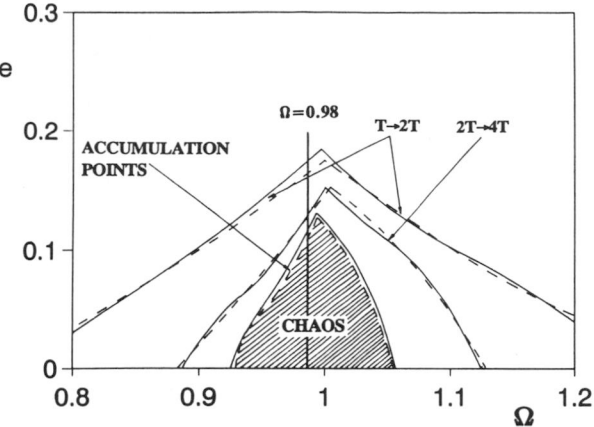

Figure 12.2 Behaviour of equations (1) for different values of e and Ω: $a = 0.077$, $b = 0$, $c = 1.0$, $B_0 = 0.045$ and $B_1 = 0.16$; analytical approximation (solid line), numerical simulation (broken line).

be obtained with an absorber mass m_a approximately 100 times smaller than the main mass (Figure 12.1).

To show the effectiveness of our method in real experimental conditions we have considered the effect of quasi-periodic noise

$$h(t) = \sum_{i=1}^{N} \alpha_i \cos(v_i t + \eta_i), \tag{5}$$

where $\alpha_i \ll B_{0,1}$ are constant and v_i and η_i are time-independent random variables, on the behaviour of equations (1). Quasi-periodic noise given by equation (5) is an approximation of the realization of the band-limited white-noise stochastic process with zero mean and a spectral density

$$s(v) = s/(v_{max} - v_{min}), \quad v \in [v_{min}, v_{max}]$$
$$0, \qquad\qquad\qquad v \notin [v_{min}, v_{max}]$$

where s is the intensity of noise and $[v_{min}, v_{max}]$ is the interval of frequencies considered and can be easily simulated experimentally (Kapitaniak 1991b).

Considering the perturbed system

$$\ddot{x} + a\dot{x} + bx + cx^3 + d(x - y) = B_0 + B_1 \cos(\Omega t) + \sum_{i=1}^{N} \alpha_i \cos(v_i t + \eta_i), \tag{6a}$$

$$\ddot{y} + e(y - x) = 0, \tag{6b}$$

we have found the interesting property that presence of noise reduces the magnitude of e necessary to obtain an appropriate periodic solution. This property is summarized in Figure 13.3, where we compare the behaviour of the system (1) for different noise intensities.

A similar controlling effect can be obtained by varying the absorber stiffness, i.e. by simultaneous changes of parameters d and e (Błażejczyk *et al.* 1993).

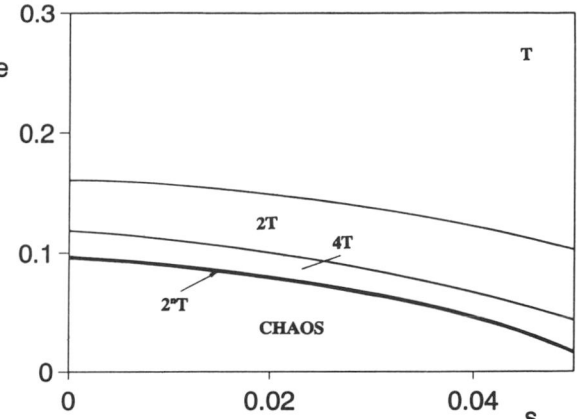

Figure 12.3 Effect of noise on the behaviour of equations (1): $\Omega = 0.98$, other parameter values as in Figure 12.2.

12.3 CONTROLLING CHUA'S CIRCUIT

An experimental verification of our method has been carried out on an example of Chua's circuit – one of the simplest chaotic systems (Madan 1993). Chua's circuit is a remarkably simple and robust electrical circuit made up of only four linear elements (one resistor, one inductor, two capacitors) and a nonlinear element known as Chua's diode.

To obtain the appropriate controlling results Chua's circuit has been coupled to a second-order linear circuit, as shown in Figure 12.4. The state equations of the whole system are

$$C_1 \frac{dv_{C_1}}{dt} = \frac{1}{R}(v_{C_2} - v_{C_1}) - g(v_{C1}),$$

$$C_2 \frac{dv_{C_2}}{dt} = \frac{1}{R}(v_{c_1} - v_{c_2}) + i_L + \frac{1}{R_x}(v_C^{(1)} - v_{C_2}),$$

$$L \frac{di_L}{dt} = -v_{C_2},$$
$$\qquad (7)$$

$$C^{(1)} \frac{dv_C^{(1)}}{dt} = -\frac{1}{R^{(1)}} v_C^{(1)} + i_L^{(1)} + \frac{1}{R_x}(v_{C_2} - v_C^{(1)}),$$

$$L^{(1)} \frac{di_L^{(1)}}{dt} = -v_C^{(1)},$$

where $v_C^{(1)}$ and $i_L^{(1)}$ are the voltage across the capacitor $C^{(1)}$ and the current though the inductor $L^{(1)}$, respectively.

In the controlling of chaos in Chua's circuit (Kapitaniak *et al.* 1993), the system parameters were chosen so that the system exhibits a chaotic attractor; specifically the so-called double scroll attractor, where $C_1 = 10$nF, $B_p = 1$ V, $C_2 = 99.34$ nF, $G_a = -0.76$ mS, $G_b = -0.41$ mS, $L = 18.46$ mH and $R = 1.64$ kΩ. For the coupled linear system we used off-the-shelf com-

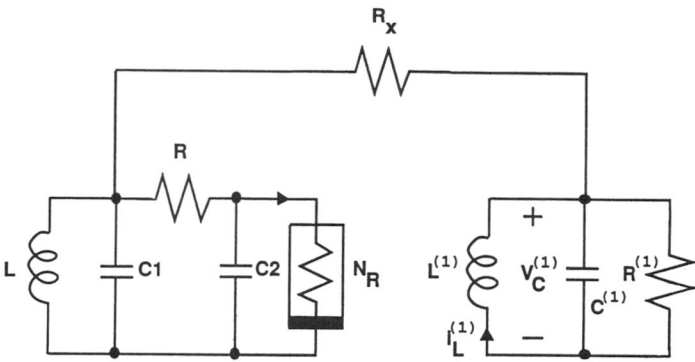

Figure 12.4   Chua's circuit coupled to a two-dimensional linear system.

ponents for the inductor and the capacitor: $L^{(1)} = 18\,\text{mH}$ with tolerance $\pm 10\%$, $C^{(1)} = 100\,\text{nF}$ with tolerance $\pm 5\%$. The value of the resistance $R^{(1)}$ was experimentally chosen to be $7.67\,\text{k}\Omega$ and resistance R_x was taken as a control parameter.

For relatively small values of coupling parameter $\varepsilon = R/R_x$ between Chua's circuit and the controlling linear system it was possible to find periodic orbits of different periods which were close to the original attractor (Kapitaniak *et al.* 1993).

Our controlling method may be straightforwardly applied to such electrical systems as microelectronics and VLSI circuits, where it is difficult, if not impossible, to access internal circuit parameters.

12.4   CONCLUSIONS

To summarize, we have shown here that chaotic behaviour can be converted to an appropriate periodic behaviour without feedback. The dynamical absorber offers a way of controlling chaos without the necessity of following a response trajectory and targeting it to the desired domain of phase space. The method can be especially useful in mechanical systems, where the feedback controllers are usually very large (sometimes larger than the controlled system). In contrast, a dynamical absorber having mass of order 1% of that of the controlled system is able, as we have shown in this example, to convert chaotic to periodic behaviour over a substantial region of parameter space. Indeed the simplicity by which chaotic behaviour may be changed in this way, and the possibility of easy access to different periodic orbits, may actually motivate the search for and exploration of chaotic behaviour in practical mechanical systems.

We have commented earlier on an electrical analogue of the mechanical configuration; it is also of some interest that this same configuration of a nonlinear system coupled to a linear oscillator is presently finding favour in simple modelling of climatic phenomena. Here the objective is better prediction rather than control of chaos. A particular example (Palmer 1993) uses a model in which a Lorenz system is coupled to a linear oscillator to develop a conceptual picture of the possible impact of El Niño and the so-called Madden–Julian oscillations on a chaotic atmosphere. The very significant influence of the coupled oscillator on the chaotic behaviour, particularly the predictability, of the Lorenz system has considerable relevance to climate forecasting.

ACKNOWLEDGEMENTS

One of us (T. K.) has been supported by KBN (Poland) under project no. 333579102.

REFERENCES

Bandyopadhyay, J. K., Ravi Kumar, V., Kulkarni, B. D. and Deshpande, P. B. (1992) On dynamical control of chaos: a study with reference to a reacting system, *Phys. Lett.*, **166A**, 197.

Błazejczyk, B., Kapitaniak, T., Wojewoda, J. and Brindley, J. (1993) Controlling chaos in mechanical systems, *Appl. Mech. Rev.*, **46**, 385.

Ditto, W. L., Rauseo, S. W. and Spano, M. L. (1991) Experimental control of chaos, *Phys. Rev. Lett.*, **65**, 3211.

Hunt, E. R. (1991) Stabilizing high-periodic orbits in a chaotic system—the diode resonator, *Phys. Rev. Lett.*, **67**, 1953.

Jackson, E. A. (1990a) The entrainment and migration controls of multiple-attractor systems, *Phys. Lett.*, **151A**, 478.

Jackson, E. A. (1990b) On the control of complex dynamical systems, *Physica*, **50D**, 341.

Kapitaniak, T. (1991a) *Chaotic Oscillations in Mechanical Systems*, Manchester, Manchester University Press.

Kapitaniak, T. (1991b) The loss of chaos in a quasiperiodically forced nonlinear oscillator, *Int. J. Bifurcation and Chaos*, **1**, 357.

Kapitaniak, T. (1992) Controlling chaos without feedback, *Chaos, Solitons and Fractals*, **2**, 519.

Kapitaniak, T., Kocarev, Lj. and Chua, L. O. (1993) Controlling chaos without feedback and control signals, *Int. J. Bifurcation and Chaos*, **3**, 324.

Madan, R. (1993) Special issue on 'Chaus's circuit: a paradigm for chaos', *J. of Circuits, Systems and Computers*, **3**, 1.

Mehta, N. J. and Henderson, R. M., (1991) Controlling chaos to generate aperiodic orbits, *Phys. Rev. A*, **44**, 4861.

Ott, E., Grebogi, C. and Yorke, J. A. (1990) Controlling chaos, *Phys. Rev. Lett.*, **64**, 1196

Palmer, T. N. (1993) Extended-range atmospheric prediction and the Lorenz model, *Bull. American Meteorological Soc.*, **74**, 49.

Shinbrot, T., Ott, E., Grebogi, C. and Yorke, Y. A. (1990) Using chaos to direct trajectories to target, *Phys. Rev. Lett.*, **65**, 3215.

Shinbrot, T., Ott. E., Grebogi, C. and Yorke, Y. A. (1992) Using chaos to target stationary states of flows, *Phys. Lett.*, **169A**, 349.

Singer, J., Wang, Y.-Z. and Bau, H. H. (1991) Controlling chaotic dynamical system, *Phys. Rev. Lett.*, **66**, 1123.

Tél, T. (1991) Controlling transient chaos, *J. Phys. A.*, **24**, L1359.

Wolf, A., Swift, J., Swinney, H. and Vastano, A. (1985) Determining Lyapunov exponents from time series, *Physica*, **16D**, 285.

J. Brindley, *Department of Applied Mathematics and Centre for Nonlinear Studies, University of Leeds, Leeds LS2 9JT, UK*

T. Kapitaniak, *Division of Control and Dynamics, Technical University of Łódź, Stefanowskiego 1/15, 90-924 Łódź, Poland*

PART IV
Engineering Applications

In some circumstances systems are designed to behave in a linear manner. However, engineers are aware that most practical systems are in fact nonlinear, in some parameter ranges at least. This means that if one varies the correct parameter and looks in sufficient detail it will be possible to view nonlinear phenomena, including chaos. Armed with new techniques it is now realistic to analyse a wide variety of engineering systems to investigate their behaviour and assess their stability. This part details a number of applications of new theories to vibration absorbers, railway vehicle dynamics, satellite rotations and other areas of study.

13 A NONLINEAR DYNAMIC VIBRATION ABSORBER FOR ROTATING MACHINERY

S. W. Shaw and C.-T. Lee

In this work we demonstrate a tuned dynamic vibration absorber system which can be used to eliminate speed fluctuations in rotating or reciprocating machinery. The primary system is modelled as a simple rotating disc, to which are attached a pair of identical absorber masses which are free to move along prescribed paths relative to the disc. The unique features of the proposed arrangement are that the absorbers are tuned to *one-half* of the order of the applied harmonic torque, and that they are effective into the fully nonlinear operating range. Due to a special symmetry, the equations of motion for this system possess a global subharmonic solution in which the absorber masses move out of phase with respect to each other and exactly cancel a purely harmonic applied torque of a given order. This is accomplished without inducing any higher harmonics, thus rendering a perfectly constant speed of rotation. The method of multiple scales is applied to obtain the corresponding approximate solution for the case of small damping, and it is determined that as the torque level is increased the angular acceleration of the disc saturates at a value which is proportional to the absorber damping level. The perturbation result is verified by numerical simulations over a large range of torque amplitudes.

13.1 INTRODUCTION

A steady rotation speed is often desired for rotating machinery in order to extend the fatigue life of components and to decrease vibration and noise levels. Centrifugal pendulum vibration absorbers (CPVA's) offer one means by which one can reduce torsional vibrations which arise from applied torques. A CPVA is essentially a mass which moves along a prescribed path relative to the rotating system. It regulates the speed of the system in a passive manner by providing a moment which counteracts some or all of the applied torque acting on the system. An important feature of the CPVA is that its tuning frequency is proportional to the system's nominal rotation rate, Ω, and it can therefore be tuned to absorb vibrations with frequencies of the form $n\Omega$ (n is referred to as the *order* of the disturbance). This is

Nonlinearity and Chaos in Engineering Dynamics
Edited by J. M. T. Thompson and S. R. Bishop,   © 1994 John Wiley & Sons Ltd

especially important in machines with reciprocating components which operate over a range of rotation speeds. These tuning arguments are based on the linearized system, and absorbers designed without considering nonlinear effects can fail miserably at moderate amplitudes of oscillation (Newland 1964; Sharif-Bakhtiar and Shaw 1992).

Modifications of the simple pendulum version of the CPVA are currently used which employ a bifilar pendulum configuration so that non-circular paths may be employed. This allows one to extend the effective range of operation by avoiding the bothersome nonlinear mistuning problems associated with the circular path of the simple pendulum (Denman 1992; Borowski *et al.* 1991). Some helicopter rotors utilize bifilar CPVAs with cycloidal paths for the absorber masses (Madden 1980), while a more recently developed design exploits the tautochronic property (i.e. constant period of oscillation, independent of amplitude) of an epicycloidal path in a centrifugal field (Denman 1985). The tautochronic absorber, which has been studied extensively by Denman, offers substantial improvements over circular paths (Denman 1992). However, a set of tautochronic absorbes tuned to a single order n is known to be not optimal, since higher-order harmonics are amplified through nonlinear effects and some residual of the primary harmonic (about 8%) persists. The higher harmonics can be dealt with by the addition of more absorbers tuned to higher orders (Borowski *et al.* 1991). It can be shown, however, that such an approach requires an infinite number of tautochronic absorbers to exactly cancel a torque which consists of a single harmonic.

In the present work we investigate an alternative configuration consisting of two absorbers which can exactly cancel a pure harmonic torque disturbance in the undamped case. This approach relies entirely on nonlinear aspects of the response, as the desired motion of the system corresponds to a subharmonic of order two in which the two absorbers, riding on half-order epicycloidal paths, move exactly out-of-phase with respect to one another. The basic idea is as follows: for an order-n torque disturbance the absorbers are taken to be tautochronic, but of order $n/2$ (this can be done for n even or odd). In this setting, the order-$(n/2)$ harmonic torques provided by the absorbers exactly cancel with each other, but their nonlinear components combine through quadratic terms so as to exactly balance the order-n harmonic applied torque, thus eliminating the torsional oscillations of the rotating inertia. This subharmonic solution is global in the sense that it exists up to a torque amplitude at which the absorber paths become singular, thus providing a maximum range of operation. In this work we focus on the basic system description and on the following two important issues: the effects of damping and the dynamic stability of the desired response.

As is typical with tuned absorbers, the level of damping in the absorber system must be kept small in order to achieve effective operation at the desired frequency, as the response of the primary system is not exactly quenched when absorber damping is present. In addition, in the present case absorber damping causes a delay in the onset of the desired subharmonic solution as the torque magnitude is increased from zero. A perturbation approach is used to investigate these effects for small levels of damping. The calculations are able to accurately predict the torque level at which the desired motion arises, the resulting response level and the dynamic stability near the onset of the subharmonic. The results are reminiscent of the saturation phenomenon which occurs in systems with two-to-one internal resonance (Nayfeh and Mook 1979). The analytical results and the effectiveness of the absorbers over a large torque range are confirmed using simulations of the fully nonlinear, three-degree-of-freedom system.

13.2   THE BASIC SYSTEM

The general system consists of a disc which is free to rotate in a plane about a fixed point and N absorber masses which move along arbitrary paths relative to the disc. Figure 13.1 shows the case for $N = 2$. The angular orientation of the disc is denoted by θ. The abosrber paths are taken to be symmetric about their vertex points, from which point an arc length variable S_i is used to specify the location of the ith absorber. The model includes viscous-type damping for the disc and each absorber. The external torque is assumed to be a combination of a constant component, T_0, and an oscillating component, $T_\theta(\theta)$. The constant torque balances the mean component of the torque which arises from the damping, thus setting the mean rotation speed of the disc, Ω, while the oscillating torque is the source of speed fluctuations. The governing equations of motion for this system are given by

$$m_i \left[\ddot{S}_i + \ddot{\theta} G_i(S_i) - \frac{1}{2} \dot{\theta}^2 \frac{dX_i}{dS_i}(S_i) \right] = - c_i \dot{S}_i, \quad i = 1, \ldots, N, \tag{1}$$

$$I_d \ddot{\theta} + \sum_{i=1}^{N} m_i \left[X_i(S_i) \ddot{\theta} + \frac{dX_i}{dS_i}(S_i) \dot{S}_i \dot{\theta} + G_i(S_i) \ddot{S}_i + \frac{dG_i}{dS_i}(S_i) \dot{S}_i^2 \right]$$

$$= \sum_{i=1}^{N} c_i G_i(S_i) \dot{S}_i - c_0 \dot{\theta} + T_0 + T_\theta(\theta), \tag{2}$$

where

$R_i(S_i)$ is the distance from the disc centre to the ith absorber position,

$X_i(S_i) = R_i^2(S_i)$,

$$G_i(S_i) = \sqrt{X_i(S_i) - \tfrac{1}{4} \left(\frac{dX_i}{dS_i}(S_i) \right)^2}$$

c_0, c_i are damping coefficients for the disc and for the ith absorber, respectively,

I_d is the moment of inertia of the disc,

m_i is the mass of the ith absorber.

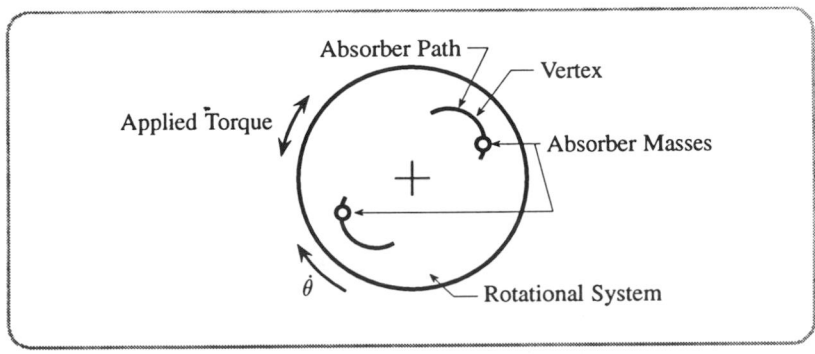

Figure 13.1   Schematic diagram of the basic system.

This system of equations can be transformed in such a manner that the oscillating part of the torque, which appears as a large nonlinearity in terms of the dependent variable θ, can be converted to an external excitation which is periodic in the *independent* variable. This is accomplished by making the reasonable assumption that the system rotates with $\dot{\theta}$ always positive, i.e. the disc never reverses direction. This allows one to convert the equations of motion via a change of variables to a form in which θ is the new independent variable. Since θ appears explicitly only in the applied torque, this plays the role of an external periodic excitation term in the new equations of motion. The transformation is realized by expressing t as a function of θ (the solution for this relationship is never actually required in the analysis). With the introduction of $\dot{\theta}$ as a new dependent variable, $\ddot{\theta}$ becomes $\dot{\theta}\dot{\theta}'$, $(\cdot)'$ d$\cdot$/dθ, thus reducing by one the order of the equations of motion. However, this transformation introduces more nonlinearities and, more importantly, renders the system non-autonomous.

We now specialize to the present case of interest. The oscillating part of the torque is taken to be a second-order harmonic, $T_\theta(\theta) = T_2 \sin(2\theta)$ (this represents the largest harmonic in a four-cylinder, in-line automotive engine, an application of interest). Two identical absorbers with first-order epicycloidal paths are to be used to counteract this torque: $N = 2$, $m_a = m_1 = m_2$, $c_a = c_1 = c_2$, $X_i(S_i) = R_0^2 - S_i^2$, where R_0 is the distance from the disc centre to the vertices of the absorbers. The transformation to the new independent variable θ is then carried out. Next, due to a special symmetry in the $N = 2$ version of the equations, it is convenient to define the following dimensionless coordinates which replace S_1 and S_2: $\xi = (S_1 - S_2)/R_0$ and $\eta = (S_1 + S_2)/R_0$. This linear transformation puts the linear part of the system in normal form. After these transformations and nondimensionalization, the equations of the motion, (1)–(2), become

$$y'\xi' + y\xi'' + y'(h_+ - h_-) + y\xi = -\hat{\mu}_a \xi', \tag{3}$$

$$y'\eta' + y\eta'' + y'(h_+ + h_-) + y\eta = -\hat{\mu}_a \eta', \tag{4}$$

$$\left(1 + 2 - \frac{\eta^2 + \xi^2}{2}\right) yy' - y^2(\eta\eta' + \xi\xi') + \frac{y}{2}[y'(\eta' + \xi') + y(\eta'' + \xi'')]h_+$$

$$+ \frac{y}{2}[y'(\eta' - \xi') + y(\eta'' - \xi'')]h_-$$

$$- \frac{y^2(\eta + \xi)(\eta' + \xi')^2}{4h_+} - \frac{y^2(\eta - \xi)(\eta' - \xi')^2}{4h_-}$$

$$= \frac{\hat{\mu}_a y}{2}[(\eta' + \xi')h_+ + (\eta' - \xi')h_-] - \hat{\mu}_\theta y + \hat{\Gamma}_0 - \hat{\Gamma}_2 \sin 2\theta, \tag{5}$$

where $y = \dot{\theta}/\Omega$, $h_+ = \sqrt{1 - \frac{1}{2}(\eta + \xi)^2}$, $h_- = \sqrt{1 - \frac{1}{2}(\eta - \xi)^2}$, $\hat{\mu}_a = c_a/m_a\Omega$, $I = I_d/m_a R_0^2$, $\hat{\mu}_\theta = c_0/m_a\Omega R_0^2$, $\hat{\Gamma}_0 = T_0/m_a\Omega^2 R_0^2$, and $\hat{\Gamma}_2 = T_2/m_0\Omega^2 R_0^2$.

For zero absorber damping, $\hat{\mu}_a = 0$, and a balance between the disk damping torque and the constant applied torque, $\hat{\mu}_\theta = \hat{\Gamma}_0$, the nondimensionalized equations have a *global* nonlinear solution which is given by

$$\xi = \sqrt{2\hat{\Gamma}_2} \sin\theta, \quad \eta = 0, \quad y = 1. \tag{6}$$

It is very important to note the features of this solution, since it represents the basis for the current design. First, we note that this solution is a subharmonic of order two, since its frequency is 1 while the external torque is of frequency 2. During this motion the absorbers move exactly out-of-phase with respect to each other ($S_1 = -S_2$), thus exactly canceling each other's first-order harmonics. However, the second-order harmonics generated by quadratic nonlinearities add in such a manner so as to exactly cancel the applied torque. This renders the disc rotation speed as a constant over a range of torque amplitudes up to $\hat{\Gamma}_2 = 1$, at which point the absorber masses reach cusps in the epicycloidal paths during their motions (where the G_i's become zero). A numerical solution of the Floquet multipliers associated with this solution shows that it is stable over the desired torque range. However, in order for this system to be of engineering use, its effectiveness must be maintained in the face of small amounts of absorber damping. This issue is tackled next.

13.3 PERTURBATION ANALYSIS

An exact, global solution of the basic system is given in the previous section for the case when $\hat{\mu}_a = 0$. However, the corresponding solution for $\hat{\mu}_a \neq 0$ is not available in closed form. Therefore, the method of multiple scales (MMS) is used to obtain a first-order approximation of this solution for $0 < \hat{\mu}_a \ll 1$ and $0 < \hat{\Gamma}_2 \ll 1$.

Let

$$\xi = \varepsilon\xi_1 + \varepsilon^2\xi_2 + \varepsilon^3\xi_3 + \cdots, \tag{7}$$

$$\eta = \varepsilon\eta_1 + \varepsilon^2\eta_2 + \varepsilon^3\eta_3 + \cdots, \tag{8}$$

$$y = 1 + \varepsilon y_1 + \varepsilon^2 y_2 + \varepsilon^3 y_3 + \cdots, \tag{9}$$

$$\frac{d}{d\theta} = D_0 + \varepsilon D_1 + \cdots, \tag{10}$$

where

$$D_p = \frac{\partial}{\partial\Theta_p}, \quad \Theta_p = \varepsilon^p\theta, \quad p = 0, 1, \ldots$$

In order to match orders for the perturbation analysis, the system parameters are scaled as follows:

$$\hat{\mu}_a = \varepsilon^2\mu_a, \quad \hat{\mu}_\theta = \varepsilon^2\mu_\theta, \quad \hat{\Gamma}_0 = \varepsilon^2\Gamma_0, \quad \hat{\Gamma}_2 = \varepsilon^2\Gamma_2.$$

These relationships are substituted into (3)–(5) and terms of order ε^p are gathered together for $p = 1, 2, 3$. The results are as follows.

First-Order, $O(\varepsilon)$:

$$D_0^2\xi_1 + \xi_1 = 0, \tag{11}$$

$$ID_0^2\eta_1 + (I + 2)\eta_1 = 0, \tag{12}$$

$$ID_0 y_1 = \eta_1. \tag{13}$$

Second-Order, $O(\varepsilon^2)$:

$$D_0^2\xi_2 + \xi_2 = -(y_1 + D_0y_1D_0 + 2D_0D_1 + y_1D_0^2)\xi_1, \tag{14}$$

$$ID_0^2\eta_2 + (I+2)\eta_2 = 2(\mu_\theta - \Gamma_0) + 2\Gamma_2\sin(2\Theta_0) - 2(\xi_1D_0\xi_1 + \eta_1D_0\eta_1)$$

$$+ 2(I+2)y_1D_0y_1 - [(I+2)y_1 + ID_0y_1D_0 + 2ID_0D_1 + (I-2)y_1D_0^2]\eta_1, \tag{15}$$

$$ID_0y_2 = -(\mu_\theta - \Gamma_0) - \Gamma_2\sin(2\Theta_0) + \eta_2 + (y_1 - y_1D_0^2)\eta_1 - ID_1y_1 + \xi_1D_0\xi_1$$

$$+ \eta_1D_0\eta_1 - (I+2)y_1D_0y_1. \tag{16}$$

Third-Order, $O(\varepsilon^3)$:

$$D_0^2\xi_3 + \xi_3 = -y_1\xi_2 - y_2\xi_1 - D_1^2\xi_1 - \mu_a D_0\xi_1 - D_1y_1D_0\xi_1 + D_0y_1\xi_1\eta_1$$

$$- D_0y_1D_1\xi_1 - D_0y_1D_0\xi_2 - D_0y_2D_0\xi_1 - 2D_0D_2\xi_1 - 2y_1D_0D_1\xi_1 - 2D_0D_1\xi_2$$

$$- y_2D_0^2\xi_1 - y_1D_0^2\xi_2, \tag{17}$$

$$ID_0^2\eta_3 + (I+2)\eta_3 = 2\mu_\theta y_1 - (I+2)(y_1\eta_2 + y_2\eta_1) - 2\xi_1D_1\xi_1 - 2\eta_1D_1\eta_1$$

$$+ 2(I+2)y_1D_1y_1 - ID_1^2\eta_1 - 2\xi_2D_0\xi_1 - 4y_1\xi_1D_0\xi_1 - (D_0\xi_1)^2\eta_1$$

$$- 2\xi_1D_0\xi_2 - (I+4)\mu_a D_0\eta_1 - (2\eta_2 + 4y_1\eta_1 + ID_1y_1 + 2\xi_1D_0\xi_1)D_0\eta_1$$

$$- \eta_1(D_0\eta_1)^2 - 2\eta_1D_0\eta_2\frac{I}{2}D_0y_1(\xi_1^2 + \eta_1^2)$$

$$+ D_0y_1[2(I+2)y_2 - ID_1\eta_1 + 2y_1D_0\eta_1 - ID_0\eta_2]$$

$$+ 2(I+2)y_1D_0y_2 - ID_0y_2D_0\eta_1 - 2ID_0D_2\eta_1 + (4-2I)y_1D_0D_1\eta_1$$

$$- 2ID_0D_1\eta_2 - \xi_1D_0^2\xi_1\eta_1 + [-\tfrac{1}{2}(\xi_1^2 + \eta_1^2) + 2y_1^2 + 2y_2 - Iy_2]D_0^2\eta_1$$

$$+ (2-I)y_1D_0^2\eta_2. \tag{18}$$

It should be noted that the equation for y is not shown at $O(\varepsilon^3)$ because it does not produce any secular terms.

Solving the equations of $O(\varepsilon)$, (11)–(13), gives:

$$\xi_1 = A_1 e^{j\Theta_0} + \text{c.c.}, \tag{19}$$

$$\eta_1 = B_1 e^{j\omega\Theta_0} + \text{c.c.}, \tag{20}$$

$$y_1 = -\frac{jB_1}{I\omega} e^{j\omega\Theta_0} + \text{c.c.}, \tag{21}$$

where $A_1 = A_1(\Theta_1, \Theta_2)$, $B_1 = B_1(\Theta_1, \Theta_2)$, $\omega = \sqrt{[(I+2)/I]}$, $j = \sqrt{-1}$, and c.c. represents the complex conjugate of the preceding terms. Substituting these results into the equations for $O(\varepsilon^2)$, (14)–(16), results in terms on the right-hand side which yield the following conditions for the removal of secular terms:

$$D_1A_1 = 0, \tag{22}$$

$$D_1 B_1 = 0. \tag{23}$$

Therefore, A_1 and B_1 are functions of Θ_2 only, i.e., $A_1 = A_1(\Theta_2)$, $B_1 = B_1(\Theta_2)$. Also note that the condition $\mu_\theta = \Gamma_0$ must hold in (16) in order for y_2 to oscillate about zero. The particular solutions of ξ_2, η_2 and y_2 are then given by:

$$\xi_2 = \frac{j}{I(2\omega + 1) + 2} A_1 B_1 e^{j(\omega + 1)\Theta_0} + \frac{j}{I(2\omega - 1) - 2} \bar{A}_1 B_1 e^{j(\omega - 1)\Theta_0} + \text{c.c.}, \tag{24}$$

$$\eta_2 = \frac{j}{3I - 2}(\Gamma_2 + 2A_1^2)e^{2j\Theta_0} + \frac{j\omega}{I + 2} B_1^2 e^{2j\omega\Theta_0} + \text{c.c.}, \tag{25}$$

$$y_2 = \frac{3}{4(3I - 2)}(\Gamma_2 + 2A_1^2)e^{2j\Theta_0} + \frac{1}{2I} B_1^2 e^{2j\omega\Theta_0} + \text{c.c.} \tag{26}$$

With the solutions of ξ_1, η_1, y_1, ξ_2, η_2 and y_2 in hand, the equations of $O(\varepsilon^3)$, (17) and (18), lead to the following condition for the removal of secular terms,

$$2I\omega D_2 B_1 = -\left[\frac{2\mu_\theta}{I\omega} + (I + 4)\mu_a\omega + j\frac{2(I + 1)A_1\bar{A}_1}{I} + j\frac{(I + 2)B_1\bar{B}_1}{I} \right] B_1. \tag{27}$$

Here the solution $B_1 = 0$ (and thus $\eta_1 = 0$) is the only possible steady-state solution, and it is stable, since the real part of the term inside the square brackers is positive for any non-zero level of damping for either the absorber or the disc.

Substituting $\eta_1 = 0$ into (17) results in secular terms which are eliminated if the following condition holds:

$$2D_2 A_1 = j\frac{3\Gamma_2\bar{A}_1}{2(3I - 2)} - \mu_a A_1 + j\frac{3A_1^2\bar{A}_1}{3I - 2}. \tag{28}$$

Defining $A_1 = \frac{1}{2}ae^{j\phi}$, where $a = a(\Theta_2)$, $\phi = \phi(\Theta_2)$, substituting these into (28), and separating the result into real and imaginary parts, results in the following equations for a and ϕ:

$$a' = \frac{3\Gamma_2 a}{4(3I - 2)} \sin 2\phi - \frac{\mu_a a}{2}, \tag{29}$$

$$a\phi' = \frac{3a}{8(3I - 2)}(2\Gamma_2 \cos 2\phi + a^2). \tag{30}$$

There are two possible steady-state solutions for a and ϕ: one is $a = a_0 = 0$ and the other is

$$a = a^* = 2\sqrt[4]{\frac{\Gamma_2^2}{4} - \frac{\mu_a^2(3I - 2)^2}{9}}, \quad \phi = \phi^* = \frac{1}{2}\left\{ \pi - \sin^{-1}\left[\frac{2\mu_a(3I - 2)}{3\Gamma_2} \right] \right\}.$$

Note that the second solution exists only for $\Gamma_2 > \Gamma_2^* = 2(3I - 2)\mu_a/3$. It is seen that a^* bifurcates from the trivial solution at a torque level corresponding to $\Gamma_2 = \Gamma_2^*$. In terms of the original system parameters, the corresponding value of T_2 is $T_2^* = 2\Omega(3I_d - 2m_a R_0^2)c_a/(3m_a)$.

The stability of these solutions can be determined by linearizing (29) and (30) about the respective steady-state solutions and checking the real parts of the resulting eigenvalues. This shows that a_0 is stable for $\Gamma_2 < \Gamma_2^*$, while a^* is stable when it exists.

The a_0 solution corresponds to the response of the linearized system; this can be seen by substituting $A_1 = B_1 = 0$ into (24)–(26), which results in a solution which is the same as that of the linearized version of (3)–(5). The presence of absorber damping renders this solution stable for small torque amplitudes. As the torque amplitude is increased, the system reaches a point where nonlinear effects begin to dominate the damping effects, and the system bifurcates into the stable motion represented by the amplitude a^*. This is the small damping continuation of the global solution given in (6).

In order to determine the effects that small levels of absorber damping have on the rotational dynamics of the disc, we consider the level of angular acceleration, $\ddot{\theta}$, which results from this solution. A non-dimensional measure of the angular acceleration is given by

$$\alpha = \frac{\ddot{\theta}}{\Omega^2} = yy' = \varepsilon^2 y_2' + O(\varepsilon^4). \tag{31}$$

The pre- and post-bifurcation amplitudes of oscillation of the angular acceleration are determined by substituting the steady-state solutions of A_1 into (26). This yields the following amplitudes of oscillation for α, denoted as α_a,

$$\alpha_a = \begin{cases} \dfrac{3\varepsilon^2 \Gamma_2}{3I-2} + O(\varepsilon^4), & \text{for } \Gamma_2 < \Gamma_2^*, \\[2mm] 2\varepsilon^2 \mu_a + O(\varepsilon^4), & \text{for } \Gamma_2 > \Gamma_2^*. \end{cases} \tag{32}$$

It is seen that α_a increases linearly with Γ_2 up to the bifurcation point, after which it saturates at the level $\alpha_a^* = 2\hat{\mu}_a$. It should be noted that for $T_2 < T_2^*$, α_a is greater than that corresponding to the same system with the absorber locked at their respective vertices. This implies that the absorbers do not function as desired at small torque levels in the damped case. Also, in the pre-bifurcation stage, $\xi = O(\varepsilon^2)$ and $\eta = O(\varepsilon^3)$ which implies that η, the nondimensionalized sum of S_1 and S_2, dominates ξ, their difference, and this results in a roughly 'in-phase' motion of the absorbers. However, in the post-bifurcation stage, $\xi = O(\varepsilon)$ while η remains at $O(\varepsilon^2)$, making $(S_1 - S_2)$ the dominant term. Thus, the absorber move roughly 'out-of-phase' with respect to each other. It is thus seen that the post-bifurcation solution is entirely consistent with the global solution in the $\mu_a \to 0$ limit.

From the design point of view, it is desired to have an early bifurcation point and a low saturation level. Therefore, it is always preferable to have the absorber damping as small as possible. The other absorber parameters, m_a and R_0, are chosen according to the geometry of the application at hand, and their influence on the bifurcation point and the saturation level must also be taken into account.

13.4   NUMERICAL SIMULATIONS

The results from numerical simulations and their comparison with the MMS calculations are presented in this section. The parameters used are: $I = 12.034$ (chosen for a particular

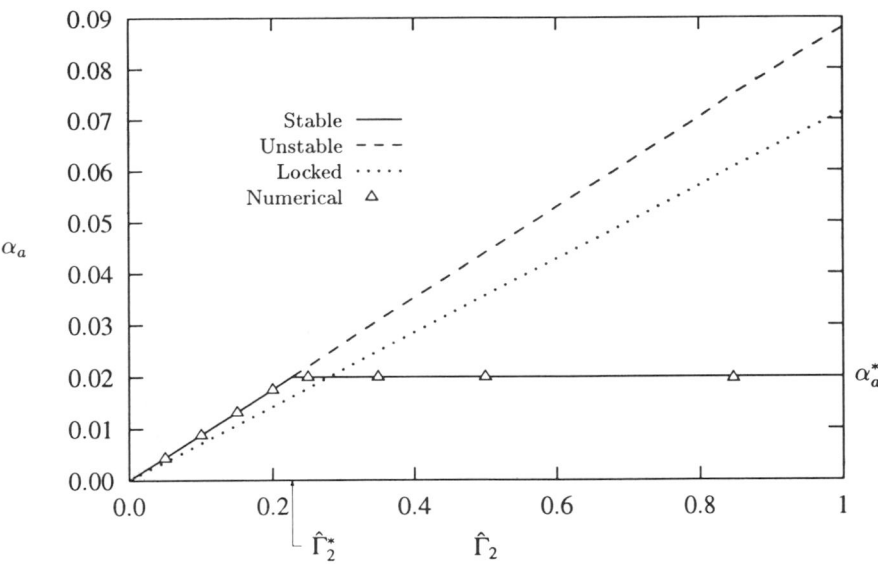

Figure 13.2 α_a as a function of the excitation amplitude; $I = 12.034$, $\hat{\mu}_a = 0.01$, $\hat{\mu}_\theta = 0.05$.

application), $\hat{\mu}_a = 0.01$ and $\hat{\mu}_\theta = 0.05$. Figure 13.2 shows that α_a increases linearly with $\hat{\Gamma}_2$, then saturates at α_a^* for $\hat{\Gamma}_2 > \hat{\Gamma}_2^*$. The case with the absorbers locked to the disc is also shown in Figure 13.2. It can be seen that the subharmonic absorbers become effective at a torque level at which the nonlinearity dominates the system response. Figure 13.3 shows the corresponding response amplitude for the 'out-of-phase' component of the absorbers.

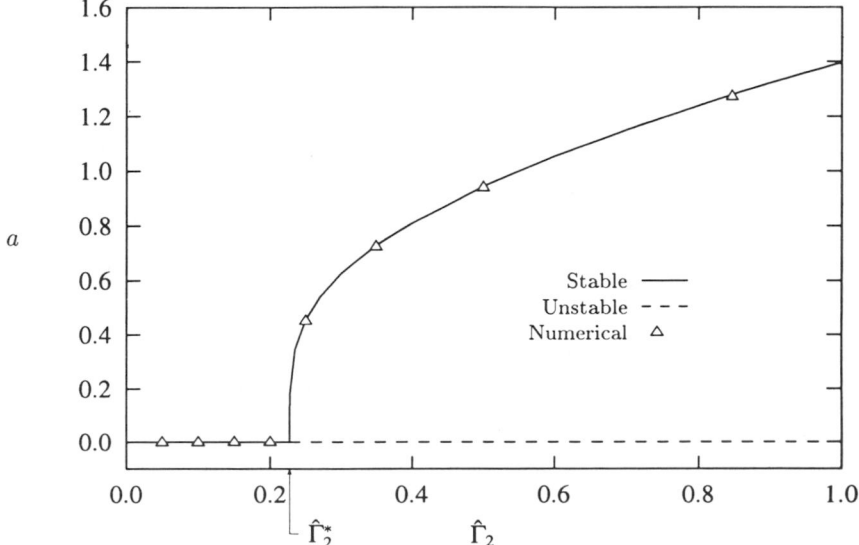

Figure 13.3 a as a function of the excitation amplitude; $I = 12.034$, $\hat{\mu}_a = 0.01$, $\hat{\mu}_\theta = 0.05$.

13.5   CONCLUSIONS

The proposed absorber configuration is able to achieve a dramatic improvement over the systems previously considered. It is able to exactly eliminate a single harmonic torque component with a finite number of absorbers (two) in the undamped case. It is also interesting to note that this system is not conceivable using extensions of the linear theory since the tuning is done using a subharmonic response. In addition, the fact that the desired response exists and is stable up to large torque levels allows for more flexibility in the system design, since one can use relatively small absorber masses to achieve the same level of performance.

Extensions of this work are under way in which we consider more realistic applied torques, in particular those of the form $T(\theta, \dot{\theta}) = T_1(\theta) + \dot{\theta}^2 T_2(\theta)$, where T_1 and T_2 are periodic in θ (this form arises quite naturally in reciprocating machinery). The main issue for this situation is quite simple to state: can one employ pairs of subharmonic absorbers to cancel corresponding individual harmonics of the torque? The equations of motion indicate that this is indeed the case, but the stability issues will be more delicate in this case. Also, residual effects will be important to consider, as the torque typically has an infinite number of harmonics, and only the dominant ones will be addressed with absorbers. Applications using specific automotive engine models are also under consideration.

ACKNOWLEDGEMENTS

This work was partially supported by grants from the US National Science Foundation. The authors are grateful to Professor Vincent T. Coppola and Dr Chrishtopher L. Lee of the University of Michigan for several suggestions which were helpful with the perturbation analysis.

REFERENCES

Borowski, V. J., Denman, H. H., Cronin, D. L., Shaw, S. W., Hanisko, J. P., Brooks, L. T., Mikulec, D. A., Crum, W. B. and Anderson, M. P. (1991) Reducing vibration of reciprocating engines with crankshaft pendulum absorbers, *SAE Computer Aided Design, Analysis and Simulation of Off-Highway Equipment* (*SP-884*), 73–79 (also *SAE Technical Paper Series*, #911876).

Denman, H. H. (1985) Remarks on Brachistochrone–Tautochrone problems, *American J. of Physics*, **53**, 781–782.

Denman, H. H. (1992) Tautochronic bifilar pendulum torsion absorbers for reciprocating engines, *J. Sound Vib.*, **159**, 251–277.

Madden, J. F. (1980) Constant frequency bifilar vibration absorber, US Patent No. 42 181 87.

Nayfeh, A. H. and Mook, D. T. (1979) *Nonlinear Oscillations*, New York, Wiley-Interscience.

Newland, D. E. (1964) Nonlinear aspects of the performance of centrifugal pendulum vibration absorbers, *J. Eng. Ind.*, **86**, 257–263.

Sharif-Bakhtiar, M. and Shaw, S. W. (1992) Effects of nonlinearities and damping on the dynamic response of a centrifugal pendulum vibration absorber, *J. Vib. Acoust.*, **114**, 305–311.

S. W. Shaw and C.-T. Lee, *Department of Mechanical Engineering and Applied Mechanics, University of Michigan, Ann Arbor, MI 48109, USA*

$\mathbf{14}$ CHAOS IN RAILWAY-VEHICLE DYNAMICS

E. Slivsgaard and H. True

We present the latest results of the analysis of the dynamics of a rolling wheelset that was started by Knudsen *et al.* (1992). For several speed ranges the wheelset oscillates chaotically, and we have selected one of these for a more detailed analysis of the periodic windows. A second-order polynomial fit is made with the well-known logistic map, and we demonstrate that it is thereby possible to determine the positions of periodic windows in the selected chaotic region with great accuracy.

14.1 INTRODUCTION

Above a certain critical speed, V_0, a rolling railway wheelset oscillates laterally on the track. In a mathematical model of the motion the oscillations are found to be related to a bifurcation. This bifurcation is either a supercritical Hopf bifurcation or a saddle–node bifurcation in connection with a subcritical Hopf bifurcation (see e.g. True 1992). At higher speeds subsequent bifurcations often lead to speed ranges of chaotic motion. In this chapter we shall examine in some detail two of the chaotic domains found by Knudsen *et al.* (1992). A description of bifurcations and chaos is presented in several textbooks. As one example of many we refer the interested reader to Guckenheimer and Holmes (1983).

The important nonlinearity in our problem arises from the contact forces in the wheel–rail contact surface. We shall assume that the wheel is conical and tapered away from the track centreline with a flange inside the track. The slope λ of the cone is 1:20. The rail heads are an arc of a circle with $r_0 = 0.4572$ m. Therefore the kinematic relations expressing the fact that the wheels and rails remain in contact are very well approximated by linear relations unless a flange hits a rail.

We adopt the model of wheel–rail contact forces described by Vermeulen and Johnson (1964). It is assumed that a finite slip velocity between the wheel and the rail is generated in the contact region. It is called the *creep*. The creep normalized by the forward speed of the car is called the *creepage*. The creep generates shear forces through friction, called *creep forces*, which have a longitudinal as well as a lateral component in the contact surface. Hertz's theory is used to calculate the contact area between a wheel and a rail, and with normal

---

Nonlinearity and Chaos in Engineering Dynamics
Edited by J. M. T. Thompson and S. R. Bishop,   © 1994 John Wiley & Sons Ltd

force N and coefficient of adhesion $\mu = 0.15$, $\mu N = 10 \, \text{kN}$. G is the shear modulus, and with a_e and b_e as the semi-axes of the contact ellipse, $G \pi a_e b_e = 65.63 \, \text{kN}$. The resultant creep force F_R is then found from

$$F_R = \mu N \begin{Bmatrix} u - \frac{1}{3}u^2 + \frac{1}{27}u^3, & u < 3, \\ 1, & u \geqslant 3, \end{Bmatrix} u = \frac{G \pi a_e b_e}{\mu N} \xi_R, \tag{1}$$

where the resultant creepage ξ_R is given by

$$\xi_R = \sqrt{(\xi_x/\Psi)^2 + (\xi_y/\Phi)^2}. \tag{2}$$

From Vermeulen and Johnson (1964) we find $\Psi = 0.542\,19$ and $\Phi = 0.602\,52$.

14.2   THE MECHANICAL MODEL

We consider the rolling motion of a suspended railway wheelset with constant speed on a straight, horizontal and 'ideal' track. The wheelset can roll without friction in a frame, which can rotate frictionless around a pivot in the floor of the carbody. The lateral motion of the wheelset is restrained by two springs with a linear characteristic, $k_1 = 18\,230 \, \text{N/m}$, as shown in figure 14.1. The mass m of the wheelset is $1022 \, \text{kg}$, the moment of inertia I is $678 \, \text{kg m}^2$, and half the track gauge b is $0.716 \, \text{m}$.

We study a model in which the coupling between the horizontal and vertical motions can be neglected. This assumption holds true in many railway-vehicle constructions. We have therefore not shown any vertical suspension. We formulate the dynamic equations of motion in a coordinate system, which moves with the constant forward speed, V, of the car along the track centreline.

The model has two degrees of freedom: lateral motion, x, and yaw motion, φ. The two equations of motion are coupled through the nonlinear creep-force terms in the equations. The flange force F_T is modelled by a very stiff linear spring, $k_0 = 14.6 \, \text{MN/m}$, with a dead band $\delta = 9.1 \, \text{mm}$.

Using Newton's equations the mathematical model becomes:

$$m \, d^2 x/dt^2 + 2k_1 x + 2F_x + F_T(x) = 0,$$
$$I d^2 \varphi/dt^2 + 2bF_y = 0, \tag{3}$$

where

$$\xi_x = \dot{x}/V - \varphi, \qquad \xi_y = b\dot{\varphi}/V + \lambda x/r_0,$$
$$F_x = (\xi_x/\Psi) F_R/\xi_R, \qquad F_y = (\xi_y/\Phi) F_R/\xi_R,$$

$$F_T(x) = \begin{cases} k_0(x - \delta), & \delta < x, \\ 0, & -\delta \leqslant x \leqslant \delta, \\ k_0(x + \delta), & x < -\delta. \end{cases} \tag{4}$$

With the substitutions $x_1 = x$, $x_2 = dx/dt$, $x_3 = \varphi$ and $x_4 = d\varphi/dt$ we have formulated a parameter-dependent nonlinear dynamical problem in $\mathbb{R}^4 \times \mathbb{R}_+$.

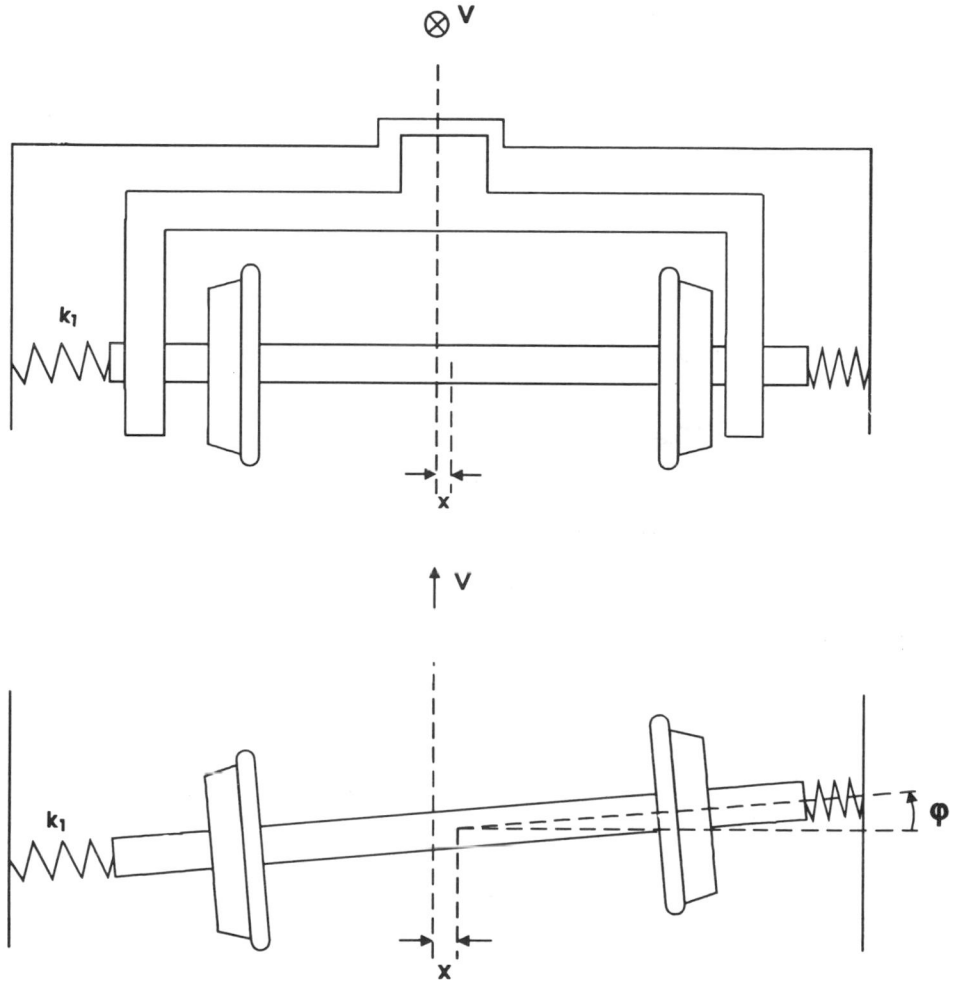

Figure 14.1 Front and top views of the wheelset model.

14.3 METHOD OF INVESTIGATION AND RESULTS

We study the dynamics of the wheelset model using the wheelset velocity, V, as the control parameter. In the numerical investigation we apply the software package PATH to follow stationary or periodic solutions dependent on the speed V. PATH was developed by Kaas-Petersen (1989) and uses the integration routine LSODA by Petzold (1983). To describe the lateral motion we choose the Poincaré section with $x_2 = 0$ and x_1 attaining its maximum value, so the Poincaré section illustrates the maximum value of the lateral motion. The results are presented as one-dimensional bifurcation diagrams.

The stationary solution remains stable, until the speed of the wheelset reaches $V_0 = 10.050$ m/s and the system undergoes a supercritical bifurcation, whereby a time-periodic solution is generated. The limit cycle grows rapidly in amplitude and at $V = 10.0555$ m/s a

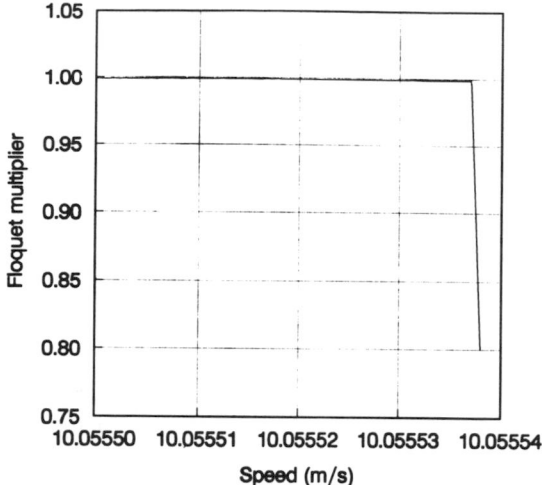

Figure 14.2 The maximum Floquet multiplier as a function of the wheelset speed just around the speed where the flange of a wheel hits the rail for the first time.

so-called 'boundary-collision' occurs, when the amplitude has become so large that the flange of the wheelset hits the rail. The elastic flange generates a lateral restoring force.

Figure 14.2 shows the maximum Floquet multiplier as a function of the speed. At speeds where there is no flange contact the Floquet multiplier has a value just below $+1$, but when flange contact occurs the value suddenly decreases. This is evidence of the expected stabilization of the motion by the flanges. When the speed grows through $V = 10.0559$ m/s a Floquet multiplier leaves the unit disc through $+1$, and a symmetry-breaking pitchfork bifurcation takes place.

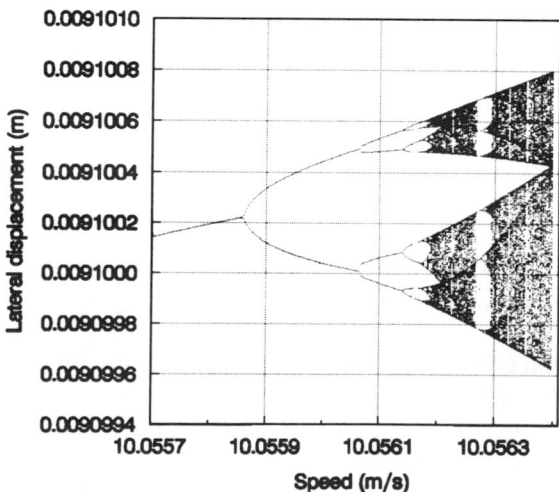

Figure 14.3 Bifurcation diagram showing the maximum value of the lateral displacement versus the speed, in the range 10.0557 m/s $< V < 10.0564$ m/s.

A bifurcation diagram for the velocity range 10.0557 m/s $< V <$ 10.0564 m/s is shown in Figure 14.3. The figure reveals the transition to chaos: first a symmetry-breaking pitchfork bifurcation creates two asymmetric limit cycles. Each of these undergoes a complete period-doubling cascade to chaos. Note, that the first period-doubling for the 'low' asymmetric limit cycle takes place exactly at the value of the speed where the flange contact of one of the rails would have ended. The two asymmetric chaotic bands change to symmetric chaos at $V \approx$ 10.0564 m/s, or in other words, the asymmetric limit cycles merge into one symmetric

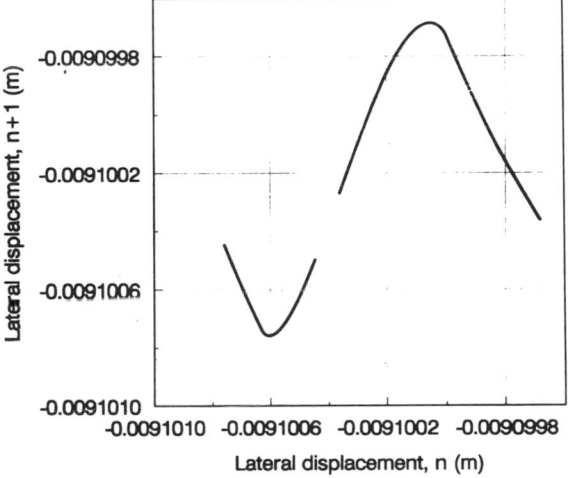

Figure 14.4 A return map for two asymmetric limit cycles at $V =$ 10.056 39 m/s. The figure shows the minimum value of the lateral displacement in the $(n + 1)$th intersection of a chaotic trajectory with the Poincaré section against the value in nth intersection.

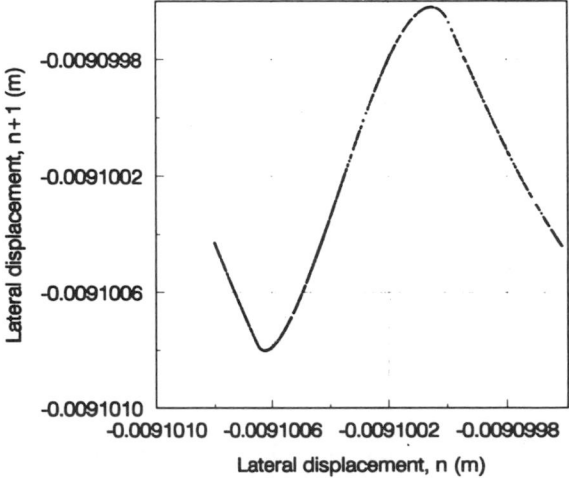

Figure 14.5 A return map for one symmetric limit cycle at $V =$ 10.056 40 m/s. The figure shows the minimum value of the lateral displacement in the $(n + 1)$th intersection of a chaotic trajectory with the Poincaré section against the value in the nth intersection.

limit cycle. The merger is illustrated in Figures 14.4 and 14.5. They show return maps of the lateral displacement for speeds just before and just after the merger. An investigation of the period-doubling cascade is presented in Section 14.4.

Figure 14.6 shows a bifurcation diagram of the first chaotic region. A crisis as described by Grebogi *et al.* (1982) is evident at $V \approx 10.18$ m/s, where three-band chaos is generated. Through increasing speed the bandwidths narrow down to a period-3 solution at $V \approx 10.20$ m/s. Notice how the centre band narrows onto flange contact. The periodic limit cycle undergoes a saddle–node bifurcation at $V \approx 10.21$ m/s, where another crisis creates large-scale chaos. At $V \approx 10.26$ m/s this chaotic attractor splits up into two asymmetric attractors. Each of them splits up into two-band chaos when the speed grows through 10.28 m/s. The four bands narrow down to two asymmetric periodic solutions which undergo a reverse period-doubling bifurcation at $V = 10.303$ m/s. As the speed is increased further the symmetry is restored at $V = 10.596$ m/s. The amplitude of the symmetric limit cycle grows monotonically with the speed.

An additional positive maximum is created at $V \approx 13$ m/s as illustrated by Knudsen *et al.* (1992). This new maximum reaches flange contact at $V = 13.81$ m/s and the second chaotic region develops. The two chaotic regions can be seen in Figure 14.7. A magnification of the low maximum on Figure 14.7 is given in Figure 14.8. The figures reveal that the second and first chaotic regions are very similar. We find that the transition to chaos for the second region is through a symmetry-breaking pitchfork bifurcation at $V = 13.813$ m/s succeeded by a complete period-doubling cascade. Figure 14.8 illustrates that a period-3 window is evident at $V = 14.35$ m/s and that the chaos disappears in two asymmetric period-2 solutions which undergo a reverse period-doubling at $V = 14.83$ m/s. Finally the symmetry is restored at $V = 17.23$ m/s.

A more detailed investigation of the two chaotic regions is given in Knudsen *et al.* (1994).

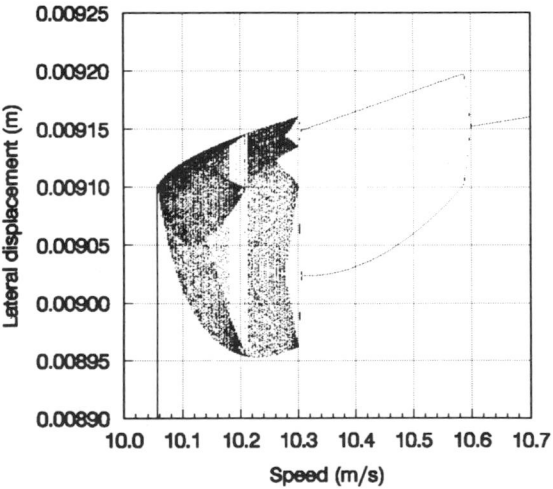

Figure 14.6 Bifurcation diagram showing the maximum value of the lateral displacement versus the speed, in the range 10.0 m/s $< V < 10.7$ m/s.

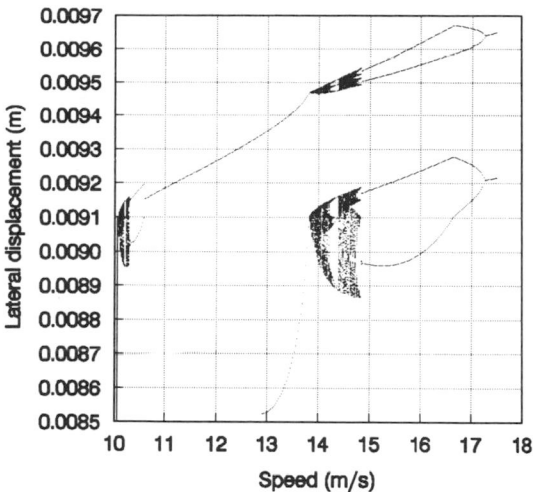

Figure 14.7 Bifurcation diagram showing the maximum value of the lateral displacement versus the speed, in the range $10.0\,\text{m/s} < V < 17.5\,\text{m/s}$. The two chaotic regions are shown.

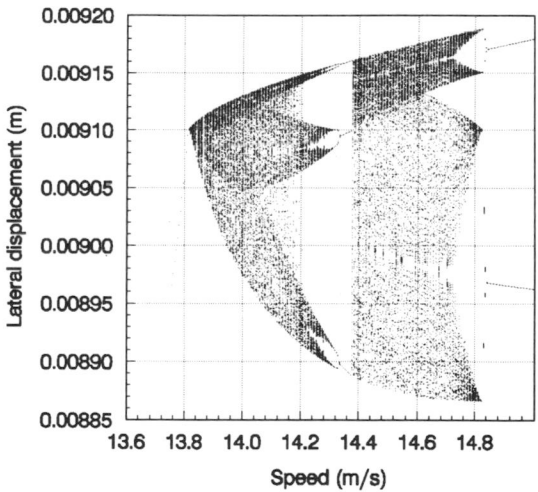

Figure 14.8 A magnification of the low maximum of Figure 15.5. The figure is a bifurcation diagram showing the maximum value of the lateral displacement versus the speed, in the range $13.6\,\text{m/s} < V < 15.0\,\text{m/s}$.

14.4 A COMPARISON WITH THE LOGISTIC MAP

Even though we have a 'boundary-collision' caused by the impact of the flange, we find that the first period-doubling cascade to chaos is characteristic with its many visible periodic windows. We therefore compare this period-doubling cascade with that of the logistic map, $x_{n+1} = ax_n(1 - x_n)$. To do this we choose some specific values of the two control parameters for the two systems, V and a: namely, the first three period-doublings, the value

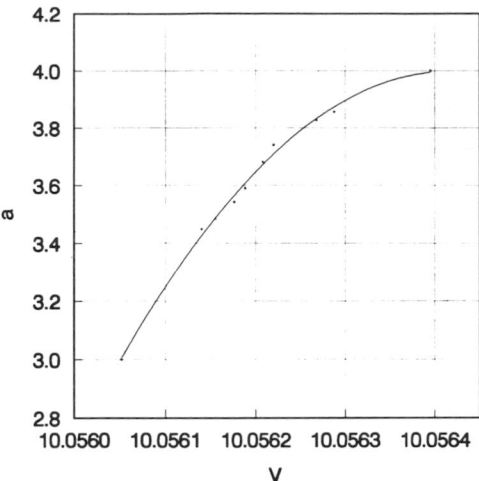

Figure 14.9 A second-order polynomial fit between the two control parameters *a* for the logistic map and *V* for the wheelset model. The dots indicate the fitting points.

where the four chaotic bands merge into two bands, the value where these bands merge into one, the period-5 window, the values where the period-3 window begins and ends respectively, and finally the value of the speed where the two asymmetric chaotic regions merge into symmetric chaos which corresponds to $a = 4$. When we make a second-order polynomial fit we get the following relation:

$$a(V) = -7\,434\,254.581 \cdot V^2 + 149\,523\,936.575 \cdot V - 751\,837\,566.439. \qquad (5)$$

Figure 14.9 shows both the second-order polynomial and the nine fitting points. A good accord is reached, and we say that we have a function which links the two systems for $10.056\,05$ m/s $\leqslant V \leqslant 10.056\,40$ m/s and $3.0 \leqslant a \leqslant 4.0$.

The logistic map has been investigated thoroughly in many articles. Using symbolic dynamics Metropolis *et al.* (1971) found an ordering of the periodic windows in the chaotic region after the period-doubling cascade. This ordering lists the values of the control parameter, *a*, for which certain superstable periodic orbits exist.

With the relation (5) it is now possible to determine the periodic windows in the chaotic region for the wheelset under consideration. We will show some examples:

Example 1 For the logistic map there is a period-4 window with the sequence RL^2 for $a = 3.96$. Solving (5) with this value of *a* yields $V = 10.056\,34$ m/s. As a check we have solved the equations of motion and found the speed at which this period-4 window exists, and the difference between the two values of *V* is 2.5 %. Furthermore a return map of the minimum value of the lateral displacement for an asymmetric attractor in the chaotic region examined shows a unimodal map with maximum just before the flange (see Figure 14.10). This indicates that the wheelset for $V = 10.056\,34$ m/s will oscillate in an asymmetric periodic motion with flange contact on one wheel and on the other wheel the following oscillation sequence: three times flange contact and one time no flange contact.

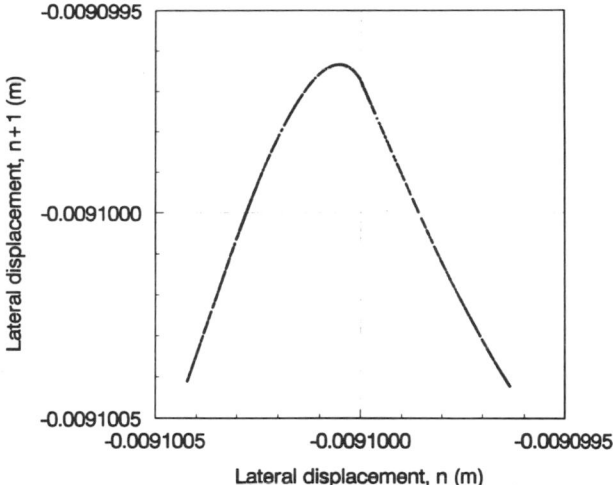

Figure 14.10 A return map for an asymmetric limit cycle at $V = 10.056\,39$ m/s. The figure shows the minimum value of the lateral displacement in the $(n+1)$th intersection of a chaotic trajectory with the Poincaré section against the value in the nth intersection. This figure is the same as the right-hand one in Figure 14.4.

Example 2 For $a = 3.99$ there exists a period-5 solution with the sequence RL^3. This corresponds to $V = 10.056\,38$ m/s. If we solve the equations of motion we find the difference between the two values of V to be 0.6%. Hereby a periodic oscillation is found with four times flange contact followed by one maximum without flange contact on one wheel and five times flange contact on the other wheel.

14.5   CONCLUSIONS

We have demonstrated the similarity between the ordering of periodic windows in a chaotic region for a specific example of a pair of unimodal maps. One of these is created through a discretization of a flow and the other is chosen to be the well-known logistic map. Bai-lin (1989) derives a recursion formula for the position of windows in the logistic map. Once obtained it applies to all maps of the same class – namely, unimodal maps.

The relation between the two systems has been approximated by a best fit of a second-order polynomial supported by nine points. Their coordinates are characteristic pairs of parameter values (e.g. bifurcations) for the logistic map and the map of the oscillating wheelset respectively. The relation makes it possible to determine the location of periodic windows in a chaotic region of the wheelset model accurately from the distribution of periodic windows in the logistic map.

ACKNOWLEDGEMENTS

The authors thank Carsten Knudsen for valuable discussions. A part of this work (H.T.) was supported by the Danish Council for Scientific and Industrial Reseach, grant no. 16-5786.M.

REFERENCES

Bai-lin, H. (1989) *Elementary Symbolic Dynamics*, Singapore, World Scientific.

Grebogi, C., Ott, E. and Yorke, J. A. (1982) Chaotic attractors in crisis, *Phys. Rev. Lett.*, **48**, 1507–1510.

Guckenheimer, J. and Holmes, P. (1983) *Nonlinear Oscillations, Dynamical Systems, and Bifurcations of Vector Fields*, New York, Springer Verlag.

Kaas-Petersen, C. (1989) *PATH – User's Guide*, Department of Applied Mathematical Studies and Centre for Nonlinear Studies, University of Leeds.

Knudsen, C., Feldberg, R. and True, H. (1992) Bifurcations and chaos in a model of a rolling railway wheelset, *Phil. Trans. R. Soc. London A*, **338**, 455–469.

Knudsen, C., Slivsgaard E., Rose, M., True, H. and Feldberg, R. (1994) Dynamics of a model of a railway wheelset, *Nonlinear Dynamics*, to appear.

Metropolis, N., Stein, M.L. and Stein, P.R. (1973) On finite limit sets for transformations on the unit interval, *Journal of Combinatorial Theory*, **15**, 25–44.

Petzold, L. (1983) Automatic selection of methods for solving stiff and nonstiff systems of ordinary differential equations, *SIAM J. Sci. Stat. Comput.*, **4**, 136–148.

True, H. (1992) Asymmetric hunting and chaos motion of railroad vehicles, *Proceedings of the 1992 ASME/IEEE Spring Joint Railroad Conference, Atlanta, Georgia, March 31–April 2*, pp. 25–40.

Vermeulen, P. J. and Johnson, K. L. (1964) Contact of non-spherical bodies transmitting tangential forces, *J. Appl. Mech.*, **31**, 338–340.

E. Slivsgaard, *Laboratory of Applied Mathematical Physics and MIDIT, The Technical University of Denmark, Bldg. 303, DK-2800 Lyngby, Denmark and The Danish State Railways, Pilestraede 58, DK-1112 Copenhagen, Denmark*

H. True, *Laboratory of Applied Mathematical Physics and MIDIT, The Technical University of Denmark, Bldg. 303, DK-2800, Lyngby, Denmark and ES-Consult, Staktoften 20, DK-2950 Vedbaek, Denmark*

15 REGULAR AND CHAOTIC ROTATIONS OF A SATELLITE IN SUNLIGHT FLUX

V. V. Beletsky and E. L. Starostin

Three-dimensional motion of a symmetrical satellite about its centre of mass under the solar radiation torques is considered. The satellite has an axially symmetrical solar stabilizer and a set of reflecting paddles arranged like a windmill. The equations in evolutionary variables are studied. The phase-space structure is investigated. By means of numerical computation of Poincaré maps, the phase trajectories are built. The regular (resonant and quasi-periodic), semi-regular (intermittent) and chaotic trajectories are distinguished. The evolution of phase portraits with change of parameters has been traced.

15.1 INTRODUCTION

Small Space Lab (SSL) is the basic spacecraft for the REGATTA missions. One-axis sun-pointing attitude orientation and stabilization, as well as spin-rate control are planned using solar-radiation pressure. For this purpose, the spacecraft is equipped with an axially symmetric solar stabilizer made from thin films and eight mirror-like solar paddles mounted symmetrically like windmill vanes (Figure 15.1).

SSL spacecrafts are intended for a variety of scientific experiments related to such fields as astrometry, plasma physics and planetary exploration.

The centre of mass of the spacecraft is assumed to move in a circular heliocentric orbit. We study the three-dimensional motion of a symmetric statellite about its centre of mass under the solar-radiation torques. No control of the motion is considered in this chapter.

15.2 MODEL AND EQUATIONS

For such a satellite, we can use the following model of radiation-pressure torque (Beletsky 1966; Beletsky and Starostin 1991):

Nonlinearity and Chaos in Engineering Dynamics
Edited by J. M. T. Thompson and S. R. Bishop,   © 1994 John Wiley & Sons Ltd

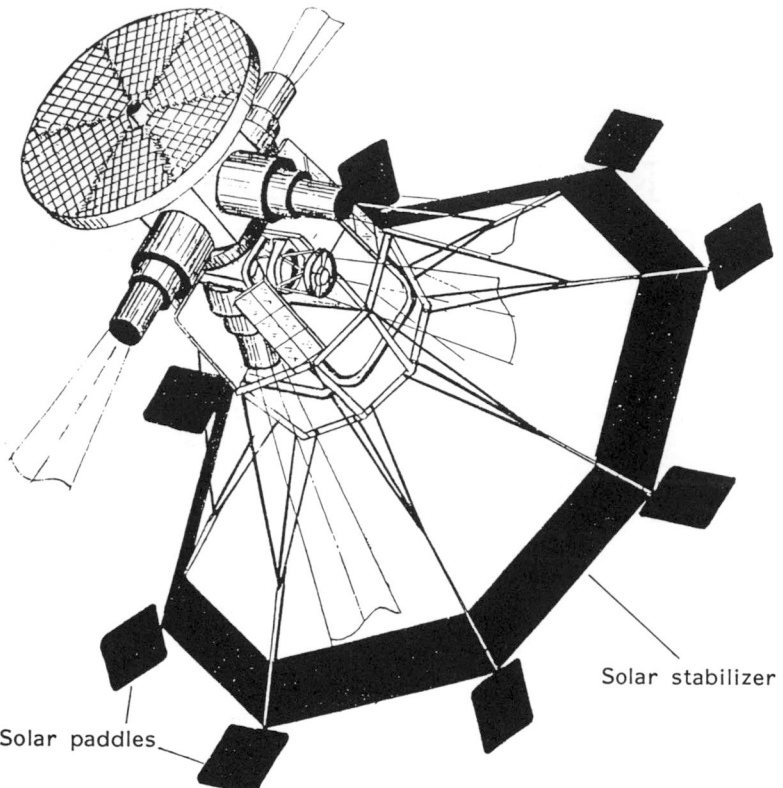

Figure 15.1   General view of Small Space Lab.

$$\boldsymbol{M} = a_s \boldsymbol{e}_r \times \boldsymbol{e}_{z'} + f_s(-a_1\gamma, \ -a_1\gamma', \ a_0\gamma'')^T, \tag{1}$$

$$a_s = p_s(R_0/R)^2(a_{s0} + a_{s1}\cos\varepsilon_s + a_{s2}\cos^2\varepsilon_s + ...),$$

$$a_{si} = \text{const}, \quad a_{si} \geqslant 0, \quad i = 0, 1, ...,$$

$$f_s = 2aS_p n_0 n_1 p_s (R_0/R)^2.$$

Here, R_0 and R are the fixed and the current orbital radius, respectively; $\boldsymbol{e}_r$ is the unit vector in the direction of the orbital radius vector $\boldsymbol{R}$; $\boldsymbol{e}_{z'}$ is the unit vector along the satellite's axis of symmetry z'; $\varepsilon_s = \angle(\boldsymbol{e}_r, \boldsymbol{e}_{z'})$; a is the distance between the centre of a paddle and the axis of symmetry; n_0 is the cosine of the angle between the normal to a paddle and the axis of symmetry; $n_1 = \sqrt{1 - n_0^2}$; S_p is the total area of the paddles; p_s is the solar-pressure constant (for the Earth orbit $R_0 = 1.496 \times 10^{11}\,\text{m}$ and $p_s = 4.64 \times 10^{-6}\,\text{Pa}$); a_0 and a_1 are the positive coefficients of the model torque; γ, γ' and γ'' are the projections of $\boldsymbol{e}_r$ on the principal axes of inertia of the satellite.

First, we put $a_{s0} = bS > 0$, $a_{sj} = 0$, $j = 1, 2, ...$, where b is the distance between the centre of pressure and the centre of mass of the satellite, both the centres lying on the axis of symmetry; S is the satellite's characteristic section area.

The torque of equation (1) consists of the 'conservative' part (the cross-product) and the 'propelling torque', projections of which are written onto principal axes of inertia of the satellite.

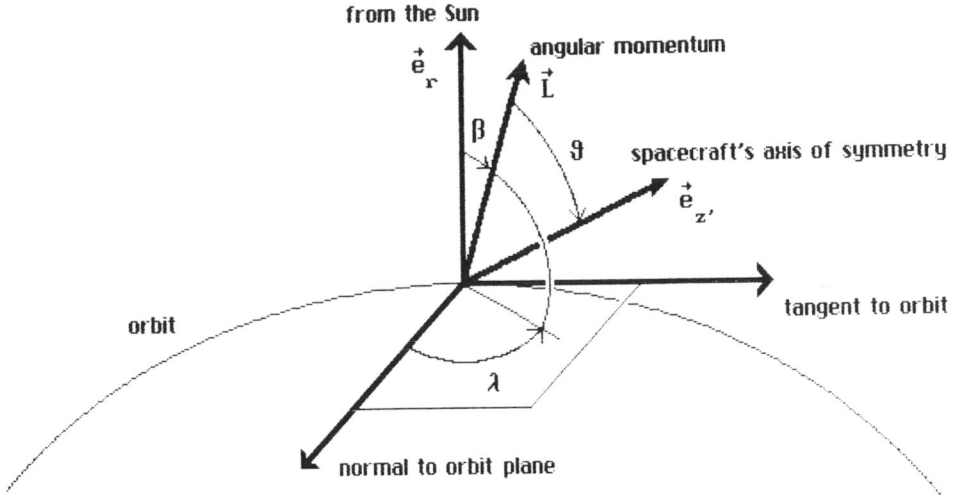

Figure 15.2   Angles ϑ, β and λ.

We employ the first-order approximation equations of motion in evolutionary variables:

$$\frac{dl}{d\tau} = f(\cos^2 \vartheta - \alpha)\cos \beta,$$

$$\frac{d\vartheta}{d\tau} = -\frac{f}{l}\sin \vartheta \cos \vartheta \cos \beta,$$

$$\frac{d\beta}{d\tau} = -\varepsilon \sin \lambda + \frac{f}{l}(\alpha - \tfrac{1}{2}\sin^2 \vartheta)\sin \beta, \tag{2}$$

$$\frac{d\lambda}{d\tau} = -\varepsilon \cot \beta \cos \lambda + \frac{1}{l}\cos \vartheta,$$

$$f = (a_0 + a_1)\frac{2aS_p n_0 n_1}{bS}, \quad \alpha = \frac{a_1}{a_0 + a_1}, \quad \varepsilon = \frac{L_0 \sqrt{\mu R}}{bSp_s R_0^2},$$

where $l = L/L_0$ is the dimensionless module of the angular momentum vector L of the spacecraft, L_0 is a fixed value of the angular momentum module; $\vartheta = \angle (L, e_{z'})$; $\beta = \angle (e_r, L)$; λ is the angle between the normal to the orbit plane and the projection of the vector L on the plane orthogonal to e_r (Figure 15.2); τ is the dimensionless time, $\varepsilon\tau = v$, v is the true anomaly of the orbit; α is the design variable; μ is the gravity constant of the Sun.

15.3   ANALYSIS OF EQUATIONS

In equations (2) the term $(1/l)\cos \vartheta$ describes a conservative effect, the terms $\sim f$ describe an effect of 'propelling' due to the screw-symmetrical paddles, and the terms with λ are due to the orbital motion.

The conservative effect reduces to precession of the angular momentum vector (at small ε approximately around the direction to the Sun with rate $d\lambda/d\tau \sim (1/l_0)\cos\vartheta_0$). The propelling effect causes profound modulations of the angular momentum vector orientation with respect to this direction (i.e. the modulations of the angle β); at the same time, the variables l and ϑ exhibit similar modulations (Figures 15.3 and 15.4). The combination

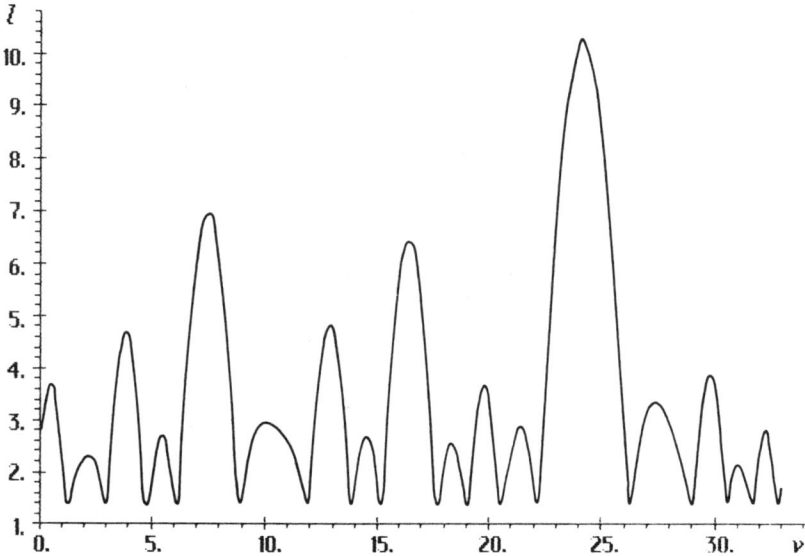

Figure 15.3   Angular momentum module l versus true anomaly v. $f = 0.15$, $\varepsilon = 0.01$; $\alpha = 0.3$, $\lambda_0 = 0.0$, $l_0 = 2.38$.

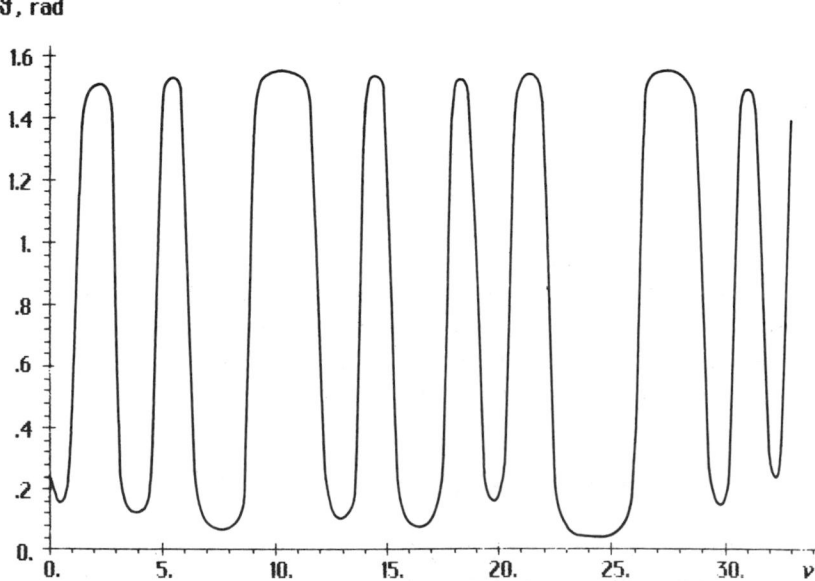

Figure 15.4   Angle ϑ versus true anomaly v, corresponding the previous figure.

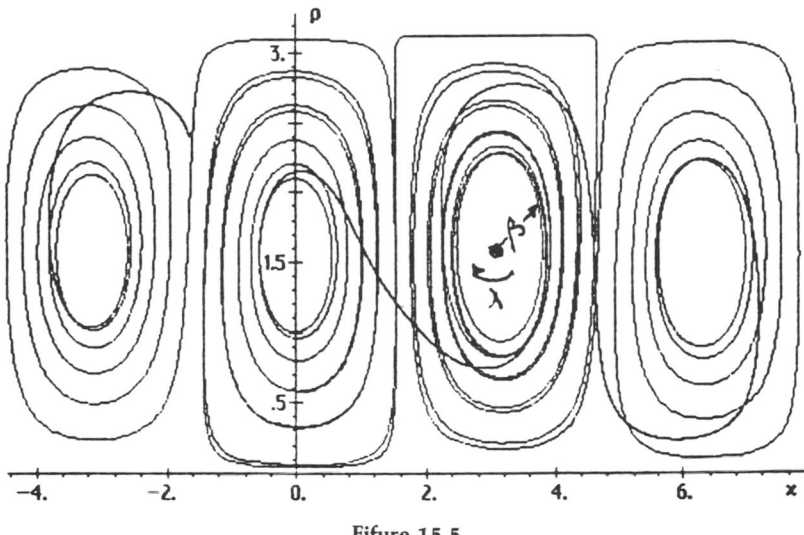

Fifure 15.5

of these two effects results in the peculiar picture of 'pulsing' precession of the vector *L* around the radius vector direction (Figure 15.5). The most interesting fact is the abrupt jumps from the motion in the vicinity of the direction *to* the Sun into the motion in the neighbourhood of the direction *from* the Sun. Such jumps are often difficult to predict. This stochasticity is inherent in the dynamical system (2) and it can be represented clearly in the appropriate phase space.

The primitive integral gives:

$$l \cot^\alpha \vartheta \sin \vartheta = C. \tag{3}$$

This enables the system order to be reduced to three, by elimination of the variable *l*, and to two by employing λ as a new independent variable. The second-order system obtained is non-autonomous and periodic in λ. In the phase space (ϑ, β), with *C* fixed, the motion may be generally described by the complex regular and chaotic trajectories, revealed by the numerical implementation of the Poincaré mapping. The plane (ϑ, β) is chosen as a surface of section and λ-mapping over 2π is studied.

For some values of the parameters, the set of trajectories may comprise only regular ones. Such cases are said to be completely integrable. An autonomous system of the third order is completely integrable if it admits of only one primitive integral.

Let us fix the parameters α and *C*. Then system (2) has two free parameters, *f* and ε. Consider behaviour of system (2) for extreme values of *f* and ε, remembering the following from integral (3).

1. If $\varepsilon^{-1} = 0$, another integral exists:

$$\sin^2 \beta \cos^2 \lambda \cos \vartheta \, \tan^{2\alpha} \vartheta = C_1.$$

and system (2) is completely integrable. There are no chaotic trajectories.
2. If $\varepsilon = 0$, then for $\varepsilon \ll 1$, it can be seen from the last of equations (2) that λ is the fast variable and equations (2) can be averaged with respect to λ, the averaged system (2)

having the primitive integral

$$\sin^2 \beta \cos \vartheta \tan^{2\alpha} \vartheta = \bar{C}_1. \tag{4}$$

and being completely integrable. The initial (non-averaged) system (2) is close to the completely integrable system; the absence of chaos (or weak chaos) can be predicted.

3. When $f = 0$, the system (2) has first integrals

$$l = l_0, \quad \vartheta = \vartheta_0, \quad \varepsilon \sin \beta \cos \lambda + \frac{1}{l_0} \cos \vartheta_0 \cos \beta = \Phi_0, \tag{5}$$

and all its trajectories are regular.

4. When $f^{-1} = 0$, the term $\varepsilon \sin \lambda$ in the third of equations (2) may be neglected. Integral (4) exists. The system is close to a completely integrable one.

It is worth noting that the consideration of items 2 and 4 may be expressed in mathematical terms.

When considering the question of whether the angle λ is a monotonic function of time or not it is noted that this condition does not always hold; i.e. the phase trajectories may be tangent to the chosen section plane (ϑ, β). Hence, when constructing the Poincaré map, the possibility should be taken into consideration that a trajectory, having come out of a point of the section $\lambda = 0$, may return and intersect the same plane once more, having not reached the section $\lambda = 2\pi$ (or $\lambda = -2\pi$).

To examine the mechanism of this in more detail we consider the system (2) when $f = 0$ once again by examining the third primitive integral (5) $\Phi(l_0, \vartheta_0; \beta, \lambda)$. Locating the critical points of the function Φ treated as a function of two variables, angles β and λ, the first partial derivatives can be calculated as:

$$\frac{\partial \Phi}{\partial \beta} = \varepsilon \cos \beta \cos \lambda - \frac{1}{l_0} \cos \vartheta_0 \sin \beta,$$

$$\frac{\partial \Phi}{\partial \lambda} = -\varepsilon \sin \beta \sin \lambda.$$

Both the derivatives become zero in two cases when:

1) $\beta = \pi i$, $i = 0, \pm 1, \pm 2, \ldots;$

    $\lambda = \pi/2 + \pi k$, $k = 0, \pm 1, \pm 2, \ldots;$ (6)

2) $\beta = \pm \beta^* + \pi m$, $m = 0, \pm 1, \pm 2, \ldots;$

    $\lambda = \pi n$, $n = 0, \pm 1, \pm 2, \ldots;$ (7)

where β^* is a root of the equation $(d\lambda/d\tau)|_{\lambda = 0} = 0$.

If one calculates the second partial derivatives Φ_{ij} and finds det Φ_{ij}, then it will turn out that in case (6) det $\Phi_{ij} = -1$ and it can be shown that in case (7) det $\Phi_{ij} > 0$. Therefore, case (6) corresponds to unstable critical points of the function Φ (saddles) and case (7) to stable centres. The trajectories that originate from the section $\lambda = 0$ in the vicinity of the stable point, never reach the neighbouring section $\lambda = 2\pi$ (-2π) (Figure 15.6).

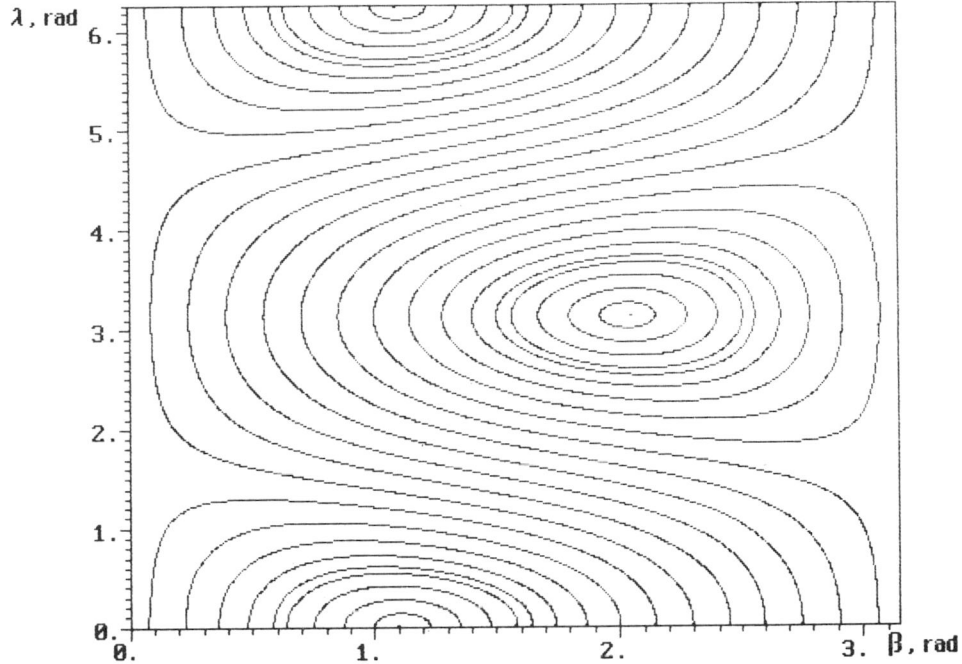

Figure 15.6   Angle λ versus angle β for various initial values of angle β. $f = 0$, $\varepsilon = 1$, $\vartheta = \pi/4$, $C = 1$.

A similar picture takes place for $f \neq 0$. The only difference is that, in the general case, there are no separatrices between stable and unstable points and trajectories may go from the vicinity of the stable point to the neighbourhood of the unstable one and vice versa. When computing the Poincaré map of the section $\lambda = 0$, a trajectory of system (2) is traced until it intersects any of the sections $\lambda = 0$, $\pm 2\pi$. By numerical integration when a trajectory is about to intersect the section plane $\lambda = 0$ (mod 2π), system (2) is substituted by an equivalent system with independent variable λ. This is done in order to find the intersection point with precision.

At $f = 0$ the stable critical points (7) constitute a curve in the (ϑ, β) plane – the derivative $d\lambda/d\tau$ vanishes on this curve. Similar curves exist in the general case, too, marked by S–S in Figures 15.9–11.

15.4   POINCARÉ MAPPING

A global analysis of the phase space was carried out and the evolution of surfaces of section were traced out with a variation of the parameters. The results obtained make it possible to specify design conditions for SSL. For astrometrical studies, the predictability of the spacecraft motion is critical, therefore the regions of chaotic motions must be avoided. On the other hand, there are scientific experiments that require the whole celestial sphere to be scanned, and in those cases chaotic rotation may be very desirable.

The following values were taken: $\alpha = 0.3$ (for all the following pictures except Figure 15.12), and $C = 1.0$.

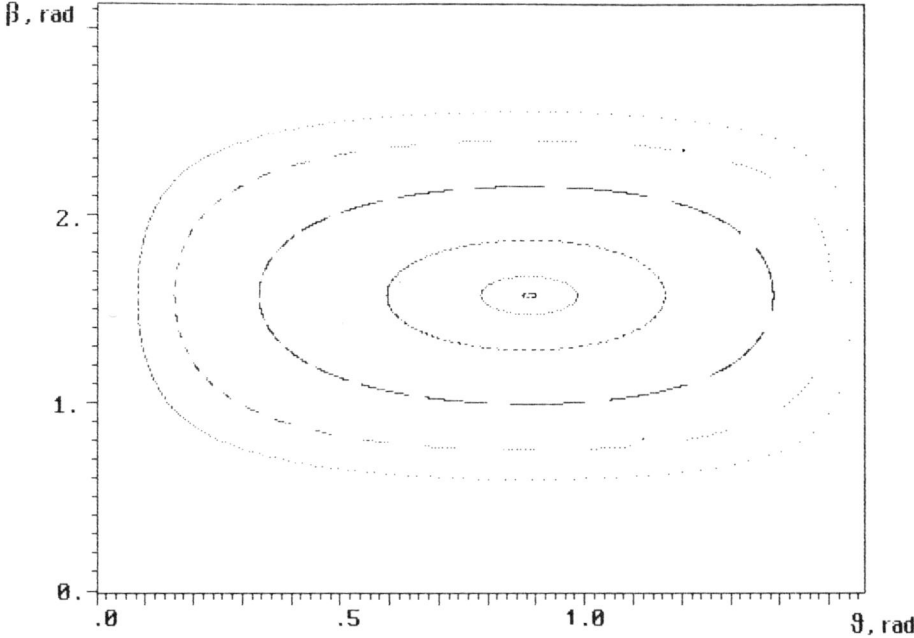

Figure 15.7   Surface of section for $f = 0.15$, $\varepsilon = 0.0$.

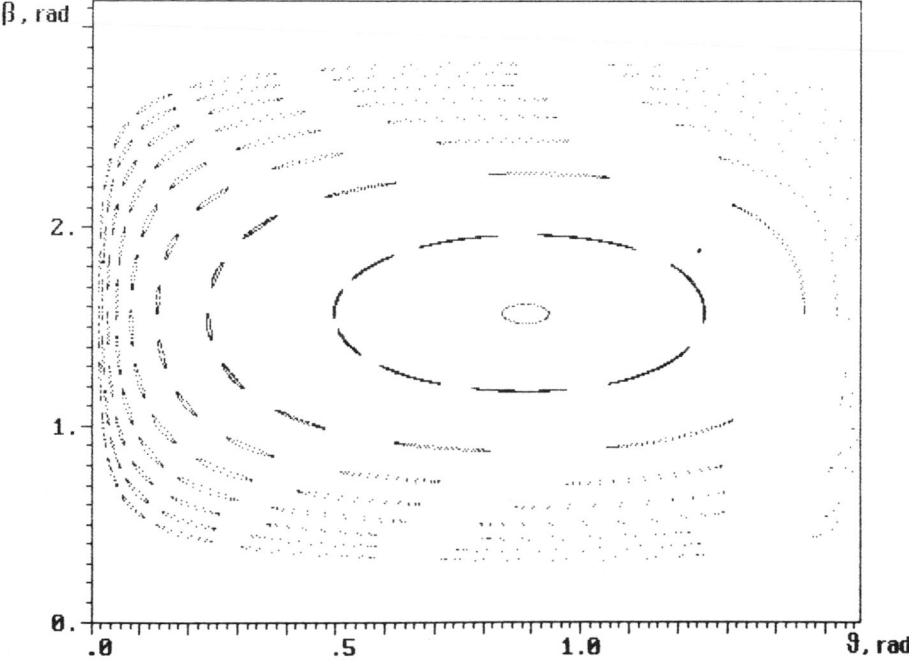

Figure 15.8   Surface of section for $f = 0.15$, $\varepsilon = 4.42 \times 10^{-5}$.

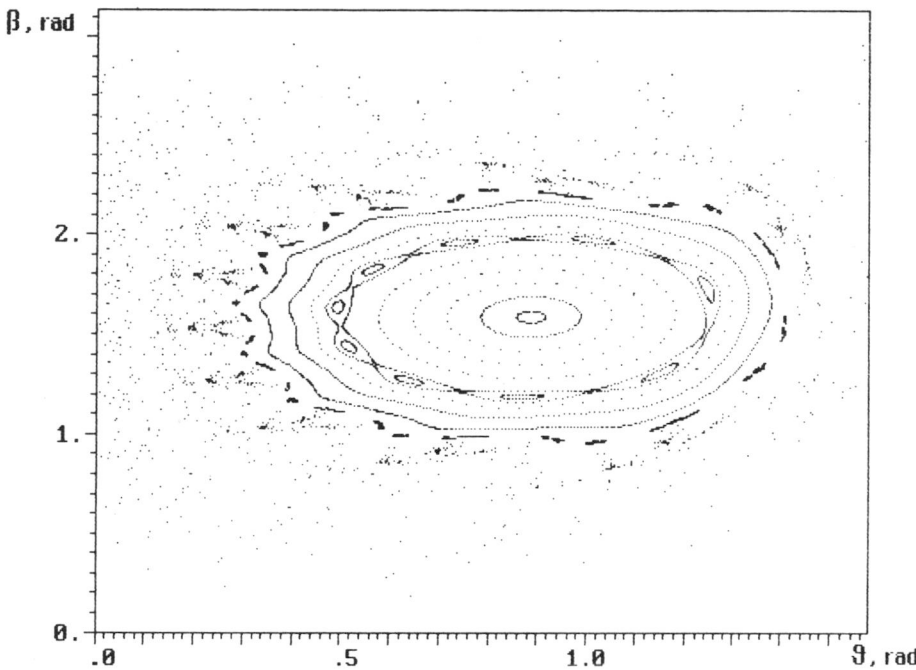

Figure 15.9   Surface of section for $f = 0.15$, $\varepsilon = 0.01$.

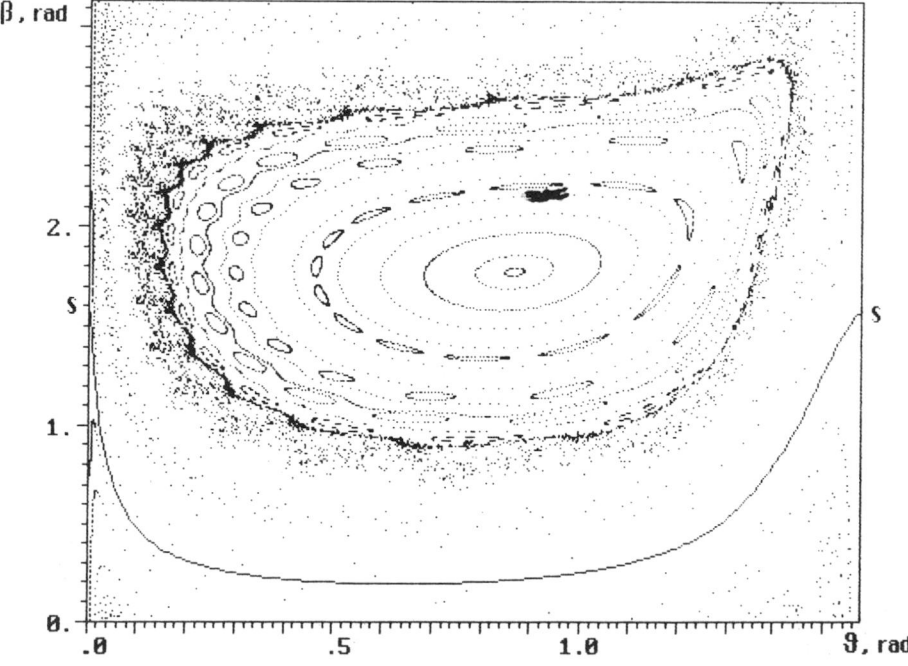

Figure 15.10   Surface of section for $f = 0.105$, $\varepsilon = 0.1$.

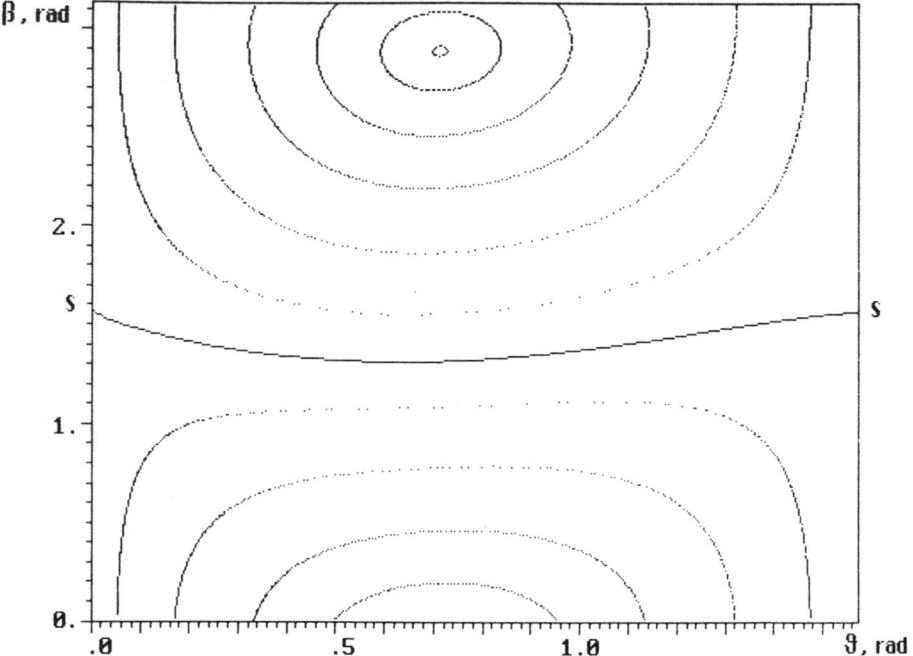

Figure 15.11   Surface of section for $f = 1.05$, $\varepsilon = 2.0$.

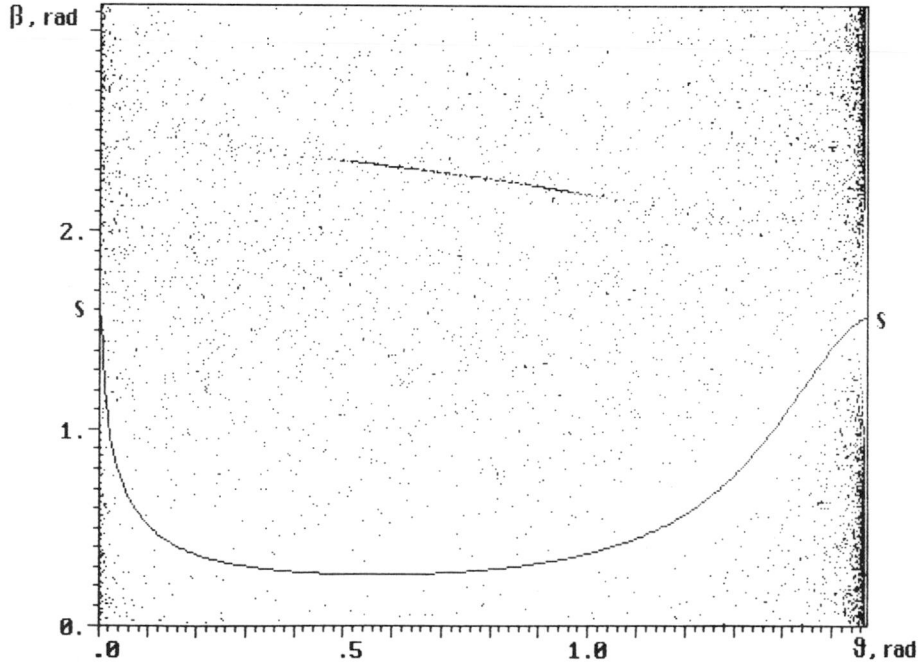

Figure 15.12   Surface of section for $f = 1.5$, $\varepsilon = 0.14$, $\alpha = 0.4225$.

β, deg

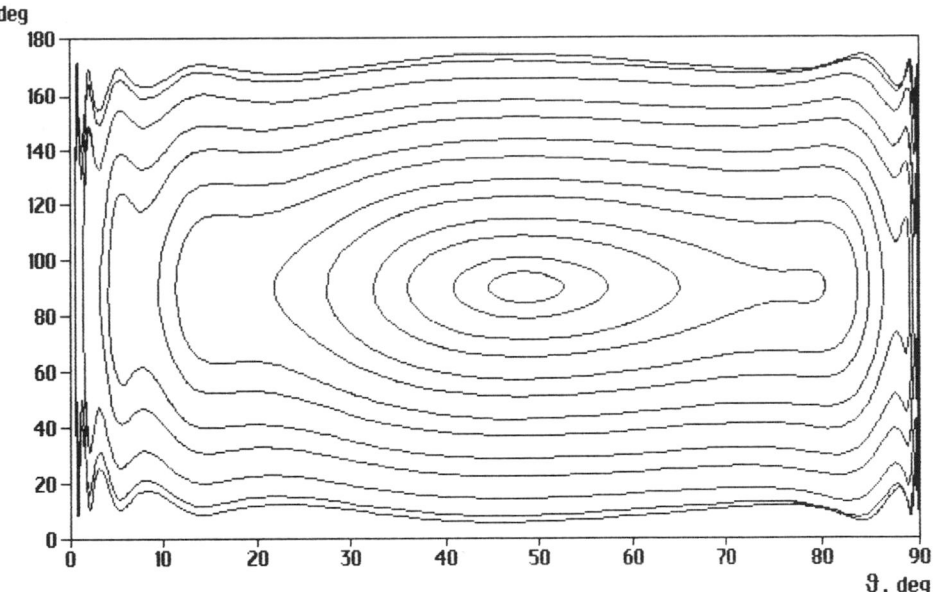

Figure 15.13   Periodic motions in the case $\alpha_{s1} > 0$, $\alpha_{sj} = 0$, $j = 0, 2, \ldots, f = 0.15$, $\varepsilon = 0.01$, $\alpha = 0.3$, $\lambda_0 = 0.0$, $l_0 = 2.38$.

Figure 15.7 is displayed for $f = 0.15$, $\varepsilon = 0.0$. There is no stochasticity detected. In Figure 15.8 ($f = 0.15$, $\varepsilon = 4.42 \times 10^{-5}$) one can see some curves splitting and chains of islands appearing. The stochasticity layers are thin and separated.

It should be noted that boundaries $\vartheta = 0$, $\vartheta = \pi/2$, $\beta = 0$, $\beta = \pi$ play the role of separatrices of the family (4); therefore, stochasticity should be expected near these boundaries first, and the 'centre' of the region is less affected by stochasticity. Figure 15.9 confirms this ($f = 0.15$, $\varepsilon = 0.01$).

Figure 15.10 ($f = 0.105$, $\varepsilon = 0.1$) exhibits development of the chaos. One can observe regularization of chaotic trajectories near the curve S–S. The nearer to the curve S–S the trajectory approaches the slower the point moves along it. Therefore, these trajectories exhibit intermittent behaviour.

Figure 15.11 ($f = 1.05$, $\varepsilon = 2.0$) reveals that the motion becomes almost completely regular. On the other hand, strong stochasticity takes place for $f = 1.5$, $\varepsilon = 0.14$ and $\alpha = 0.4225$ (Figure 15.12). Only one trajectory is presented in this figure. As α increases, global stochasticity occurs.

Let us now revert to the radiation-pressure torque expression (1). Put $a_{s1} > 0$, $a_{sj} = 0$, $j = 0, 2, \ldots$ (Beletsky *et al.* 1992). Then the last of equations (2) is found to be

$$\frac{d\lambda}{d\tau} = -\varepsilon \cot \beta \cos \lambda + \frac{1}{l}(\tfrac{3}{2} \sin^2 \vartheta - 1) \cos \beta.$$

Numerical experiments reveal that in this case all trajectories turned out to be periodic (Figure 15.13). No stochasticity was found at any set of parameters. The motion of the satellite is easy to predict.

REFERENCES

Beletsky, V. V. (1965) *Motion of an Artificial Satellite about its Center of Mass*, Jerusalem, Israel Program for Scientific Translations.

Beletsky, V. V. and Starostin, E. L. (1991) Regular and chaotic rotions of the satellite in sunlight flux, Keldysh Inst. of Appl. Math., preprint No 68 [in Russian].

Beletsky, V. V., Prokofyeva, E. V. and Starostin, E. L. (1992) Spacecraft rotational dyanmics in light flux, Keldysh Inst. of Appl. Math., preprint No 47 [in Russian].

V. V. Beletsky and E. L. Starostin, *M. V. Keldysh Institute of Applied Mathematics, Russian Academy of Sciences, 4 Miusskaya Square, 125047 Moscow, Russia*

16 DYNAMICS AND CHAOS OF RESONANTLY EXCITED STRUCTURES WITH CYCLIC SYMMETRY

A. K. Bajaj, S. Samaranayake and O. D. I. Nwokah

The present work investigates the nonlinear vibratory response of a simplified model of cyclic structures to periodic excitations. This cyclic structure, in its linear approximation, possesses pairwise double-degenerate natural frequencies with orthogonal normal modes. When the linear coupling stiffness is $O(\varepsilon)$, the cyclic system consists of n weakly coupled identical nonlinear oscillators so that all the oscillators are in internal resonance. The dynamic response of this system for strong as well as weak coupling cases is studied using the method of averaging. The amplitude equations are carefully analysed using local bifurcation theory, and the effects of the forcing amplitude and frequency, and the modal damping, on the steady-state responses are determined.

16.1 INTRODUCTION

Mechanical and structural systems are inherently nonlinear and have many different sources of nonlinearities. Nonlinearities lead to a wide range of interesting phenomena, especially when the structures are lightly damped, possess internal resonances, and are excited near resonant frequencies (Nayfeh and Balachandran 1989; Tousi and Bajaj 1985).

An important class of structural systems found in engineering applications are those with cyclic symmetry. They arise naturally, for example, in large circular space antennae, bladed-disc assemblies and circular magnetic storage devices. The literature on vibrations in bladed-disc assemblies is quite extensive. Most of the studies have been restricted to the prediction of linear vibratory response, with the more recent works being specifically concerned with the mode-localization phenomenon (e.g. see Wei and Pierre 1988). Tobias and Arnold (1957) and Tobias (1957) concern the earliest of studies on the effects of stiffness nonlinearities on the vibration of structures with cyclic symmetry. They examined the free and forced oscillations

Nonlinearity and Chaos in Engineering Dynamics
Edited by J. M. T. Thompson and S. R. Bishop,   © 1994 John Wiley & Sons Ltd

in thin discs undergoing axial vibrations. They experimentally observed both 'standing-wave' and 'travelling-wave' solutions, along with the jump from one type of motion to another as the frequency of external excitation was varied.

Recently, Vakakis (1992) has conducted an extensive study of the free and forced responses of a nonlinear periodic structure. For the free vibratory response, he has shown that the system can only possess mutually orthogonal 'similar' nonlinear normal modes, and that these pairs of nonlinear modes have distinct 'backbone' curves. For a harmonically excited structure, Vakakis (1992) and Samaranayake *et al.* (1992) have found that a stable branch of 'travelling-wave'-type solutions bifurcates from a branch of 'standing-wave'-type motions. The results of the latter authors show that the coupled-mode periodic motions may also bifurcate to chaotic amplitude-modulated motions.

Several researchers have investigated weakly coupled structures with cyclic symmetry. Wei and Pierre (1988) investigated the occurrence of mode localization in linear models. Ashwin *et al.* (1990) and more recently, Ashwin and Swift (1992) have studied the dynamics of weakly coupled identical oscillators.

In the present work, a model of multi-degree-of-freedom cyclic systems with cubic nonlinearities is studied for its resonant response. The method of averaging is used to derive the amplitude or the averaged equations for cases of both strong and weak coupling. In the strong coupling case, the averaged equations are very similar to those that have arisen in previous studies of systems with $O(2)$ symmetry (e.g. Bajaj and Johnson 1992). A careful computer-assisted bifurcation analysis for constant and non-constant solutions of the averaged equations is performed. It is shown that, over some frequency interval, there exist 'travelling-wave'-type solutions which arise due to pitchfork bifurcation from the 'standing-wave' or normal-mode solutions. For lower damping, the 'travelling-wave'-type solutions give rise to either multiple branches of 'travelling-wave' solutions or to a branch of limit-cycle and chaotic solutions.

In the weak coupling case, the system is equivalent to n weakly coupled identical nonlinear oscillators and, again, the method of averaging is used to predict the response. The interest here is focused on the effect of the linear coupling coefficient on the transition from localized to extended response of the system.

16.2 THE CYCLIC SYSTEM AND EQUATIONS OF MOTION

Consider a cyclic system consisting of n identical particles of mass m each. Each particle is attached to its neighbours by nonlinear extensional springs (constants k_1 and εk_2) and is hinged to the ground by a nonlinear torsional spring (stiffnesses T_1 and εT_2). Let x_i be the displacement of the ith particle. Then, the quations of motion for the lightly damped cyclic structure can be written in the form (Samaranayake *et al.* 1992):

$$x_i'' + (1 + 2k)x_i - k(x_{i+1} + x_{i-1}) + \varepsilon f_i/m\bar{\omega}^2 = 0, \quad i = 1, 2, \ldots, n, \tag{1}$$

where

$$f_i = k_2\{(x_i - x_{i+1})^3 + (x_i - x_{i-1})^3\} + T_2(x_i^3/a^4) + \bar{d}x_i'/\bar{\omega}, \tag{2}$$

where a prime denotes derivative with respect to the non-dimensional time τ, $\tau = \bar{\omega}t$, $\bar{\omega} = (T_1/ma^2)^{1/2}$ and $k = k_1/m\bar{\omega}^2$. Note that when the linear coupling stiffness k_1 is also $O(\varepsilon)$, the weak coupling case, each oscillator in equations (1) has the same linear natural frequency.

As is well known, the strongly coupled linear cyclic system (when $\varepsilon = 0$) has mostly pairwise degenerate natural frequencies, $\omega'_j s$, with orthogonal normal modes (Vakakis 1992; Samaranayake *et al.* 1992). The mode shapes of the *n*-degree-of-freedom cyclic system are denoted by U_0, U^c_j and U^s_j. The modes U^c_j and U^s_j are, respectively, the 'cos-*j*' and 'sin-*j*' modes, and correspond to the same natural frequency ω_j. In the case of weak coupling, k_1 is $O(\varepsilon)$, all the *n* natural frequencies are clustered in a small neighbourhood of the natural frequency of the individual subsystem.

16.3   STRONGLY COUPLED SYSTEM

In order to study nonlinear interaction between the pairs of modes with identical linear natural frequencies, an external harmonic excitation with amplitude of $O(\varepsilon)$ is introduced on the structure. We assume that the excitation frequency Ω is close to one of the degenerate natural frequencies, ω_j and the excitation is spatially distributed like one of the corresponding two modes, that is, $F^{ex} = \varepsilon \mu U^c_j \cos \Omega \tau$. Then, only the 'cos-*j*' and 'sin-*j*' modes contribute to the response and the steady-state monofrequency response of the system can be expressed in the form

$$x(\tau) = \{B_j(\tau) U^c_j \cos(\Omega_j \tau + \beta_j(\tau)) + C_j(\tau) U^s_j \cos(\Omega_j \tau + \gamma_j(\tau))\}/e,$$

where

$$e = \left\{ \sum_{i=1}^{n} \cos^2 i\alpha_j \right\}^{1/2} = \left\{ \sum_{i=1}^{n} \sin^2 i\alpha_j \right\}^{1/2} = (n/2)^{1/2}, \tag{3}$$

and where we have introduced Ω_j to be the frequency of response.

Then, following the method of averaging (Hale 1969), the averaged equations governing the evolution of amplitudes and phases B_j, β_j, C_j and γ_j are given by (Samaranayake *et al.* 1992):

$$B' = \varepsilon \{pBC^2 \sin 2(\beta - \gamma) - dB - \sigma \sin \beta\},$$

$$B\beta' = \varepsilon \{B\{3pB^2 + [2 + \cos 2(\beta - \gamma)]pC^2 - \lambda\} - \sigma \cos \beta\},$$

$$C' = \varepsilon \{pB^2 C \sin 2(\gamma - \beta) - dC\}, \tag{4}$$

$$C\gamma' = \varepsilon C \{[2 + \cos 2(\gamma - \beta)]pB^2 + 3pC^2 - \lambda\},$$

where

$$p = (k_2 L + T_2 M/a^4)/2m\bar{\omega}\Omega_j n^2,$$

$$d = \bar{d}/2m\bar{\omega}^2, \quad \sigma = \mu/2m\bar{\omega}\Omega_j \quad \text{and} \quad \lambda = (\omega_j^2 - \Omega_j^2)/2\Omega_j \varepsilon.$$

Constant solutions of equations (4) represent steady-state periodic motions of the cyclic structure. Non-constant steady-state solutions are also possible and they represent amplitude- and phase-modulated motions of the system. Considering equations (4), we first note that all the nonlinearities are combined through only one coefficient *p*. Secondly, the development is valid for all degenerate pairs of modes with identical natural frequencies, ω_j. The particular mode combination and the corresponding natural frequency are identified only through the frequency number *j*. The nonlinearity, *p*, the force, σ, and the frequency mistuning, λ, all depend on *j*. The nonlinear coefficient *p* can be absorbed by the change of parameters $\lambda_j/p \rightarrow \lambda_j$, $d/p \rightarrow d$ and $\tau p \rightarrow \tau$. Thus, all the degenerate frequencies exhibit qualitatively the same nonlinear dynamical response. Finally, note that equations (4) are a special case of the one-parameter family of equations studied in the literature in the context of systems with $O(2)$ symmetry (see Bajaj and Johnson 1992). Thus, the cyclic systems (systems with D_n symmetry) also lead to the canonical averaged equations for systems with $O(2)$ symmetry and are expected to exhibit a complex-amplitude dynamics similar to the one discovered for stretched strings (Bajaj and Johnson 1992).

Equations (4) have two types of solutions: the single-mode motions which occur when $C = 0$, and the coupled-mode motions which occur when both B and C are other than zero. These correspond, respectively, to a 'standing-wave' and a 'travelling- wave' pattern for the structure.

Figure 16.1 shows the various single-mode and coupled-mode steady-state constant solu-

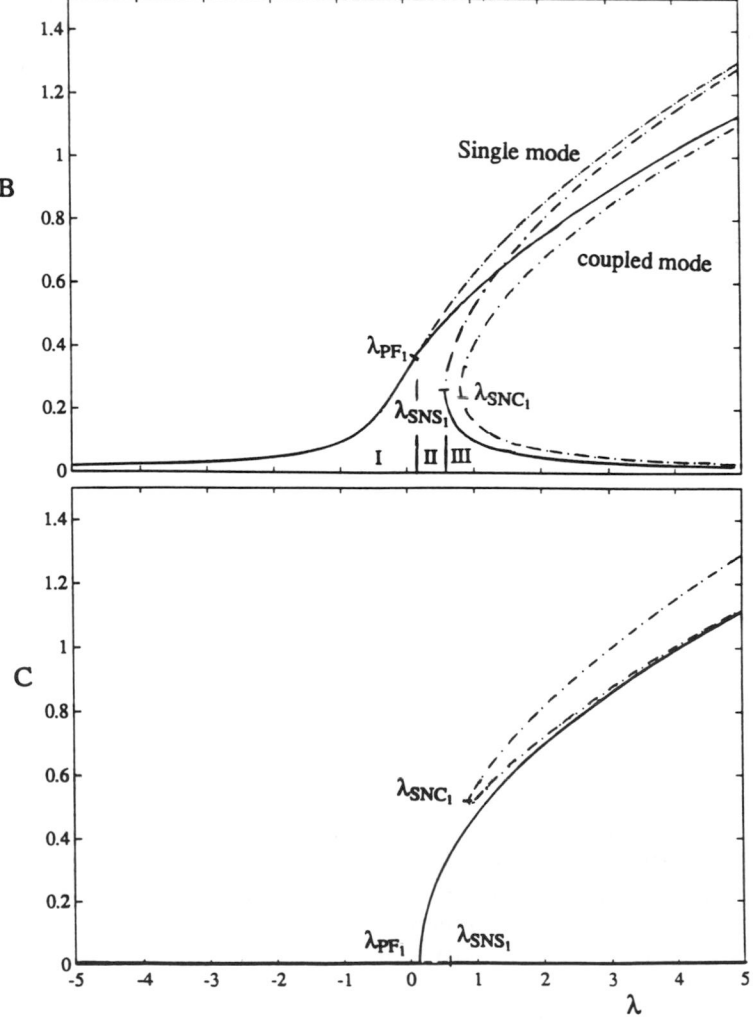

Figure 16.1   Steady-state constant amplitude response; $\sigma = 0.1$, $d = 0.0$.

tions B and C for zero damping as a function of the excitation frequency λ_0. The force amplitude is held constant at $\sigma = 0.1$ and the nonlinear parameter $p = 1$. Over the interval I, there exists only one single-mode solution. Over the interval II, a stable coupled-mode solution coexists with an unstable single-mode solution. Over the interval III, a stable single-mode solution coexists with a stable coupled-mode solution along with other unstable solutions.

Figure 16.2 shows the various bifurcation sets in the (σ, λ_0) plane for the constant solutions at a fixed damping, $d = 0.12$. They correspond to a pitchfork bifurcation (PF) from a single-mode to the coupled-mode solution, saddle–node bifurcations in the single-mode (SNS) and the coupled-mode (SNC) branches, and to a Hopf bifurcation (HB) in a coupled-mode branch. These sets show that for force levels below $\sigma \approx 0.035$, only the single-mode solution exists and it is unique. At a force in the interval $0.035 \leqslant \sigma \leqslant 0.045$, there exist three single-mode responses over the frequency interval $(\lambda_{SNS_1}, \lambda_{SNS_2})$. As the force is further increased, the upper branch of single-mode solutions undergoes a pitchfork bifurcation and now a coupled-mode branch exists over the frequency interval $\lambda_{PF_1} \leqslant \lambda_0 \leqslant \lambda_{PF_2}$. The upper pitchfork point λ_{PF_2} quickly approaches the turning point SNS_2 as σ increases, with both $\to \infty$ as $\sigma \to \infty$. For still higher force amplitudes, first, the coupled-mode solutions undergo Hopf bifurcation at frequencies λ_{HB_1} and λ_{HB_2}, and then the coupled-mode branch develops turning points at λ_{SNC_1} and λ_{SNC_2}. The lower turning point and Hopf point are very close to each other, whereas the coupled-mode constant solution between the frequencies λ_{PF_1} and λ_{SNC_2} always remains stable.

A representative set of response curves for $\sigma = 0.1$ and $d = 0.12$, which are very similar to those for a stretched string (Bajaj and Johnson 1992), are shown in Figure 16.3. These were obtained using AUTO (Doedel 1986) and are given in terms of the two Cartesian

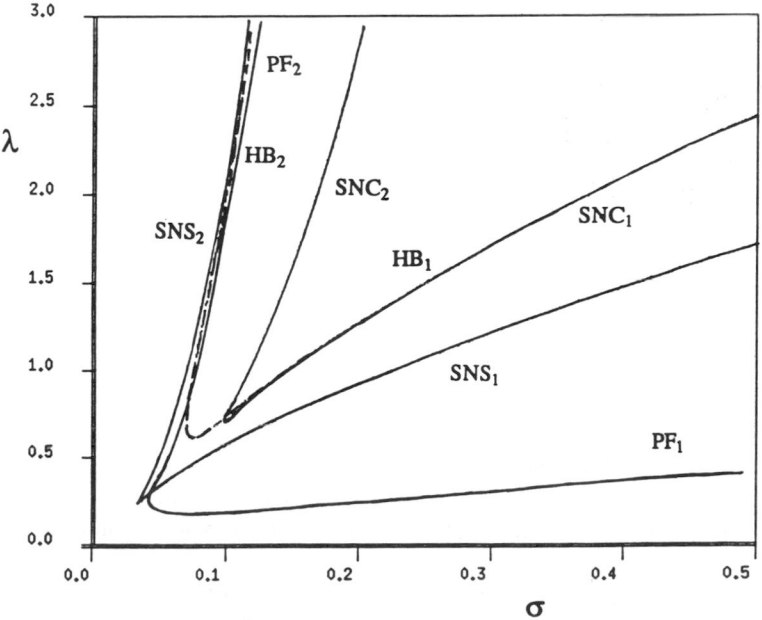

Figure 16.2 Bifurcation sets for the constant solutions; $d = 0.12$.

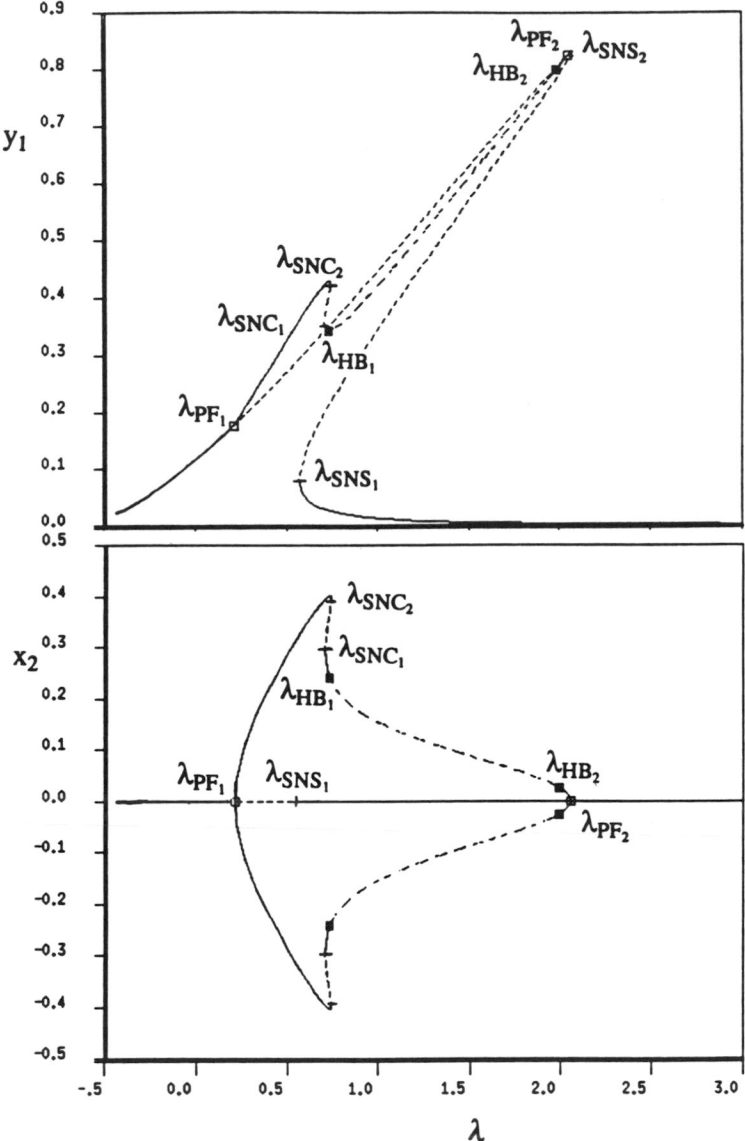

Figure 16.3   Steady-state responses for y_1 and x_2; $\sigma = 0.1$, $d = 0.12$.

components y_1, and x_2 ($x_1 = B \cos \beta, y_1 = B \sin \beta, x_2 = C \cos \gamma, y_2 = C \sin \gamma$). There is a wide frequency interval $\lambda_{HB_1} \leqslant \lambda_0 < \lambda_{HB_2}$ where the coupled-mode constant solution is unstable by a Hopf bifurcation and we expect limit-cycle solutions near the bifurcation frequencies λ_{HB_1} and λ_{HB_2}.

Figure 16.4 shows a qualitative bifurcation diagram in this Hopf interval. At $\lambda_{HB_1} = 0.732$, a limit-cycle solution (denoted as P_1 solution) arises due to a Hopf bifurcation in the coupled-mode constant solution branch. It undergoes a period-doubling bifurcation to a

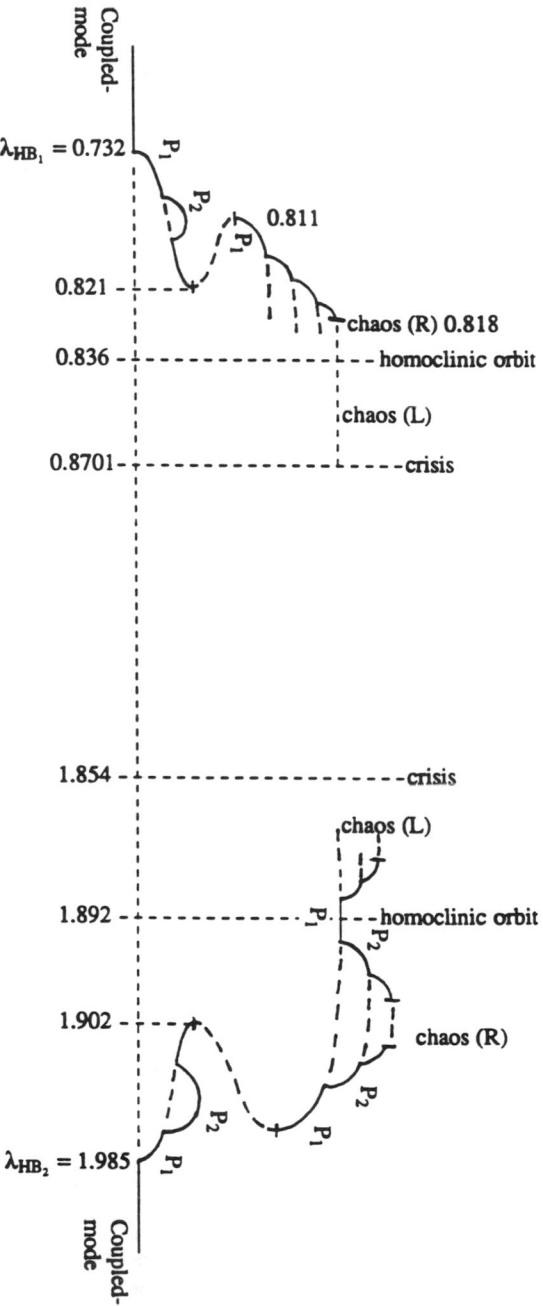

Figure 16.4   Qualitative bifurcation diagram for coupled-mode motions; $\sigma = 0.1$, $d = 0.12$.

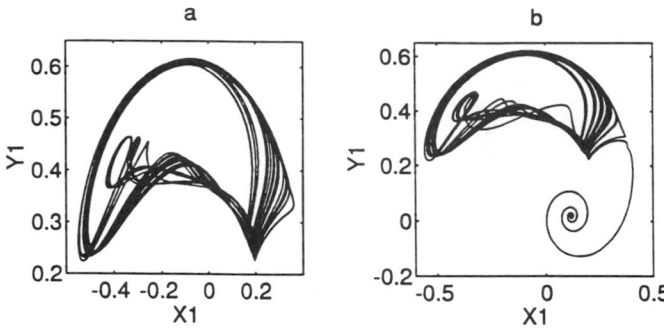

Figure 16.5 Phase plots displaying 'crisis'; $\sigma = 0.1$, $d = 0.12$: (a) $\lambda_0 = 0.87$ (chaos), (b) $\lambda_0 = 0.8707$ (transient chaos).

solution with twice the period (P_2 solution), and then reverses to a P_1 solution. Around $\lambda_0 = 0.821$, the P_1 solution seems to jump to a chaotic solution encircling only a single coupled-mode equilibrium point (Rössler type). This solution, when continued for lower frequencies, arises due to period-doubling bifurcations from a limit-cycle solution which also jumps back, for $\lambda_0 < 0.811$, to the P_2 solution. This branch of solutions arises, at $\lambda_0 = 0.811$, due to a saddle–node bifurcation, where a stable periodic (P_1) and an unstable periodic solution are created. This unstable periodic solution continues up to $\lambda_0 = 0.821$ where it annihilates the stable P_1 solution, thus causing the jump from solutions branch arising at the Hopf bifurcation point. The chaotic solution merges, via a homoclinic orbit (Guckenheimer and Holmes 1983), with its twin around $\lambda_0 = 0.836$ and results in a symmetric chaotic attractor. This chaotic solution encircles three equilibrium points (Lorenz type). The Lorenz-type chaotic solution undergoes a 'crisis' (Grebogi *et al.* 1983) around $\lambda_0 = 0.8701$, and after long 'chaotic transients', settles down to the lower stable single-mode constant solution. Figures 16.5a and 16.5b, respectively, show the chaotic attractor for $\lambda_0 = 0.87$ and the transients for $\lambda_0 = 0.8707$. A somewhat similar behaviour is observed for frequencies beginning with the right Hopf frequency λ_{HB_2}, with the 'crisis' arising near $\lambda_0 = 1.854$.

16.4   WEAKLY COUPLED SYSTEM

In this case, the linear coupling spring k is $O(\varepsilon)$ and the response of the system (1) to harmonic external excitation with frequency Ω can be assumed to be

$$\boldsymbol{x}(\tau) = \sum_{j=1}^{n} A_j \boldsymbol{U}_j \cos(\Omega\tau + \alpha_j), \qquad (5)$$

where $\boldsymbol{U}_j = [0, 0, \ldots, 0, 1, \ldots, 0]^{\mathrm{T}}$ with 1 only in the jth place, A_j is the amplitude of motion of the jth particle, and α_j is its phase. Then, proceeding with the averaging process, the time-evolution of the amplitudes and phases, A_j and α_j, is governed by the equations (Samaranayake *et al.* 1993)

$$A_j' = -\varepsilon\{\kappa[A_{j-1}\sin(\alpha_j - \alpha_{j-1}) + A_{j+1}\sin(\alpha_j - \alpha_{j+1})] + dA_j + \mu_j\sin\alpha_j\}, \qquad (6)$$

$$A_j\alpha_j' = \varepsilon\{-\kappa[A_{j-1}\cos(\alpha_j - \alpha_{j-1}) + A_{j+1}\cos(\alpha_j - \alpha_{j+1}) - 2A_j]$$
$$+ pA_j^3 - \lambda A_j + \mu_j\cos\alpha_j\}, \quad j = 1, 2, \ldots, n,$$

where

$$\kappa = k/2m\bar{\omega}^2\Omega, \qquad d = \bar{d}/2m\bar{\omega}^2, \qquad p = T_2/4m\bar{\omega}^2\Omega,$$

$$\mu_j = \mu U_j\Gamma/2m\bar{\omega}^2\Omega, \qquad \lambda = (1 - \Omega^2)/2\Omega\varepsilon.$$

The solutions of the $2n$ averaged equations (6) are a function of the non-dimensional physical parameters κ, p and d, which represent, respectively, the coupling constant, the nonlinearity in an individual subsystem, and the damping in an individual oscillator. The parameters λ and μ_j represent the frequency and the amplitude of external excitation of the jth particle.

The equations (6) represent a class of systems which have a very complex and rich dynamics. It is being explored systematically by the authors, with some preliminary results for a three-mass system presented in Samaranayake *et al.* (1993). The solutions or the system response depends crucially on the nature of excitation, as reflected in the symmetry properties of the averaged equations. Variation in the coupling constant can then lead to symmetry-breaking bifurcations. Consider, for example, a three-mass system with identical in-phase forces acting on the three masses. One set of solutions is then clearly the in-phase identical motions of the particles which are unaffected by the coupling spring. If the individual subsystems are Duffing type with three solutions over a frequency interval, there are three solutions of the coupled system which preserve the symmetry of the equations of motion. The stability of these solutions does, however, depend on the coupling parameter. Further careful consideration of the steady-state solutions for $\kappa = 0$ shows that seven additional solutions exist and four out of the ten solutions are stable. As κ is slowly increased, some of these solutions undergo change in stability and possible new solutions arise. Also, one of these ten solutions represents a localized response in that one particle has an amplitude of response much larger than that of the other two particles. Clearly, the number of possible solutions grows very quickly with the dimension of the system.

ACKNOWLEDGEMENT

This work is supported in part by the financial support of Southern University/US Army Research Office under the grant B840-101-0930/DAAL03-91-G-0247.

REFERENCES

Ashwin, P. and Swift, J. W. (1992) The dynamics of n weakly coupled identical oscillators, *J. Non. Sci.*, **2**, 69–108.

Ashwin, P., King, G. P. and Swift, J. W. (1990) Three identical oscillators with symmetric coupling, *Nonlinearity*, **3**, 585–603.

Bajaj, A. K. and Johnson, J. M. (1992) On the amplitude dynamics and crisis in resonant motion of stretched strings, *Phil. Trans. Roy. Soc. Lond. A*, **338**, 1–41.

Doedel, E. (1986) AUTO: software for continuation and bifurcation problems in ordinary differential equations, *Report*, Dept. Appl. Math. Cal. Inst. of Tech.

Grebogi, G., Ott, E. and Yorke, J. A. (1983) Crises, sudden changes in chaotic attractors, and transient chaos, *Physica D*, **7**, 181–200.

Guckenheimer, J. and Holmes, P. (1983) *Nonlinear Oscillations, Dynamical Systems, and Bifurcation of Vector Fields*, New York, Springer Verlag.

Hale, J. K. (1969) *Ordinary Differential Equations*, New York, Wiley-Interscience.

Nayfeh, A. H. and Balachandran, B. (1989) Modal interaction in dynamical and structural systems, *Appl. Mech. Rev.*, **42**, 175–201.

Samaranayake, S., Bajaj, A. K. and Nwokah, O. D. I. (1992) Amplitude modulated dynamics and bifurcations in the resonant response of a structure with cyclic symmetry, *Nonlinear Vibrations* (eds, R. A. Ibrahim, N. Sri Namachchivaya, A. K. Bajaj), DE-Vol. 50, ASME, pp. 139–150.

Samaranayake, S., Bajaj, A. K. and Nwokah, O. D. I. (1993) Resonant vibrations in weakly coupled nonlinear structures with cyclic symmetry, *14th Biennial ASME Conference on Mechanical Vibrations and Noise, Alburquerque, NM, Sept. 1993.*

Tobias, S. A. (1957) Free undamped non-linear vibrations of imperfect circular disks, *Proc. Inst. Mech. Eng.*, **171**, 691–701.

Tobias, S. A. and Arnold, R. N. (1957) The influence of dynamical imperfection on the vibration of rotating disks, *Proc. Inst. Mech. Eng.*, **171**, 669–690.

Tousi, S. and Bajaj, A. K. (1985) Period-doubling bifurcations and modulated motions in forced mechanical systems, *ASME J. App. Mech.*, **52**, 446–452.

Vakakis, A. F. (1992) Dynamics of a nonlinear periodic structure with cyclic symmetry, *Acta Mech*, **95**, 197–226.

Wei, S. -T. and Pierre, C. (1988) Localization phenomena in mistuned assemblies with cyclic symmetry. Part I: free vibration, *ASME J. Vib., Acous., Str. Rel. in Des*, **110**, 429–438.

A. K. Bajaj, S. Samaranayake and O. D. I. Nwokah, *School of Mechanical Engineering, Purdue University, West Lafayette, IN 47907-1288, USA*

17 NONLINEARITY AND CHAOS IN THE FINITE DYNAMICS OF CABLE MODELS

F. Benedettini, G. Rega and A. Salvatori

The finite forced dynamics of an elastic suspended cable are analysed through different models and various approaches. Numerical and geometrical techniques are used to investigate the bifurcations and chaotic phenomena in a single-degree-of-freedom (SDOF) model. Analytical techniques are used to obtain steady-state regular solutions in a multi-degree-of-freedom model able to account for the experimentally complex behaviour of the real system.

17.1 INTRODUCTION

The finite forced dynamics of an elastic suspended cable are studied through different models. First, use of a single-DOF model allows us to see the main features and richness of planar response of the mechanical system associated with the occurrence of both odd and even nonlinear terms. Numerical and geometrical techniques are combined to provide a deep insight into nonlinear and chaotic phenomena in the light of the system's global dynamics.

Secondly a more accurate representation of the real structure is obtained through a 4-DOF model, based also on the results of previous analytical and experimental models. Possible classes of steady-state regular motions under multiple internal resonance conditions are identified, and some sample results highlighting the complex behaviour of the system are presented.

Nonlinearity and Chaos in Engineering Dynamics
Edited by J. M. T. Thompson and S. R. Bishop,   © 1994 John Wiley & Sons Ltd

17.2   BIFURCATION AND CHAOS IN A SINGLE-DEGREE-OF-FREEDOM CABLE MODEL

17.2.1   Continuum cable model and discrete system for planar dynamics

A heavy elastic cable suspended between two supports at the same level is considered, whose initial static equilibrium configuration lies in the x–y plane (Figure 17.1). The three-dimensional (3D) dynamic configuration occupied under the action of external loads is described by the displacement components u, v, w. The following assumptions are made: (i) parabolic static equilibrium configuration, (ii) moderately large rotations in the cable motion, (iii) initial tension–axial rigidity ratio $H/EA \ll 1$. Assuming the Lagrangian strain as strain measure and applying the extended Hamilton principle, the 3D equations of motion are:

$$EA[u' + y'v' + 1/2(v'^2 + w'^2)]' - \mu_u \dot{u} + p_u = m\ddot{u},$$

$$\{Hv' + EA(y' + v')[u' + y'v' + 1/2(v'^2 + w'^2)]\}' - \mu_v \dot{v} + p_v = m\ddot{v},$$

$$\{Hw' + EAw'[u' + y'v' + 1/2(v'^2 + w'^2)]\}' - \mu_w \dot{w} + p_w = m\ddot{w}, \qquad (1)$$

where the prime and dot denote $\partial/\partial x$ and $\partial/\partial t$, respectively.

The system's planar dynamics are described by the first two equations with $w = 0$. In the absence of an external load in the u direction, and neglecting corresponding inertia and viscous forces, the inplane horizontal displacement can be expressed in terms of the vertical one through a standard condensation procedure, and only one partial integrodifferential equation in the v component is obtained. This is accurate for studying suspended cables for which $H/EA = O[(d/l)^2]$ and the components u, v are respectively $u = O[\varepsilon d^2/l]$ and $v = O[\varepsilon d]$, ε being a small parameter of the order of amplitude. In technical applications (overhead transmission lines), such situations occur for sag-to-span ratios up to about 1/20.

By introducing non-dimensional quantities, representing the displacement through the single-eigenfunction of the linearized problem and a time function $q(t)$, and considering frequency harmonic forcing with given spatial distribution, application of the Galerkin method

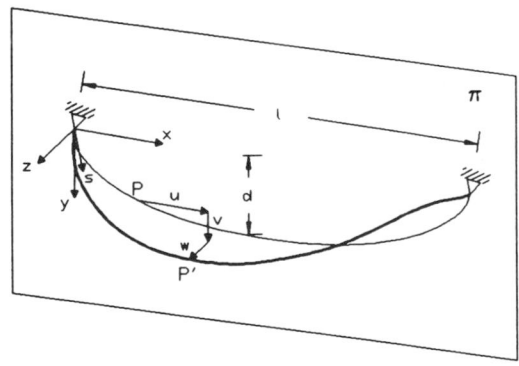

Figure 17.1   Finite dynamics of a suspended cable.

leads to a single ordinary differential equation (ODE):

$$\ddot{q} + \mu\dot{q} + q + c_2 q^2 + c_3 q^3 = P\cos\Omega t \qquad (2)$$

that retains both the quadratic and cubic interaction terms occurring in the original continuum model, associated with the initial curvature and stretching of the cable axis, respectively.

Though the Hamiltonian system associated with equation (2) has three equilibrium points, corresponding to an asymmetric two-well potential, for the numerical case of technical interest considered here ($c_2 = 35.952$, $c_3 = 534.53$), the oscillator has a single equilibrium point and one-well potential. The associated cable vibrating with the first symmetric mode under uniform forcing belongs to the class of initially pre-stressed cables which do not undergo compression.

The presence of both even and odd nonlinear terms in equation (2) gives rise to a rich and complex set of responses associated with the meaningful resonance conditions for the system, namely the primary one, the order-1/2 and -1/3 subharmonics and the order-2 and -3 superharmonics.

17.2.2 Behaviour charts, response curves and bifurcation diagrams

Several numerical techniques are used to examine the steady-state behaviour of the nonlinear system (1), namely, (i) point-by-point computer simulations with prescribed initial conditions

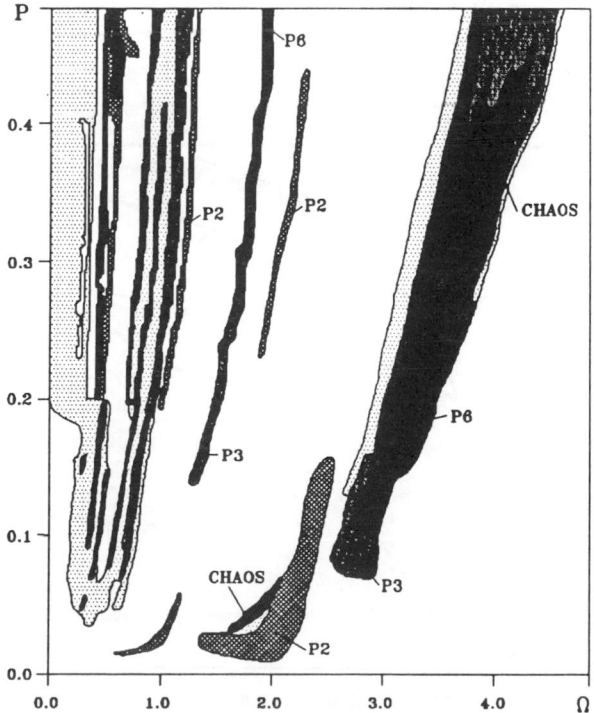

Figure 17.2 Chart of behaviour in excitation-parameter plane.

(i.c.) (ii) path-following algorithm giving fixed points of mappings corresponding to the flow, (iii) cell-mapping procedures showing basins of attraction of coexisting solutions.

A behaviour chart (zero i.c.) in the forcing-parameter space (Ω, P) is shown in Figure 17.2. The kind of response was evaluated on the basis of its Poincaré map and confirmed locally through several quantitative measures (FFT, Lyapunov exponents, fractal dimension). Well-

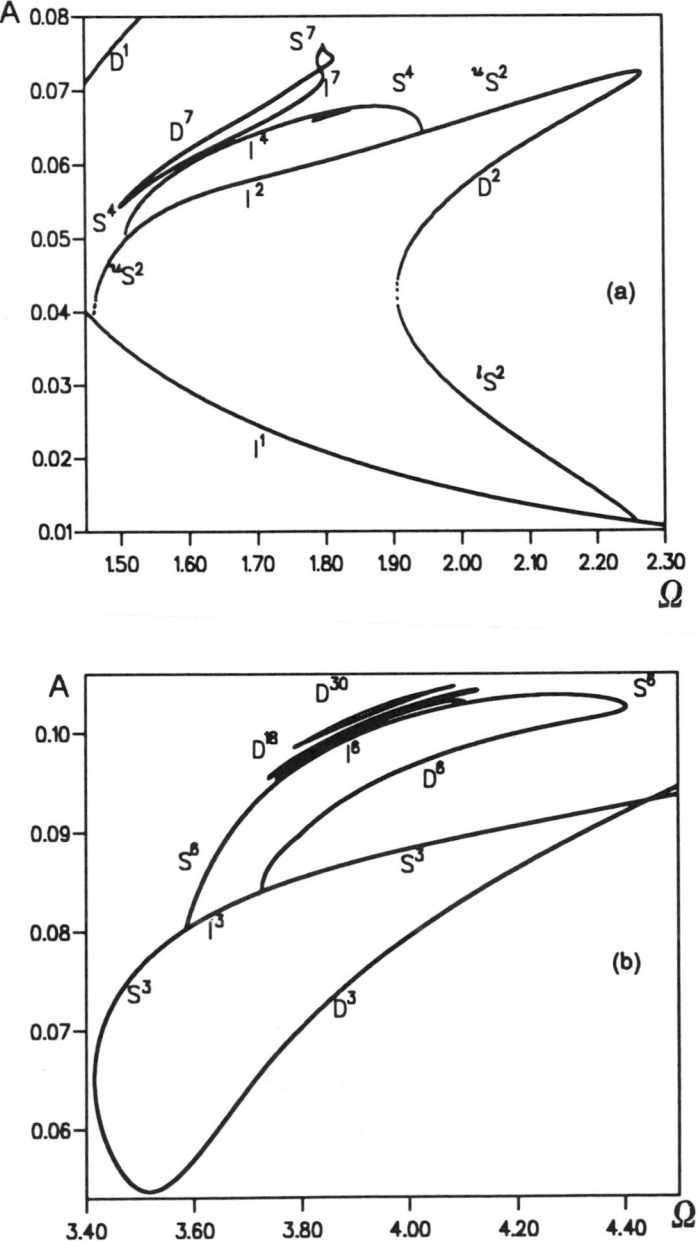

Figure 17.3   Frequency–response curves in the (a) $\frac{1}{2}$- and (b) $\frac{1}{3}$-subharmonic ranges.

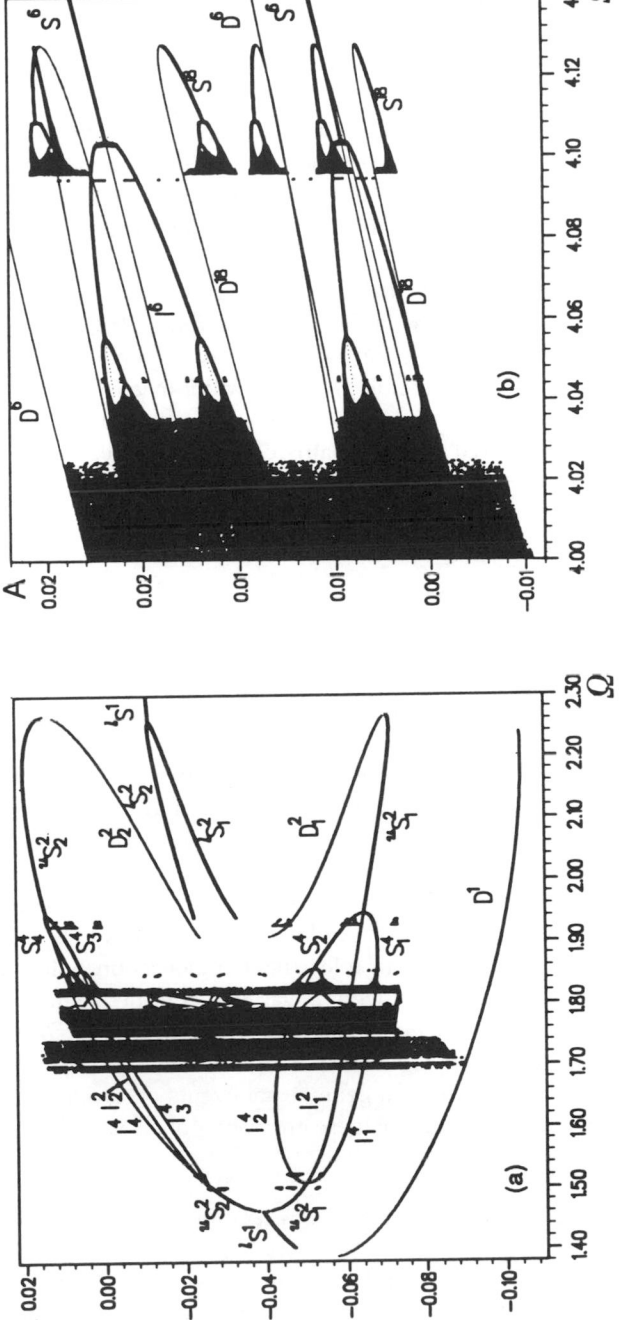

Figure 17.4 Bifurcation diagrams in the (a) $\frac{1}{2}$- and (b) $\frac{1}{3}$-subharmonic ranges.

established zones of regular and chaotic response occur in the neighbourhood of 1/2- and 1/3-subharmonic resonances, while the picture in the superharmonic region is more sensitive to variations of control parameters.

Frequency–response curves at forcing amplitude values for which chaos occurs in the first two regions are reported in Figure 17.3. This shows magnifications of overall response curves, aimed at focusing attention on the period-2 (P-2) solution bifurcated via period-doubling in the neighbourhood of $\Omega \cong 2$ from the non-resonant (lower) P-1 solution (Figure 17.3a), and on the P-3 solution which arose via saddle–node bifurcation in the neighbourhood of $\Omega \simeq 3$ and coexisting with lower P-1 (Figure 17.3b). In the respective regions, these are the two heading solutions of dominant cascades to different chaotic attractors via various transition mechanisms.

Several competing stable attractors are seen to occur. Besides the main responses (P-1, P-2, P-4 in Figure 17.3a, P-1, P-3, P-6 in Figure 17.3b), there also exist some branches of higher-periodicity solutions responsible for the occurrence of secondary (sometimes incomplete) cascades to chaos and/or of periodic windows within chaos zones.

Bifurcation diagrams giving overall or local pictures of the existing attractors are shown in Figure 17.4. Whether compared with the paths of the main direct (D_j^m) and inverse (I_j^m) saddles corresponding to P-m unstable solutions with $j = 1, \ldots, m$ different image points, they are of great help in understanding the geometrical mechanisms through which global bifurcations occur, mostly as far as sudden modifications of size of a chaotic attractor are concerned (crises).

17.2.3   Attractor and basin bifurcations: a selection of geometrical events

Satisfactory insight into the attractor and basin bifurcations occurring when varying a control parameter is obtained primarily by following the evolution of attractor–basin phase portraits. The numerical results with path-following, cell-mapping and computer simulation algorithms are examined in the light of a geometrical description of system dynamics, which allows understanding of the global attractor structure and of the attractor and basin metamorphoses (Ueda 1991). It is based on the construction of global invariant manifolds of the saddle points of the mappings – which play a fundamental role in basin organization (Soliman and Thompson 1991) – and on the relevant evolution with a varying control parameter. Global stable (unstable) manifolds of a direct (inverse) saddle are referred to as $W_k^{s(u)}(D)(W_k^{s(u)}(I))$, with the subscript k distinguishing between the two branches of the manifolds separated by the saddle. In what follows, a selection of meaningful geometrical events occurring for the oscillator (2) are discussed in the light of the behaviour of these invariant global manifolds (Rega *et al.* 1994).

Competing attractors and heteroclinic tangles

Figure 17.5 refers to a frequency value in the neighbourhood of 1/2-subharmonic resonance where three attractors coexist, the sinks $^uS^1$, $^uS_j^2$, $^lS_j^2$, corresponding to the upper P-1 (light-grey basin), upper P-2 (white), lower P-2 (dark grey) solutions, respectively. The direct saddles D^1 and D^2 corresponding to the unstable hysteresis branches of the P-1 and P-2 solutions are located on the basin boundaries.

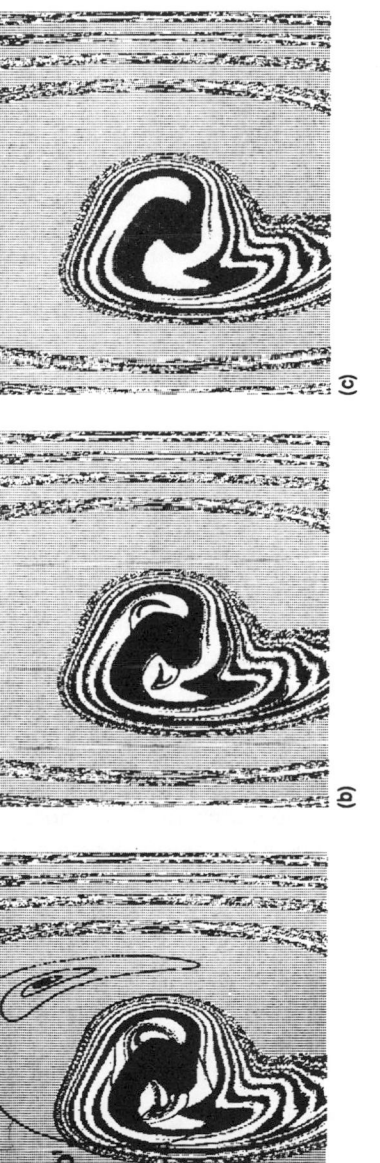

Figure 17.5 Attractor–basin portrait with manifolds of (a) D^1, (b) D_j^2, (c) I^1.

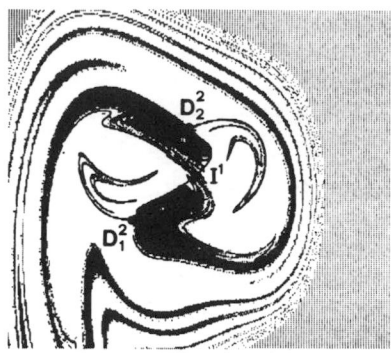

Figure 17.6   Competing P-2 attractors.

$W_i^u(D^1)$ (Figure 17.5a) is seen to be born strongly heteroclinically tangled with the two branches of both $W^s(D_1^2)$ and $W^s(D_2^2)$, i.e. $W_i^u(D^1) \cap W^s(D^2) \neq 0$, being $W^s(D^2) \equiv W^s(D_1^2) \cup W^s(D_2^2)$ (Figure 17.5b); this latter agrees quite well with the boundary separating the dark-grey basin from the two unjoined subdomains corresponding to the upper P-2 solution. $W_i^u(D^1)$ is accumulated onto the two branches of both $W^u(D_1^2)$ and $W^u(D_2^2)$, so that $W^u(D^2) \equiv W^u(D_1^2) \cup W^u(D_2^2)$ is contained within the closure of $W_i^u(D^1)$, i.e. $\overline{W_i^u(D^1)} \supseteq W^u(D^2)$. Equally, both $W^s(D_1^2)$ and $W^s(D_2^2)$ are accumulated on the inside of the basin boundary formed by $W^s(D^1)$, which is thus contained within the closure of $W^s(D^2)$, i.e. $\overline{W^s(D^2)} \supseteq W^s(D^1)$. Such a picture produces a high degree of inter-twining between the white and dark-grey basins such as to render them inaccessible from inside the D^1 saddle (Grebogi *et al.* 1987). The upper P-2 basin (white) is increasing in size with the decreasing control parameter while the lower one (dark grey) is decreasing in size. This latter appears joined, but it is actually separated in two subdomains, too, by the stable manifold $W^u(I^1)$ (Figure 17.5c), I^1 being the saddle originating at period-doubling bifurcation of lower P-1 ($^1S^1 \rightarrow I^1 + {}^1S^2$).

In Figure 17.6, I^1 is located at about the position where the previously unjoined subdomains are going to join and the previously joined ones are going to part. In its neighbourhood, repeated folding of $W_u^s(D_1^2)$ and $W_1^s(D_2^2)$ occurs; accordingly, infinitely thin fingers of the whilte basin are accumulated onto I^1 and its stable manifold. Two main heteroclinic tangles develop, $W_u^s(D_1^2) \cap W_i^u(I^1) \neq 0$ and $W_1^s(D_2^2) \cap W_u^u(I^1) \neq 0$, and entail inaccessibility of I^1 from either basin, locally fractal basin boundary, and rapid erosion of the dark-grey basin that will vanish at the saddle–node annihilation of D_j^2 and $^1S_j^2$) (see Figure 17.4a).

Boundary crisis and transient chaos

Figure 17.7a shows homoclinic tangency between $W^s(D^1)$ and $W_i^u(D^1)$, after which infinite homoclinic crossings of the manifolds rapidly develop and entail jumps of the basin boundary inward into the white evolving region. The steady-state chaotic attractor contained within the closure of $W_i^u(D^1)$ (Figure 17.7b) and its basin, suddenly disappear by collision with D^1 and its stable manifold, and after this global bifurcation the upper P-1 basin occupies the whole i.c. space (Figure 17.8). However, $W^s(D^1)$ delineates a region with the same shape as that of the previously steady chaotic basin, while starting points on $W_i^u(D^1)$ escape to the

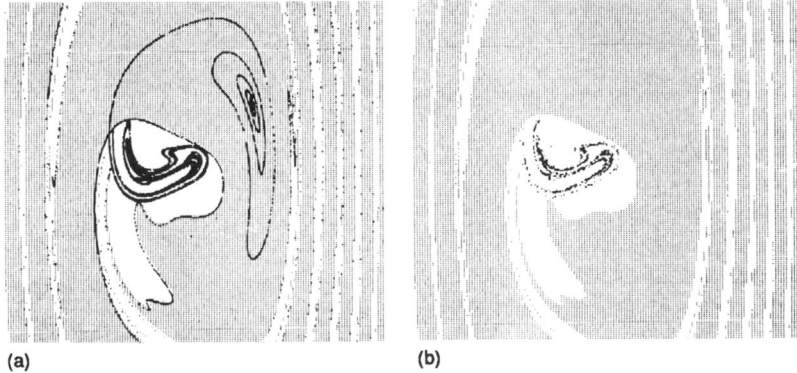

Figure 17.7   (a) Manifolds of D^1 and (b) chaotic attractor, at boundary crisis.

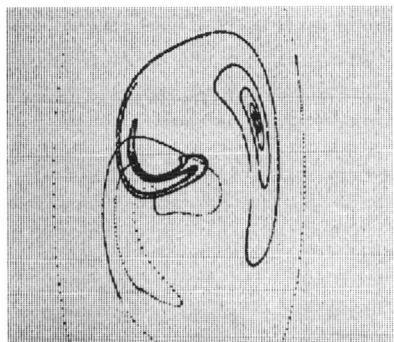

Figure 17.8   Transient chaos.

upper sink $^uS^1$ after visiting that region for a rather long time. The formerly chaotic topological structure still has notable attractive properties, though no longer being trapped inside the stable manifold $W^s(D^1)$ (basin boundary) and thus only transient.

Crises mechanisms

Figure 17.4b shows transition to chaos from a P-6 response sampled with a three times iterated mapping. This solution undergoes a main period-doubling cascade ending in a chaotic attractor made up of six pieces, which cannot join, being the relevant subdomains permanently disconnected by those of stable P-3 solution (Figure 17.3b). The evolution occurs via successive interior crises, at each of which – due to collision with an inverse saddle I^n and the associated homoclinic tangles of the relevant manifolds – the $2n$ pieces of a chaotic attractor merge two-by-two giving rise to an n-piece chaotic attractor. Prior to ending in chaos of the main sequence, further cascades to motions of disturbed periodicity occur (see, e.g., the one originating from a P-18 solution). Each of the relevant saddles (D^{18}) first causes destruction through boundary crisis of the attractor ensuing from the corresponding sink (S^{18}) and then sudden widening through interior crisis of another kind of n-piece chaotic attractor ensuing from S^6.

17.3 THREE-DIMENSIONAL FINITE DYNAMICS OF MULTI-DEGREE-OF-FREEDOM CABLE MODELS

17.3.1 Towards more realistic cable models

An important question arises about the reliability of a planar model in describing the complex dynamics of an actual cable. Analytical models with two or three DOF have recently been proposed and used to analyse modal interactions occurring for the cable in conditions of internal resonance. The following cases of interaction have been considered: between symmetric and antisymmetric modes in the planar motion, between symmetric in-plane and out-of-plane modes in the spatial motion (Rao and Iyengar 1991), and between symmetric in-plane and both symmetric and anti-symmetric out-of-plane modes in the spatial motion, this latter case involving two simultaneous internal resonances (Lee and Perkins 1992). In all cases some suggestions about the possibility of occurrence of chaotic motions can be inferred.

At the same time, experiments conducted on a simple discrete model of a cable made up of a string carrying two equal masses (Benedettini and Moon 1991) have shown the occurrence of meaningful zones of quasi-periodic and chaotic motions and the important role played by the coupling between antisymmetric in-plane and the out-of-plane modes. By way of example, a quasi-periodic 3D motion obtained for a nylon wire carrying eight masses is illustrated in Figure 17.9 through different dynamic measures.

Previous studies highlight that in some areas the results obtained using the single-DOF model are able to capture all, or most of, the behaviour without losing any interesting phenomena, but in others a planar description is too crude to match the actual dynamics, so that a more complete analytical model is needed.

Since for an actual cable the first in-plane mode is the symmetric or the antisymmetric one, depending on the sag-to-span ratio and/or the mechanical properties, it seems suitable to consider at least two harmonics in the description of both the planar and the out-of-plane component in order to gather the meaningful interactions which are likely to occur.

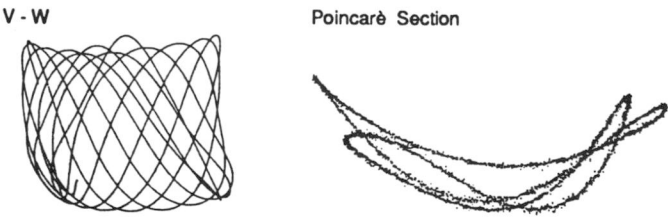

Figure 17.9   Quasi-periodic 3D motion (experimental).

17.3.2 A four-degree-of-freedom model

Starting from equations (1) and eliminating the in-plane horizontal displacement as in Section 17.2.1 leads to two integro-differential equations:

$$\{Hv' + (EA/l)(y' + v') \int_0^l [y'v' + 1/2(v'^2 + w'^2)]dx\}' - \mu_v \dot{v} + p_v = m\ddot{v},$$

$$\{Hw' + (EA/l)w' \int_0^l [y'v' + 1/2(v'^2 + w'^2)]dx\}' - \mu_w \dot{w} + p_w = m\ddot{w}. \tag{3}$$

A 4-DOF model (Benedettini *et al.* 1993) is obtained by describing both v and w through a space-time decoupling using a two-term truncated series of linearized eigenfunctions, one symmetric, $f_1(f_3)$, and one antisymmetric, $f_2(f_4)$:

$$v(x, t) = q_1(t)f_1(x) + q_2(t)f_2(x); \qquad w(x, t) = q_3(t)f_3(x) + q_4(t)f_4(x). \tag{4}$$

In the same way the external forces are decoupled considering time-varying loads with assigned spatial distributions, symmetric $\psi_1(\psi_3)$ and antisymmetric $\psi_2(\psi_4)$:

$$p_v(x, t) = \psi_1(x)p_1(t) + \psi_2(x)p_2(t); \qquad p_w(x, t) = \psi_3(x)p_3(t) + \psi_4(x)p_4(t). \tag{5}$$

The following nondimensionalized discrete form of the equations of motion is obtained:

$$\ddot{q}_1 + \mu_1\dot{q}_1 + \lambda_1^2 q_1 + \sum_{i=1}^4 c_i q_i^2 + \sum_{i=1}^4 d_{1i}q_1 q_i^2 = c_1^e p_1(t),$$

$$\ddot{q}_k + \mu_k\dot{q}_k + \lambda_k^2 q_k + c_{k1}q_k q_1 + \sum_{i=1}^4 d_{ki}q_k q_i^2 = c_k^e p_k(t), \qquad k = 2, 4, \tag{6}$$

$$\ddot{q}_3 + \mu_w\dot{q}_3 + \lambda_3^2 q_3 + c_{31}q_3 q_1 + \sum_{i=1}^4 d_{3i}q_3 q_i^2 = c_3^e p_3(t),$$

where $\lambda_j = \omega_j/\omega_1$, modal dampings are considered, and the various coefficients depend on the elasto-geometric properties of the cable and the assumed spatial shapes. Under the assumption of equal damping, the second and fourth equations are identical, the eigenfunctions describing the nth-order planar and non-planar antisymmetric modes being equal.

Different combinations of internal resonances can be considered, with stronger or weaker interaction among the involved DOF in the system response. Moreover, the occurrence of specific conditions of external, primary and/or sub- (super-) harmonic resonances, can give a major role to the involved, symmetric or antisymmetric, planar or non-planar, DOF or to a subset of them.

17.3.3 Perturbation analysis at multiple internal resonances

The following situation is studied. (i) Cable at first crossover point under multiple internal resonance (i.r.) conditions involving all four modes; indeed, at this point, both planar modes and the antisymmetric non-planar mode have the same frequency (1:1 i.r.), while the symmetric non-planar has half the frequency (2:1 i.r.). (ii) Harmonic external excitation with in-plane and out-of-plane components having same the frequency near the first crossover frequency; this means simultaneous primary resonance for the planar and the antisymmetric non-planar modes and 1/2-subharmonic resonance for the symmetric non-planar mode.

An approximate solution of equations (6) is found using the method of multiple time scales. A second-order approximation is pursued by expanding the solution in a three-term power series of ε: $q_j(t, \varepsilon) = \sum_{n=1}^{3} \varepsilon^n q_{jn}(T_0, T_1, T_2)$, $j = 1, 2, 3, 4$.

The external and the various i.r. conditions are controlled by the detunings σ, ρ_k, ρ_3:

$$\Omega = 1 + \varepsilon\sigma = 1 + \varepsilon(\sigma_1 + \varepsilon\sigma_2), \tag{7a}$$

$$\lambda_k = 1 + \varepsilon\rho_k = 1 + \varepsilon(\rho_{k1} + \varepsilon\rho_{k2}), \qquad k = 2, 4, \tag{7b}$$

$$2\lambda_3 = 1 + \varepsilon\rho_3 = 1 + \varepsilon(\rho_{31} + \varepsilon\rho_{32}), \tag{7c}$$

equation (7b) taking into account perfect tuning of λ_2 and λ_4. Furthermore, the damping coefficients and the excitation amplitudes are expanded as $\mu_j^* = \varepsilon\mu_j$, $P_j^* = \varepsilon^2 P_j$, $j = 1, 2, 3, 4$.

The perturbation procedure involving sequential solutions of systems of ODEs and elimination of secular producing terms at various ε-orders, is followed. Eventually, the equations governing the slow-time-scale modulation of the complex amplitudes of motion A_j to second nonlinear order are obtained as $\dot{A}_j = \varepsilon D_1 A_j + \varepsilon^2 D_2 A_j$, $j = 1, 2, 3, 4$, where $D_i = \partial/\partial T_i$. Approximate steady-state periodic solutions to equations (6) are found from the singular points of an autonomous form of previous equations. For a given set of system parameters, the four real amplitudes of motion a_j and the four phases are obtained by solving a system of eight differential nonlinear equations which admit qualitatively different classes of solutions, partially trivial (involving one to three active DOF) or non-trivial (four DOF).

Their stability is studied by considering perturbations δA_j to the steady-state, trivial or non-trivial, complex amplitudes A_j^*, $j = 1, 2, 3, 4$; a system of linearized ODEs is obtained, whose solutions are sought in the form $\delta A_j = (B_{jre} + iB_{jim})e^{i\vartheta_j}$. Eventually, the autonomous system $\{\dot{B}\} = [J]\{B\}$ is obtained, where $\{B\}$ is an eight-component vector containing real and imaginary parts of the complex amplitude of the perturbations δA and the matrix $[J]'$ depends on the stationary solution whose local stability is analysed.

The time evolution of each perturbation depends on the sign of the eigenvalues Λ_n, $n = 1, 2, \ldots, 8$ of $[J]$. When varying a control parameter, zeroing of one eigenvalue gives a bifurcation condition of the solution between stable and unstable branches along an equilibrium path. At the relevant point, either the same class of motion with different amplitude and/or period or a new class of motion is established.

The stability of each class is evaluated with respect to the complete set of possible perturbations. From both mathematical and physical points of view, it is worth distinguishing between perturbations on active and non-active modes and also considering separately the single perturbations on the non-active modes. Indeed, the composition in active modes of a considered class governs the degree of coupling of the eigenvalue problem to be solved in the relevant stability analysis. For a class of steady motions in which there exist m active and p non-active modes ($m + p = 4$), the kind of functional dependence on the steady amplitudes a_j in the off-diagonal terms of the matrix $[J]$ is such that it is never necessary to solve an 8×8 coupled eigenvalue problem in all four perturbations, but just one coupled subproblem of order $2m \times 2m$ and p uncoupled subproblems of order 2×2. This circumstance reduces the algorithmic effort of the whole stability analysis and allows us to identify the key perturbation for instability at the bifurcation point. This is also useful in the search for the new class of motion – likely to contain just the DOF responsible for instability of the former class – that is established at the relevant bifurcation point.

17.3.4   Classes of steady-state regular motions

Attention is focused on the case in which only external symmetric in-plane excitation is present. The possible, complete or incomplete, classes of motion that are mechanically meaningful and mathematically possible according to the developed 4-DOF model are presented with reference to the dictating i.r. conditions.

(i) *Absence of i.r.* The planar symmetric dynamics studied with the 1-DOF model occur. The unimodal symmetric solution (a_1) is also the only possible one-component motion for the 4-DOF model.

(ii) *Pure 1:1 i.r.* The three different classes of motions (a_1, a_2), (a_1, a_4), (a_1, a_2, a_4) containing only one or both equal antisymmetric components can be activated. Their stable existence depends strongly on the damping ratio μ_2/μ_4, the first or second class being activated respectively on whether or not $\mu_4 > \mu_2$. The third class corresponds to the condition $\mu_2 = \mu_4$, which gives rise to the situation of a non-hyperbolic fixed point for which the performed linearized stability analysis does not make sense.

(iii) *Pure 2:1 i.r.* The 3D symmetric bimodal motion (a_1, a_3) is activated.

(iv) *Simultaneous 1:1 and 2:1 i.r.* The three different classes (a_1, a_2, a_3), (a_1, a_3, a_4), (a_1, a_2, a_3, a_4) can be activated. They undergo the same damping conditions as the corresponding ones without the a_3 component in the pure 1:1 i.r. condition.

The size of the ranges of stable existence for the incomplete (uni-, bi-, trimodal) solutions in control parameter space has to be evaluated.

By way of example, the results obtained for the case with $\mu_4 > \mu_2$ are presented. In Figure 17.10a the response at a given excitation frequency ($\Omega = 0.975$) with excitation amplitudes varying in a low range is shown. The unimodal planar solution is stable only for very low excitation values, where it is the only possible motion. It becomes unstable via pitchfork bifurcation at a forcing level where a bimodal symmetric non-planar solution originates, according to the key role for instability played by the δa_3 perturbation, as shown by the decoupled stability problem for the unimodal solution which is now described by four 2×2 elementary subproblems. The remaining separate perturbations do not limit the stability range of the unimodal solution. In more general terms, referring to subsets of perturbations can give an incorrect stability picture for a given class of motion, whether or not the subset contains the key perturbation for instability. The circumstance of a newly born solution containing just the component associated with the key perturbation for the previous solution, still occurs at a higher forcing value, where a trimodal (a_1, a_2, a_3) solution bifurcates from the bimodal (a_1, a_3) one via pitchfork bifurcation. The dependence of solutions on forcing frequency and the agreement with numerical results are shown in Figure 17.10b, which refers to a forcing amplitude for which either the (a_1) or the (a_1, a_3) motions occur. For higher forcing amplitudes there is coexistence of the bimodal and trimodal solutions,

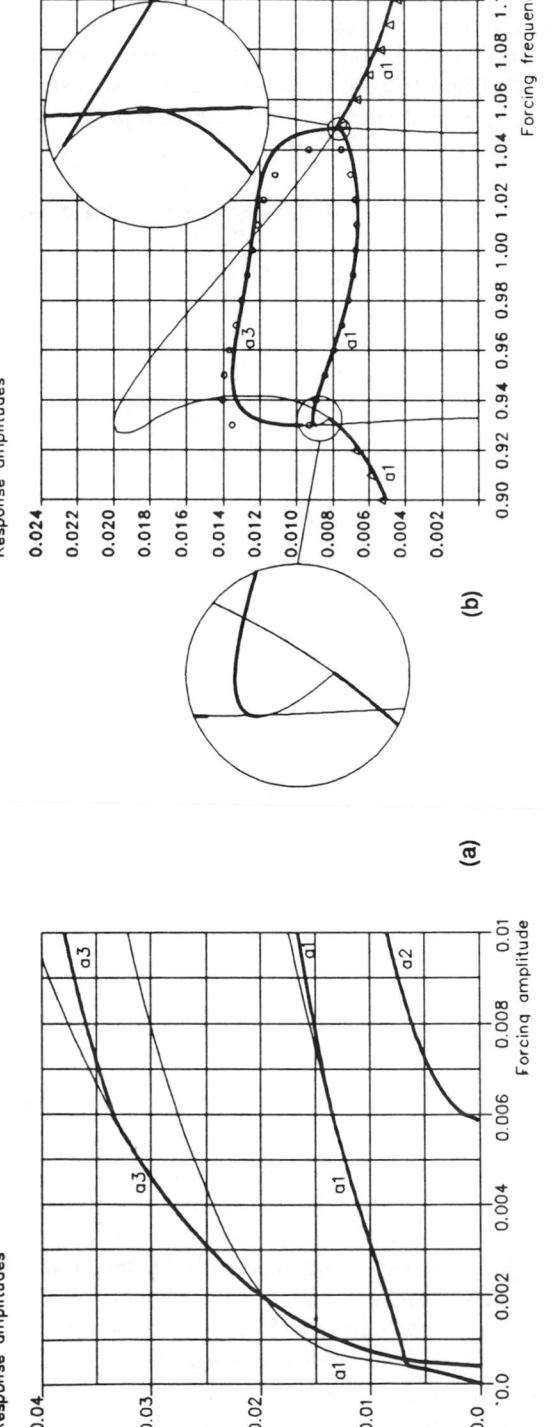

Figure 17.10 Classes of steady-state regular motions.

which throws some light on the likely complexity of the global picture of system response associated with the occurrence of multiple attractors and their basins of attraction, which is computationally hard to obtain for a eight-dimensional problem.

17.4 CONCLUDING REMARKS

A single-DOF model with odd and even nonlinearities representative of the cable planar dynamics has been considered. It exhibits many fundamental characteristics of the behaviour of nonlinear dynamical systems: multiple competing attractors, strange chaotic attractors in various regions of control-parameter space, different local and global bifurcation mechanisms, regular and fractal basin boundaries.

A multi-DOF model has subsequently been developed and studied through analytical techniques, to provide useful insight into the complex behaviour of the actual mechanical system which has several competing classes of regular motions.

REFERENCES

Benedettini, F. and Moon, F. C. (1991) Experimental dynamics of a hanging cable carrying concentrated masses, *Atti DISAT, University of L'Aquila*, **1**, to appear in *Int. J. Bif. Chaos.*

Benedettini, F., Rega, G. and Alaggio, R. (1994) Nonlinear oscillations of a four-degree-of-freedom model of a suspended cable under multiple internal resonance conditions, to appear in *J. Sound Vib.*

Grebogi, C., Ott, E. and Yorke, J. A. (1987) Basin boundary metamorphoses: changes in accessible boundary orbits, *Physica*, **24D**, 243–262.

Lee, C. L. and Perkins, N. C. (1992) Three-dimensional oscillations of suspended cables involving simultaneous internal resonances, *Proc. ASME Wint. Ann. Meet.*, **144**, 59–67.

Rao, G. V. and Iyengar, R. (1991) Internal resonance and non-linear response of a cable under periodic excitation, *J. Sound Vib.*, **149**, 25–41.

Rega, G., Salvatori, A. and Benedettini, F. (1993) Numerical and geometrical analysis of bifurcation and chaos for an asymmetric elastic nonlinear oscillator, to appear in *Nonlin. Dyn.*

Soliman, M. S. and Thompson, J. M. T. (1991) Basin organization prior to a tangled saddle-node bifurcation, *Int. J. Bif. Chaos*, **1**, 107–118.

Ueda, Y. (1991) Survey of regular and chaotic phenomena in the forced Duffing oscillator, *Chaos, Sol. and Fract*, **1**, 199–231.

F. Benedettini, G. Rega and A. Salvatori, *Dipartimento di Ingegneria delle Strutture, Acque e Terreno, Università dell' Aquila, Monteluco di Roio, 67040 L'Aquila, Italy*

18 CHAOTIC OSCILLATIONS OF A FLUID-CONVEYING VISCOELASTIC TUBE

A. Steindl and H. Troger

The planar oscillations of a slender initially bent tube with circular cross-section converying incompressible fluid are studied by a numerical analysis. A large-displacement but small-strain beam model for the tube, which is geometrically nonlinear and physically linear, is used. By varying two parameters, namely, the initial curvature and the flow rate, values of the initial curvature are found where a homoclinic orbit exists. In this parameter region for quasi-statically increasing flow rate, after loss of stability of the continuously deforming equilibrium configuration, a transition to chaotic tube motions via periodic motions with various periods is found. In fact a Silnikov-type scenario is present with all its interesting properties.

18.1 INTRODUCTION

Slender fluid-conveying tubes are an important technical system with extremely complicated dynamic behaviour. There have been many investigations, both theoretical and experimental: for example Bajaj and Sethna (1984), Benjamin (1961), Copeland and Moon (1992), Païdoussis and Issid (1974), Rousselet and Herrmann (1981), Steindl (1992) and Sugiyama et al. (1985).

In Champneys (1991) the planar motion in a horizontal plane of a pair of elastically joined articulated pipes conveying fluid is, studied. Basically, a fluid-conveying planar double pendulum with elastic joints is treated. The interesting behaviour obtained by a numerical analysis depends, as one would expect, first on the flow rate and secondly on the unloaded stress-free configuration of the elastic double pendulum. Here, as well as the reflectionally symmetric straight configuration we also consider unloaded unstrained configurations that break the reflectional symmetry. The amount of this non-symmetry is expressed by the angle ψ measured between the first pipe and the second one for which the second joint is unstrained.

Nonlinearity and Chaos in Engineering Dynamics
Edited by J. M. T. Thompson and S. R. Bishop,   © 1994 John Wiley & Sons Ltd

Hence, a codimension-2 problem is studied (Troger and Steindl 1991) where the bifurcation parameter is the flow rate and the angle ψ is the second parameter.

It has been shown in Bajaj and Sethna (1984) that the reflectionally symmetric straight configuration ($\psi = 0$) loses its stability by a Hopf bifurcation to a flutter motion. For non-symmetric pipe configurations ($\psi \neq 0$) it is shown in Champneys (1991) that there exists a certain small range of values of ψ where a very complex dynamical behaviour of the double pendulum is obtained. This is basically due to the occurrence of a homoclinic orbit (Guckenheimer and Holmes 1983). Since the conditions of the Silnikov scenario are fulfilled, we find all the features of the dynamics obtainable from it, as nicely described in Glendinning and Sparrow (1984) and Hirschberg and Knobloch (1992): for example, it is shown that a countable infinity of period-doubling cascades are present in this problem. Of course, in performing a numerical analysis of a physical system only a very few could be detected.

In the present chapter we extend the work of Champneys (1991) especially from the mechanical modelling point of view by dealing with a more realistic physical model. That is, we first study a continuous pipe model, hence, discarding the possibility of fixing stiffnesses of elastic joints as we please. Secondly, in Champneys (1991) the motion of the double pendulum takes place in a horizontal plane, thus eliminating the influence of gravity. We assume that the motion of the tube takes place in a vertical plane, thus also including the effect of gravity. The tube is clamped at one end and free at the other end. We give special care to the derivation of the equations of motion of the tube fluid system, especially in order to derive consistent expressions for the dissipation due to internal damping. Director rod theory is used (Buzano *et al.* 1985; Simo 1985) which allows us in a geometrically exact way to derive the nonlinear kinematic relations describing the deformation of the tube. Under the assumption of large displacements but small strains the linear Kelvin–Voigt law of viscoelasticity is introduced.

We also treat a two-parameter problem. Of course, as primary bifurcation parameter again the flow rate is used. The second parameter, which in Champneys (1991) was the angle ψ at which the second joint was unstrained, is now in our analysis the initially bent unstrained configuration of the tube. We assume that this unstrained configuration is of circular form as is quite natural for a piece of tube which has been stored for some time round a drum.

The method of analysis of the complex motion follows that given in Champneys (1991). However, since our mechanical model is continuous and therefore described by a set of partial differential equations a proper dimension-reduction method must be applied. Here we apply the method of 'inertial manifolds' (Foias *et al.* 1988; Brown *et al.* 1990). For choice of the appropriate dimension of the critical variable set, results of numerical simulations are used.

18.2   MECHANICAL MODEL AND EQUATIONS OF MOTION

For the derivation of the equations of motion we follow the presentations given in Steindl and Troger (1992) where equations for the three-dimensional motion of a tube are given. We start with this set describing the three-dimensional motion because it allows us immediately to restrict the equations to planar motions only. In addition this approach has the advantage that we are able to check if the planar oscillations become unstable and, hence, oscillations out of the plane of bending could occur.

In the mechanical model used we adopt the assumptions made in Lundgren *et al.*

(1979). For the fluid flow they are: (i) the fluid motion is an incompressible 'slug' flow with constant velocity U relative to the tube; (ii) the variation in the pressure of the fluid across a cross-section due to lateral accelerations is neglected. For the deformation of the tube the following assumptions are made: (a) the tube is of uniform annular cross-section; (b) the tube is long compared with its diameter; (c) effects of rotatory inertia and shear deformation are neglected; (d) the centre line of the tube is inextensible; (e) plane sections before deformation remain plane during deformation (Bernoulli–Euler theory). However, we do not assume that the tube is initially straight, and in addition we also introduce internal (material) damping.

Due to assumption (b) we can expect that even for large deformations – and there will be no restriction on the magnitude of deformation – we still have small strains allowing us to introduce the linear Kelvin–Voigt material law of viscoelasticity.

We derive the equations of motion in three steps (Steindl and Troger 1992; Buzano *et al.* 1985).

1. We consider the deformation of the tube. It is given, first, by the radius vector $r(s, t)$ defining the location of a point on the axis of the tube and, second, by the rotation matrix $B(s, t)$ defining the orientation of the cross-section of the tube, given by the vectors (t_1, t_2, t_3) with respect to a space-fixed triad (e_1, e_2, e_3) (Figure 18.1). From the rotation matrix B we can calculate the skew-symmetric infinitesimal rotation matrix $\hat{\Omega}$ in the form $(( )' = \partial/\partial s, \text{ or } (\dot{ }) = \partial/\partial t)$:

$$\hat{\Omega}(s, t) = B^{\mathrm{T}} B' = \begin{pmatrix} 0 & -\Omega_3 & \Omega_2 \\ \Omega_3 & 0 & -\Omega_1 \\ -\Omega_2 & \Omega_1 & 0 \end{pmatrix}, \tag{1}$$

where Ω_1 and Ω_2 measure the curvature of the rod about the axes t_1 and t_2, and Ω_3 measures the twist about t_3. Hence, the kinematic relations are

$$r' = Be_3,$$
$$B' = B\hat{\Omega}. \tag{2}$$

2. From the principles of linear and angular momentum we obtain the equations of motion:

$$T' = T \times \Omega + F \times e_3,$$
$$F' = F \times \Omega + B^{\mathrm{T}}(-(m_{\mathrm{T}} + m_{\mathrm{F}})g e_3 + \delta \dot{r} + m_{\mathrm{T}} \ddot{r} + m_{\mathrm{F}}(\ddot{r} + 2\dot{r}'U + r''U^2)), \tag{3}$$

where F is the force and T is the moment in the cross-section in the $\{t_i\}$ system; m_{T} and

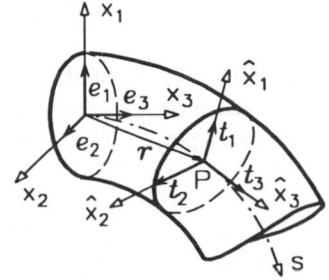

Figure 18.1 The deformation of the tube is described by the position of a point on the centreline given by the radius vector $r(s, t,)$ and by the orientation of the tripod (t_1, t_2, t_3) with respect to the space-fixed tripod (e_1, e_2, e_3).

m_F are the masses of the tube and the fluid per unit length, respectively; U is the velocity of the fluid flow; and δ is an external damping coefficient.

3. Finally, a relationship between the moment vector T and the curvatures Ω must be formed. This relationship is supplied by the material law. Due to the inextensibility constraint, that is, assumption (d) given above, the axial strain is zero and, hence, no constitutive relationship relating the strain to the section-force F exists. However, the relationship between the moment vector $T = (T_1, T_2, T_3)^T$ and the deformation vector consisting of the components of Ω is given by

$$T_1 = EJ(\Omega_1 + \alpha_1\dot{\Omega}_1),$$
$$T_2 = EJ(\Omega_2 - \Omega_{20} + \alpha_1\dot{\Omega}_2), \tag{4}$$
$$T_3 = GJ_T(\Omega_3 + \alpha_3\dot{\Omega}_3),$$

where EJ and GJ_T are the bending and torsional stiffnesses, the α_i are material damping coefficients, and Ω_{20} is the initial curvature of the planar deformed unstrained tube in the (e_1, e_3) plane.

The whole set of equations of motion is given by (2), (3) and (4). They are supplemented by boundary conditions at both ends of the tube ($s = 0$ and $s = l$):

$$r(0) = 0, \quad B(0) = E, \quad F(l) = 0, \quad T(l) = 0. \tag{5}$$

Before we proceed we introduce non-dimensional quantities in the following way:

$$\tilde{s} = \frac{s}{l}, \quad \tilde{r} = \frac{r}{l}, \quad \tilde{\Omega} = l\Omega, \quad v = \sqrt{\frac{EJ}{m_T + m_F}} \frac{1}{l^2},$$

$$\tilde{t} = vt, \quad \beta = \frac{m_F}{m_F + m_T}, \quad \tilde{F} = \frac{l^2 F}{EJ}, \quad \tilde{\delta} = \delta v, \tag{6}$$

$$\gamma = \frac{l^3 g(m_T + m_F)}{EJ}, \quad \gamma_3 = \frac{GJ_T}{EJ}, \quad \rho = \sqrt{\frac{m_F}{EJ}} lU, \quad \tilde{\alpha}_i = \alpha_i v.$$

From now on all quantities in the equations are dimensionless; however, for convenience, we drop the tildes.

18.3   STATIC PLANAR DEFORMATION OF THE FLUID-CONVEYING INITIALLY BENT TUBE

As first step in the study of the planar tube motion we calculate the static deformation of the curved tube. We assume that bending takes place in the (e_1, e_3) plane. Hence, the rotation matrix B depends only on one angle, which we designate by ϑ. It is given by

$$B = R_2(\vartheta) = \begin{pmatrix} \cos\vartheta & 0 & \sin\vartheta \\ 0 & 1 & 0 \\ -\sin\vartheta & 0 & \cos\vartheta \end{pmatrix}. \tag{7}$$

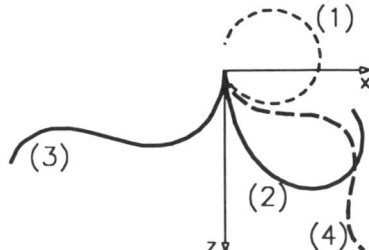

Figure 18.2 Planar states of the initially curved tube ($\Omega_{20} = 6.0$) in a vertical plane for various values of the flow rate ρ: (1) initial, unstrained state (gravitational acceleration $g = 0.0$, $\rho = 0$); (2) shape of the fluid-filled tube ($g = 9.81 \, \text{m/s}^2$, $\rho = 0$); (3) shape of the tube at stability boundary ($\rho = \rho_H = 5.9$) (see Figure 18.3); (4) shape of the tube where the homoclinic orbit occurs ($\rho = \rho_h \approx 8.5$).

The equations of motion reduce to the planar steady-state equilibrium conditions

$$\bar{r}_1' = \sin \bar{\vartheta},$$
$$\bar{r}_3' = \cos \bar{\vartheta},$$
$$\bar{\vartheta}' = \bar{\Omega}_2 = \Omega_{20} + \bar{T}_2,$$
$$\bar{T}_2' = -\bar{F}_1,$$
$$\bar{F}_1' = -\bar{F}_3 \bar{\Omega}_2 + \cos \bar{\vartheta} q_1 - \sin \bar{\vartheta} q_3, \tag{8}$$
$$\bar{F}_3' - \bar{F}_1 \bar{\Omega}_2 + \sin \bar{\vartheta} q_1 + \cos \bar{\vartheta} q_3,$$
$$q_1 = \varrho^2 \bar{r}_1'' = \varrho^2 \cos \bar{\theta} \bar{\theta}',$$
$$q_3 = \varrho^2 \bar{r}_3'' - \gamma = -\varrho^2 \sin \bar{\vartheta} \bar{\vartheta}' - \gamma.$$

Variables describing the statically deformed tube are designated with an overbar. Ω_{20} refers to the unstrained state. $\bar{T}_2$ is the bending moment about t_2.

In Figure 18.2 four configurations of the tube are shown for the deformation in a vertical plane where the influence of gravity is taken into account. Note that the static deformation of the tube occurs in a wide range.

18.4   CALCULATION OF THE STABILITY BOUNDARY

The configurations calculated from (8) are only stable up to a critical flow rate ϱ_c, which can be calculated by linearization of the equations of motion about the considered state. We obtain two sets of equations: one for the linearized motion in the plane of bending and one for the small oscillations out of the plane of bending.

With $\vartheta = \bar{\vartheta} + \Theta$, $r = \bar{r} + R$, $F = \bar{F} + \hat{F}$, $B = \bar{B} + \hat{B}$ where $\bar{B}$ is given by (7) and $\Omega_2 = \bar{\Omega}_2 + U_2$ where Θ, U_2, $\hat{F}$, $\hat{B}$, T and R designate small deviations from the static equilibrium; the differential equations for small oscillations in the plane of bending are

$$R_1' = \cos \bar{\vartheta} \Theta,$$
$$R_3' = -\sin \bar{\vartheta} \Theta,$$

$$\Theta' = U_2, \quad T_2 = U_2 + \alpha_1 \dot{U}_2,$$

$$T_2' = -\hat{F}_1,$$

$$\hat{F}_1 = -\hat{F}_3 \bar{\Omega}_2 - \bar{F}_3 U_2 + \cos\bar{\vartheta} Q_1 - \sin\bar{\vartheta} Q_3 - \Theta(\sin\bar{\vartheta} q_1 + \cos\bar{\vartheta} q_3),$$

$$\hat{F}_3 = \hat{F}_1 \bar{\Omega}_2 + \bar{F}_1 U_2 + \sin\bar{\vartheta} Q_1 + \cos\bar{\vartheta} Q_3 + \Theta(\cos\bar{\vartheta} q_1 - \sin\bar{\vartheta} q_3),$$

$$Q_1 = \ddot{R}_1 + 2\sqrt{\beta \varrho} \dot{R}_1 + \varrho^2 R_1'' + \alpha_e \dot{R}_1,$$

$$Q_3 = \ddot{R}_3 + 2\sqrt{\beta \varrho} \dot{R}_3 + \varrho^2 R_3'' + \alpha_e \dot{R}_3,$$

For small oscillations out of the plane of bending we obtain

$$R_2' = \hat{B}_3 \sin\bar{\vartheta} - \hat{B}_1 \cos\bar{\vartheta},$$

$$\hat{B}_1' = \Omega_1 \cos\bar{\vartheta} + \Omega_3 \sin\bar{\vartheta},$$

$$\hat{B}_3' = \Omega_3 \cos\bar{\vartheta} - \Omega_1 \sin\bar{\vartheta},$$

$$T_1 = \Omega_1 + \alpha_1 \dot{\Omega}_1, \quad T_3 = \gamma_3(\Omega_3 + \alpha_3 \dot{\Omega}_3),$$

$$T_1' = \bar{T}_2 \Omega_3 - T_3 \bar{\Omega}_2 + \hat{F}_2,$$

$$T_3' = T_1 \bar{\Omega}_2 - \bar{T}_2 \Omega_1,$$

$$\hat{F}_2 = -\bar{F}_1 \Omega_3 + \bar{F}_3 \Omega_1 + (\ddot{R}_2 + 2\sqrt{\beta \varrho} \dot{R}_2' + \varrho^2 R_2'' + \alpha_e \dot{R}_2) - \hat{B}_{3q1} + \hat{B}_{1q3}.$$

The quantities $\bar{R}_2$, $\hat{B}_1$, $\hat{B}_3$, T_1, T_3 and $\hat{F}_2$ are skew-symmetric under a reflection in the (x_1, x_3) plane.

The stability boundary in the ϱ, Ω_{20} parameter space for in-plane instability is given in Figure 18.3. At the whole boundary the loss of stability occurs at a pure imaginary pair of eigenvalues. We note that about $\Omega_{20} \approx 4.5$ there exists a domain of initial curvatures of the

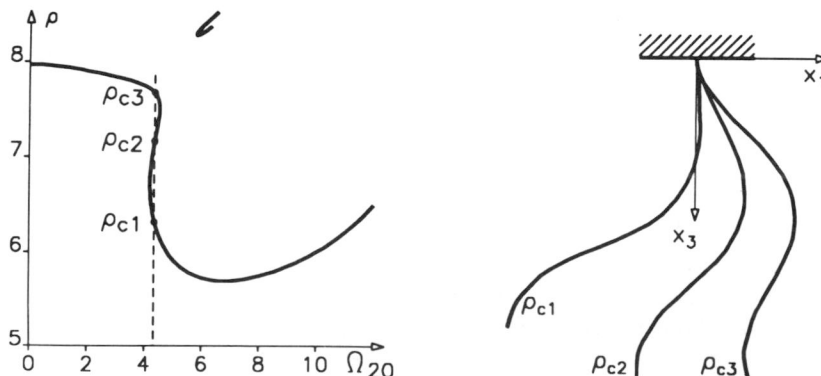

Figure 18.3 In the left-hand diagram the stability boundary in the (ϱ, Ω_{20}) parameter space for the in-plane motion is shown. At loss of stability a purely imaginary pair occurs, leading to a supercritical Hopf bifurcation. The small curve located at $\rho = 8.5$ and $\Omega_{20} = 6$ gives parameter values for which homoclinic orbits exist. In the right-hand diagram tube configurations at loss of stability are shown corresponding to the three critical states ρ_{ci}.

unstrained tube where for increasing flow rate ϱ at ϱ_{c1} a loss of stability by a supercritical Hopf bifurcation occurs. Increasing the flow rate beyond ϱ_{c1} the amplitude of the periodic motion increases but then gradually decreases to yield again a static equilibrium at ϱ_{c2}. In the interval $[\varrho_{c2}, \varrho_{c3}]$ we have a stable static equilibrium. For $\varrho = \varrho_{c3}$, again a supercritical Hopf bifurcation occurs. However, no further bifurcations could be found in the interval $(\varrho_{c2}, \varrho_{c2})$.

18.5. INERTIAL MANIFOLD REDUCTION

The reduction of the infinite-dimensional problem to a low-dimensional problem in critical variables is performed making an approximate inertial manifold calculation following Brown *et al.* (1990). First, we discretize by means of a 10-degree-of-freedom finite-difference scheme to obtain a 20th-order system. We designate those four variables corresponding to the eigenvalues with the largest real part as critical variables by p, and decompose the 20th-order system in the following form:

$$\frac{dp}{dt} = Ap + PF(p + q), \tag{9a}$$

$$\frac{dq}{dt} = Bq + QF(p + q), \tag{9b}$$

$p \in \mathbb{R}^4$, $q \in \mathbb{R}^{16}$. $P: \mathbb{R}^{20} \to \mathbb{R}^4$ denotes the projection to the first four modes and $Q - I - P$ is the projection onto the orthogonal complement. We note that the eigenvalues of the critical modes are both complex-conjugate pairs, having the largest real parts: one pair with a positive and the other with a negative real part. The eigenvalues of B are all strictly in the left half-plane separated by a gap from those corresponding to the critical modes. We now replace (9) by

$$\frac{dp}{dt} = Ap + PF(p + \phi_a(p)).$$

For ϕ_a, which is an approximation of the inertial, manifold, an iterative scheme is set up in the following way. Setting $dq/dt = 0$ in (9b) we obtain the mapping

$$T(q) = - B^{-1} QF(p + q).$$

The inertial manifold ϕ is given as its fixed point. We approximate it by ϕ_a which follows from the following iteration. Starting with $\phi_0 \equiv 0$, which is the linear approximation used in the traditional Galerkin approach, we explicitly obtain

$$\phi_1 = - B^{-1} QF(p), \quad \phi_2 = - B^{-1} QF(p + \phi_1(p)), \dots$$

In our calculations we used ϕ_2. We further note by comparing the configuration corresponding to ϱ_{c3} in Figure 18.3 and the configuration where the homoclinic orbit occurs in Figure 18.2 that they are quite similar in shape. Hence, the modes to calculate the critical system in the inertial manifold reduction near the homoclinic orbit are taken at the stability boundary at ϱ_{c3}.

18.6   THE HOMOCLINIC ORBIT

The main task to prove chaotic motions of the tube is to show the existence of a homoclinic orbit. We proceed as in Champneys (1991). That is, we check the period of the primary oscillatory orbit following from the Hopf bifurcation at the stability boundary increasing the flow rate beyond the stability boundary for various values of the initial curvature Ω_{20}. Around $\Omega_{20} = 6$ we found that the period tends to increase strongly for increasing flow rate ϱ. Interestingly the wiggly curve obtained in Champneys (1991) and Glendinning and Sparrow (1984) is also found (Figure 18.4). Hence, we are encouraged to expect that a scenario similar to that described in Glendinning and Sparrow (1984) occurs also for the curved tube.

However, our situation is different from that described in Champneys (1991) and Glendinning and Sparrow (1984) where a classical Silnikov scenario with one real eigenvalue and a pair of complex eigenvalues is found. Here the critical system is four-dimensional because both the stable and the unstable manifold in the neighbourhood of the equilibrium are two-dimensional and carry a stable and an unstable focus.

In Figure 18.5 projections of the homoclinic orbit to two different three-dimensional subspaces of the four-dimensional phase space are depicted showing that the local behaviour

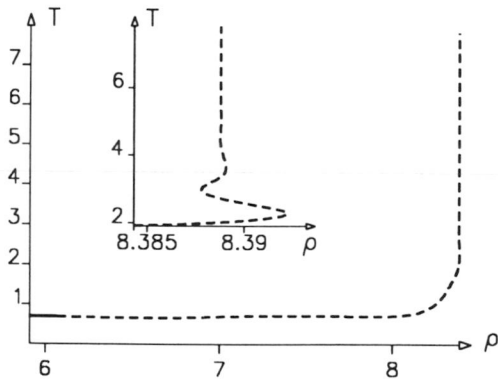

Figure 18.4   Period T of the oscillatory motion in its dependence on the flow rate ρ for the primary branch bifurcating off the deformed equilibrium for an initially unconstrained configuration $\Omega_{20} = 6$. The wiggly character of the curve becomes evident only for a smaller scale in the detailed part.

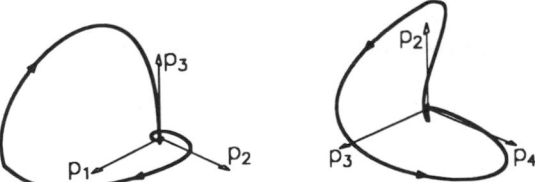

Figure 18.5   Projections of the homoclinic orbit in the four-dimensional phase space to two different three-dimensional subspaces showing the stable and unstable local focal behaviour.

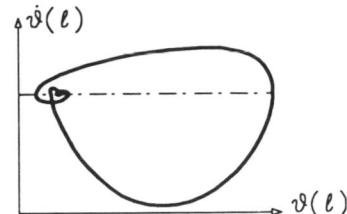

Figure 18.6   Projection of the motion of the tangent to the end of the tube in the phase plane.

of the trajectories in the neighbourhood of the equilibrium is of the focus type, both in the stable and unstable manifold. Physically speaking, the tube starts to oscillate with increasing amplitude with frequency ω_u then making a large amplitude move to return back to the equilibrium oscillating now with decreasing amplitude with frequency ω_s. This is clearly shown in Figure 18.6 where the motion of the tangent to the end of the tube is projected into its phase plane. Finally, we remark that for the calculation of the homoclinic orbit a boundary value problem is formulated prescribing the position of the tube for $t \to -\infty$ and $t \to +\infty$.

18.7   CONCLUSIONS

In Glendinning and Sparrow (1984) section 3.2.4 and in Hirschberg and Knobloch (1992) it is shown that from the existence of a homoclinic orbit with the wiggly curve similar to that shown in Figure 18.4 the existence of an infinite number of horseshoe maps can be proved. Hence, a very complex and complicated behaviour of the tube can result, though this has been located only in a very narrow parameter region.

ACKNOWLEDGEMENT

This research project has been supported by the Austrian Science Foundation (FWF) under project P07003.

REFERENCES

Bajaj, A. K. and Sethna, P. R. (1984) Flow induced bifurcations to three-dimensional oscillatory motions in continuous tubes, *SIAM J. Applied Math.*, **44**, 270–286.

Benjamin, T.B. (1961) Dynamics of a system of articulated pipes conveying fluid, I. Theory; II. Experiments, *Proceedings of the Royal Society London A*, **261**, 457–486; 487–499.

Brown, H. S., Jolly, M. S., Kevrekidis, I. G. and Titi, E. S. (1990) Use of approximate inertial manifolds in bifurcation calculations. In: *Continuation and Bifurcations: Numerical Techniques and Applications*, D. Roose *et al.* (eds.) Dordrecht, Kluwer, pp. 9–23.

Buzano, E., Geymonat, G. and Poston, T. (1985) Post-bucking behavior of a non-linearly hyperelastic thin rod with cross-section invariant under the dihedral group D_n, *Archive of Rational Mechanics and Analysis*, **89**, 307–388.

Champneys, A. R. (1991) Homoclinic orbits in the dynamics of articulated pipes conveying fluid, *Nonlinearity*, **4**, 747–774.

Copeland, G. S. and Moon, F. C. (1992) Chaotic flow-induced vibration of a flexible tube with end mass, AMD, Vol. 152, *Stability and Control of Pipes Conveying Fluid*, (M. P. Paidoussis and N. S. Namachchivaya, eds.), Book no. 400729, New York, ASME, pp. 63–77.

Foias, C., Jolly, M. S., Kevrekidis, I. G., Sell, G. R. and Titi, E. S. (1988) On the computation of inertial manifolds, *Physics Letters*, **131A**, 433–437.

Glendinning, P. and Sparrow, C. (1984) Local and global behavior near homoclinic orbits, *J. of Statistical Physics*, **35**, 645–696.

Guckenheimer, J. and Holmes, P. (1983) *Nonlinear Oscillations, Dynamical Systems, and Bifurcations of Vector Fields*, Applied Math. Sciences Vol. 42, Berlin, Springer-Verlag.

Hirschberg, P. and Knobloch, E. (1992) Silnikov–Hopf-Bifurcation, Preprint.

Lundgren, T. S., Sethna, P. R. and Bajaj, A. K. (1979) Stability boundaries for flow induced motions of tubes with an inclined terminal nozzle, *J. of Sound and Vibration*, **64**, 553–571.

Païdoussis, M. P. and Issid, N. T. (1974) Dynamic stability of pipes conveying fluid, *Journal of Sound and Vibration*, **33**, 267–294.

Rousselet, J. and Herrmann, G. (1981) Dynamic behavior of continuous cantilevered pipes conveying fluid near critical velocities, *Journal of Applied Mechanics*, **48**, 945–947.

Simo, J. C. (1985) A finite strain beam formulation. The three-dimensional dynamic problem. Part I. *Computer Methods in Appl. Mech. and Engineering*, **49**, 55–70.

Steindl, A. (1992) Hopf/steady-state mode interaction for a fluid conveying elastic tube with D_4-symmetric support, *Int. Series of Num. Math.*, **104**, Basel, Birkhäuser Verlag, pp. 305–315.

Steindl, A. and Troger, H. (1992) Nonlinear three-dimensional oscillations of an elastically constrained fluid conveying viscoelastic tube with O(2)-symmetry, AMD, Vol. 152, *Stability and Control of Pipes Conveying Fluid* (M. P. Paidoussis and N. S. Namachchivaya, eds), Book no. G00729, New York, ASME, pp. 47–62.

Sugiyama, Y., Tanaka, Y., Kishi, T. and Kawagoe, H. (1985) Effect of a spring support on the stability of pipes conveying fluid, *Journal of Sound and Vibration*, **100**, 257–270.

Troger, H. and Steindl, A. (1991) *Nonlinear Stability and Bifurcation Theory, An Introduction for Engineers and Applied Scientists*, Vienna and New York, Springer-Verlag.

A. Steindl and H. Troger, *Institut für Mechanik, Technische Universität Wien, Wiedner Haupstrasse 8-10/325, A-1040 Wien, Austria*

19 STRONGLY RESONANT HOPF BIFURCATIONS AND VORTEX-INDUCED VIBRATIONS

W. F. Langford and K. Zhan

Vortex-induced vibrations of engineering structures can have destructive effects, especially at principal (or 1:1) resonance. This chapter presents a new approach to the study of this resonance, in the context of a general model for coupled oscillators at 1:1 resonance. Previous studies using coupled oscillator models, as reviewed here, have tended to ignore the distinction between semisimple and non-semisimple double eigenvalues. The role of this distinction in determining the behaviour of solutions of the system of governing equations is clarified in this chapter. It is shown that, in the absence of special symmetry or constraints, the non-semisimple case is the one to be expected for systems in 1:1 resonance. A general normal-form procedure for the non-semisimple case enables one to explore the complex dynamics and bifurcation behaviour which can occur in the neighbourhood of 1:1 resonance. These tools and ideas apply not only to vortex-induced vibrations, but in general to coupled oscillators at principal resonance.

19.1 INTRODUCTION

Flow-induced vibrations are important in the design of engineering structures, because they may lead to destructive effects; see Blevins (1990) for a comprehensive account. Some examples are the effects of winds on towers, chimneys and power lines and of river and ocean currents on bridges and drilling platforms, and the flow around tubes in the heat exchangers of nuclear reactors. Typically, the presence of the structure can lead to the formation of periodic vortices (von Kármán vortex shedding) in the fluid flow, which in turn induces a periodic transverse force on the structure. The shedding frequency is a function of the Reynolds number, which depends on the diameter of the cylinder and on the free-stream velocity. The structure itself is often elastic with its own natural frequency, and its response to the periodic fluid force is known as *vortex-induced vibration*. A large-amplitude vibration may result when the natural frequencies of the cylinder and of the fluid vortices are in resonance.

Nonlinearity and Chaos in Engineering Dynamics
Edited by J. M. T. Thompson and S. R. Bishop,   © 1994 John Wiley & Sons Ltd

This phenomenon has been modelled by a pair of coupled nonlinear second-order ordinary differential equations; some classical examples are reviewed in the next section. Seriously destructive effects are most likely to occur when the natural frequency of the structure is at or near strong resonance with the vortex-shedding frequency, that is, in the ratio $n:m$ where n and m are small integers. The present chapter focuses on systems near $1:1$ resonance where the two frequencies are nearly equal (also called *principal resonance*), for which very strong effects are possible. Previous studies of mathematical models of this case are reviewed and compared with a new approach, using the non-semisimple double Hopf bifurcation theory of van Gils *et al.* (1990). The general theory of double Hopf bifurcation is well developed in the generic *non-resonant* case; that is, when the ratio of the pure imaginary eigenvalues (characteristic frequencies) is irrational, see for example, Iooss and Langford (1980) and Guckenheimer and Holmes (1983). Only recently have detailed results been obtained for the strongly resonant cases where the imaginary eigenvalues are in the ratios $1:1$ and $2:1$, see van Gils *et al.* (1990) and Leblanc and Langford (1993), respectively. The analysis assumes a prior reduction to a four-dimensional centre manifold, and then explores three-parameter unfoldings of the equivalent normal forms for these vector-field singularities; see Vanderbauwhede (1989) for the general procedure. The normal form can be expressed as a four-dimensional real system or as a two-dimensional complex system which is more convenient to analyse. In the strongly resonant cases, the normal form has an $SO(2)$ symmetry which permits us to reduce the dimension for most of the dynamics on the centre manifold from four to three. This permits richer dynamics than in the non-resonant case, which has $SO(2) \times SO(2)$ symmetry, and thus reduces to a pair of amplitude equations with only two-dimensional dynamics. The singularity is simplified by means of a singular change of coordinates (Krupa transformation), and a blow-up involving both state variables and parameters. This reveals the existence of a wide variety of interesting local dynamics, including multiple periodic orbits, hysteresis, quasi-periodic motion, invariant 2-tori, 3-tori and evidence of chaos. In the three-parameter unfolding of the $1:1$ resonance case, one finds codimension-2 bifurcations including Bogdanov–Takens bifurcation, the Z_2-symmetric cusp, non-resonant double Hopf bifurcation, and a steady-state/Hopf mode interaction. The codimension-1 bifurcations include the standard Hopf, saddle–node and pitchfork bifurcations.

These general results are applied to a generalization of the Hartlen–Currie model (see equation (1) below) of vortex-induced vibrations. Previous analyses, using perturbation methods, have not completely described the dynamics in these strongly resonant cases. The present approach predicts new complex dynamical behaviour, see Langford and Zhan (1992). Furthermore, the normal-form analysis reveals flaws in conventional models such as (1), and suggests a more stable and generic class of models for the study of vortex-induced vibrations.

19.2 COUPLED OSCILLATOR MODELS

The class of *coupled oscillator models* for vortex-induced vibrations has been widely used and studied in the engineering literature. These are two-degree-of-freedom models in which the motion of the fluid wake is modelled as a self-excited oscillation, similar to the van der Pol equation, while the structure is modelled as a damped harmonic oscillator, and the two equations are coupled by linear and/or nonlinear terms. The parameters in the equations are determined either from physical principles or empirically by matching to experimental data.

The earliest coupled oscillator model appears to be that of Hartlen and Currie (1970):

$$y'' + 2\beta y' + y = aC_L,$$
$$C_L'' + (\gamma C_L'^2 - \alpha\omega_0)C_L' + \omega_0^2 C_L = by',$$

$$(1)$$

where y is the displacement of the cylinder, C_L is the amplitude of the lift force of the fluid on the structure, ω_0 is the frequency ratio, α and β are linear damping coefficients, $\gamma C_L'^3$ is a nonlinear damping function and a and b are coupling parameters. The first equation in (1) models the oscillation of the structure and the second models and fluid wake oscillations. Note that in this model, the two oscillators are only linearly coupled through the terms on the right-hand sides of the equations.

This model was modified by Currie *et al.* (1974), by the addition of a nonlinear restoring force in the structure oscillator (intended to introduce hysteresis):

$$y'' + 2\beta y' + y(1 - \varepsilon y^2) = aC_L,$$
$$C_L'' + (\gamma C_L'^2 - \alpha\omega_0)C_L' + \omega_0^2 C_L = by'.$$

$$(2)$$

Another modification of the Hartlen–Currie model was proposed by Landl (1975) in order to allow an isolated stable solution away from the main branch of the response curve. This has the form

$$y'' + 2\beta y' + y = aC_L,$$
$$C_L'' + (\alpha - \gamma C_L'^2 + \delta C_L'^4)C_L' + \omega_0^2 C_L = by',$$

$$(3)$$

where α, γ, δ are fitted to experimental data.

A model proposed by Skop and Griffin (1973) is described by differential equations which are like the Hartlen–Currie model, but with the addition of a cubic restoring term C_L^3 and nonlinear coupling of y in the C_L equation.

A somewhat different type of model was proposed by Iwan and Blevins (1974); the flow-field model derived from vortex field theory, and has the form

$$y'' + 2\beta y' + y = a_1 C_L'' + a_2 C_L',$$
$$C_L'' + (\gamma C_L'^2 - \alpha\omega_0)C_L' + \omega_0^2 C_L = b_1 y'' + b_2 y'.$$

$$(4)$$

This differs significantly from the previous models in that the coupling terms on the right-hand sides involves first and second derivatives. However, the second-derivative coupling terms can be eliminated by substitution, and it is found that the predictions of this model are very similar to the Hartlen–Currie model, at least locally.

A combination model, which is intended to model 'galloping' as well as vortex-induced vibration, was presented by Corless and Parkinson (1988). Although we do not consider galloping in this chapter, the model does fit into our general framework. The equations have the form

$$y'' + (r_1 - r_3 y'^2 + r_5 y'^4 - r_7 y'^6)y' + y = aC_L,$$
$$C_L'' + (\gamma C_L'^2 - \alpha\omega_0)C_L' + \omega_0^2 C_L = b_1 y'' + b_2 y'.$$

$$(5)$$

The model of Tamura and Amano (1983) differs from the above models in that instead of the lift force C_L, the angular position θ is used as the second variable; however, the form of the model is not unlike those above:

$$y'' + 2\beta y' + y = a\theta,$$

$$\theta'' + (\gamma\theta'^2 - \alpha\omega_0)\theta' + \omega_0^2\theta = b_1 y'' + b_2 y'. \tag{6}$$

All of these models are two-degree-of-freedom coupled oscillators, with linear and/or nonlinear coupling and weak nonlinearities when the amplitudes are small. In the cases of interest here, the linear damping coefficients α and β are small and the frequency ratio ω_0 is near 1. These similarities suggest that it may be useful to conduct a general mathematical analysis of systems of differential equations of this form, to determine the typical behaviour of the solutions. A general model which includes all of the above cases is

$$y'' + \delta_1 y' + y + k_1 C_L + c_1 C_L' + f_1(y, y', C_L, C_L') = 0,$$

$$C_L'' + \delta_2 C_L' + (1 + \mu)C_L + k_2 y + c_2 y' + f_2(y, y', C_L, C_L') = 0. \tag{7}$$

Here δ_1 and δ_2 are the damping parameters and μ is the detuning parameter, all of which are assumed small. The linear coupling coefficients k_1, k_2, c_1, c_2 are not necessarily small, and the nonlinear functions f_1 and f_2 are assumed quadratic or higher order in their arguments.

Next let us consider the numerical values of the parameters used in the classical models. In the original model of Hartlen and Currie (1970), the following data were used:

$$y'' + 0.00288y' + y = 0.002\omega_0^2 C_L,$$

$$C_L'' + (\gamma C_L'^2 - 0.02\omega_0)C_L' + \omega_0^2 C_L = 0.4y'. \tag{8}$$

Since ω_0 is near 1, it is clear that the coupling is not symmetric; the effect of the structure on the wake oscillator is much stronger (0.4) than is the effect of the wake on the structure (0.002). Similarly for the modified model of Currie *et al.* (1974), we find

$$y'' + 0.002y' + y(1 - 0.002y^2) = 0.0016\omega_0^2 C_L,$$

$$C_L'' + (\gamma C_L'^2 - 0.01\omega_0)C_L' + \omega_0^2 C_L = y'. \tag{9}$$

For the model of Iwan and Blevins (1974), neglecting the second-derivative terms, we find

$$y'' + 0.034y' + y = 0.00015C_L'$$

$$C_L'' + (\gamma C_L'^2 - 0.0995\omega_0)C_L' + \omega_0^2 C_L = 0.6303\omega_0 y', \tag{10}$$

where again, the structure has a dominating effect on the wake. The same conclusion holds for the model of Tamura and Amano (1983), and for the combination model of Corless and Parkinson (1988) for which the relevant data is given as:

$$y'' + (r_1 - r_3 y'^2 + r_5 y'^4 - r_7 y'^6)y' + y = 3.74 \times 10^{-4} C_L,$$

$$C_L'' + (\gamma C_L'^2 - \alpha\omega_0)C_L' + \omega_0^2 C_L = -2.35y'' + 0.855y'. \tag{11}$$

On the other hand, in case the viscous damping is small and the mass of the structure is light compared to that of the fluid, the coefficients k_2 and c_2 in the general model (7) may be assumed to be small while k_1 and c_1 are of order 1. Under these conditions, the structure is strongly affected by the fluid oscillator and the fluid is weakly affected by the structure.

From this survey of classical models, one way may conclude that the linear coupling coefficients are not uniformly small. In fact, the coefficient for the effect of the structure on the wake is typically $O(1)$ in these models, while the influence of the wake oscillator on the structure is small. It is also possible for strong linear coupling to act in the opposite direction, or, indeed, in both directions. Therefore, in the general model (7), no restrictive assumptions will be made regarding the magnitudes of the linear coupling coefficients. This is in contrast to most previous analyses of these models by perturbation methods, which assume that the linear coupling coefficients are uniformly small.

19.3   NON-SEMISIMPLE 1:1 RESONANCE

We consider a general system of two coupled oscillators of the form (7), which have weak damping and nearly equal natural frequencies (when uncoupled); that is to say, δ_1, δ_2 and μ are all assumed small. Transform the linear part of (7) to a system of first-order equations, defining

$$x_1 = y, x_2 = y', x_3 = C_L, x_4 = C'_L \tag{12}$$

to obtain the linear system

$$
\begin{aligned}
x'_1 &= x_2, \\
x'_2 &= -\delta_1 x_2 - x_1 - k_1 x_3 - c_1 x_4, \\
x'_3 &= x_4, \\
x'_4 &= -\delta_2 x_4 - (1+\mu)x_3 - k_2 x_1 - c_2 x_2.
\end{aligned}
\tag{13}
$$

The characteristic polynomial of this system, the roots of which are the eigenvalues λ, is

$$(\lambda^2 + \delta_1 \lambda + 1)(\lambda^2 + \delta_2 \lambda + 1 + \mu) - (c_1 \lambda + k_1)(c_2 \lambda + k_2). \tag{14}$$

As stated previously, we are interested in the case of 1:1 resonance, that is, when the two natural frequencies of the coupled oscillators are equal, or nearly so. For the linear system, this equality corresponds to eigenvalues which are double pure-imaginary complex conjugates; that is, the four roots of the characteristic polynomial are of the form

$$\lambda = \pm \omega i, \pm \omega i. \tag{15}$$

This will occur when the following conditions hold:

$$
\begin{aligned}
\delta_1 + \delta_2 &= 0, \\
\mu\delta_1 - c_2 k_1 - c_1 k_2 &= 0, \\
(\mu - c_1 c_2 + \delta_1 \delta_2)^2 + 4(k_1 k_2 - c_1 c_2 + \delta_1 \delta_2) &= 0,
\end{aligned}
\tag{16}
$$

provided

$$2 + \mu + \delta_1 \delta_2 - c_1 c_2 > 0.$$

The study of 1:1 resonance then corresponds to describing the dynamics of systems with parameter values near those for which (16) is satisfed. Note the fact that *three* equations must be satisfied by the parameters corresponds to the statement that 1:1 resonance has codimension 3; see van Gils *et al.* (1990). Under these conditions, the eigenvalues of (13) are given by $\pm \omega_i$ with

$$\omega^2 = 1 + (\mu + \delta_1 \delta_2 - c_1 c_2)/2. \tag{17}$$

Whenever multiple eigenvalues occur, it is important to distinguish between the two possible cases, of semisimple or non-semisimple eigenvalues, that is, whether the corresponding Jordan block is diagonal (over the complex field) or has the form of a semisimple plus a nilpotent matrix. The latter case is called non-semisimple. These two types of Jordan blocks for the 2×2 complex case are respectively as follows:

$$\begin{bmatrix} i\omega & 0 \\ 0 & i\omega \end{bmatrix} \quad \begin{bmatrix} i\omega & 1 \\ 0 & i\omega \end{bmatrix}. \tag{18}$$

It is well known that in a general family of matrices, when double eigenvalues occur, the non-semisimple case is typical and the semisimple case is exceptional; see Arnold (1983) § 30. A more precise statement, which follows from Arnold (1983), is the following:

> In a generic two-real-parameter family of matrices, there are only matrices with simple eigenvalues and, for some isolated values of the parameters, matrices with one 2×2 non-semisimple Jordan block. If in a family there are matrices with a more complicated Jordan structure, then we can remove them by an arbitrarily small perturbation of the family.

If we add to this statement the condition that the double eigenvalue has zero real part, as here, then we would expect to find a non-semisimple block (18) in generic three-parameter families, which satisfy (16) at isolated values. The case of semisimple double eigenvalues occurs generically only in families with more parameters (see Arnold 1983). As pointed out by Arnold, if non-generic behaviour is found to occur in a problem, then it is likely that one of two events has occurred: either in the idealization of the phenomenon something essential was eliminated which has changed the structure qualitatively, or there are some special properties such as symmetry or Hamiltonian structure of the problem.

For the present model equations, one can verify directly that the non-semisimple case is the relevant one, and there are no special properties which would lead to the semisimple case. If λ is a double semisimple eigenvalue, then the matrix $\lambda I - A$, where A is the matrix of coefficients in (13), has a two-dimensional kernel (or eigenspace), while if λ is non-semisimple the kernel is one-dimensional. A direct calculation shows that, if the eigenvalue $\lambda = i\omega$ is given by (17), then the corresponding eigenvector $v = (v_1, v_2, v_3, v_4)$ satisfies

$$v_2 = i\omega v_1,$$

$$v_4 = i\omega v_3, \tag{19}$$

$$(k_2^2 + c_2^2 \omega^2)v_1 + (k_2 - c_2 \omega i)(1 + \mu - \omega^2 + \delta_2 \omega i)v_3 = 0,$$

which defines a one-dimensional eigenspace in general. (An obvious exception is $c_1 = c_2 = k_1 = k_2 = 0$; that is, uncoupled oscillators.) Therefore, the appropriate analysis is one which is based on the assumption of a non-semisimple double eigenvalue.

19.4   THE NORMAL FORM FOR NON-SEMISIMPLE 1:1 RESONANCE

A general study of non-semisimple 1:1 resonance, based on Poincaré–Birkhoff normal-form theory, was presented by van Gils *et al.* (1990), to which the reader is referred for the details. The result is that a general nonlinear system with linear part as above, having non-semisimple imaginary eigenvalues, can be reduced locally to a system of the form

$$\rho' = 2\rho[\alpha + u + F_1],$$
$$u' = \mu_1 - u^2 + v^2 + F_3, \qquad (20)$$
$$v' = \mu_2 - 2uv + F_4,$$

where α, μ_1 and μ_2 are unfolding parameters which can be related to the physical parameters in (13) (see Langford and Zhan 1992), ρ is the amplitude of the periodic solution, u and v give the amplitude of the component of the solution in the space of the generalized eigenvector, and the F_j are functions of a special form, namely functions of the invariant polynomials up to any finite order. The leading term in each F_j is linear in ρ, and the values of these coefficients of ρ suffice to determine most of the local dynamics and bifurcation behaviour of the system, as shown by van Gils *et al.* (1990). See Langford and Zhan (1992) for the calculation of these coefficients for vortex-induced vibration models. The details need not be repeated here. However, the conclusion of that paper is that, for most of the classical coupled-oscillator models for vortex-induced vibrations, the leading terms of F_3 and F_4 are degenerate. This degeneracy is removed by the general model (7) presented here.

19.5   CONCLUSIONS

There is an apparent gap in the literature on coupled oscillators in resonance, where the case of 1:1 non-semisimple resonance is concerned. For example, the classic reference of Nayfeh and Mook (1979) explores in detail subharmonic and super-harmonic resonances, as well as combination resonances and parametric excitation; but there is no treatment there of the non-semisimple 1:1 principal resonance. Steen and Davis (1982) studied 1:1 resonance in a pair of coupled oscillators, but only for nonlinear coupling which effectively forces the semisimple (non-generic) case. Semisimple eigenvalues can be generic in the presence of symmetry, as in the case studied by Bajaj and Sethna (1982). However, no such symmetry is present in the case of vortex-induced vibrations at principal resonance, or for that matter in many other cases of coupled oscillators in mechanics. This chapter highlights this gap, and introduces a powerful tool, that of non-semisimple 1:1 Hopf bifurcation theory, to fill the gap.

REFERENCES

Arnold, V. I. (1983) *Geometrical Methods in the Theory of Ordinary Differential Equations*, New York, Springer-Verlag.

Bajaj, A. K. and Sethna, P. R. (1982) Bifurcations in three-dimensional motions of articulated tubes, *Trans. ASME, J. Appl. Mech.*, **49**, 606–618.

Blevins, R. D. (1990) *Flow-Induced Vibration*, Second Edition, New York, Van Nostrand Reinhold.

Corless, R. M. and Parkinson, G. V. (1988) A model of combined effects of vortex-induced oscillation and galloping, *J. Fluids and Struct.*, **2**, 203–220.

Currie, I. G., Hartlen, R. T. and Martin, W. W. (1974) The response of circular cylinders to vortex shedding. In: *Flow-Induced Structural Vibrations*, (E. Naudrecher, ed.), New York, Springer-Verlag, pp. 128–142.

Guckenheimer, J. and Holmes, P. (1983) *Nonlinear Oscillations, Dynamical Systems and Bifurcations of Vector Fields*, Berlin, Springer-Verlag.

Hartlen, R. T. and Currie, I. G. (1970) Lift oscillator model of vortex-induced vibration, *J. Eng. Mech., Proc. ASME*, **41**, 577–591.

Iooss, G. and Langford, W. F. (1980) Conjectures on the routes to turbulence via bifurcations, *Annals New York Acad. Sci.*, **351**, 489–505.

Iwan, W. D. and Blevins, R. D. (1974) A model for vortex induced oscillation of structures, *J. Appl. Mech.*, **41**, 581–586.

Landl, R. (1975) A mathematical model for vortex-excited vibrations of bluff bodies, *J. Sound Vib.*, **42**, 219–234.

Langford, W. F. and Zhan, K. (1992) Dynamics of strong 1:1 resonance in vortex-induced vibration. In: *ASME Int. Symp. on Flow-Induced Vib. and Noise, Vol. 7, Fundamental Aspects of Fluid-Structure Interactions* (M. P. Paidoussis and P. B. Abraham, eds.), New York, ASME, pp. 117–127.

LeBlanc, V. G. and Langford, W. F. (1993) Bifurcation of periodic orbits in 1:2 resonance: a singularity theory approach, *Fields Institute for Research in Mathematical Sciences Report* FI93–DS15, Waterloo, Canada.

Nayfeh, A. H. and Mook, D. T. (1979) *Nonlinear Oscillations*, New York, Wiley-Interscience.

Skop, R. A. and Griffin, C. M. (1973) A model for the vortex-excited resonant response of bluff cylinders, *J. Sound Vib.*, **27**, 225–233.

Steen, P. and Davis, S. H. (1982) Quasi-periodic bifurcation in nonlinearly coupled oscillators near a point of strong resonance, *SIAM J. Appl. Math.*, **42**, 1345–1368.

Tamura, Y. and Amano, A. (1983) Mathematical model for vortex-induced oscillations of continuous systems with circular cross section, *J. Wind Eng. Indu.*, **14**, 431–442.

Vanderbauwhede, A. (1989) Center manifolds, normal forms and elementary bifurcations, *Dynamics Reported*, **2**, 89–170.

van Gils, S. A., Krupa, M. and Langford W. F. (1990) Hopf bifurcation with nonsemisimple 1:1 resonance, *Nonlinearity*, **3**, 825–850.

W. F. Langford and K. Zhan, *Department of Mathematics and Statistics, University of Guelph, Guelph, Ontario N1G 2W1, Canada*

PART V
Random Vibration

The nonlinear dynamical equations of motion governing mathematical or experimental studie of engineering systems, as we have seen, can exhibit many interesting phenomena; multiple solutions are seen to be the norm. Environmental forcing, such as wave, wind or earthquake excitation can rarely be described as regular and so the assessment of the effects of random fluctuations in the forcing excitation becomes an important issue for the engineer. Clearly, when tackling such issues the first problem is to correctly model the forcing and the state of the system. This being done, the problem is then to determine how this non-regular forcing might cause the system to behave in an undesirable way.

The response of a nonlinear system in a randomly fluctuating environment has been traditionally studied under the assumption that the force can be modelled as a random process exhibiting a wide-band power spectrum. This hypothesis enables the well-known Fokker–Planck–Kolmogorov equation to be implemented which governs the diffusion of the probability of a Markov process. When this assumption cannot be justified it has been shown that a variation of the input bandwidth has a significant effect on the shape of the probability distribution of the response amplitude (see Koliopulos and Bishop 1993, and the references cited therein). In engineering systems, fatigue damage and certain other types of failure depend upon the statistical behaviour of the peaks of the response process and thus multiple levels of response play a vital role in determining the structural life-span.

The subject of random vibration is a vast area of research (indeed an earlier IUTAM Symposium was dedicated to this field, see the reference by Hennig 1983), and certainly too large to cover in detail here. The following chapters, however, relate specifically to studies that additionally focus attention on nonlinear behaviour apparent in their deterministic counterparts.

There are a number of ways in which a random element can be incorporated within a nonlinear system. A survey – which also includes some recent results – of nonlinear systems in the presence of multiplicative and additive weak noise is given in the chapter by *Namachchivaya and Doyle*. This chapter deals with the question of stability as well as stochastic bifurcation which is also the subject discussed in the chapter by *Ariaratnam*.

The phenomenon of stochastic resonance forms the content of the chapter by *Dykman et al.* Here the response of a nonlinear system is enhanced by the addition of external noise and the work is seen to be bridging the gap between deterministic and stochastic dynamics. The chapter by *Belyaev* complements these more theoretical studies with a study of high-frequency vibrations in a system consisting of substructures modelled as a random-heterogeneous continuum.

REFERENCES

Hennig, K. (ed.) (1983) *Proceedings of the IUTAM Symposium on Random Vibrations and Reliability,* Berlin, Akademie-Verlag.
Koliopulos, P. K. and Bishop, S. R. (1993) Quasi-harmonic analysis of the behaviour of a hardening Duffing oscillator subjected to filtered white noise, *Nonlinear Dynamics,* **4**, 279–288.

20 STOCHASTIC DYNAMICS

N. Sri Namachchivaya and M. M. Doyle

This chapter gives an overview of the methods of analysis used to determine the response, stability and stochastic bifurcations in nonlinear systems in the presence of multiplicative and additive weak noise. As applications, two examples of second-order nonlinear systems exhibiting codimension-one and codimension-two bifurcations in the presence of noise are presented. Some recent results are cited in the references.

20.1 INTRODUCTION

Recently, in the study of the dynamics of mechanical systems, much effort has been devoted to the determination of the system behaviour in the presence of uncertainty or randomness. Such randomness may enter the system in a variety of ways. A mechanical component, as part of larger system, may be subjected to stochastic loading conditions due to random vibrations transmitted from other machinery. Environmental changes, such as earthquakes, can also contribute to the random excitation of the system. Alternatively, the randomness may stem from material properties such as the distribution of imperfections or defects. A temporal random variation is known as a *stochastic process* while a random variation which is spatially distributed is called a *random field*. Most of the discussion in this chapter is motivated by the study of random temporal variations. Often these fluctuations are modelled by a zero-mean δ-correlated Gaussian white noise because of the wealth of mathematical analysis available to handle these problems. Unfortunately, this assumption does not always provide a good description of the fluctuations that occur naturally. In particular, the δ-correlation of white noise is an idealization of the correlation of real processes which often have finite correlation times.

A class of problems in which stochastic excitation plays an important role is that of flow-induced oscillations or fluid-structure interaction. One example of this type of problem is a pipe conveying fluid where the effect of the turbulent flow is to induce vibrations in the pipe. These vibrations may eventually lead to catastrophic failure of the structure. Another example in this class of problems is the effect of turbulence in the flow across a structure, such as a bridge or power line. In the context of machinery, rotating drive shafts are often

Nonlinearity and Chaos in Engineering Dynamics
Edited by J. M. T. Thompson and S. R. Bishop,   © 1994 John Wiley & Sons Ltd

subjected to a stochastic axial load generated by other machine components. This dynamic load can give rise to transverse vibrations in the shaft. After prolonged use, even small vibrations lead to material fatigue and eventually failure. Another problem in the area of rotating system is the study of helicopter blade dynamics under atmospheric turbulence as examined by Lin *et al.* (1979). The effects of stochastic excitation are also evident in the study of vehicle dynamics; in particular, vibration isolation systems, such as the shock absorption system in a car.

In the study of dynamical systems excited by a stochastic process, several properties of the response are sought. The excitation-parameter values at which the system loses stability are of primary importance. The concept of stability for systems with a random component will be discussed in more detail in Section 20.2. Closely related to the concept of stability is the determination of parameter values at which bifurcations occur, although the exact definition of a stochastic bifurcation is still the subject of debate. Response statistics such as first- and second-order moments, as well as the probability density function, are vital in describing the behaviour of the system. In engineering practice, it is often necessary to determine and quantify the reliability of a machine component. A useful measure of this reliability is the first time to fail, or first passage time, of a stochastically excited system.

Many mechanical and structural systems under random parameter variations or fluctuations in the environment as described above can be written as

$$\dot{x} = f(x; \eta) + g(x; \eta) \xi_t, \quad x \in \mathbb{R}^n \tag{1}$$

where $\xi_t = \xi(\omega, t)$ represents stationary stochastic processes (white noise, coloured noise, etc.) and $\mathbb{R}$ represents the set of real numbers. In the case of additive noise, $g_{ij}(x; \eta)$ is a constant. Let $x = \tilde{x}(t, \eta)$ be a stationary solution of equation (1). Then, one asks the following questions:

- How do we find $\tilde{x}(t, \eta)$?
- Is $\tilde{x}(t, \eta)$ stable in some sense?
- What is stochastic bifurcation?
- How do we determine the bifurcating solutions?

We shall attempt to answer some of these questions in this chapter.

20.1.1   Stochastic differential equations and diffusion

In order to illustrate the analysis of a stochastically excited system, consider an Itô stochastic differential equation corresponding to equation (1) in $\mathbb{R}$:

$$dx = \mu(x)dt + \sigma(x)dw_t \tag{2}$$

where

$$\mu(x) = f(x; \eta) + \tfrac{1}{2}g_x(x; \eta)g(x; \eta), \quad \sigma(x) = g(x; \eta),$$

and w_t is a Wiener process with unit variance. Let x be the state variable representing the energy or amplitude of a second-order nonlinear problem. Associated with equation (2), we define $\{x_t, a \leqslant t \leqslant b\}$ and denote the transition function and transition density function respectively by $P(x, t|x_0, t_0)$ and $p(x, t|x_0, t_0)$, i.e.

$$P\{x - t < x | x_{t_0} = x_0\} = P(x, t|x_0, t_0) = \int_{-\infty}^{x} p(u, t|x_0, t_0) du.$$

Now defining $t = t_0 + \tau$, and depending on whether $p(x, t|x_0, t_0)$ is considered as a function of the pair (x, t) or the pair (x_0, t_0), we can write the Kolmogorov *backward* and *forward* diffusion equations as

$$\frac{\partial(p)}{\partial \tau} = L(p) = \mu(x_0) \frac{\partial(p)}{\partial x_0} + \frac{1}{2} \sigma^2(x_0) \frac{\partial^2(p)}{\partial x_0^2} \quad \text{and} \quad \frac{\partial(p)}{\partial \tau} = L^*(p)$$

respectively, with $p(x, t_0|x_0, t_0) = \delta(x - x_0)$ and L is the standard differential generator.

In order to understand the bevhaiour of the diffusion process $x(t)$, we shall introduce two canonical measures, namely, the *scale* and *speed* measures. A diffusion is said to be on its natural scale if $\mu(x)$ is identically zero. For any diffusion not on its natural scale, it is possible to find a transformation $S(x)$ such that the new drift term (using the Itô differential rule) is identically zero, i.e.

$$\mu_s(S(x)) = \mu(x) \frac{\partial S}{\partial x} + \frac{1}{2} \sigma^2(x) \frac{\partial^2 S}{\partial x^2} \equiv 0 \tag{3}$$

and the Itô equation reduces to

$$dS(x) = \sigma(x) s(x) dw_t \quad \text{where } s(x) dx = dS.$$

The transformation satisfying equation (3) is called the *scale measure* and is given by

$$S(x) = \int^x s(\eta) d\eta, \quad s(\eta) = \exp\{-B(\eta)\}, \quad B(\eta) = \int^\eta \frac{2\mu(\xi)}{\sigma^2(\xi)} d\xi.$$

Defining the *speed measure* M in the form $dM = m(x) dx$ where $m(x) = [\sigma^2(x) s(x)]^{-1}$ yields the reduced differential operator

$$L(\cdot) = \frac{1}{2} \frac{\partial}{\partial M} \left[\frac{\partial(\cdot)}{\partial S} \right].$$

The solutions of various equations can now be given in terms of scale and speed measures. For example, the stationary density function which solves $L^*p(x) = 0$ is

$$p_{st}(x) = m(x)[c_1 S(x) + c_2]$$

where c_1 and c_2 are determined by boundary and normality conditions.

20.2   STOCHASTIC STABILITY

One of the primary concerns in the analysis of dynamical systems is the determination of the stability of the steady-state solutions. This analysis becomes more difficult when these systems are excited by a stochastic process. The stability of a linear stochastic system can be defined in several ways. The weakest, or least conservative, definition is that of stability in distribution. A more conservative estimate of the stability boundary is described by stability in probability. Thus, if a system is stable in probability, it is also stable in distribution. The last two definitions of stability in the stochastic sense are stability in the rth mean and almost-sure stability, or stability with probability one. If a dynamical system excited by noise is stable according to either of these definitions, it is stable in distribution and in probability, as well. However, rth mean and almost-sure stability do not imply each other, i.e. a system can be almost-surely stable while its 2nd moments grow exponentially.

20.2.1 Moment stability

Moment stability, as the name implies, refers to boundedness of the nth-order moment. An equilibrium solution $\bar{x}$ is asymptotically stable in the nth moment if

$$\lim_{t \to \infty} E[|x(t; x_0) - \bar{x}|^n] = 0.$$

This automatically implies boundedness of *all lower-order moments*. Consider the linear system described by the equations of motion

$$\ddot{q}_i + \omega_i^2 q_i + \beta_{ij} \dot{q}_j + f(t) k_{ij} q_j = 0, \quad i = 1, 2, \ldots, n \tag{4}$$

where $f(t)$ is a stationary stochastic process with zero mean value. Such non-gyroscopic multi-degree-of-freedom systems were studied by Ariaratnam and Srikantaiah (1978). Assuming $S(\omega)$ has significant values only in the neighbourhood of a particular frequency ω_0, the second-moment stability conditions are given as follows: When $\omega_0 = 2\omega_i$, $i = 1, 2, \ldots, n$,

$$\beta_{ii} > \frac{k_{ii}}{2\omega_i^2} S(2\omega_i).$$

When $\omega_0 = |\omega_i \pm \omega_j|$, $i, j = 1, 2, \ldots, n$, $i \neq j$,

$$\frac{\beta_{ii} \beta_{jj}}{\beta_{ii} + \beta_{jj}} > \pm \frac{k_{ij} k_{ji}}{4\omega_i \omega_j} S(|\omega_i \pm \omega_j|).$$

Similar results for gyroscopic systems are given in Sri Namachchivaya and Ariaratnam (1987). Finally, the mean-square stability of a non-conservative system was obtained by Sri Namachchivaya and Tien (1990) and the results were applied to systems with a follower force.

20.2.2 Sample stability

The equilibrium solution is said to be almost-surely stable if (see Kozin 1969)

$$P\left\{ \lim_{\|x_0 - x_\mu^*\| \to 0} \sup_{t \geq t_0} \|x(t) - x_\mu^*\| = 0 \right\} = 1.$$

This definition is strengthened to almost-sure asymptotic stability if the equilibrium solution is almost-surely stable and there exists an $\varepsilon > 0$ and $\delta > 0$ such that $\| x_0 - x_\mu^* \| < \delta$ implies

$$\lim_{T \to \infty} P \left\{ \sup_{t \geqslant t_0 + T} \| x(t) - x_\mu^* \| \geqslant \varepsilon \right\} = 0.$$

Since moment stability does not imply almost-sure stability and vice versa, it is generally acknowledged that it is important to establish both stability boundaries for applications of stochastic models to real phenomena. Kozin and Sugimoto (1977), with extensions by Arnold (1984), established a characterization between moment stability and almost-sure stability for linear Itô stochastic differential equations when the process is ergodic on the entire surface of the n-sphere. It was shown that the region of sample stability is the limit of the regions of pth moment stability for p approaching zero. The application of this method is difficult due to the problem of determining conditions for arbitrary pth moment stability in multi-dimensional multiplicative systems. For this reason, moment stability calculations are generally limited to mean-square stability.

The almost-sure stability of a dynamical system excited by noise is determined by the sign of the maximal Lyapunov exponent. To illustrate this, consider the linear stochastic system in $\mathbb{R}^n$:

$$\dot{x}(t) = A(\xi(t))x(t), \quad x(0) = x_0, \quad t \geqslant 0,$$

$$d\xi = \chi_0(\xi)dt + \sum_{i=1}^{r} \chi_i(\xi) \circ dw_i. \tag{5}$$

In this case, $\xi(t)$ is a stationary ergodic diffusive process satisfying the above Stratonovich equation and $\chi_i(\xi)$ are nonlinear functions. Assume there exists a smooth positive density u which solve $G^*u = 0$ where G is the differential generator of $\xi(t)$. According to Oseledec's (1968) multiplicative ergodic theorem, the Lyapunov exponents of the solution of equation (5), $x(t; x_0)$, for the the initial condition $x_0 (x_0 \neq 0)$ are of the form

$$\lambda(x_0) = \lim_{t \to \infty} \sup \frac{1}{t} \ln \| x(t; x_0) \|$$

where $\lambda(x_0)$ takes on one of $r(1 \leqslant r \leqslant n)$ fixed or non-random values $\lambda_1 < \cdots < \lambda_r$. Which λ_i is realized depends on the initial conditions x_0. The multiplicities of the Lyapunov exponents sum to the dimension of the system, n. Associated with each λ_i there exists a random linear space E_i, known as an Oseledec space, such that $E_1 \oplus E_2 \oplus \cdots \oplus E_r = \mathbb{R}^n$, and $\lambda(x_0) = \lambda_i$ if and only if $x_0 \in E_i$ with probability one. The dimension of each Oseledec space E_i is given by the multiplicity of the associated Lyapunov exponent, λ_i.

A method of calculating the maximum Lyapunov exponent of equation (5) was given by Kliemann and Arnold (1983). The essence of the method is to project the system in $\mathbb{R}^n$ onto the surface of an n-dimensional sphere such that the SDE for $\ln \| x \|$ can be written explicitly as a function of only n-1 independent angles. To this end, introduce the following transformation in $\mathbb{R}^n$: $s = x/\| x \| \in \mathbf{S}^{n-1}$, $\rho = \ln \| x \| \in \mathbb{R}^+$ where $\mathbf{S}^{n-1}$ represents a sphere of dimension n-1. This yields

$$\dot{s} = h(s, \xi) = [A(\xi) - Q(s, \xi)]s, \quad \dot{\rho} = Q(s, \xi) = s^{\mathrm{T}} A(\xi)s.$$

The new problem is given by

$$\dot{\rho} = Q(s, \xi), \quad d\begin{pmatrix} s \\ \xi \end{pmatrix} = \begin{pmatrix} h(s, \xi) \\ \chi_0(\xi) \end{pmatrix} dt + \sum_{i=1}^{r} \begin{pmatrix} 0 \\ \chi_i(\xi) \end{pmatrix} \circ dw_i,$$

where the pair (s, ξ) is a diffusive process with generator

$$L = G + h \frac{\partial}{\partial s}.$$

It was then shown that the top Lyapunov exponent is given by

$$\lambda = \int Q(s, \xi) dv \qquad \text{a.s.}$$

where v is the invariant stationary probability measure of (s, ξ). Khaşminskii (1967) used a version of this method for obtaining necessary and sufficient conditions for almost-sure stochastic stability of linear Itô equations when the noise is white. An extension of these results, which includes all possible singularities that can exist in one-dimensional diffusion, was presented by Nishioka (1976). Necessary and sufficient conditions for sample stability of systems subjected to white-noise excitation have also been obtained by Kozin and Prodromou (1971) and Mitchell and Kozin (1974). The case in which the system is excited by a real noise process has been investigated by Arnold *et al.* (1986), Pardoux and Wihstutz (1988), Pinsky (1991) and, more recently Sri Namachchivaya (1991a).

The studies mentioned above deal primarily with two-dimensional systems. Recent advances in stochastic stability have facilitated the calculation of Lyapunov exponents for multi-dimensional systems. This is possible provided the system can be reduced to a one-dimensional diffusion problem. Results for three- and four-dimensional systems have been presented by Ariaratnam and Xie (1992) in which the sample stability of coupled oscillators was examined. Various codimension-two dynamical systems were studied by Sri Namachchivaya and Talwar (1994) where averaging was applied to obtain a set of approximate Itô equations for amplitudes and phases. However, in order to completely decouple the amplitude and phase equations, certain restrictive conditions on the manner in which the noise entered the equations were imposed.

Recently, Sri Namachchivaya and Van Roessel (1993) employed a perturbative approach, based on the work of Arnold *et al.* (1986), in order to obtain an asymptotic approximation for the Lyapunov exponent and rotation numbers of the double oscillator in the form of equation (4). In order to make the problem tractable, it was assumed that the infinitesimal generator $G(\xi)$ associated with the noise process $\xi(t)$ has an isolated simple zero eigenvalue. Again, assuming $S(\omega)$ has significant values only in the neighbourhood of ω_0, the almost-sure stability boundaries are as follows:

When $\omega_0 = 2\omega_i$, $i = 1, 2$,

$$\beta_{ii} > \frac{k_{ii}}{\omega_i} S(2\omega_i).$$

When $\omega_0 = |\omega_1 \pm \omega_2|$,

$$(\beta_{11} + \beta_{22}) + (\beta_{11} - \beta_{22}) \coth \left[\frac{\pm 2(\beta_{11} - \beta_{22}) \omega_1 \omega_2}{k_{12} k_{21} S(|\omega_1 \pm \omega_2|)} \right] < \frac{\pm k_{12} k_{21}}{2\omega_1 \omega_2} S(|\omega_1 \pm \omega_2|).$$

This investigation was extended by Doyle and Sri Namachchivaya (1994) to include more general linear systems with all possible singularities and removing the restrictive conditions imposed by the use of the method of averaging.

There exist numerous physical systems in which a stochastic excitation is coupled with a harmonic loading. In such situations, almost-sure stability results do not exist except for a class of systems which have only one critical mode (see Sri Namachchivaya (1991a)). Extensions for systems with two critical modes are given by Talwar and Sri Namachchivaya (1994).

20.2.3   First-passage problem

Consider the domain of attraction D of an equilibrium point x_l bounded by Γ. Suppose that at time t_0 the state of the system corresponds to some point x_0 within D. In the first-passage problem, we are interested in the time T it takes for the trajectory starting at x_0 to reach the boundary Γ of D for the first time, i.e.

$$T(x_0) = \min\{t : x(t) \in \Gamma \,|\, x(t_0) = x_0\}, \quad x_0 \in D.$$

Define the probability that the trajectory has not reached the boundary Γ during the time interval τ as

$$P(x_0, \tau) = P\{x(t_0 + \tau) \in D \,|\, x(t_0) = x_0\} = \int_D p(u, t_0 + \tau \,|\, x_0, t_0)\,du.$$

Then it is natural to define $P(x_0, \tau) = \Pr\{\tau < T(x_0)\}$ and this is governed by Kolmogorov's backward equation with the initial and boundary conditions

$$P(0, x_0) = 1 \quad \text{for} \quad x_0 \in D \quad \text{and} \quad P(\tau, x_c) = 0 \quad \text{for} \quad x_c \in \Gamma.$$

The distribution function of the first-passage time $\Pr\{\tau = T\} = 1 - P(\tau, x_0)$ and the corresponding Pontryagin equation for the nth moment is given by

$$L[M_n(x_0)] = -n M_{n-1}(x_0) \quad \text{and} \quad M_n(x_c) = 0.$$

Bergman and Spencer (1992) have examined the first-passage problem for two- and three-dimensional linear and nonlinear systems subjected to additive white noise by solving the related backward Kolmogorov equations in the presence of absorbing boundaries by the finite-element method giving the probability of failure versus time for all initial conditions in the safe domain. Moments of first-passage times have similarly been computed by solving the Pontryagin–Vitt equations. An instability theorem based on the boundary behaviour of the amplitude process is presented by Sri Namachchivaya (1989).

20.3   ANALYSIS OF HIGHER-ORDER SYSTEMS

Although a large amount of literature exists for use in analysing nonlinear deterministic dynamical systems, much work remains to be done to develop consistent methods for studying nonlinear stochastic systems. At present, the two methods most commonly used in obtaining mean-square statistics of the response are stochastic linearization and closure techniques. It is often necessary to simplify the equations as much as possible before proceeding with the analysis in order to make the problem tractable. Two reduction techniques, namely, stochastic averaging and stochastic normal forms, have proven helpful in reducing the dimensionality of the nonlinear systems. In this section, we briefly describe each of these techniques.

20.3.1 Equivalent linearization and closure techniques

Historically, the earliest formulation of statistical linearization was carried out by Booton (1954), Kazakov (1956) and Caughey (1963). The philosophy behind this concept is to replace an nth-order nonlinear model governed by equation (1) with $g(x:\eta) = 1$ under the assumption that the system is stable and an ergodic probability measure exists, by an equivalent linear system

$$\dot{y} = Ay + \xi_t \tag{6}$$

where for the stationary case the matrix A is constant. It is clear that an approximate linear form (6), can be obtained by adding and subtracting Ax in equation (1) and choosing A such that the difference, $f(x; \eta) - Ax$, is as small as possible according to some suitable criterion. If one chooses to minimize the difference in the mean-square sense, i.e.

$$\min_{A} \{ \| f(x; \eta) - Ax \|^2 \} \Rightarrow A = E\{f(x)x^{\mathrm{T}}\} E\{xx^{\mathrm{T}}\}^{-1} \tag{7}$$

and the expectations are taken with respect to the stationary measure of (1). However, this stationary measure is unknown. If it were known, there would be no need to linearize in the first place. Thus the statistics used to define A are taken from the linear system (6) and A is calculated through the expectations given in equation (7) with x replaced by the new variable y. For vector polynomial functions f one can obtain the expression for A in terms of joint moments of the excitation process. It is clear that there are two approximations, namely, linearization and approximation of statistics. Even though the method of linearization has various approximations, this method is being widely used by engineers due to the lack of implementable alternative procedures. The relationship to parameter estimation for linear systems in this context is discussed by Kozin (1988). A recent survey on this topic is given by Socha and Soong (1991).

The goal of the closure techniques is to obtain a statistical description of the response x (see, for example, Lin and Wu (1983) and Ibrahim *et al.* (1985)). This method can be easily applied if the random excitations are Gaussian white noises or filtered Gaussian white-noise processes. Using the multi-dimensional Itô equations corresponding to equation (1) and Itô's differential rule, we obtain the following equations for the statistical moments of x:

$$\frac{d}{dt} E[\phi(x, t)] = E\left[\frac{\partial \phi}{\partial t} + \left(\mu_j \frac{\partial \phi}{\partial x_j} + \frac{1}{2} \sigma_{jl} \sigma_{kl} \frac{\partial^2 \phi}{\partial x_j \partial x_k} \right) \right]$$

where ϕ is x_i, $x_i x_r$, $x_i x_r x_s$, ... The moments can be written in terms of the cumulants as follows:

$$E[x_j] = \kappa_1[x_j], \quad E[x_j x_k] = \kappa_2[x_j x_k] + \kappa_1[x_j]\kappa_1[x_k],$$

$$E[x_j x_k x_l] = \kappa_3[x_j x_k x_l] + 3\{\kappa_1[x_j]\kappa_2[x_k x_l]\}_s + \kappa_1[x_j]\kappa_1[x_k]\kappa_1[x_l],$$

$$E[x_j x_k x_l x_m] = \kappa_4[x_j x_k x_l x_m] + 3\{\kappa_2[x_j x_k]\kappa_2[x_l x_m]\}_s$$

$$+ 4\{\kappa_1[x_j]\kappa_3[x_k x_l x_m]\}_s + 6\{\kappa_1[x_j]\kappa_1[x_k]\kappa_2[x_l x_m]\}_s + \kappa_1[x_j]\kappa_1[x_k]\kappa_1[x_l]\kappa_1[x_m],$$

$$\vdots$$

where $\kappa_j[\cdot]$ is the jth cumulant and $\{\cdot\}_s$ indicates a symmetrizing argument with respect to all its arguments. In order to obtain a closed system, a scheme for truncating the equations is often adopted. In this procedure, all cumulants of order higher than n are set to zero. The statistical moments of order higher than n can then be calculated in terms of lower-order moments using the above relations. Thus, applying cumulant-neglect closure (see, for example, Lin and Wu (1983)) to the set of moment equations yields a set of nonlinear moment equations of the form

$$\dot{\bar{x}} = F(\bar{x}, \lambda)$$

where $\bar{x} = \{\langle x_1 \rangle, \langle x_2 \rangle; \langle x_1^2 \rangle, \langle x_1 x_2 \rangle, \langle x_2^2 \rangle; \cdots; \langle x_1^n \rangle, \langle x_1^{n-1} x_2 \rangle, \cdots, \langle x_2^n \rangle\}$. Since the above equations are autonomous deterministic equations, we have a classical problem of analysis, i.e. to find the zeros of $F(\bar{x}, \lambda)$. To this end, one uses the basic tools of analysis such as the implicit function theorem, Lyapunov–Schmidt reduction, etc.

20.3.2 Stochastic averaging and stochastic normal forms

The method of stochastic averaging involves the convergence of parametrized sequence of processes $\{x^\varepsilon(t)\}$ to a limit process $\{x^0(t)\}$, i.e. the averaging method finds a process $\{x^0(t)\}$ such that $x^\varepsilon(t) \to x^0(t)$ in the sense of weak convergence of measures as $\varepsilon \to 0$. It is important that the limit process $\{x^0(t)\}$ obtained by this procedure be much more mathematically tractable than the true physical process and the parameter value ε corresponding to the physical process be small enough to yield a good approximation. Four variations of the stochastic averaging theorem are available and a complete description of these has been compiled by Sri Namachchivaya (1990).

Consider equation (1) with $f(x; \eta) = \varepsilon^2 F(x, \eta)$ and $g(x; \eta)\xi_t = \varepsilon G(x, \xi_t; \eta)$ where ξ_t is a stationary stochastic process with zero mean whose correlation function satisfies the strong mixing condition, i.e. $R(\tau)$ tends to zero sufficiently fast as τ increases. The solution of this equation can be uniformly approximated in a weak sense on a time interval of $O(1/\varepsilon^2)$ by a Markov process which is continuous with probability 1 and satisfies the Itô stochastic differential equation with drift and diffusion terms given as

$$m_i(x) = M_t\left\{ F_i(x, t) + \sum_{j=1}^n \int_{-\infty}^0 E\left[\frac{\partial G_i}{\partial x_j}(x, t, \xi_t) G_j(x, t+\tau, \xi_{t+\tau})\right] d\tau \right\}$$

$$[\sigma\sigma^\mathsf{T}]_{ij} = M_t\left\{ \int_{-\infty}^\infty E[G_i(x, t, \xi_t), G_j(x, t+\tau, \xi_{t+\tau})] d\tau \right\}$$

where M_t is the averaging operator defined as

$$M_t(\cdot) = \lim_{T \to \infty} \frac{1}{T} \int_{t_0}^{t_0 + T} (\cdot)\, \mathrm{d}t.$$

The method of stochastic averaging has been extended to include rapidly decaying stable modes coupled with rapidly oscillating modes by Papanicolaou and Kohler (1976). This method was extended further to include nonlinear terms by Sri Namachchivaya and Lin (1988). Further extension of the principle of averaging for Itô stochastic differential equations was done by Khas'minskii (1968).

The method of normal forms, originally developed for deterministic dynamical systems has been extended to include stochastic excitations (see Coulett *et al.* (1985) and Sri Namachchivaya and Lin (1991)), with the objective of obtaining an optimal reduction of dimensionality of the system while retaining its essential dynamic characteristics. Similar to the deterministic case, the crucial step in the normal-form computation is to find the so-called resonant terms which cannot be eliminated through a change of variables. Subsequent to the reduction of the dimensionality, the associated stochastic normal form is obtained using a Markovian approximation.

Consider the system described by equation (1) with explicit linear operator $A(\eta)$ and $f(x; \eta)$ containing only nonlinear terms. In order to reduce the dimensionality of the problem, we begin by decomposing the state vector x and the linear operator $A(\eta)$ into critical and stable components, i.e. $x = \{x_c, x_s\}$ and $A = \{A_c, A_s\}$ where the eigenvalues of A_c are purely imaginary or zero and the eigenvalues of A_s have negative real parts. The corresponding nonlinear and time-dependent terms are then $f = \{f_c, f_s\}$ and $F = \{F_c, F_s\}$, respectively. The system can then be written in first-order form as

$$\dot{x}_c = A(\eta)_c x_c + f_c(x_c, x_s, \eta) + \sigma F_c(x_c, x_s, \eta; \xi(t)),$$

$$\dot{x}_s = A(\eta)_s x_s + f_s(x_c, x_s, \eta) + \sigma F_s(x_c, x_s, \eta; \xi(t)). \tag{8}$$

Consider the near-identity transformations

$$x_c = y_c + W_c(y_c, y_s, \eta) + \sigma U_c(y_c, y_s, \eta, \xi(t)) + \sigma^2 V_c(y_c, y_s, \xi(t)),$$

$$x_s = y_s + W_s(y_c, y_s, \eta) + \sigma U_s(y_c, y_s, \eta, \xi(t)) + \sigma^2 V_s(y_c, y_s, \xi(t)),$$

where $W = \{W_c, W_s\}$ is a homogeneous vector polynomial of degree k and $U = \{U_c, U_s\}$ and $V = \{V_c, V_s\}$ are vector polynomials with time-dependent coefficients. Substituting these expressions into equations (8) yields

$$\dot{y}_c = A_c y_c + g_c(y_c, \eta) + \sigma G_c(y_c, \eta, \xi(t)) + \sigma^2 H_c(y_c, \eta, \xi(t)),$$

$$\dot{y}_s = A_s y_s + g_s(y_c, y_s, \eta) + \sigma G_s(y_c, y_s, \eta, \xi(t)) + \sigma^2 H_s(y_c, y_s, \eta, \xi(t)),$$

where W, U and V are such that g_c, G_c and H_c are as simple as possible and take values in the critical subspace E_c.

The above lower-dimensional equations are replaced by diffusive Markov processes whose transition probabilities at time intervals $\Delta t (\Delta t \gg \tau_{cor})$ are approximately the same as those of the original processes. The drift and diffusion coefficients which completely describe the

Markov processs are obtained using a method which deals with asymptotic behaviour of the solutions of the lower-dimensional system when $\tau_{cor} \to 0$ where τ_{cor} is the correlation time. It was shown by Sri Namachchivaya and Lin (1991) that the approximation of the solution of the lower-dimensional equations by a Markov process gives rise to the same drift and diffusion terms as those obtained in the extended stochastic averaging technique employed by Sri Namachchivaya and Lin (1988).

20.4 BIFURCATIONS OF NONLINEAR SYSTEMS

When a linear stochastic system loses stability, the oscillations will increase without limit in the mean-square or almost-sure sense. In real physical systems, however, the amplitude cannot become arbitrarily large. In order to determine the response of an actual system, it is necessary to examine the bifurcation behaviour of the corresponding nonlinear system and to calculate the stationary amplitudes and their densities. In order to understand this behaviour, three approaches are proposed: (1) the method of extrema (i.e., finding the *most probable value* of a variable described by the maximum of an otherwise broad stationary distribution), (2) the method of Lyapunov exponents, (3) bifurcations of autonomous nonlinear moment equations derived using non-Gaussian closure. A brief description of each of these methods is given below.

The extrema of the density function $p_{st}(x)$ are important for several reasons. The number and location of the extrema are the most distinguishing features of $p_{st}(x)$ and contain essential information on the stationary behaviour of the system. The extrema are the continuation of the deterministic steady states. Furthermore, it may be noted that as $\varepsilon \to 0$ the extrema tend to the deterministic steady states of the system. If the process x_t is ergodic, we can interpret the density function $p_{st}(x)$ as a measure for that part of the time an arbitrary sample path spends in an infinitesimal vicinity of x. This motivates the usual identification of the extrema: the maxima, where the process spends relatively much time, are the stable steady states, and the minima, which the process leaves rather quickly, are the unstable states.

The method of Lyapunov exponents which yields the almost-sure stability of stochastically excited systems was discussed in Section 20.2. In this section, we consider the nonlinear stochastic system described by equation (1) with initial conditions $x(0) = x_0$ and η is a control parameter. In this case, the invariant probability of the stationary solution and the corresponding Lyapunov exponents $\lambda_i(\eta)$ will also depend on η. Furthermore, the Lyapunov exponents and the random linear spaces E_i are the stochastic analogues of the real parts of the eigenvalues and the generalized eigenspaces, respectively. Thus, one can define the point of bifurcation as η_c, where $\lambda_i(\eta_c) = 0$, and the new stationary solution can be obtained, in principle, by examining the stochastic normal form defined on the stochastic centre manifold near the bifurcation point (see, for example, Boxler (1989) and references therein).

Consider the non-autonomous system of differential equations

$$\dot{x} = \mu x + \omega_0 y - x^3 + \alpha_0 g(t)x, \quad \dot{y} = -\omega_0 x + \mu y - x^2 y, \tag{9}$$

where μ, ω_0 and α_0 are real quantities (μ, $\omega_0 > 0$) and $g(t)$ is a stationary stochastic process with zero mean. In the absence of stochastic excitation, i.e. when $\alpha_0 = 0$, equation (9) has the solution $x = y = 0$ which loses stability via a Hopf bifurcation as μ increases through zero. The bifurcating solution exists for $\mu > 0$ and is orbitally asymptotically stable.

In the case where $\alpha_0 \neq 0$, stochastic averaging yields the set of autonomous equations

$$da = \left[\left(\alpha + \frac{\gamma}{2}\right)a - \frac{a^3}{2}\right]dt + \sqrt{\gamma a}\,dW_a, \quad d\phi = -\eta dt + \gamma dW_\phi$$

with parameters

$$\alpha = \mu + \tfrac{1}{2}S_{gg}(2\omega_0), \quad \gamma = \tfrac{1}{8}[S_{gg}(0) + S_{gg}(2\omega_0)], \quad \eta = \tfrac{1}{8}\psi_{gg}(2\omega_0),$$

and spectral densities

$$S_{gg}(\omega) = 2\int_0^\infty R_{gg}(\tau)\cos\omega\tau\,d\tau, \quad \psi_{gg}(\omega) = 2\int_0^\infty R_{gg}(\tau)\sin\omega\tau\,d\tau,$$

where $R_{gg}(\tau) = E[g(t)g(t + \tau)]$. The Lyapunov exponent, determined from the linear system, is given as $\lambda = \alpha = \mu + \tfrac{1}{8}S_{gg}(2\omega_0)$. Hence, the almost-sure stability boundary is described by

$$\mu_c = -\tfrac{1}{8}S_{gg}(2\omega_0).$$

The steady-state solution $p_{st}(a)$ describing the amplitude of the response and the nth stationary moment are

$$p_{st}(a) = 2\left(\frac{1}{2\gamma}\right)^\nu \frac{a^{2\nu-1}}{\Gamma(\nu)}\exp\left(-\frac{a^2}{2\gamma}\right), \quad M_n^s = (2\gamma)^{n/2}\frac{\Gamma\left(\nu + \dfrac{n}{2}\right)}{\Gamma(\nu)},$$

provided $\nu = \alpha/\gamma > 0$. Assuming the excitation has a broad-band spectrum with a constant spectral density S_0, the most probable value of the response amplitude is

$$a = \sqrt{2\mu - \frac{S_0}{8}}, \quad \mu > \frac{S_0}{16}.$$

As a second exmaple, we consider a stochastically perturbed codimension-two bifurcation associated with non-semisimple double zero eigenvalues. The normal form corresponding to this problem represents the parametrically perturbed van der Pol–Duffing oscillator as shown by Sri Namachchivaya (1991b):

$$\dot{x} = y, \quad \dot{y} = (\mu_0 + \xi_t)x + \mu_2 y \pm x^3 - x^2 y.$$

Using the appropriate scaling, i.e. $\mu_0 = \varepsilon^2 v_0$, $\mu_2 = \varepsilon^2 v_2$, $x = \varepsilon u$, $y = \varepsilon^2 v$, $\tau = \varepsilon t$, we have

$$u' = v, \quad v' = v_0 u \pm u^3 + \varepsilon[(v_2 - u^2)v] + \varepsilon^{1/2}v_0 u\xi_t,$$

which can be interpreted as a Stratonovich or an Itô equation since the correction term is identically zero. Introducing $H = v^2/2 + P_{1,2}$, $P_{1,2} = -v_0 u^2/2 \pm u^4/4$ and $G(u) = (v_2 - u^2)$, the Itô equations for u and H can be written as

$$du = \sqrt{Q_{1,2}}dt, \quad dH = \varepsilon\{Q_{1,2}(u)G(u) + \tfrac{1}{2}v_0^2 u^2\}dt + \varepsilon^2\sqrt{Q_{1,2}}(v_0 u)dw_t,$$

where $Q_{1,2} = 2(H - P_{1,2}(u))$ and we consider only the following two cases: (1) $H = v^2/2 + P_1$, $v_0 < 0$ and the Hamiltonian levels of interest lie within $H = (0, v_0^2/4)$; (2) $H = v^2/2 + P_2$,

$v_0 > 0$ and the Hamiltonian levels of interest lie within $H = (-v_0^2/4, 0)$. Applying the theory of Khas'minskii (1968), we obtain the one-dimensional Itô equation

$$dH = \left(\frac{A(H)}{\Lambda(H)} \right) dt + \left(\frac{\sigma_{HH}}{\sqrt{\Lambda(H)}} \right) dw$$

where

$$A(H) = \frac{2}{15} \left(\frac{v_0}{2-m} \right)^{5/2} \left\{ \left[5(v_2/v_0)(2-m)^2 - 4(m^2 - m + 1) + \frac{15}{2} S_0 v_0 \right] E(m) \right.$$

$$\left. - 2[5(v_2/v_0)(2-m)(1-m) - (2-m)(1-m)]F(m) \right\}$$

$$\sigma_{HH}^2 = \frac{4}{15} \left(\frac{v_0}{2-m} \right)^{5/2} \left\{ [2S_0 v_0^2(m^2 - m + 1)]E(m) - S_0 v_0^2(2-m)(1-m)F(m) \right\}$$

$$\Lambda(H) = F(m) \sqrt{\frac{2-m}{v_0}}, \quad T(H) = 2\Lambda(H), \quad H = \frac{v_0^2(m-1)}{(2-m)^2}.$$

In the above equations, $F(m)$ and $E(m)$ are complete elliptic integrals of the first and second kinds, respectively. By solving the corresponding Fokker–Planck equation, the stationary probability density is obtained as

$$w_{st}(H) = c e^{-\psi}, \quad \psi = 2 \int \frac{B(H)}{\sigma_{HH}^2(H)} dH + \ln(\Lambda(H))$$

For details of this calculation, refer to Sri Namachchivaya (1991b). The same problem has been solved by Tien *et al.* (1993) using elliptic stochastic averaging. Assuming the conditional density $w_{st}(u|H)$ is inversely proportional to velocity, the density in (u, v) can be written as $w(u, H(u, v)) = w(H(u, v))w(u|H(u, v))$ as in Sri Namachchivaya (1991b). In addition, extrema of the density and first-passage time to exit various energy levels were also obtained.

20.5 CONCLUSIONS

The development of mathematical methods to aid in the study of the response of *lower-dimensional* nonlinear dynamical systems excited by noise has produced results in a wide variety of fields and deepened our understanding of the mechanisms of instability in such systems. However, the answers to several important questions are still incomplete. For example, how does one analyse higher-dimensional systems? Why is noise beneficial in some nonlinear systems but harmful in others? What is the effect of noise on the stable and unstable periodic orbits? What is the effect of noise on the dynamics of a system which is operating in a chaotic regime? These issues are the subject of current and future study.

ACKNOWLEDGEMENTS

The first author would like to acknowledge the support of the Air Force Office of Scientific Research through grant 93-0063 monitored by Dr. Spencer Wu.

REFERENCES

Ariaratnam, S. T. and Srikantaiah, T. K. (1978) Parametric instabilities in elastic structures under stochastic loading, *J. Struct. Mech.*, **6** (4), 349–365.

Ariaratnam, S. T. and Xie, W. C. (1992) Lyapunov exponents and stochastic stability of coupled linear systems under real noise excitation, *ASME J. Appl. Mech.*, **59** (3), 664–673.

Arnold, L. (1984) A formula connecting sample and moment stability of linear stochastic systems, *SIAM J. Appl. Math.*, **44** (4), 793–802.

Arnold, L., Papanicolaou, G. and Wihstutz, V. (1986) Asymptotic analysis of Lyapunov exponent and rotation number of the random oscillator and applications, *SIAM J. Appl. Math.*, **46** (3), 427–450.

Bergman, L. A. and Spencer, Jr., B. F. (1992) Robust numerical solution of the transient Fokker-Planck equation for nonlinear dynamical systems, *Proceedings of the IUTAM Symposium on Nonlinear Stochastic Mechanics, Turin Italy*, 49–60.

Booton, R. C. (1954) Nonlinear control systems with random inputs, *IRE Trans., Circuit Theory, CT-1*, **1**, 9–19.

Boxler, P. (1989) A stochastic version of center manifold theory, *Prob. Thy. Rel. Fields*, **83**, 509–545.

Caughey, T. K. (1963) Equivalent linearization techniques, *J. Acoust. Soc. Amer.*, **35** (11), 1706–1711.

Coulett, P. H., Elphick, C. and Tirapequi, E. (1985) Normal form of a Hopf bifurcation with noise, *Phys. Letters*, **111A**, 277–282.

Doyle, M. M. and Sri Namachchivaya, N. (1994) Almost-sure asymptotic stability of a general four-dimensional dynamical system driven by real noise *J. Stat. Phys.* **75** (314), 525–552.

Ibrahim, R. A., Soundarajan, A. and Heo, H. (1985) Stochastic response of nonlinear dynamic systems based on a non-Gaussian closure, *ASME J. Appl. Mech.*, **52** (4), 965–970.

Kazakov, I. E. (1956) Approximate probability analysis of the operational precision of essentially nonlinear feedback control systems, *Auto. and Remote Control*, **17**, 423–450.

Khas'minskii, R. Z. (1967) Necessary and sufficient conditions for the asymptotic stability of linear stochastic systems, *Thy. Prob. Appl.*, **12** (1), 144–147.

Khas'minskii, R. Z. (1988) On the principles of averaging for Itô stochastic differential equations, *Kybernetica (Prague)*, **4**, 260–279.

Kliemann, W. and Arnold, L. (1983) Lyapunov exponents of linear stochastic systems, Technical Report 93, Institute for Dynamical Systems, University of Bremen.

Kozin, F. (1969) A survey of stability of stochastic systems, *Automatica*, **5**, 95–112.

Kozin, F. (1988) The method of statistical linearization for nonlinear stochastic vibrations in nonlinear stochastic dynamic engineering systems, *Proceedings of the IUTAM Symposium, Innsbruck, Austria*, pp. 45–56.

Kozin, F. and Prodromou, S. (1971) Necessary and sufficient conditions for almost-sure sample stability of linear Itô equations, *SIAM J. Appl. Math.*, **21** (3), 413–424.

Kozin, F. and Sugimoto, S. (1977) Relations between sample and moment stability for linear stochastic differential equations, *Proceedings of Conferences on Stochastic Differential Equations and Applications*, pp. 145–162.

Lin, Y. K. and Wu, W. F. (1983) Applications of cumulant closure to random vibration problems, *ASME Ran. Vib. AMD*, **65**, 113–125.

Lin, Y. K., Fujimori, Y. and Ariaratnam, S. T. (1979) Rotor blade stability in turbulent flows, *AIAA Journal*, Part I: **17** (6), 545–552; Part II: **17** (7), 673–678.

Mitchell, R. R. and Kozin, F. (1974) Sample stability of second order linear differential equations with wide-band noise coefficients, *SIAM J. Appl. Math.*, **27** (4), 571–604.

Nishioka, K. (1976) On the stability of two-dimensional linear stochastic systems, *Kodai Mathematics Seminar*, **27**, 211–230.

Oseledec, V. I. (1968) A multiplicative ergodic theorem. Lyapunov characteristic numbers for dynamical systems, *Trans. Moscow Math. Soc.*, **19**, 197–231.

Papanicolaou, G. C. and Kohler, W. (1976) Asymptotic analysis of determinate and stochastic equations with rapidly varying components, *Comm. Math. Phys.*, **46**, 217–232.

Pardoux, E. and Wihstutz, V. (1988) Lyapunov exponent and rotation number of two-dimensional linear stochastic systems with small diffusion, *SIAM J. Appl. Math.*, **48** (2), 442–457.

Pinsky, M. A. and Wihstutz, V. (1991) Lyapunov exponents for white and real noise driven two-dimensional systems, *Lect. Appl. Math.*, **27**, 201–213.

Socha, L. and Soong, T. T. (1991) Linearization in analysis of nonlinear stochastic systems, *Appl. Mech. Rev.*, **44** (10), 399–422.

Sri Namachchivaya, N. (1989) Instability theorem based on the nature of the boundary behaviour for one-dimensional diffusion, *Struct. Mech. Arch.*, **14** (3), 131–142.

Sri Namachchivaya, N. (1990) Stochastic bifurcation, *Appl. Math. Comp.*, **38**, 101–159.

Sri Namachchivaya, N. (1991a) Almost-sure stability of dynamical systems under combined harmonic and stochastic excitations, *J. Sound Vib.*, **151** (1), 77–90.

Sri Namachchivaya, N. (1991b) Co-dimension two bifurcations in the presence of noise, *ASME J. Appl. Mech.*, **58** (1), 259–265.

Sri Namachchivaya, N. and Ariaratnam, S. T. (1987) Stochastically perturbed linear gyroscopic systems, *Mech. Struct. Mach.*, **15** (3), 323–345.

Sri Namachchivaya, N. and Lin, Y. K. (1988) Application of stochastic averaging for system with high damping, *Prob. Engin. Mech.*, **3** (3), 185–196.

Sri Namachchivaya, N. and Lin, Y. K. (1991) Method of stochastic normal forms, *Int. J. Nonlin. Mech.*, **26** (6), 931–943.

Sri Namachchivaya, N. and Talwar, S. (1994) Maximal Lyapunov exponents for stochastically perturbed co-dimension two bifurcations, *J. Sound Vib.*, **169**, 349–372.

Sri Namachchivaya, N. and Tien, W. M. (1990) Stochastically excited linear nonconservative systems, *Mech. Struct. Mach.*, **18** (4), 459–481.

Sri Namachchivaya, N. and Van Roessel, H. J. (1993) Maximal Lyapunov exponent and rotation numbers for two coupled oscillators driven by real noise, *J. Stat. Phys.*, **71** (3, 4), 549–567.

Talwar, S. and Sri Namachchivaya, N. (1994) Lyapunov exponents for two-degree of freedom systems under periodic and stochastic excitations, in preparation.

Tien W. M., Sri Namachchivaya, N. and Coppola, V. T. (1993) Stochastic averaging using elliptic functions to study nonlinear stochastic systems, *Nonlinear Dynamics*, **4**, 373–387.

N. Sri Namachchivaya and M. M. Doyle, *Nonlinear Systems Group, Talbot Laboratory, University of Illinois, Urbana, IL 61801, USA*

21 SOME ILLUSTRATIVE EXAMPLES OF STOCHASTIC BIFURCATION

S. T. Ariaratnam

The chapter illustrates some concepts of bifurcation in stochastic dynamical systems through simple one- and two-dimensional examples. The role of the maximal Lyapunov exponent in determining the point of dynamical bifurcation and the stochastic stability of the bifurcating solution is explained. An alternative concept, prevalent in the physics literature, which regards stochastic bifurcation as implying a qualitative change in the response probability distribution is also discussed.

21.1   INTRODUCTION

We first consider a *deterministic* dynamical system described by the ordinary differential equation

$$\dot{x} = f(x, \alpha), \qquad x(0) = x_0 \in \mathscr{R}^n, \tag{1}$$

where $f(x, \alpha)$ is continuously differentiable in x and the parameter α. Let $x_s(t, \alpha)$ denote the steady-state solution for a specific value of α. As the parameter α is increased, suppose that this solution becomes unstable at some value α_c of α and a further (bifurcating) solution appears for α beyond the value α_c. The stability of the original solution $x_s(t, \alpha)$ and of the new bifurcating solution can be examined by investigating the stability of the trivial solution of the equation resulting from linearization of equation (1) around each of these solutions. Now suppose that the parameter α, instead of being a constant, is subjected to a random fluctuation. The resulting dynamical system is then governed by the *stochastic* differential equation

$$\dot{x} = f(x, \alpha + \sigma \xi(t)), \qquad x(0) = x_0 \in \mathscr{R}^n, \tag{2}$$

where $\xi(t)$ is a stochastic process and σ is an intensity parameter. The stationary solution

Nonlinearity and Chaos in Engineering Dynamics
Edited by J. M. T. Thompson and S. R. Bishop,   © 1994 John Wiley & Sons Ltd

$x_s(t, \alpha)$ of equation (2), if it exists, now corresponds to the steady-state solution of the deterministic equation (1). The question then arises as to what one understands by a bifurcation of the stochastic response. In analogy with the deterministic case, it is natural to linearize equation (2) around the reference solution $x_s(t, \alpha)$ and examine the stability, in some sense, of the resulting linear stochastic differential equation

$$\dot{v} = A(x_s(t, \alpha), \xi(t)) v, \quad v(0) = v_0 \in \mathcal{R}^n, \tag{3}$$

where $A(x_s, \xi)$ is the $n \times n$ Jacobian matrix $(\partial f_i / \partial x_j)$ evaluated at $x = x_s(t, \alpha)$. In equation (3), the random processes $x_s(t, \alpha)$ and $\xi(t)$ appear as coefficients. The stability of the trivial solution of equation (3) implies the stability of the reference solution $x_s(t, \alpha)$. The average exponential growth of solutions of equation (3) are given, with probability one, by the so-called Lyapunov exponents λ defined by

$$\lambda = \lim_{t \to \infty} \frac{1}{t} \log \| v(t, v_0) \|. \tag{4}$$

According to a theorem of Oseledec (1968), there will in general be n such Lyapunov exponents depending on the random initial vector v_0 which, under certain mild ergodicity properties of the random matrix $A(t)$, are *deterministic real numbers*. The trivial solution of equation (3) will be stable with probability one (w.p.1) if the largest Lyapunov exponent is negative, and unstable w.p.1 if it is positive. Thus, the vanishing of the largest Lyapunov exponent determines the value α_c of α at which stochastic bifurcation to a new solution can occur. The stability of this new solution must be determined in a similar manner by evaluating the largest Lyapunov exponent of the linearization of equation (2) around this solution. This is the criterion of stochastic bifurcation advocated by dynamic systems theorists (Arnold 1988) and is in accord with that for the corresponding deterministic system when the stochastic fluctuation of the parameter is set to zero.

On the other hand, many physicists, chemists and biologists (Horsthemke and Lefever 1984; Stratonovich 1967) have adopted a different criterion which is based on a change in form of the stationary probability distribution of the response. According to the latter concept, a bifurcation is considered to occur when this probability distribution undergoes a qualitative change from a unimodal to a bimodal or multi-modal distribution. The two criteria yield different values for the critical bifurcation parameter α_c.

In this chapter the two concepts are illustrated through some examples of one- and two-dimensional dynamical systems, and their relative significance is discussed.

21.2   ONE-DIMENSIONAL EXAMPLES

Consider the deterministic system whose local behaviour is described by

$$\dot{x} = \alpha x - \beta x^n, \quad x_0 \in \mathcal{R}^+, \tag{5}$$

where $\alpha \in \mathcal{R}$ is the bifurcation parameter, $\beta \in \mathcal{R}^+$ and $n \in \mathcal{N}$. The trivial solution $x_s = 0$ is stable for $\alpha < 0$ and unstable for $\alpha > 0$. A non-trivial bifurcating solution given by $x_s^{n-1} = \alpha / \beta$ exists for $\alpha > 0$; hence $\alpha_c = 0$. When $n = 2, 3$ we have respectively a transcritical and a pitchfork bifurcation at $\alpha = \alpha_c$.

Now suppose that α, β are changed to $\alpha + \sigma_1 \xi_1(t)$, $\beta + \sigma_2 \xi_2(t)$ where $\xi(t)$, $\xi_2(t)$ are independent physical white-noise processes of unit intensity. The system is now described by the Stratonovich stochastic differential equation

$$dx = (\alpha x - \beta x^n)dt + \sigma_1 x \circ dW_1 - \sigma_2 x^n \circ dW_2 \tag{6}$$

whose Itô form is

$$dx = \left[\left(\alpha + \frac{1}{2}\sigma_1^2\right)x - \beta x^n + \frac{1}{2}nx^{2n-1}\sigma_2^2\right]dt + \sigma_1 x dW_1 - \sigma_2 x^n dW_2 \tag{7}$$

where $W_1(t)$, $W_2(t)$ are indepenent standard Wiener processes. Linearization of equation (7) around a reference stationary solution $x_s(t)$ leads to the Itô equation

$$dv = [\alpha + \tfrac{1}{2}\sigma_1^2 - n\beta x_s^{n-1} + \tfrac{1}{2}n(2n-1)\sigma_2^2 x_s^{2n-2}]v dt + \sigma_1 v dW - n\sigma_2 x_s^{n-1} v dW_2. \tag{8}$$

If $\rho = \log v$, then $\rho(t)$ satisfies the Itô equation

$$d\rho = [\alpha - n\beta x_s^{n-1} + \tfrac{1}{2}n(n-1)\sigma_2^2 x_s^{2n-1}]dt + \sigma_1 dW_1 - n\sigma_2 x_s^{n-1} dW_2. \tag{9}$$

From the definition (4), the Lyapunov exponent of the linearized system is

$$\lambda = \alpha - n\beta \langle x_s^{n-1}\rangle + \tfrac{1}{2}n(n-1)\sigma_2^2 \langle x_s^{2n-2}\rangle \tag{10}$$

where $\langle \cdot \rangle$ denotes expectation. For the trivial solution $x_s = 0$, $\lambda := \lambda_0 = \alpha$. Hence the trivial solution is stable w.p.1 for $\alpha < 0$ and unstable w.p.1 for $\alpha > 0$. A stochastic bifurcation to a non-trivial solution therefore occurs at $\alpha := \alpha_c = 0$. In order to determine the Lyapunov exponent $\lambda := \lambda_s$ corresponding to the non-trivial solution x_s, it is necessary to first determine the moments $\langle x_s^{n-1}\rangle$ and $\langle x_s^{2n-2}\rangle$ appearing in equation (10). To this end, we set up the Itô equation governing $\log x_s$:

$$d(\log x_s) = [\alpha - \beta x_s^{n-1} + \tfrac{1}{2}(n-1)\sigma_2^2 x_s^{2n-2}]dt + \sigma_1 dW_1 - \sigma_2 x_s^{n-1} dW_2. \tag{11}$$

Since $x_s(t)$ is a stationary process, $d\langle \log x_s\rangle/dt = 0$ and hence

$$\alpha - \beta \langle x_s^{n-1}\rangle + \tfrac{1}{2}(n-1)\sigma_2^2 \langle x_s^{2n-2}\rangle = 0 \tag{12}$$

which, when substituted in equation (10), gives

$$\lambda_s = -(n-1)\alpha \tag{13}$$

showing that the non-trivial solution $x_s(t)$, which exists only when $\alpha > 0$, is stable w.p.1.

The stationary probability density $p(x)$ of $x_s(t)$ is governed by the Fokker–Planck equation

$$\frac{1}{2}\frac{d^2}{dx^2}[\sigma^2(x)p] - \frac{d}{dx}[m(x)p] = 0 \tag{14}$$

where, from equation (7),

$$m(x) = (\alpha + \tfrac{1}{2}\sigma_1^2)x - \beta x^n + \tfrac{1}{2}nx^{2n-1}\sigma_2^2,$$

$$\sigma^2(x) = \sigma_1^2 x^2 + \sigma_2^2 x^{2n}.$$

The solution for $p(x)$ is obtained as

$$p(x) = C_n \frac{x^{\gamma}}{(\sigma_1^2 + \sigma_2^2 x^{2n-2})^{\delta_n}} \exp\left[-\frac{2\beta}{(n-1)\sigma_2\sigma_2} \tan^{-1}\left(\frac{\sigma_2 x^{n-1}}{\sigma_1} \right) \right] \qquad (15)$$

where

$$\gamma = \frac{2\alpha}{\sigma_1^2} - 1, \quad \delta_n = \frac{1}{2} + \frac{\alpha}{(n-1)\sigma_1^2},$$

and C_n is the normalizing constant. The most probable value of x, denoted by $\hat{x}$, obtained by setting $dp/dx = 0$ or by solving for x from the equation

$$m(x) - \frac{1}{2}\frac{d}{dx}[\sigma^2(x)] = 0 \qquad (16)$$

is either the trivial value $\hat{x} = 0$ or the non-trivial solution given by

$$\hat{x}^{n-1} = \frac{[2n(\alpha - \frac{1}{2}\sigma_1^2)\sigma_2^2 + \beta^2]^{1/2} - \beta}{n\sigma_2^2} \qquad (17)$$

which exists only when $\alpha \geqslant \sigma_1^2/2$. For $\alpha < 0$, $p(x)$ is the Dirac delta function. The density $p(x)$ for $\alpha > 0$, $\sigma_2 = 0$ is shown in Figure 21.1. When $0 < \alpha < \sigma_1^2/2$, $p(x)$ has an integrable singularity at $x = 0$; when $\alpha = \sigma_1^2/2$ it has a single peak at $x = 0$ and is a monotonic decreasing function; when $\alpha > \sigma_1^2/2$ the peak is shifted away from $x = 0$. Thus a qualitative change in the probability distribution occurs when $\alpha = \sigma_1^2/2$. In the physics literature, $\alpha_c^* = \sigma_1^2/2$ is usually taken as the bifurcation value of the parameter α. Figure 21.2 shows the stochastic bifurcation diagram for the system.

From the viewpoint of dynamical systems theory, a bifurcation from the trivial solution occurs when $\alpha = \alpha_c$. However, for values of α in the range $\alpha_c < \alpha < \alpha_c^*$, it is evident from Figure 21.1 that in any fixed interval of time the response remains close to $x = 0$ for a large fraction of this time, and significant deviations from $x = 0$ are observable only when $\alpha > \alpha_c^*$. Hence α_c^* may also be regarded as a bifurcation value of the parameter α in the sense of

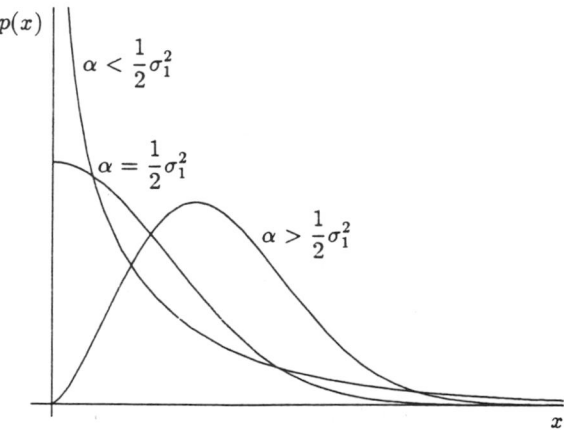

Figure 21.1   Probability density function.

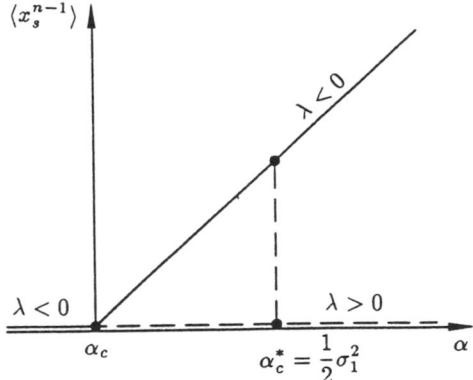

Figure 21.2 Bifurcation diagram for $\dot{x} = \alpha x - \beta x^n + \sigma_1 x \xi(t)$.

bifurcation of probability density. It is worth noting that the random fluctuations have no effect on α_c but do have a beneficial effect on α_c^* by delaying the onset of bifurcation as α is increased.

21.3 TWO-DIMENSIONAL EXAMPLE

Consider now a two-dimensional system governed by the equation

$$\ddot{x} - 2\alpha\dot{x} + \beta|\dot{x}|^\nu \dot{x} + x = 0, \quad \nu > 0, \tag{18}$$

in which $\beta \in \mathscr{R}^+$ is a fixed constant and $\alpha \in \mathscr{R}$ is a variable parameter. When $\alpha > 0$, the system exhibits a Hopf bifurcation from the trivial solution to a periodic limit cycle of amplitude $(\alpha/\gamma)^{1/\nu}$ for small values of α, β where

$$\gamma = \frac{\beta\Gamma\left(\dfrac{3+\nu}{2}\right)}{\sqrt{\pi}\,\Gamma\left(\dfrac{5+\nu}{2}\right)}. \tag{19}$$

Equations of the form (18) often arise in certain problems in structural dynamics such as the motion of offshore platforms under wind and wave actions (Brouwers 1986). We now suppose that the damping coefficient α and the stiffness coefficient are both subjected to random fluctuations so that the resulting equation of motion is

$$\ddot{x} - [2\alpha + \sigma_2\xi_2(t)]\dot{x} + \beta|\dot{x}|^\nu\dot{x} + [1 + \sigma_1\xi_1(t)]x = 0. \tag{20}$$

If $\xi_1(t)$, $\xi_2(t)$ may be approximated by independent 'physical' white-noise processes, this equation may be written in the Stratonovich form,

$$dx_1 = x_2 dt,$$

$$dx_2 = (2\alpha x_2 - \beta|x_2|^\nu x_2 - x_1)dt - \sigma_1 x_1 \circ dW_1 + \sigma_2 x_2 \circ dW_2,$$

where $W_1(t)$, $W_2(t)$, are independent standard Wiener processes. Transforming to polar coordinates (a, θ) by the relations

$$x_1 = a \cos \theta, \quad x_2 = -a \sin \theta,$$

results in

$$da = [(2\alpha + \beta a^\nu |\sin^\nu \theta|)a \sin^2 \theta] dt - \sigma_1 a \sin \theta \cos \theta \circ dW_1 - \sigma_2 a \sin^2 \theta \circ dW_2,$$

$$d\theta[1 + (\alpha + \beta a^\nu |\sin^\nu \theta|) \sin \theta \cos \theta] dt - \sigma_1 \cos^2 \theta \circ dW_1 - \sigma_2 \sin \theta \cos \theta \circ dW_2. \quad (21)$$

For small values of α, β, σ_1 and σ_2, asymptotic solutions for a, θ may be obtained by applying the well-known stochastic averaging procedure to the right-hand sides of equations (21). This results in the following uncoupled Itô equation for the averaged amplitude a:

$$da = \left(\alpha + \frac{3}{16}\sigma_1^2 + \frac{5}{16}\sigma_2^2 - \gamma a^\nu \right) a \, dt + \left(\frac{\sigma_1^2 + 3\sigma_2^2}{8} \right)^{1/2} a \, dW \quad (22)$$

where $W(t)$ is the standard Wiener process. If $a_s(t)$ denotes the stationary solution of equation (22), then writing $a = a_s + v, v = e^\rho$, the linearized Itô equation around $a_s(t)$ is obtained as

$$d\rho = [\alpha + \tfrac{1}{8}(\sigma_1^2 + \sigma_2^2) - \gamma(v + 1)a_s^\nu] dt + \sigma dW \quad (23)$$

where

$$\sigma = (\sigma_1^2 + 3\sigma_2^2)^{1/2}/2\sqrt{2}.$$

From equation (23), the Lyapunov exponent of the linearized system is

$$\lambda = \alpha + \frac{1}{8}(\sigma_1^2 + \sigma_2^2) - \gamma(v + 1)\langle a_s^\nu \rangle. \quad (24)$$

For the trivial solution $a_s(t) = 0$,

$$\lambda := \lambda_0 = \alpha + \frac{1}{8}(\sigma_1^2 + \sigma_2^2). \quad (25)$$

In order to obtain $\langle a_s^\nu \rangle$ for the non-trivial solution, we set up the following Itô equation for $\log a_s$:

$$d(\log a_s) = [\alpha + \tfrac{1}{8}(\sigma_1^2 + \sigma_2^2) - \gamma a_s^\nu] dt + \sigma dW \quad (26)$$

and, since $d\langle \log a_s \rangle/dt = 0$ for a stationary solution, we obtain

$$\langle a_s^\nu \rangle = \frac{1}{\gamma}[\alpha + \tfrac{1}{8}(\sigma_1^2 + \sigma_2^2)] \quad (27)$$

provided $\alpha > -(\sigma_1^2 + \sigma_2^2)/8$. Substituting for $\langle a_s^\nu \rangle$ in the expression (24), the Lyapunov exponent corresponding to the non-trivial solution is

$$\lambda := \lambda_s = -v[\alpha + \tfrac{1}{8}(\sigma_1^2 + \sigma_2^2)]. \quad (28)$$

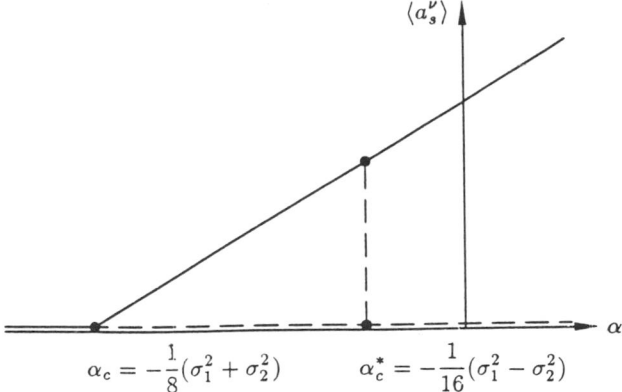

Figure 21.3 Bifurcation diagram for $\ddot{x} - [2\alpha + \sigma_2\xi_2(t)]\dot{x} + \beta|x|^\nu\dot{x} + [1 + \sigma_1\xi_1(t)]x = 0$.

Hence the trivial solution becomes unstable w.p.1 when $\alpha > \alpha_c = -(\sigma_1^2 + \sigma_2^2)/8$ and the non-trivial stationary solution, which exists only when $\alpha > \alpha_c$, is stable for these values of α. Thus a dynamical bifurcation occurs at $\alpha = \alpha_c$.

The probability density of $a_s(t)$, found from the Fokker–Planck equation corresponding to equation (22), is

$$p(a) = Ca^k \exp\left(-\frac{2\gamma}{\nu\sigma^2}a^\nu\right)$$

where

$$k = \frac{16\alpha + \sigma_1^2 - \sigma_2^2}{8\sigma^2}, \quad C = \frac{\nu}{\Gamma(k + 1/\nu)}\left(\frac{2\gamma}{\nu\sigma^2}\right)^{(k+1/\nu)},$$

provided $\alpha > \alpha_c = -1/8(\sigma_1^2 + \sigma_2^2)$.

This probability density is similar to that shown in Figure 21.1. It has an integrable singularity at $a = 0$ for $\alpha_c < \alpha < \alpha_c^*$ and single peak at $a = [(\alpha - \alpha_c^*)/\gamma]^{1/\nu}$ for $\alpha > \alpha_c^* = -(\sigma_1^2 - \sigma_2^2)/16$. A change in the form of the probability density occurs when $\alpha = \alpha_c^*$. As remarked by Stratonovich (1967), fully developed oscillations begin to be noticable only for $\alpha > \alpha_c^*$. The bifurcation diagram is shown in Figure 21.3. It may be noted that both α_c and α_c^* are affected by the addition of noise to the system parameters.

21.4 CONCLUSIONS

The effect of stochastic perturbation of the parameters of some simple dynamical systems exhibiting bifurcational behaviour has been examined. The important role of the Lyapunov exponent in determining the stability of the trivial and the non-trivial stationary solutions and the point of dynamical bifurcation has been shown through explicit calculation of the appropriate Lyapunov exponents. It was found that, even though the trivial solution loses its stability at a dynamical bifurcation point, significant departures from the trivial state can be observed only for parameter values beyond that at which a qualitative change in the probability distribution occurs. In any physical system, in addition to parametric stochastic fluctuations, there is also present stochastic disturbances of an additive nature. In such systems, an abrupt bifurcation does not occur, but rather a gradual transition to higher and higher response amplitudes as the bifurcation point is crossed. Hence, from a physical viewpoint, the concept of bifurcation of probability density measure may be more realistic.

ACKNOWLEDGEMENT

The research for this paper was supported (in part) by Grant No. A-1815 from the Natural Sciences and Engineering Research Council of Canada.

REFERENCES

Arnold, L. (1988) Lyapunov exponents of nonlinear stochastic systems. In: *Proc. IUTAM Symposium on Nonlinear Stochastic Dynamic Engineering Systems, Innsbruck, Austria,* F. Ziegler and G. I. Schuëller (Eds.), Berlin, Springer-Verlag, pp. 181–201.

Brouwers, J. J. H. (1986) Stability of a nonlinearly damped second-order system with randomly fluctuating restoring coefficient, *Int. J. Non-Linear Mechanics,* **21**, 1–43.

Horsthemke, W. and Lefever, R. (1984) *Noise-Induced Transitions,* Berlin, Springer-Verlag.

Oseledec, Y. I. (1968) A multiplicative ergodic theorem – Lyapunov characteristic numbers for dynamical systems, *Trans Moscow Math. Soc.,* **19**, 197–231 (English translation).

Stratonovich, R. L. (1967) *Topics in the Theory of Random Noise, II,* New York, Gordon and Breach, p. 184.

S. T. Ariaratnam, *Solid Mechanics Division, University of Waterloo, Waterloo, Ontario N2L 341, Canada*

22 STOCHASTIC RESONANCE

M. I. Dykman, D. G. Luchinsky, R. Mannella,
P. V. E. McClintock, S. M. Soskin, N. D. Stein and N. G. Stocks

We review stochastic resonance (SR), a counter-intuitive phenomenon in which the signal due to a weak periodic force in a nonlinear system can be *enhanced* by the addition of external noise. A theoretical approach based on linear response theory (LRT) is described. It is pointed out that, although the LRT theory of SR is by definition restricted to the small-signal limit, it possesses substantial advantages in terms of simplicity, generality and predictive power. We outline the application of LRT to overdamped motion in a bistable potential, the most commonly studied form of SR. Two new forms of SR, predicted on the basis of LRT and subsequently observed in analogue electronic experiments, are described.

22.1 INTRODUCTION

One of the most active current growth areas of nonlinear dynamics lies in the relatively unexplored region separating the two major divisions of the subject: that is, within the interface separating 'deterministic' nonlinear dynamics (e.g. Thompson and Stewart 1986), where externally applied forces are precisely known (e.g. periodic), from stochastic nonlinear dynamics where the system under study fluctuates under the influence of a random force (e.g. Moss and McClintock 1989). *Stochastic resonance* (SR), in which the signal due to a weak periodic force in a nonlinear system can, remarkably, be amplified by the addition of external noise, provides an example of a phenomenon in this interface region. It arises through a tripartite interaction between nonlinearity, fluctuations and a periodic force, and it cannot occur unless all three of these features are simultaneously present.

The notion of SR was originally introduced (Nicolis 1982; Benzi *et al.* 1982) in relation to the earth's ice-age cycle. The phenomenon was subsequently demonstrated in an electronic circuit (Fauve and Heslot 1983) and in a ring laser (McNamara *et al.* 1988). Following this latter paper, there has been a veritable explosion of activity leading to the observation

Nonlinearity and Chaos in Engineering Dynamics
Edited by J. M. T. Thompson and S. R. Bishop,   © 1994 John Wiley & Sons Ltd

of SR or associated phenomena in a wide variety of contexts, including passive optical systems (Dykman *et al.* 1991), electron spin resonance (Gammaitoni *et al.* 1991a), sensory neurons (Longtin *et al.* 1991), and a magneto-elastic strip (Spano *et al.* 1992). These references are merely illustrative: a fuller bibliography can be found within the proceedings of a recent conference on SR (Moss *et al.* 1993).

In this chapter we introduce SR and set it within the context of classical linear response theory (LRT). We emphasize that the LRT perception of the phenomenon is very general. Not only does it provide a good description of SR in systems with static bistable potentials (conventional SR) but it also leads on naturally to the prediction of new forms of SR in quite different kinds of systems: see Wiesenfeld (1993). In Section 22.2 we describe this LRT approach and in Section 22.3 we show how it may be applied to conventional SR. Sections 22.4 and in 22.5 describe two quite new forms of SR — associated with fluctuational transitions between coexisting periodic attractors, and for underdamped nonlinear oscillators in the absence of bistability — that were predicted on the basis of LRT and subsequently observed in electronic experiments. In Section 22.6 we summarize the results, discuss future directions, and draw conclusions.

22.2   LINEAR RESPONSE THEORY OF STOCHASTIC RESONANCE

We shall define SR as an increase of the amplitude of a periodic signal in a nonlinear system resulting from the addition of external noise at the input; often, the signal/noise ratio at the output will also increase, an effect that meets the stricter definition of SR used by some authors. In both cases, the signal decreases again for sufficiently strong noise, giving rise to a resonance-like curve when the amplitude is plotted against noise intensity, thereby accounting for the terminology.

The theory of SR has been perceived as difficult, because of the need to treat stochastic and periodic forces together in a highly nonlinear system. It has mostly been developed with the simplifying assumption of a discrete two-state model (in the case of bistable systems) or, in the case of continuous systems, has been based on an approximate or numerical solution of the Fokker–Planck equation for a periodically driven system, sometimes with contradictory results (Benzi *et al.* 1982; Nicolis 1982; Presilla *et al.* 1989; Gammaitoni *et al.* 1989; McNamara and Wiesenfeld 1989; Fox 1989; Hu Gang *et al.* 1990; Jung and Hanggi 1990, 1991).

The alternative approach to SR introduced by Dykman *et al.* (1990a, b), based on LRT, is quite different. According to LRT (see e.g. Landau and Lifshitz 1980), if a system with coordinate q is driven by a weak force $A \cos \Omega t$, a small periodic term $\delta \langle q(t) \rangle$ will appear in the ensemble-averaged value of the coordinate, oscillating at the same frequency Ω:

$$\delta \langle q(t) \rangle = a \cos(\Omega t + \phi), \qquad A \to 0, \tag{1}$$

$$a = A|\chi(\Omega)|, \qquad \phi = -\arctan[\mathrm{Im}\, \chi(\Omega)/\mathrm{Re}\, \chi(\Omega)] \tag{2}$$

where $\chi(\Omega)$ is the *susceptibility* of the system. The function $\chi(\Omega)$ contains virtually everything needing to be known about the response of the system to a weak driving force. It gives both the *amplitude* a of the signal and its *phase lag* ϕ relative to the driving force. The occurrence of a delta-shaped spike at frequency Ω in the spectral density of fluctuations

(SDF), $Q(\omega)$, of the system

$$Q(\omega) = \lim_{\tau \to \infty} (4\pi\tau)^{-1} \left| \int_{-\tau}^{\tau} dt q(t) \exp(i\omega t) \right|^2 \tag{3}$$

follows immediately from (1) on account of the principle of the decay of correlations

$$\langle q(t)q(t') \rangle \to \langle q(t) \rangle \langle q(t') \rangle \text{ for } |t - t'| \to \infty.$$

The *intensity* $\frac{1}{4}a^2$ (i.e. area) of the spike can be found from (2). Following Fauve and Heslot (1983) and McNamara *et al.* (1988), the signal/noise ratio in SR is often defined as the ratio R of the area of the spike to the value $Q^{(0)}(\Omega)$ of the SDF at frequency Ω but in the absence of the driving force. From (1)–(3), this quantity may be expressed in terms of the susceptibility as

$$R = \frac{1}{4}A^2 |\chi(\Omega)|^2 / Q^{(0)}(\Omega) \qquad (A \to 0). \tag{4}$$

Consequently, the evolution of $\chi(\Omega)$, or of $\chi(\Omega)$ and $Q^{(0)}(\Omega)$, with increasing noise intensity shows immediately whether or not SR in the signal or in the signal/noise ratio, respectively, is to be expected at frequency Ω in any given system.

In the particular case of systems that are in thermal equilibrium, or quasi-equilibrium, the susceptibility at frequency Ω can be obtained very simply from the fluctuation dissipation relations (Landau and Lifshitz 1980),

$$\operatorname{Re} \chi(\Omega) = \frac{2}{T} P \int_0^\infty d\omega_1 Q^{(0)}(\omega_1) \omega_1^2 (\omega_1^2 - \Omega^2)^{-1},$$

$$\operatorname{Im} \chi(\Omega) = \frac{\pi\Omega}{T} Q^{(0)}(\Omega), \tag{5}$$

where P implies the Cauchy principal part, and T is the temperature (noise intensity) in energy units. It is interesting to note that a knowledge of $Q^{(0)}(\omega)$ and its evolution with T is then sufficient in itself to predict whether or not the system in question will exhibit SR: this would be true even where the underlying dynamics was unknown, and the information about $Q^{(0)}(\omega)$ had been acquired by experiment.

22.3 STOCHASTIC RESONANCE IN STATIC DOUBLE-WELL POTENTIALS

The initial tests of the above ideas were performed (Dykman *et al.* 1990a, b) through the measurement of SDFs and the investigation of SR in an electronic model of the damped double-well Duffing oscillator,

$$\ddot{q} + 2\Gamma\dot{q} + U'(q) = A \cos \Omega t + f(t),$$

$$U(q) = -\frac{1}{2}q^2 + \frac{1}{4}q^4, \qquad \langle f(t) \rangle = 0, \qquad \langle f(t)f(t') \rangle = 4\Gamma T \delta(t - t'). \tag{6}$$

The results are shown in Figure 22.1, where the scaled signal/noise ratio $\tilde{R} = 6.51 \times 10^{-4} R$

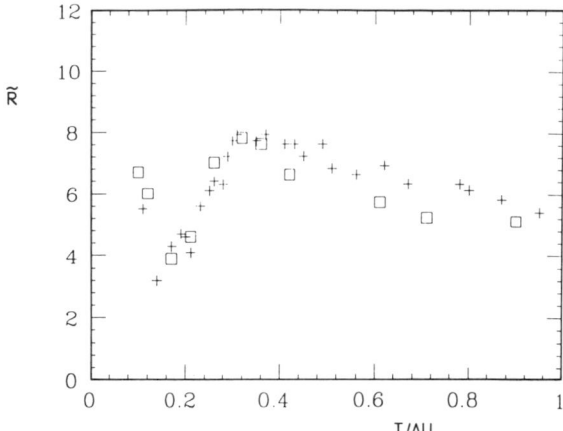

Figure 22.1 Signal/noise ratio $\tilde{R}$ for the electronic model of the double-well oscillator (6), as a function of scaled noise intensity $T/\Delta U$. Direct measurements of $\tilde{R}$ (squares) are compared with results calculated from measured spectra $Q^{(0)}(\omega)$ (crosses) using the fluctuation dissipation relations (5). (After Dykman *et al* 1990a.)

is plotted as a function of scaled noise intensity $T/\Delta U$, $\Delta U(=\frac{1}{4}$ for the potential in (6)) being the height of the potential barrier between the wells. The square data points represent direct measurements of $\tilde{R}$, obtained from the heights of the delta spikes in $Q(\omega)$; the crosses are also obtained experimentally, but in a completely different way, from Equations (2), (4) and (5) using measurements of $Q^{(0)}(\omega)$ in the absence of the periodic force. The fact that the agreement is excellent, within the experimental error, without the use of any adjustable parameters, can be regarded as a direct confirmation of the validity of the LRT perception of small-signal SR.

For the particular parameters used for the measurements of Figure 22.1, the magnitude of the rise in $\tilde{r}$ is relatively modest; much larger increases can be obtained for lower frequencies Ω and larger damping constants Γ. Nonetheless, it is clear that there is a range of $T/\Delta U$ within which $\tilde{R}$ rises with increasing T, i.e. there is a manifestation of SR.

Of course, for a theory of SR, one would also need to be able to calculate $Q^{(0)}(\omega)$, rather than having to measure it experimentally. Although this has been done (Dykman *et al.* 1988) for the underdamped system (6), the most detailed experimental and theoretical studies relate to the overdamped system, which has been widely used as the standard system for investigations of SR,

$$\dot{q} + U'(q) = A\cos\Omega t + f(t)$$

$$U(q) = -\frac{1}{2}q^2 + \frac{1}{4}q^4, \qquad \langle f(t)\rangle = 0, \qquad \langle f(t)f(t')\rangle = 2D\delta(t-t') \qquad (7)$$

where $f(t)$ is now a zero-mean Gaussian noise of intensity D (the way in which the overdamped limit is taken to obtain (7) from (6) is discussed by e.g. Risken 1989). Like (6), (7) for $A=0$ is also a thermal equilibrium system so that, in order to find the susceptibility $\chi(\omega)$, it is only necessary to calculate the SDF $Q^{(0)}(\omega)$ in the absence of the periodic force for substitution in the fluctuation dissipation relations (5) with T replaced by D. In the limit of weak noise,

$D \ll \Delta U$, both $Q^{(0)}(\omega)$ and $\chi(\omega)$ can be obtained analytically (Dykman and Krivoglaz 1979, 1984; Dykman *et al.* 1989) as a sum of partial contributions from fluctuations about the equilibrium positions q_n and from inter-well transitions.

$$Q^{(0)}(\omega) = \Sigma_{n=1,2} w_n Q_n^{(0)}(\omega) + Q_{tr}^{(0)}(\omega), \quad \chi(\omega) = \Sigma_{n=1,2} w_n \chi_n(\omega) + \chi_{tr}(\omega). \tag{8}$$

Here w_n is the population of the nth stable state and, for the model (7), $w_1 = w_2 = \frac{1}{2}$, $Q_1^{(0)}(\omega) = Q_2^{(0)}(\omega)$ and $\chi_1(\omega) = \chi_2(\omega)$. The SDF for the intra-well vibrations $Q_n^{(0)}(\omega)$ is obtained by expanding $U(q)$ about the equilibrium position; $Q_{tr}^{(0)}(\omega)$ can be written down in terms of the transition probabilities $W_{nm}^{(0)}$, defining the probability of an $n \to m$ transition in the absence of periodic forcing.

These calculations result in explicit analytic predictions for $R(D)$ and $\phi(D)$. The latter is of particular interest in view of the results of earlier calculations and experiments, with Nicolis (1982), McNamara and Wiesenfeld (1989) and Hu Gang *et al.* (1990) on the one hand and Gammaitoni *et al.* (1990, 1991a, b) on the other. An analogue electronic experiment was performed (Dykman *et al.* 1992a) to measure $\phi(D)$ for comparison with the LRT theoretical predictions, yielding the results shown by the data points in the main section of Figure 22.2; the inset shows a plot of R/A^2 as a function of D in the range near the minimum where other theories fail. In both cases, the agreement between experiment and theory is very satisfactory, providing further confirmation of the validity of the LRT approach to SR. The dashed line shows the prediction of earlier (two-state) theories (e.g. Nicolis 1982; McNamara and Wiesenfeld 1989) that do not include the effect of intra-well motion.

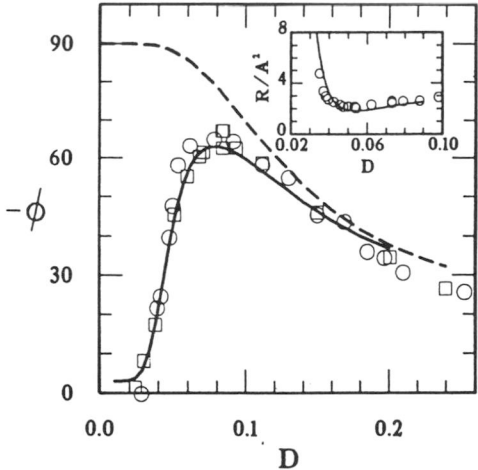

Figure 22.2 The phase shift $-\phi$ (degrees) between the periodic force and the response measured (data points) for an electronic model of the overdamped double-well system (7) with $\Omega = 0.1$ and $A = 0.04$ (circles) and $A = 0.2$ (squares). The solid curve represents the LRT theory based on (5); the dashed curve represents the prediction of earlier two-state theories. Inset: the normalized signal/noise ratio as a function of noise intensity D, showing that the LRT theory works well near the minimum. (After Dykman *et al.* 1992a.)

22.4   STOCHASTIC RESONANCE FOR PERIODIC ATTRACTORS

It is clear from the above discussion that any system whose susceptibility $\chi(\Omega)$ increases with noise intensity may be expected to display SR when driven by a weak periodic force of frequency Ω. Dykman and Krivoglaz (1979) had noticed such an effect in the imaginary part of the susceptibility of a periodically driven nonlinear oscillator with coexisting periodic attractors. It was therefore obvious that SR was to be expected in systems of this kind, and that it would be likely to have some unusual and characteristic features distinguishing it from conventional SR. The new phenomenon has been sought and recently found and investigated (Dykman *et al.* 1993b, c). The results have implications for a large class of passive optically bistable systems and, in particular, for optically bistable microcavities.

The system that we consider is the nearly resonantly driven, underdamped, single-well Duffing oscillator with additive noise,

$$\ddot{q} + 2\Gamma\dot{q} + \omega_0^2 q + \gamma q^3 = F\cos(\omega_F t) + f(t),$$

$$\Gamma, |\delta\omega| \ll \omega_F, \; \gamma\delta\omega > 0, \; \delta\omega = \omega_F - \omega_0, \; \langle f(t) \rangle = 0, \; \langle f(t)f(t') \rangle = 4\Gamma T\delta(t - t'). \tag{9}$$

Note that the force $F\cos(\omega_F t)$ is not very weak; neither is it so strong that the system becomes chaotic or displays subharmonics. Within a certain parameter range, (9) is characterized by two coexisting periodic attractors of different amplitude and phase. Weak noise $f(t)$ causes occasional transitions between them. For appropriate noise intensity, these transitions can become coherent on average with an additional weak periodic trial force $A\cos(\Omega t + \phi)$ added to (9), provided that Ω is close to ω_F, leading to stochastic amplification, i.e. SR. It can be shown that the system responds strongly to the trial force, not only at Ω but also at $|2\omega_F - \Omega|$; the relevant susceptibilities can be calculated by an extension of the Dykman and Krivoglaz (1979) theory.

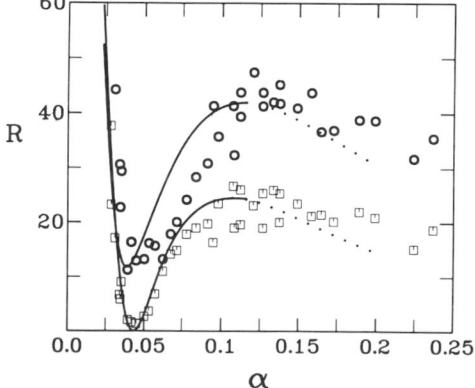

Figure 22.3   The signal/noise ratio R of the response of the system (9) to a weak trial force at frequency Ω, as a function of noise reduced intensity $\alpha = 3\gamma T/32\omega_F^3\Gamma$, in experiment and LRT theory: at the trial frequency Ω (circle data and associated curve); and at the mirror-reflected frequency $|2\omega_F - \Omega|$ (squares). For noise intensities near and beyond the maxima in $R(\alpha)$, the asymptotic theory is only qualitative and so the curves are shown dotted. (After Dykman *et al.* 1993c.)

Measurements of the signal/noise ratio R in an analogue electronic experiment (data points) are compared with the theoretical predictions in Figure 22.3. Although the results are very similar to those found in conventional SR (Moss *et al.* 1993), it must be emphasized that SR for periodic attractors also possesses a number of features that are entirely different. The most important of these are, first, that it is a high-frequency phenomenon. Stochastic amplification takes place not at a low frequency comparable to the inter-attractor hopping rate, as in conventional SR, but at the much higher frequency Ω comparable to ω_0. Secondly the stochastic enhancement of the signal at the mirror-reflected frequency $|\Omega - 2\omega_F|$ has no analogue in conventional SR. Other differences, and the relationship to phenomena in nonlinear optics, are discussed by Dykman *et al.* (1993c).

22.5 STOCHASTIC RESONANCE IN MONOSTABLE SYSTEMS

Until recently, it was the almost universal assumption (Moss *et al.* 1993, and references therein) that the stochastic amplification, or, SR could occur only as the result of nearly periodic fluctuational transitions between coexisting attractors, corresponding to the minima of a static bistable potential. The high-frequency SR of Section 22.4 extends the picture to encompass periodic attractors, but it still requires bistability. The LRT picture of SR, however, does not involve any such requirement: any system, bistable or otherwise, in which the susceptibility $\chi(\Omega)$ increases with noise intensity would, in view of (1) and (2), be expected to display SR.

One monostable system in which SR is to be anticipated on these grounds is the underdamped, single-well, Duffing oscillator subject to a constant field,

$$\ddot{q} + 2\Gamma\dot{q} + U'(q) = A \cos \Omega t + f(t),$$

$$U(q) = \tfrac{1}{2}q^2 + \tfrac{1}{4}q^4 + Bq, \quad \Gamma \ll 1, \quad \langle f(t) \rangle = 0, \quad \langle f(t)f(t') \rangle = 4\Gamma T\delta(t - t'), \tag{10}$$

which is known (Stocks *et al.* 1993a) to have extremely sharp zero-dispersion peaks (ZDPs) in its SDFs provided that $|B| > 8/(7)^{3/2}$ so that the variation of the oscillator's eigenfrequency $\omega(E)$ with energy E possesses an extremum (Dykman *et al.* 1990c). The ZDPs rise exponentially fast with increasing noise intensity T. Thus, because (10) for $A = 0$ is a system of the thermal equilibrium type, to which (5) is applicable, $\chi(\Omega)$ may also be expected to rise extremely fast provided that Ω is chosen to be in the close vicinity of the ZDP.

Experimental results (Stocks *et al.* 1993b) obtained from an analogue electronic model of (10) with $B = 2$ are shown by the circle data of Figure 22.4a. The quantity plotted is the square of the stochastic amplification factor

$$S(T) = a(T)/a(0)$$

where $a(T)$ is the amplitude of the signal for noise intensity T. The fact that $S(T)$ rises very rapidly (from the value of unity that it would take in the absence of stochastic amplification) provides a clear signature of SR. The fuller curve is the LRT prediction, based on (1) and (5) using the expressions for $Q^{(0)}(\omega)$ given by Dykman *et al.* (1990c). It is in very satisfactory agreement with the data. The phase shift $\phi(T)$ has also been measured and calculated, as shown by the circle data and associated curve in Figure 22.4b. Here, too, experiment and

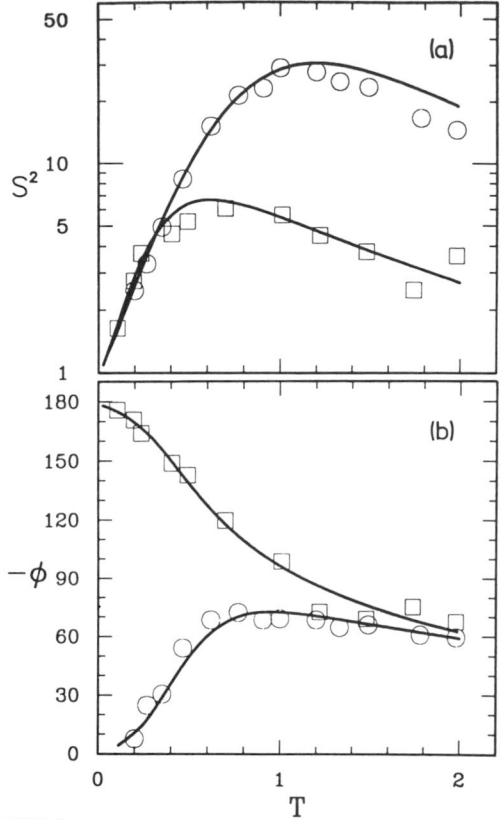

Figure 22.4   (a) The squared stochastic amplification factor S^2 measured as a function of noise intensity T for the electronic model of the system (10) with $|B| = 2$, $A = 0.02$ (circle data), compared with the LRT theory (curve) based on the fluctuation dissipation relations (5). (b) The phase shift ϕ between the periodic force and the response, measured (circle data) under the same conditions and compared with the LRT theoretical curve. (After Stocks *et al.* 1993b, where the square data points and associated curve, for $B = 0$, are also discussed.)

LRT theory agree well. A physically motivated discussion (together with an explanation of the experimental and theoretical results obtained for $B = 0$ shown by the square data points and associated curves) of this new form of SR has been given by Stocks *et al.* (1993b). It can be demonstrated (Stocks *et al.* 1992) on the basis of LRT that, for sufficiently small Γ in (10), substantial increases, not only in the signal, but also in the signal/noise ratio R are to be expected.

22.6   CONCLUSION

We conclude that LRT provides a good description, not only of conventional SR, but also of the other new forms of SR that can be predicted on that basis. These include SR for periodically modulated noise (Dykman *et al.* 1992b) which has not been considered here, as well as the SR for periodic attractors and monostable systems discussed above. Although LRT is, by definition, applicable only in the small-signal limit, in combination with the corresponding physical picture of SR it provides a valuable clue as to the general type of behaviour to be expected even for larger amplitudes of the trial force.

REFERENCES

Benzi, R., Parisi, G., Sutera, A. and Vulpiani, A. (1982) Stochastic resonance in climatic change, *Tellus*, **34**, 10–16.

Dykman, M. I. and Krivoglaz, M. A. (1979) Theory of fluctuational transitions between stables states of a nonlinear oscillator, *Sov. Phys. JETP*, **50**, 30–37.

Dykman, M. I. and Krivoglaz, M. A. (1984) Theory of nonlinear oscillator interacting with a medium. In: *Soviet Physics Reviews*, I. M. Khalatnikov (ed.), New York, Harwood, Vol. 5, pp. 265–441.

Dykman, M. I., Mannella, R., McClintock, P. V. E., Moss, F. and Soskin, S. M. (1988) Spectral density of fluctuations of a double-well Duffing oscillator driven by white noise, *Phys. Rev. A*, **37**, 1303–1313.

Dykman, M. I., Krivoglaz, M. A. and Soskin, S. M. (1989) Transition probabilities and spectral density of fluctuations of noise driven bistable systems. In *Noise in Nonlinear Dynamical Systems*, F. Moss and P. V. E. McClintock (eds.), Cambridge, Cambridge University Press, Vol. 2, pp. 347–380.

Dykman, M. I., Mannella, R., McClintock, P. V. E. and Stocks, N. G. (1990a) Comment on stochastic resonance in bistable systems, *Phys. Rev. Lett.*, **65**, 2606.

Dykman, M. I., McClintock, P. V. E., Mannella, R. and Stocks, N. G. (1990b) Stochastic resonance in the linear and nonlinear responses of a bistable system to a periodic field, *Sov. Phys. JETP Lett.*, **52**, 141–144.

Dykman, M. I., Mannella, R., McClintock, P. V. E., Soskin, S. M. and Stocks, N. G. (1990c) Noise-induced narrowing of peaks in the power spectra of underdamped nonlinear oscillators, *Phys. Rev. A*, **42**, 7041–7049.

Dykman, M. I., Velikovich, A. L., Golubev, G. P., Luchinsky, D. G. and Tsuprikov, S. V. (1991) Stochastic resonance in an all-optical passive bistable system, *Sov. Phys. JETP Lett.*, **53**, 193–197.

Dykman, M. I., Mannella, R., McClintock, P. V. E. and Stocks, N. G. (1992a) Phase shifts in stochastic resonance, *Phys. Rev. Lett.*, **68**, 2985–2988.

Dykman, M. I., Luchinsky, D. G., McClintock, P. V. E., Stein, N. D. and Stocks, N. G. (1992b) Stochastic resonance for periodically modulated noise intensity, *Phys. Rev. A*, **46**, R1713–1716.

Dykman, M. I., Luchinsky, D. G., Mannella, R., McClintock, P. V. E., Stein, N. D. and Stocks, N. G. (1993a) Stochastic resonance: linear response and giant nonlinearity, *J. Stat. Phys.*, **70**, 463–478.

Dykman, M. I., Luchinsky, D. G., Mannella, R., McClintock, P. V. E., Stein, N. D. and Stocks, N. G. (1993b) Nonconventional stochastic resonance, *J. Stat. Phys.*, **70**, 479–499.

Dykman, M. I., Luchinsky, D. G., Mannella, R., McClintock, P. V. E., Stein, N. D. and Stocks, N. G. (1993c) High frequency stochastic resonance in periodically driven systems, *Sov. Phys. JETP Lett.*, **58**, 145–151.

Fauve, S. and Heslot, F. (1983) Stochastic resonance in a bistable system, *Phys. Lett.*, **97A**, 5–7.

Fox, R. F. (1989) Stochastic resonance in a double well, *Phys. Rev. A*, **39**, 4148–4153.

Gammaitoni, L., Marchesoni, F., Menichella-Saetta, E. and Santucci, S. (1989) Stochastic resonance in bistable systems, *Phys. Rev. Lett.*, **62**, 349–352.

Gammaitoni, L., Marchesoni, F., Menichella-Saetta, E. and Santucci, S. (1990) Reply to Comment on Stochastic resonance in bistable systems, *Phys. Rev. Lett.*, **65**, 2607.

Gammaitoni, L., Martinelli, M., Pardi, L. and Santucci, S. (1991a) Observation of stochastic resonance in bistable electron-paramagnetic-resonance systems, *Phys. Rev. Lett.*, **67**, 1799–1802.

Gammaitoni, L., Marchesoni, F., Martinelli, M., Pardi, L. and Santucci, S. (1991b) Phase shifts in bistable EPR systems at stochastic resonance, *Phys. Lett.*, **158A**, 449–452.

Hu Gang, Nicolis, G. and Nicolis, C. (1990) Periodically forced Fokker–Planck equation and stochastic resonance, *Phys. Rev. A*, **42**, 2030–2041.

Jung, P. and Hanggi, P. (1990) Resonantly driven Brownian motion: basic concepts and exact results, *Phys. Rev. A*, **41**, 2977–2988.

Jung, P. and Hanggi, P. (1991) Amplification of small signals via stochastic resonance, *Phys. Rev. A*, **44**, 8032–8042.

Landau, L. D. and Lifshitz, E. M. (1980) *Statistical Physics*, 3rd edn., Part 1, New York, Pergamon.

Longtin, A., Bulsara, A. and Moss, F. (1991) Time interval sequences in bistable systems and the noise-induced transmission of information by sensory neurons, *Phys. Rev. Lett.*, **67**, 656–659.

McNamara, B. and Wiesenfeld, K. (1989) Theory of stochastic resonance, *Phys. Rev. A*, **39**, 4854–4869.

McNamara, B., Wiesenfeld, K. and Roy, R. (1988) Observation of stochastic resonance in a ring laser, *Phys. Rev. Lett.*, **60**, 2626–2629.

Moss, F. and McClintock, P. V. E. (eds.) (1989) *Noise in Nonlinear Dynamical Systems*, Cambridge, Cambridge University Press, 3 vols.

Moss, F., Bulsara A. and Shlesinger, M. F. (eds.) (1993) Proceedings of the NATO ARW: Stochastic resonance in Physics and Biology, *J. Stat. Phys.*, **70** (1/2), special issue.

Nicolis, C. (1982) Stochastic aspects of climatic transitions – response to a periodic forcing, *Tellus*, **34**, 1–9.

Presilla, C., Marchesoni, F. and Gammaitoni, L. (1989) Periodically time-modulated bistable systems: Nonstationary statistical properties, *Phys. Rev. A*, **40**, 2105–2113.

Risken, H. (1989) *The Fokker–Planck Equation*, 2nd edn., Berlin, Springer-Verlag.

Spano, M. L., Wun-Fogle, M. and Ditto, W. L. (1992) Experimental observation of stochastic resonance in a magnetoelastic ribbon, *Phys. Rev. A*, **46**, 5253–5256.

Stocks, N. G., Stein, N. D., Soskin, S. M. and McClintock, P. V. E. (1992) Zero-dispersion stochastic resonance, *J. Phys. A* **25**, L1119–1125.

Stocks, N. G., McClintock, P. V. E. and Soskin, S. M. (1993a) Observation of zero-dispersion peaks in the fluctuation spectrum of an underdamped single-well oscillator, *Europhys. Lett.*, **21**, 395–400.

Stocks, N. G., Stein, N. D. and McClintock, P. V. E. (1993b) Stochastic resonance in monostable systems, *J. Phys. A*, **26**, L385–390.

Thompson, J. M. T. and Stewart H. B. (1986) *Nonlinear Dynamics and Chaos*, New York, Wiley.

Wiesenfeld, K. (1993) Signals from noise: stochastic resonance pays off, *Physics World*, **6** (February), 23–24.

M. I. Dykman, *Department of Physics, Stanford University, Stanford, CA 94305, USA*

D. G. Luchinsky, *AU-Russian Institute for Metrological Service, Andreevskaya nab 2, 117965 Moscow, Russia*

R. Mannella, *Dipartimento di Fisica, Università di Pisa, Piazza Torricelli 2, 56100 Pisa, Italy*

P. V. E. McClintock and N. D. Stein, *School of Physics and Materials, University of Lancaster, Lancaster LA1 4YB, UK*

S. M. Soskin, *Institute of Semiconductor Physics, pr. Nauki 45, 252038 Kiev, Ukraine*

N. G. Stocks, *Department of Engineering, University of Warwick, Coventry CV4 7AL. UK*

23 NONLINEAR HIGH-FREQUENCY VIBRATIONS OF COMPLEX ENGINEERING STRUCTURES

A. K. Belyaev

Complex structures are modelled by a random–heterogeneous continuum. The field of high-frequency vibration is studied by means of analysis of the framework's displacement. It is shown that considerable spatial decay of high-frequency vibration can be explain by: (i) the secondary systems' resonant absorption, (ii) dispersion and multiple scattering, (iii) material damping and dry friction between the structural members. The effect of high-frequency vibration saturation is observed and used for explanation of full-scale vibration testing of a large-sized spacecraft.

23.1 INTRODUCTION

Such complex engineering structures as buildings, ships, aircraft and spacecraft are considered. A detailed description of their vibrations is made difficult, first by the complexity of the structure's shape, then by assembly of separate substructures, and finally, by the presence of various secondary systems. Even if it were possible to obtain an 'exact' boundary-value problem for the actual complex structure and even if one could succeed in solving this problem, the very interpretation of this 'exact' solution would present great difficulty. The field of vibration of an actual complex structure (under, say, wide-band excitation) is a complicated function of time and spatial coordinates since a great number of modes are excited. According to engineering experience only the first, second and rarely the third global resonances take place in actual complex structures. Beyond the region of these few global resonances the vibrations localize within the substructures (e.g. Bendiksen 1987; Pierre and Dowell 1987; Cornwell and Bendiksen 1989; Pierre 1990; Cha and Pierre 1991) and the conventional methods of structural dynamics cannot be applied. A deterministic modelling does not result in an adequate modelling of complex structures because of many inherent

Nonlinearity and Chaos in Engineering Dynamics
Edited by J. M. T. Thompson and S. R. Bishop,   © 1994 John Wiley & Sons Ltd

uncontrolled factors (uncertainties, Ibrahim 1987). They arise from stiffness, mass and damping fluctuations caused by variations in material properties as well as variations resulting from manufacturing and assembly. The study of essential heterogeneity, complexity and uncertainties of actual complex structures naturally leads to the concept of random-heterogeneous media. For simplicity the one-dimensional case will be analysed; however, the generalization to the three-dimensional case can be done by analogy (e.g. Belyaev 1991).

The chapter is arranged as follows. In Section 23.2 the boundary problem is obtained with the help of the Hamiltonian variational principle since the mechanical parameters of the structure are perturbed in a deterministic sense. The Dyson integral equation originally applied in quantum electrodynamics is used in Section 23.3 to find the mean field of high-frequency vibration of essentially heterogeneous media. The main reasons for considerable spatial decay of high-frequency vibration and some nonlinear effects are listed in Section 23.4. The chapter closes with some examples of applications and conclusions in Section 23.5.

23.2   BOUNDARY PROBLEM FOR HIGH-FREQUENCY VIBRATION IN A COMPLEX STRUCTURE

A complex structure consisting of N substructures is schematically depicted in Figure 23.1. Let us consider a typical substructure L_n and take into account that vibrations localize within the substructure at high frequencies. Then the absolute displacement $u_n(x, t)$ of the substructure points can be represented in the form of modal analysis

$$u_n(x, t) = \sum_{k=1}^{\infty} u_{nk}(x) q_{nk}(t) + u(x, t) \tag{1}$$

where u_{nk} are the normal modes and q_{nk} are the generalized coordinates. The normal modes are specified to be orthonormal within the substructure and to vanish on the boundaries of substructures. Thus the function $u(x, t)$ coincides with the actual displacement of framework (primary structure) and may be referred to as the displacement of the framework. The local vibration of the substructure is represented in equation (1) by the standard sum of modal analysis and the vibration transmission from the substructure to the adjacent ones is described by the function $u(x, t)$. Hence the vibration propagation in complex structures can be studied by means of analysis of the field $u(x, t)$. The expressions for kinetic K and

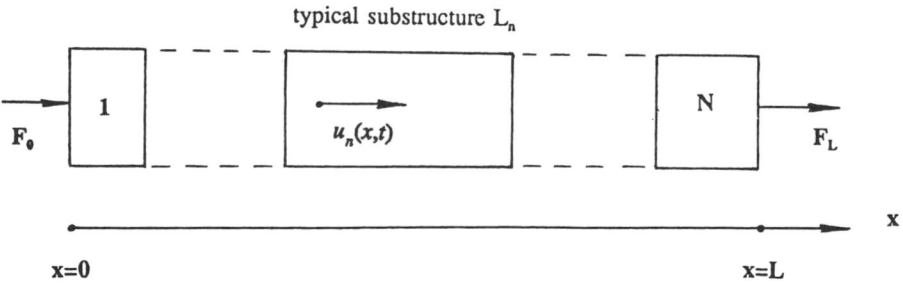

typical substructure L_n

F_0 1 $u_n(x,t)$ N F_L

x=0   x=L   x

Figure 23.1   Mechanical model of complex structure.

potential P energy and for the external loads' work W are derived using equation (1):

$$K = \frac{1}{2} \sum_{n=1}^{N} \int_{L_n} \mu \dot{u}_n^2 dx = \frac{1}{2} \int_{L} \mu \dot{u}^2 dx + \sum_{n=1}^{N} \sum_{k=1}^{\infty} \left(\dot{q}_{nk} \int_{L_n} \mu \dot{u} u_{nk} dx + \frac{1}{2} \dot{q}_{nk}^2 \right), \tag{2}$$

$$P = \frac{1}{2} \sum_{n=1}^{N} \int_{L_n} EA(u_n')^2 dx = \frac{1}{2} \int_{L} EA(u')^2 dx + \frac{1}{2} \sum_{n=1}^{N} \sum_{k=1}^{\infty} \omega_{nk}^2 q_{nk}^2, \tag{3}$$

$$W = \sum_{n=1}^{N} \sum_{k=1}^{\infty} p_{nk} q_{nk} + F_0 u(0) + F_L u(L). \tag{4}$$

Here $\mu(x)$ is the mass density per unit length, $E(x)$ is Young's modulus, $A(x)$ is the cross-sectional area, ω_{nk} are the eigenfrequencies of the substructure L_n, p_{nk} are the generalized forces, and the external forces F_0 and F_L are applied at the ends $x = 0$ and $x = L$ respectively.

The system of normal modes u_{nk} is known to be complete; hence, in order to obtain a unique boundary problem for u and q_{nk} the following additional condition was imposed:

$$\sum_{n=1}^{N} \sum_{k=1}^{\infty} q_{nk} \int_{L_n} EA u' u_{nk}' dx = 0. \tag{5}$$

Under this condition the terms in equation (3) for potential energy are orthogonal; thus the framework and secondary systems drive each other only kinematically.

The boundary problem obtained by means of the Hamiltonian variational principle is given as follows:

$$x \in L_n, \quad (EA u')' - \mu \left(\ddot{u} + \sum_{k=1}^{\infty} \ddot{q}_{nk} u_{nk} \right) = 0, \quad n = 1, 2, \ldots, N, \tag{6}$$

$$\ddot{q}_{nk} + \omega_{nk}^2 q_{nk} = - \int_{L_n} \mu u_{nk} \ddot{u} dx, \quad k = 1, 2, \ldots, \infty, \tag{7}$$

$$x = 0, \quad EA u' = F_0; \quad x = L, \quad EA u' = - F_L. \tag{8}$$

The generalized Jenkins model (Palmov 1976) is used to take into account the inherent non-linear material damping and the mechanical energy dissipation due to dry friction between the structural members. Its rheological model (Figure 23.2) is composed of an infinite number of Jenkins elements, each consisting of an elastic element $E dh$ in series with a dry damper. Each dry damper has a maximum allowance load $Eh dh$ characterized by the density of the yield-strength distribution $R(h)$, where h is the dimensionless yield strength. The generalized Jenkins model is a universal model which is valid for the description of the internal hysteresis in materials caused by nonlinear stress–strain behaviour (amplitude-dependent internal damping). This model is suitable for describing friction in sliding or fretting of joints, supports or other parts of the structure during their relative motion.

The spectral representations

$$u(x, t) = \int_{-\infty}^{\infty} u(x, \omega) e^{i\omega t} d\omega, \quad q_{nk}(x, t) = \int_{-\infty}^{\infty} q_{nk}(x, \omega) e^{i\omega t} d\omega,$$

$$F_0(t) = \int_{-\infty}^{\infty} F_0(\omega) e^{i\omega t} d\omega \tag{9}$$

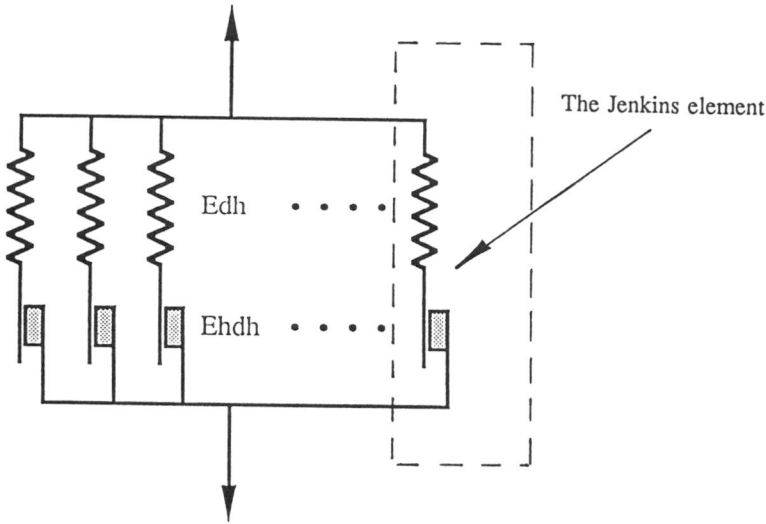

Figure 23.2   The generalized Jenkins model.

are suitable for description of the deterministic and random vibrations. It allows us to introduce the complex Young's modulus for harmonics of frequency ω (for details see Belyaev 1993):

$$\hat{E} = E(1 + i\chi)^2; \quad \chi(a) = \int_0^1 \frac{1}{2} \eta \sqrt{1 - \eta^2} R\left(\frac{\pi a \eta}{4}\right) \frac{\pi a}{4} \, d\eta \tag{10}$$

where a is the deformation amplitude. The correspondence principle in the theory of dynamic plasticity allows us to derive the following boundary problem for harmonic ω:

$$x \in L_n, \quad (\hat{E}Au')' + \mu\omega^2\left(u + \sum_{k=1}^{\infty} u_{nk}q_{nk}\right) = 0, \quad n = 1, 2, \ldots, N, \tag{11}$$

$$(-\omega^2 + \hat{\omega}_{nk}^2)q_{nk} = \omega^2 \int_{L_n} \mu u_{nk} u \, dx, \quad k = 1, 2, \ldots, \infty, \tag{12}$$

$$x = 0, \quad \hat{E}Au' = F_0; \quad x = L, \quad \hat{E}Au' = -F_L, \tag{13}$$

$\hat{\omega}_{nk} = \omega_{nk}(1 + i\chi)$ being the 'complex eigenfrequencies'. Having obtained q_{nk} from equation (12) and substituted it into equation (11) we have the following equation for $u(x, \omega)$ (Belyaev 1993):

$$x \in L_n, \quad (\hat{E}Au')' + \mu\omega^2 s(\omega)u = 0, \quad n = 1, 2, \ldots, N, \tag{14}$$

$$s(\omega) = 1 + \omega^2 \sum_{k=1}^{\infty} \frac{\beta_{nk}}{-\omega^2 + \hat{\omega}_{nk}^2}. \tag{15}$$

Close consideration is required for the spectrum $s(\omega)$ since this function characterizes the generalized spectral properties of the whole structure. As seen from equation (15), $s(\omega)$ is formed by an infinite number of resonance curves corresponding to, a SDOF-system,

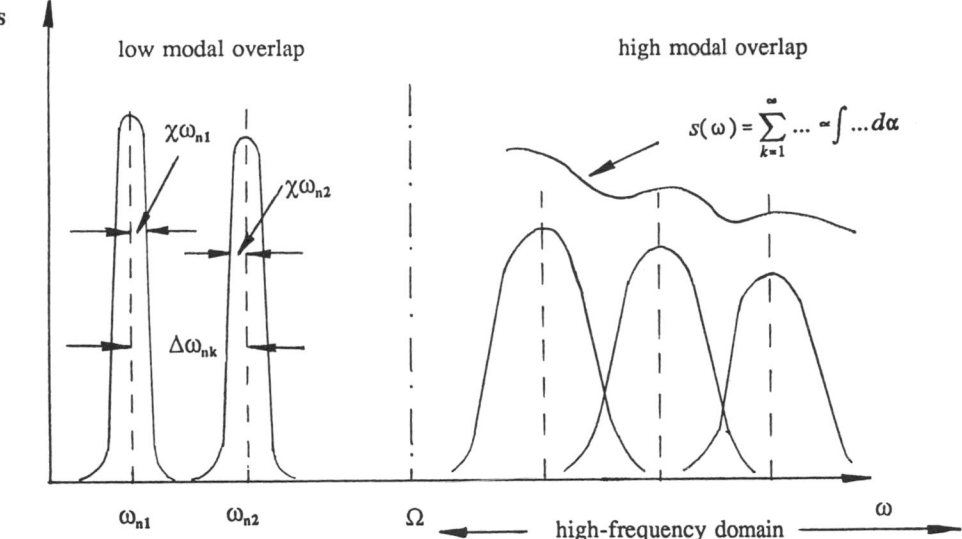

Figure 23.3   Frequency domains.

(Figure 23.3). The width of each resonance curve is $\chi\omega_{nk}$ at the 'half-power' level. If the resonant width is large compared to the eigenfrequency separation $\Delta\omega_{nk}$ (high modal overlap),

$$\Delta\omega_{nk} = \omega_{nk+1} - \omega_{nk} \leqslant \chi(\omega_{n+1} + \omega_{nk}) \quad \text{or} \quad \Delta\omega_{nk}/2\omega_{nk} \leqslant \chi, \tag{16}$$

then the resonance curves in equation (15) merge and the sum in (15) can be replaced by an integral:

$$s(\omega) = 1 + \omega^2 \int_0^\infty \frac{\Phi(\alpha)\,d\alpha}{-\omega^2 + \alpha^2(1 + i\chi)^2} \tag{17}$$

where $\Phi(\alpha)$ is a locally smooth function of the eigenfrequency distribution. Hence, each complex structure has a critical frequency Ω which depends on the relative density of natural frequency spectrum and the structural damping (see equation (16) and Figure 23.3) and is therefore specific for each complex structure. In the high-frequency domain $(\omega > \Omega)$ the structure behaves as a mechanical system with the continuous spectrum of eigenfrequencies that is typical for an infinite continuum. The denotations

$$s(\omega) = [1 - i\kappa(\omega)]^2, \quad \kappa(\omega) = \chi\omega^2 \int_0^\infty \frac{\alpha^2 \Phi(\alpha)\,d\alpha}{(-\omega^2 + \alpha^2)^2 + 4\chi^2\alpha^4} \tag{18}$$

will be used hereafter. For small value of χ and locally smooth function $\Phi(\alpha)$ the integral, equation (18) can be estimated with the help of the random vibration theory (Bolotin 1984), to give

$$\kappa(\omega) = \tfrac{1}{2}\pi\omega\Phi(\omega). \tag{19}$$

23.3 MEAN FIELD OF HIGH-FREQUENCY VIBRATION

By means of the new independent variable y and dependent variable V:

$$y = \langle v \rangle \int_0^x \frac{d\zeta}{v(\zeta)}; \quad u(x) = (EA\mu)^{-1/4} V(y), \quad v = \sqrt{EA/\mu}, \tag{20}$$

equation (14) can be written in the form

$$\frac{d^2 V}{dy^2} + [\lambda^2 + \varepsilon(y)] V = \delta(y - y_0), \quad \lambda^2 = \frac{\omega^2 s(\omega)}{\langle v \rangle^2 (1 + i\chi)^2},$$

$$\varepsilon(y) = -(EA\mu)^{-1/4} \frac{d^2}{dx^2} (EA\mu)^{1/4}, \tag{21}$$

where $\langle \ \rangle$ denotes the mean value. The widely used asymptotical methods cannot be applied to study equation (21) since actual complex structures should be modelled by essentially heterogeneous random media. For such media ε may considerably exceed λ^2, equation (21), hence conventional asymptotical methods and perturbation theories fail. Let us consider the random function $\varepsilon(y)$ as a normal one, then the Dyson method (Sobzyk 1985) can be applied. In fact, this method is a more sophisticated application of the asymptotic analysis. The small parameter is present in the Dyson theory in the form of the ratio of the scale of heterogeneity to the wavelength. The operator $L_0 = d^2/dy^2 + \lambda^2$ of equation (21) in the case $\varepsilon = 0$ has the inverse operator

$$\varphi(y) = f(y) M_0(y) = \int G_0(y, \rho) f(\rho) d\rho, \quad G_0(y, \rho) = -\exp(-i\lambda|y - \rho|)/2i\lambda, \tag{22}$$

where G_0 is the Green's function of equation (21) in the case $\varepsilon = 0$. As shown by Sobzyk (1985) (see also Belyaev 1993) the mean field of vibration $\langle V(y) \rangle$ is the solution of the Dyson integral equation

$$\langle V(y) \rangle = G_0(y, y_0) + \int \int G_0(y, \rho_1) Q(\rho_1, \rho_2) \langle V(\rho_2, y) \rangle d\rho_1 d\rho_2 \tag{23}$$

where Q is the kernel of an integral operator which is analogous to the quantum mechanical mass operator of quantum field theory.

Statistical homogeneity of the field $\varepsilon(y)$ is assumed; hence the correlation function B of the random field $\varepsilon(y)$ has the form

$$B(y_1, y_2) = \langle \varepsilon(y_1)\varepsilon(y_2) \rangle = B(y_1 - y_2) = \sigma^2 l \delta_l(y_1 - y_2). \tag{24}$$

In equation (24) an essentially heterogeneous random medium is modelled by a delta-shaped correlation function localized in the region $|\rho| < l$ (Figure 23.4) where σ is the standard deviation of the heterogeneity (intensity) and l is an integral radius of correlation (scale). The heterogeneity is supposed to be a small-scale ($|\lambda l| \ll 1$, $|y| \gg l$) and intensive one (no restriction on σ). Under such assumptions the solution of equation (23) is given (for details see Belyaev 1993) as

$$\langle V(y) \rangle = -\exp[-i\lambda(1 - i\xi)|y - y_0|]/2i\lambda; \quad \xi = 0.25\sigma^2 l |\lambda|^{-3}. \tag{25}$$

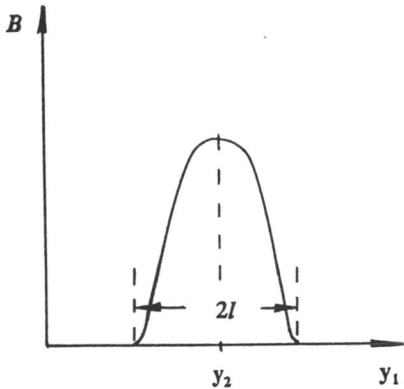

Figure 23.4   Delta-shaped correlation function of an essentially heterogeneous random medium.

23.4   CAUSES OF SPATIAL DECAY OF HIGH-FREQUENCY VIBRATION

Equation (25) is in fact an equation for the amplitude of deformation a since λ is a function of u. One obtains with the help of the logarithmic derivative from equations (20) and (25)

$$\frac{1}{a}\frac{da}{dx} = \operatorname{Im}\lambda(1 - i\xi) = \frac{\omega}{\langle v \rangle}\operatorname{Im}\frac{(1 - i\kappa)(1 - i\xi)}{(1 + i\chi)} = -\frac{\omega}{\langle v \rangle}(\kappa + \xi + \chi(a)) \qquad (26)$$

since κ and χ do not depend on a. If we take a standard distribution function of the theory of internal damping (Palmov 1976), i.e. $R(h) = \beta H h^{\beta - 1}(H > 0, \beta > 0)$, then equation (10) gives $\chi = g a^{\beta}$ where $g = \beta H (\pi/4)^{\beta} B((\beta + 1)/2; 3/2)$ and $B(\; ; \;)$ is the Eulerian beta-function. Then the solution of equation (26) is

$$a(x) = \left\{ \left[a(0)^{-\beta} + \frac{g}{\kappa + \xi} \right] \exp(\beta\omega(\kappa + \xi)x/\langle v \rangle) - \frac{g}{\kappa + \xi} \right\}^{-1/\beta}. \qquad (27)$$

The analysis of this equation indicates the main reasons for the considerable spatial absorption of high-frequency vibrations in complex structures. For large x the exponent in equation (27) prevails and we have

$$a(x) = a(0)\left[1 + \frac{\chi[a(0)]}{\kappa + \xi} \right]^{-1/\beta} \exp(-\omega(\kappa + \xi)x/\langle v \rangle). \qquad (28)$$

If $\chi \ll \kappa + \xi$, equation (28) can be rewritten in the form

$$a(x) = a(0)\exp(-\omega(\kappa + \xi)x/\langle v \rangle). \qquad (29)$$

Hence the inequality $\chi \ll \kappa + \xi$ can be understood as a condition of the validity of the linear theory. According to experimental data (Palmov 1976) $\kappa + \xi \approx 0.2$–0.4. As seen from equation (29) the parameters κ and ξ are the characteristics of the vibration decay. The parameter $\kappa(\omega)$ characterized the resonant absorption of vibration by structural members since $\kappa(\omega)$ is

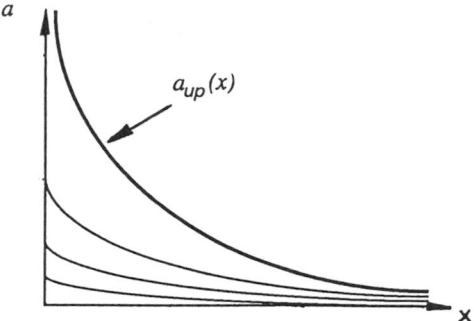

Figure 23.5 The field of actual deformation and its majorant.

determined by the function of the eigenfrequencies distribution $\Phi(\omega)$, equation (19). This means that the secondary systems act as dynamic absorbers with respect to the framework. In the case of high modal overlap the considerable resonant absorption in the whole high-frequency domain is observable. On the contrary, the parameter ξ describes not the absorption, but dispersion and multiple scattering of the propagating wave. For essentially heterogeneous media this provides a considerable spatial attenuation.

If the condition of linearity ($\chi \ll \kappa + \xi$) is violated, then one should use equation (27), which has an interesting special property. For any $a(0)$, i.e. for any amplitude of the external force, the following inequality holds:

$$a(x) < \left\{ \frac{g}{\kappa + \xi} \left[\exp\left(\frac{\beta\omega(\kappa + \xi)x}{\langle v \rangle} \right) - 1 \right] \right\}^{-1/\beta} = a_{up}(x). \tag{30}$$

This formula introduces the upper limit of the vibration in the structure $a_{up}(x)$ even if the excitation power grows beyond any bounds. This upper limit depends on the spatial distance from the external excitation source (x) and integral mechanical characteristics of the structure ($\langle v \rangle$, κ, χ, β, g) and it does not depend on the intensity of external excitation. Hence, for any complex structure one can plot a universal curve $a_{up}(x)$ and the amplitudes of actual deformation are located below this curve as depicted in Figure 23.5.

23.5   APPLICATIONS AND CONCLUSIONS

The existence of the majorant for the field of actual deformation $a_{up}(x)$ was used for predicting the vibration of a railway-noise-excited tall building (Belyaev and Ziegler 1991) since no information about the external loads was needed.

The property of the vibration saturation was also used to explain the failure in vibro-testing an extended spacecraft. It is generally accepted practice to incorporate into the design of spacecraft some special devices (e.g. explosive bolts) which function as a release mechanism to separate two subassemblies of the craft. These firings are widely used in space engineering. For instance, on a typical Gemini mission, over a hundred pyrotechnic firings took place (Bucciarelli and Ashkinazi 1973). Mounted directly on the structure, a pyrotechnic device generates a considerable high-frequency dynamic environment which can affect sensitive

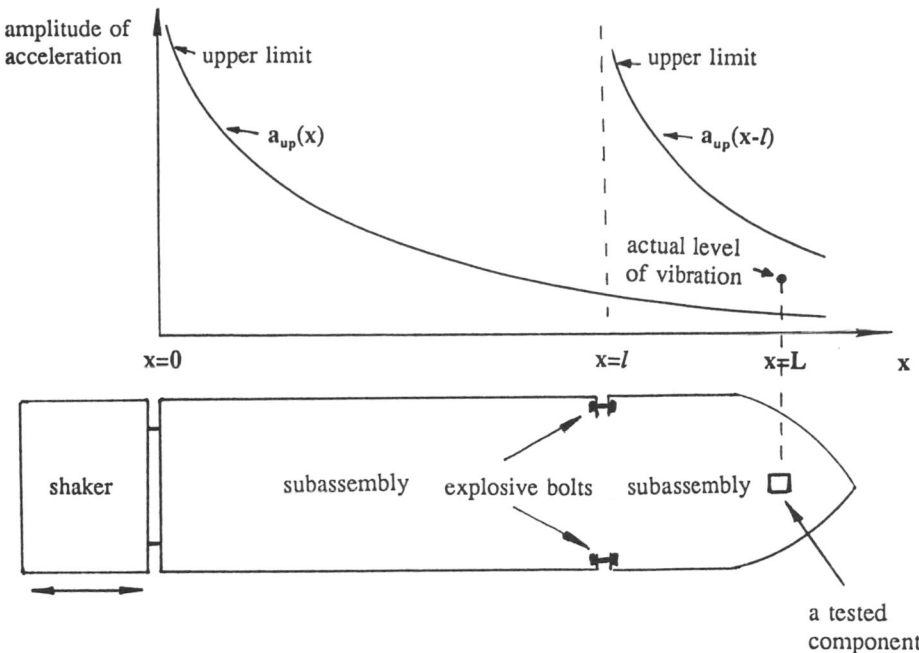

Figure 23.6 Vibration testing of a large-sized spacecraft.

equipment carried aboard the structure in the neighbourhood of the pyrotechnic. One observed failure in vibro-testing that was designed to evaluate the sensitivity of both structural hardware and various pieces of sensitive electrical equipment. A powerful electrodynamic shaker could not produce in fullscale testing the same vibration level which was caused by the firing of small explosive bolts. If the vibration exciter is mounted rather far from the tested component this desired level exceeds $a_{up}(x)$, i.e. 'the nature-allowed level'. Hence, as seen from Figure 23.6 the desired level of response cannot be excited by the shaker even if it had limitless power. Such poorly arranged vibration testing may and did result in a dynamic buckling of the moving coil of the electrodynamic shaker (Belyaev 1994).

Summarizing, one can say that for each actual complex structure a domain of high frequencies is observed. Considerable spatial decay of vibration at high frequencies is caused by (i) dispersion and multiple scattering in essentially heterogeneous structures; (ii) inherent material damping and dry friction between the components of the structure (because of their nonlinear nature one can obtain a majorant for the field of actual deformation); (iii) resonant absorption of vibration by the secondary systems. The secondary systems, especially sensitive structural members, may be damaged since they absorb mechanical energy at their eigenfrequencies.

ACKNOWLEDGEMENT

The research was done under the auspices of the Alexander von Humboldt Foundation, Germany.

REFERENCES

Belyaev, A. K. (1991) Vibrational state of complex mechanical structures under broad-band excitation, *Int. J. Solids and Structures*, **27**, 811–823.

Belyaev, A. K. (1993) High-frequency vibration of extended complex structures, *Probabilistic Engineering Mechanics*, **8**, 15–24.

Belyaev, A. K. (1994) Failure of vibration testing caused by dynamical buckling of shaker coil, *Proceedings of the 5th Int. Conference on Recent Advances in Structural Dynamics, University of Southampton, England* (N. S. Ferguson and H. F. Wolf, eds.).

Belyaev, A. K. and Ziegler, F. (1991) Traffic-noise excited uniaxial waves in complex structures, In: *Trends in Applications of Mathematics to Mechanics*, W. Schneider, H. Troger and F. Ziegler (eds.), New York, Longman, pp. 108–117.

Bendiksen, O. O. (1987) Mode localization phenomenon in large space structures, *AIAA J.*, **25**, 1241–1248.

Bolotin, V. V. (1984) *Random Vibrations of Elastic Systems*, The Hague, Nijhoff.

Bucciarelli, L. L. and Ashkinazi, J. (1973) Pyrotechnic shock synthesis using nonstationary broad band noise, *ASME J. Appl. Mech.*, **40**, 429–432.

Cha, P. D. and Pierre, C. (1991) Vibration localization by disorder in assemblies of monocoupled, multimode component systems, *ASME J. Appl. Mech.*, **58**, 1072–1081.

Cornwell, P. J. and Bendiksen, O. O. (1989) Localization of vibrations in large space reflectors, *AIAA J.*, **27**, 219–226.

Ibrahim, R. A. (1987) Structural dynamics with parameter uncertainties, *Appl. Mech. Rev.*, **40**, 309–328.

Palmov, V. A. (1976) *Vibration of Elastoplastic Bodies*, Mascow, Nauka, [in Russian].

Pierre, C. (1990) Weak and strong vibration localization in disordered structures: a statistical investigation, *J. Sound and Vibration*, **139**, 111–132.

Pierre, C. and Dowell, E. H. (1987) Strong mode localization in nearly periodic disordered structures, *AIAA J.*, **27**, 227–241.

Sobzyk, K. (1985) *Stochastic Wave Propagation*, Amsterdam, Elsevier.

A. K. Belyaev, *Department of Mechanics and Control Processes, State Technical University of St. Petersburg, Polytechnicheskaya 29, 195251, St. Petersburg, Russia*

PART VI
Mathematical Techniques

It is generally agreed that Henri Poincaré, around 1900, was the first person to view dynamical systems from a topological perspective at a time when analysis was seen to be the norm. His novel ideas were to view not just a single solution of a system but a whole ensemble of trajectories to give a complete, geometrical picture of the system behaviour within the full phase space. It is also known that Poincaré was aware that non-periodic solutions of deterministic equations existed long before the seminal paper of Edward Lorenz in 1963. Since this time we have discovered that much knowledge of the qualitative behaviour of nonlinear systems can be gained by topological approaches to the dynamics. On the other hand, vibration analysis is probably one of the most important analytical methods in the engineer's toolkit. This term covers a range of techniques which utilize perturbation and averaging methods to determine the vibrational responses of complex structures. Vibration theory has become a standard course for undergraduate study and is covered by many authors including Thomson (1993) and Nayfeh and Mook (1979).

TOPOLOGICAL METHODS

The use of knot theory in dynamical systems is a relatively new area of study applying powerful topological theorems to understand qualitative changes in nonlinear systems. Many of the ideas have apparent restrictions to three-dimensional flows but this does not diminish their contribution to our knowledge. Studying elements within the flow allows us to determine the allowable bifurcation structure of the system without having to complete the full and detailed numerical calculations. This knot-theoretic approach is introduced by *Holmes and Ghrist* who advocate an examination of the gluing bifurcation while concepts of braid theory are applied to driven oscillators in the chapter by *McRobie and Thompson*.

TIME-SERIES ANALYSIS

It is often the case that systems from engineering, biology and other scientific fields are sufficiently complex that a simple mathematical model is not available. However, it is likely that these same systems can be monitored in some way to produce a series of data points which are representative of the behaviour. The first question that naturally arises is 'from such data is it possible to quantify the state of the system?' In many cases the recorded data might appear random but from our knowledge of chaotic systems we know that such behaviour can result from very simple models (May 1976). Thus the second question often asked is 'from examination of the data from a single time series alone is it possible to reveal an underlying low-dimensional chaotic system?' These are just two of the questions that have sparked off a field of study in time-series analysis. This new area of research utilizes techniques which have been in practice for some time; for instance most experimental or real data incorporates noise which must be separated as far as possible from the underlying signal of interest (in some cases the noise may be the dominant factor!). The second task involves the construction of a suitable state or phase space within which the data can be embedded. Two of the techniques used for reconstructing the phase space are the method of using delay coordinates (see Packard *et al.* 1980) and a systematic approach using singular-value decomposition (see Broomhead and King 1986). If the time series contains low-dimensional dynamical structure then this will be apparent in the reconstructed phase space and the next issue is then to find approximations to an equivalent low-dimensional system. This approximation can be carried out on a local level or a global level, and indeed some methods such as radial basis function approximations are able to capture behaviour at both levels. Once an approximate system has been evaluated predictions can be made regarding future events as parameters are varied.

Filtering techniques used in the analysis of time series are introduced in the chapter by *Davies and Stark* while the wider issue of using time series is addressed in the later chapter by *Abarbanel* in Part VII.

STABILITY AND THE HARMONIC BALANCE METHOD

A question of prime importance to the engineer is whether a particular solution (harmonic, subharmonic etc.) to a dynamical system is achievable; this incorporates questions of existence and the additional requirement that the solution be stable. Quantifying stability in nonlinear systems is not a straightforward task and a chapter is included here by *Schiehlen* which puts forward some new ideas based on stability numbers in which the simple and double pendulums are used as illustrations.

One of the most well documented analytical techniques for the study of nonlinear systems is the method of harmonic balance. A solution is effectively assumed to have a certain form containing harmonic components. This solution is then substituted into the governing differential equation and coefficients are first equated to give algebraic equations, which can then be solved for the amplitudes of the harmonic components. Approximations are usually made along the way so that there are restrictions on the applicability of the solution, but the method invariably provides useful guides as to the behaviour and stability of the system with insight into the form of the solution.

The method of harmonic balance forms the basis for the chapters by *Szemplińska-Stupnicka and Rudowski* and *Narayanan and Sekar*. In 'the first of these, analytical estimates of the first period-doubling bifurcation can be used to provide a lower bound for the eventual escape of a driven oscillator from a potential well. Furthermore, by considering the basin of attraction, it is possible to establish the minimum impulsive load that would cause the system to escape from the well. In the second of these two chapters an efficient and speedy implementation of harmonic balancing is described which identifies periodic solutions, estimates their stability, and allows further bifurcational analysis to be performed.

REFERENCES

Broomhead, D. S. and King, G. P. (1986) Extracting qualitative dynamics from experimental data, *Physica*, **20D**, 217–239.

Lorenz, E. N. (1963) Deterministic non-periodic flows, *J. Atmos. Sci.*, **20**, 130–141.

May, R. M. (1976) Simple mathematical models with very complicated dynamics, *Nature*, **26**, 459–467.

Nayfeh, A. H. and Mook, D. T. (1979) *Nonlinear Oscillations*, New York, Wiley.

Packard, N. H., Crutchfield, J. P., Farmer, J. D. and Shaw, R. S. (1980) Geometry from time series, *Phys. Rev. Lett.*, **45**, 712–715.

Thomson, W. T. (1993) *Theory of Vibration with Applications* (4th edn.), London, Chapman and Hall.

24 KNOTTING WITHIN THE GLUING BIFURCATION

P. Holmes and R. Ghrist

After a brief review of the uses of *knots* and *templates* in bifurcation theory for three-dimensional flows, we begin an examination of the class of homoclinic bifurcations known as *gluing bifurcations* from a knot-theoretic point of view. We present a topological classification of the periodic orbits which appear in the gluing bifurcations associated with a *saddle* fixed point: that is, an equilibrium point having all real eigenvalues in the linearization. For this classification, the global twisting of the flow about the orbits is of fundamental importance. We use the associated template to show the predominance of *torus knots* among the orbits created in these bifurcations.

24.1 KNOTS, TEMPLATES, AND THE TOPOLOGY OF BIFURCATIONS

In a three-dimensional flow, a periodic orbit defines a knot, by the uniqueness of solutions of ordinary differential equations (ODEs). Adopting this viewpoint enables us to import a vast body of results and techniques from theoretical knot and link theory to understand the topology of periodic orbits. This in turn lends itself naturally to the bifurcation problem for a parametrized family of flows: given a class of orbits, from what bifurcations did they arise, and in what order? Many problems in nonlinear oscillations are amenable to such an analysis; in particular, many 'chaotic' systems exhibit infinitely many knot types (e.g. the Lorenz system (Birman and Williams 1983a) and the suspended Smale horseshoe map (Holmes and Williams 1985; Ghrist and Holmes 1993)), providing a rich topological structure for analysis.

The knot-theoretic approach to bifurcations is very natural. For example, consider a saddle-node bifurcation of periodic orbits. Intuitively, one can think of the two orbits growing closer and closer together until they coalesce at the bifurcation point and then disappear. From the uniqueness of solutions, it then follows that these orbits have the same *knot type* (that is, as embeddings of S^1, they are *ambient isotopic*). Furthermore, they must be *linked* with all other coexisting orbits in precisely the same manner—again, in order to maintain uniqueness of solutions. Therefore, knowledge of topological data immediately yields

Nonlinearity and Chaos in Engineering Dynamics
Edited by J. M. T. Thompson and S. R. Bishop,   © 1994 John Wiley & Sons Ltd

bifurcation invariants. Other local bifurcations have analogous topological interpretations, e.g. a period-multiplying bifurcation corresponds to a *cabling* of the knot (Holmes 1986).

The strategy for examining systems via knot-theoretical techniques is as follows: given a three-dimensional flow, 'embed' it in a parametrized family for which certain elements have well-understood behaviour, e.g. a certain class of knots is unique, or perhaps non-existent. Then, varying parameters back to the original system, one traces bifurcations back, keeping track of knotting and linking to 'match up' with orbits in the original system. This approach has unearthed surprising behaviour in a family of Hénon maps (Holmes and Williams 1985; Holmes 1986, 1989). Other systems which have been examined include the (geometric) Lorenz system (Birman and Williams 1983a) and the perturbed pendulum (Josephson junction) equation (Holmes 1987). In general, these methods are also of use in understanding the ordering of bifurcations in the creation of horseshoes (Holmes and Williams 1983) in chaotic systems. Knot and braid theory has also found its way into the analysis of experimental time-series data (Mindlin *et al.* 1991), and, in connection with numerical simulation, in bifurcation studies of periodically forced oscillators (McRobie and Thompson 1993a, b). Here we will focus on results for a specific class of bifurcations. Our aim is to completely describe the knot and link types of all periodic orbits created in the global 'gluing bifurcations' which occur near a degenerate homoclinic orbit in a dissipative three-dimensional flow. We then envisage the use of continuation methods to follow these orbits far (in parameter space) from their origin and thus reveal global genealogies. Although we will be brief in our review of the necessary theory, the interested reader may consult Ghrist and Holmes (1993) for a thorough introduction. We wish to stress that although the first papers in the subject appeared over ten years ago (Birman and Williams 1983a, b) there is a great deal left to be done – the subject is still in its infancy.

In order for such an approach to be useful, however, one must be able to efficiently extract topological information from a system. Thus, we turn to the dimension-reducing tool of Birman and Williams (1983a, b): the *template*. For three-dimensional flows which satisfy certain conditions, one identifies orbits which have the same future behaviour; in effect, one collapses each fibre in a strong stable foliation down to a point. The result projects the three-dimensional flow to a *semi-flow* on a *branched two-manifold* which is the template. An example will best serve to illustrate. For the geometric Lorenz system (Guckenheimer and Williams 1979), collapsing out the strong stable direction yields the template $\mathcal{T}$ pictured in Figure 24.1. This is essentially what one sees in numerical experiments on the Lorenz equations: cf. Figure 3 of Lorenz (1963). There are two *branches*, x and y, which meet at the branchline I. The three-dimensional flow is now a two-dimensional semi-flow since one cannot 'flow backwards' at the branchline. The utility of templates lies in the following theorem (Birman and Williams 1983a, b).

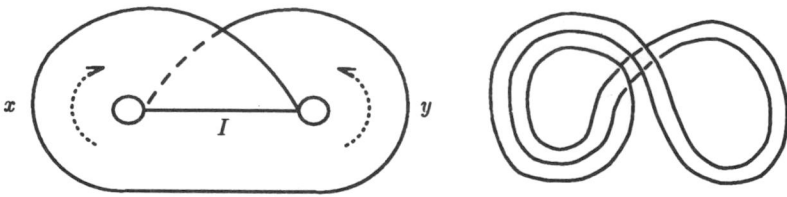

Figure 24.1   The Lorenz template and a periodic orbit.

Theorem 1 (the template theorem) *Given a flow on a three-manifold having a hyperbolic chain-recurrent set, there exists a template such that (with perhaps one or two exceptions) the collapsing map is one-to-one on the union of periodic orbits via an ambient isotopy (i.e. a continuous deformation preserving all knotting and linking information).*

Organizing knots on the template is simplified by the following symbolic tool: given a periodic orbit on $\mathcal{T}$, we may describe it by a finite word in the symbols x and y by recording the order in which the orbit travels along the x and y branches. For example, Figure 24.1 shows the Lorenz template with the periodic orbit $xxyxy(=x^2yxy)$. (*Exercise*: is this knotted?) Of course, cyclic permutations (shifts) of a word do not change the orbit: $x^2yxy = xyxyx$.

We may therefore by Theorem 1 describe all orbits in our original system symbolically. Upon noting that the branchline is a natural choice for a Poincaré section, we may extract the (one-dimensional) first-return map, which for the Lorenz system is an expanding map of the interval with a single discontinuity (see Guckenheimer and Holmes 1983, § 2.3, § 6.4).

In summary, we may use the structure inherent in the template to examine the topology of the orbits. We may then turn to the one-dimensional return map and invoke the tools of *kneading theory* and *symbolic dynamics* to relate various knots to bifurcations in the system (Ghrist and Holmes 1993; Holmes and Williams 1985). Such an analysis is typically a blend of topological visualization and symbolic manipulations. As an example, we re-prove a proposition from Birman and Williams (1983a) which will be of use in classifying knotted orbits in the gluing bifurcation.

First, we recall the definition of a particular family of knots. A *torus knot* is a knot which can be arranged to fit on the surface of the standardly embedded torus $T^2 = S^1 \times S^1 \subset \mathbb{R}^3$. We say the torus knot is of type (p, q) for $p, q \in \mathbb{Z}$ if it winds around the longitudinal direction p times and around the meridional direction q times. We require p, q to be relatively prime and note that, except for the relation $(p, q) \sim (q, p)$, these two integers completely classify and distinguish all torus knots (Rolfsen 1977). We note that the family of torus knots is a very restrictive class, of which a great deal is known. Identifying torus knots in flows has been useful in bifurcation analyses (Holmes and Williams 1985; Holmes 1986).

Consider again the Lorenz template $\mathcal{T}$ and the symbolic description of knots. There are infinitely many (but not all!) knot types coexisting on $\mathcal{T}$, some of which are torus knots. Consider an *evenly distributed* word in x and y, that is, a word composed of syllables of only the forms xy^k, xy^{k+1} or $x^ky, x^{k+1}y$ (cf. Section 24.2 below): for example, x^2yxy and xy^3xy^4 are evenly distributed while x^3y^2 and xy^3xy^5 are not. In Birman and Williams (1983a), the following is stated.

Proposition 1 *On $\mathcal{T}$, an evenly distributed word corresponds to a torus knot.*

Proof Say the given word has p x's and q y's and assume that $p > q$. (If $q > p$, flip $\mathcal{T}$ about the vertical axis and proceed by symmetry.) Then, since the word is evenly distributed, there are no consecutive y's: each loop about the y-branch is immediately followed by a loop about the x-branch. As such, we may fit the orbit within an unbranched subtemplate $\mathcal{S} \subset \mathcal{T}$ and isotope it as per Figure 24.2. What remains after these visual gymnastics is a ribbon containing the orbit which fits on the torus T^2 such that the orbit winds p times longitudinally and q times meridionally: a (p, q) torus knot. $\qquad\square$

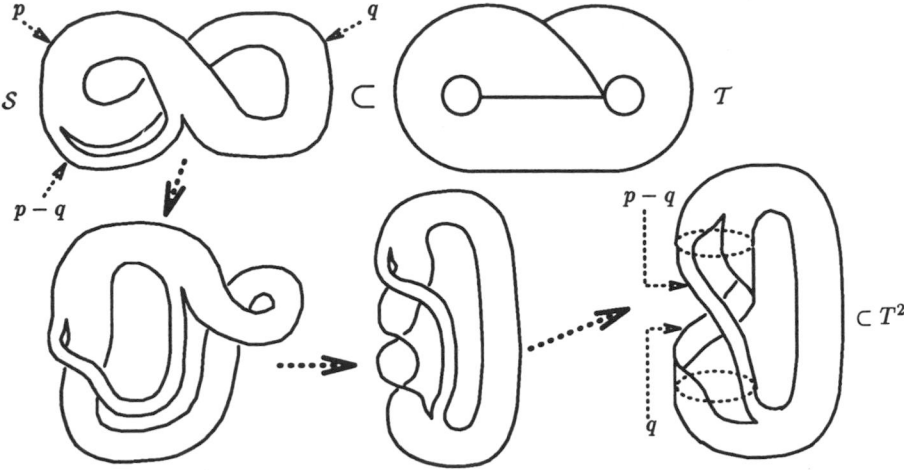

Figure 24.2 The subtemplate $\mathscr{S}$ fits on T^2. The labels p, q and $p-q$ refer to the number of strands on each ribbon.

24.2 THE GLUING BIFURCATION

The bifurcation problem has been examined from a knot-theoretic viewpoint so far only in the cases of local bifurcations such as saddle–node and period-multiplying bifurcations (Ghrist and Holmes 1993; Holmes 1988). Global bifurcations are more subtle in that one requires information along an entire trajectory of the flow: e.g. in homoclinic bifurcation theory, one examines orbits which bifurcate from a homoclinic connection. Our goal is to introduce the knot-theoretic viewpoint to the study of global bifurcations by means of a simple example of a homoclinic bifurcation: the so-called *gluing bifurcation* (Coullet *et al.* 1984; Gambaudo *et al.* 1988; Glendinning 1988).

Consider a system $\dot{x} = f(x)$ (where f is C^1 with uniformly Lipshitz partials) having two orbits, Γ_x and Γ_y, homoclinic to a hyperbolic fixed point p. For the gluing bifurcation, it is necessary that the unique unstable eigenvalue $\lambda_u (> 0) \in \mathbb{R}$ be *weak* in the sense that $\lambda_u < |\text{Re}\,\lambda_i|$ for all other stable eigenvalues $\lambda_i (\text{Re}\,\lambda_i < 0)$. Finally, to permit linearization near p, we require no first-order resonances among the eigenvalues:

$$\text{Re}(\lambda_i) \neq \lambda_u + \text{Re}(\lambda_j) \quad \forall i, j. \tag{1}$$

We will be interested in a neighbourhood W of $\Upsilon = \Gamma_x \cup \Gamma_y$ for C^1 perturbations of such *critical* systems. One describes the periodic orbits emerging in the unfolding of a critical system symbolically, as with orbits on the template. Assuming Υ is bounded away from other fixed points and taking W to be a small neighbourhood of Υ, orbits in W cannot 'double back' and must follow Υ monotonically. In fact, the weak-eigenvalue assumption implies that the flow is contracting and hence that there is an attractor contained in W. Thus, labelling Γ_x, Γ_y with the letters x, y we can assign to any periodic orbit γ in W a finite acyclic word in x and y given by what order it travels around each loop: this is the orbit's *signature*. Just as in the case of template orbits, we have a shift-equivalence class of words, since beginning at a different point yields a cyclic permutation of the word.

In a gluing bifurcation, it is possible to have periodic orbits with non-trivial words, e.g.

x^2yxy. Which words are possible is at the heart of understanding the unfolding. It is a fact stated in Gambaudo *et al.* (1988) and Glendinning (1988) that any periodic orbit bifurcating from Υ in a critical system has a *rotation-compatible* signature. That is, it may be obtained as the symbol sequence of a point $z \in I$ iterated by a rotation map $\rho_\theta : z \leftrightarrow (z + \theta) \bmod 1$ with the partition $I_x = (0, 1 - \theta]$, $I_y = (1 - \theta, 1]$ for some $\theta \in [0, 1)$. Given such a word, the unique θ for that word is called the *rotation number*. Note: to compute the rotation number from a given rotation-compatible word, take the number of y's and divide by the total length of the word: e.g. $x^2yxy \Rightarrow \theta = 2/5$. In short, we can describe rotation-compatible words as precisely the 'evenly distributed' words of Proposition 1: we may then define the rotation number uniquely by the above formula.

The main result concerning the gluing bifurcation is the following (see Coullet *et al.* 1984; Gambaudo and Tresser 1988).

Theorem 2 *For every sufficiently small C^1 perturbation of a critical system there are at most two periodic orbits in a small neighbourhood W of Υ. Any periodic orbits which may be present are attracting and have rotation-compatible signatures. Finally, if there are two periodic orbits, then their associated rotation numbers are Farey neighbours.*

We recall that two rational numbers p/q and p'/q' are *Farey neighbours* if $|pq' - qp'| = 1$.

The fact that at most two periodic orbits exist follows from the observations that the attractor lies within the closure of the one-dimensional unstable manifold $W^u(p)$, of which there are two 'sides' (separated by p). Although Theorem 2 severely restricts the number and type of periodic orbits near Υ, there is nevertheless enough latitude within many such systems to allow for bifurcation diagrams with infinitely many curves (Gambaudo *et al.* 1984). In such a contracting system, unlike the expanding Lorenz flow, at most two periodic orbits can coexist, but as parameters vary, we will see that infinitely many periodic orbits (and knot types) may be created and annihilated, depending upon the global structure of the system.

We complete our review by recalling the universal features of the unfolding of a gluing bifurcation (Coullet *et al.* 1984; Glendinning 1988). Let

$$\dot{x} = f(x; \alpha, \beta) : x \in \mathbb{R}^3; \alpha, \beta \in \mathbb{R} \tag{2}$$

be a continuous two-parameter family of systems for which $f(x, 0, 0)$ is a critical system. Construct a cylindrical ($S^1 \times I$) Poincaré section Σ to the fixed point p (assumed fixed under variation of parameters) transversal to $W^s_{loc}(p)$. The local unstable manifold $W^u_{loc}(p)$ is one-dimensional: set up local coordinates about p having $W^u_{loc}(p)$ as the u-axis. At $\alpha = \beta = 0$, the two homoclinic orbits first hit Σ in two points, each with zero u-coordinate. Turning on the bifurcation parameters α, β will send the 'first' intersections of $W^u(p)$ with Σ to the points with u-coordinates $\mu(\alpha, \beta)$, $-\nu(\alpha, \beta)$. Assuming these are locally invertible functions, we may choose μ and ν to be the unfolding parameters for the system. Having done this, one can show:

Proposition 2 *For (μ, ν) sufficiently small, the parameter space has the structure appearing in Figure 24.3, where the notation 'x:y' denotes coexistence of two orbits. The periodic orbit signatures are completely determined except within the two 'wedges' defined to lowest order by*

$$-c_1\nu^\delta < \mu < c_2\nu^\delta; \nu > 0 \quad -c_3\mu^\delta < \nu < c_4\mu^\delta; \mu > 0 \tag{3}$$

where $c_i > 0$ and $\delta = -\lambda_s\lambda_n > 1$.

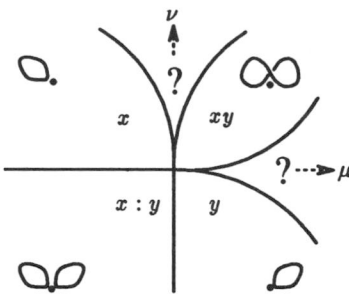

Figure 24.3   The parameter plane unfolding.

It is precisely these undetermined wedges of Proposition 2 which make this bifurcation interesting.

The gluing bifurcation takes its name from the fact that, as one varies parameters from the third quadrant to the first quadrant through the origin, the two 'simple' coexistent periodic orbits x and y are glued together in the double homoclinic connection Υ which then breaks to form a *single* periodic orbit xy.

24.3   TOPOLOGICAL CLASSIFICATION OF THE REAL SADDLE

We will be interested in examining a three-dimensional critical system having a *(real) saddle* fixed point p: i.e. all the eigenvalues of the linearization are real. This implies that there is no 'spiralling' of $W^u(p)$ into p. As such, there are two distinct topological configurations depending upon which sides of $W^s_{loc}(p)$ the homoclinic orbits re-enter: these are the *figure-of-eight* and the *butterfly*, given in Figure 24.4. Our goal is to provide a topological classification of the periodic orbits which may appear in these configurations, with emphasis on the latter. For these systems, we assume the existence of a strong stable foliation (reported in Glendinning (1988) to be a generic condition in these cases) and proceed to collapse such foliations out, leaving a template. The associated semi-flows on such templates are *contracting*, whereas *expanding* templates have been the norm in the literature. More precisely, we have a two-parameter (μ, ν) family of templates, each of which holds, by Theorem 2, at most two closed orbits.

It is noted that the templates we construct for these systems must be embedded in $\mathbb{R}^3$. Thus, we must take into account the 'twist' of the flow around the homoclinic connections,

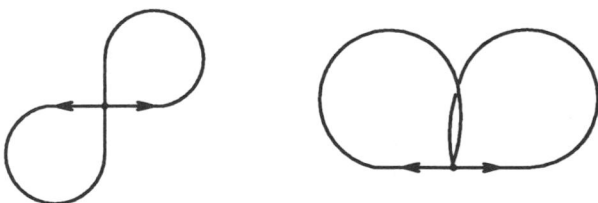

Figure 24.4   The figure-of-eight and butterfly configurations.

which will correspond to certain branches of the template being likewise twisted. At this time, we ignore more exotic, knotted embeddings of the template branches.

24.3.1   Classification of the butterfly

The templates associated with the butterfly configuration are similar to their expanding cousin, the Lorenz template. Temporarily disregarding extraneous twisting, there are three cases to consider: *untwisted, singly twisted* and *doubly twisted**. These appear in Figure 24.5 in this order. Taking the branchline as a Poincaré section, the first return map is discontinuous and one-dimensional. By performing a local analysis at the fixed point, and assuming an affine global return map along the branches (cf. the derivation in Gambaudo *et al.* (1986) and Glendinning (1988)), the return map in a sufficiently small neighbourhood of the degenerate homoclinic loop ($\mu = v = 0$) is equivalent to

$$f_{\mu,v}:\xi \longmapsto \begin{cases} \mu - a|\xi|^{\delta} & : \xi \leqslant 0, \\ -v + b\xi^{\delta} & : \xi > 0, \end{cases} \tag{4}$$

where $\delta > 1$, μ and v are the unfolding parameters as above, and the signs of a, b depend upon the orientability of the x, y branches respectively (negative if twisted, positive if untwisted). We note that δ is the ratio of the weakest stable eigenvalue to the unstable eigenvalue, and $\delta > 1$ corresponds to our assumption that the latter be weak (see Section 24.2). Representatives of these maps appear below their template counterparts in Figure 24.5. In contrast to the Lorenz first return map, these are *contracting* maps of the interval with a single discontinuity.

Case 1: untwisted $(a, b > 0)$

It is stated in Glendinning (1988) and Gambaudo and Tresser (1985) that for $\mu, v > 0$ this system has at most one periodic orbit, based on the theory of circle maps (one views the map

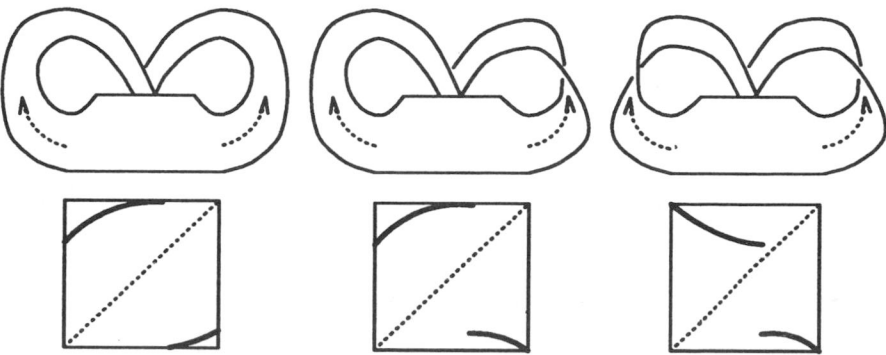

Figure 24.5   The three templates (above) with induced maps (below).

---

*The expanding counterparts to these three systems appear in Afraimovich *et al.* (1982) as *oriented, semi-oriented* and *non-oriented* respectively.

as a monotone injective map of the circle with a single discontinuity). The structure in other regions of the parameter plane is trivial and covered by the general theory (Glendinning 1988). It is possible to find periodic orbits with non-trivial knot types; however, since the signature of any orbit must be a rotationcompatible word, and since these are precisely the words which are evenly distributed, we may call upon Proposition 1. Note that although the *dynamics* of this butterfly system differ greatly from that of the Lorenz, the associated templates are isotopic (we can deform one to the other). Therefore, we have:

Corollary 1   *Any periodic orbit appearing in the unfolding of an untwisted butterfly saddle is a torus knot. If the rotation number of the signature is $\theta = p/(q+p)$, then the corresponding knot type is (p, q).*

For example, the orbit with signature $x^2 yxy$ is a $(2, 3)$ torus knot: a *trefoil*. It is very significant that only torus knots appear: in the expanding case, one has a *much* wider array of knot types (Birman and Williams 1983a). Corollary 1 is reminiscent of the Morgan–Wada theorem, which states that non-singular zero-entropy flows in S^3 yield only torus knots, with perhaps additional cablings and connected sums thereof (Morgan 1978; Wada 1989). Though this theorem does not apply to the systems we are examining, the weak unstable eigenvalue assumption nevertheless yields a 'tame' family of flows, whose periodic orbits we expect to be related to simple classes of knots. And again, we stress that embedding this template in a more complicated fashion may not yield torus knots: see the remarks below.

The unfolding of the bifurcation in this case has been stated in Gambaudo *et al.* (1986), Procaccia *et al.* (1987) and Turaev and Silnikov (1987). We recall the unfolding diagram along with the stable orbits* (4) numerically computed along a slice of constant $\nu > 0$ in Figure 24.6. Note the existence of 'tongues' of orbits of fixed rotation number and that the rotation number is a continuous monotonic function of each parameter. Those (measure-zero) regions

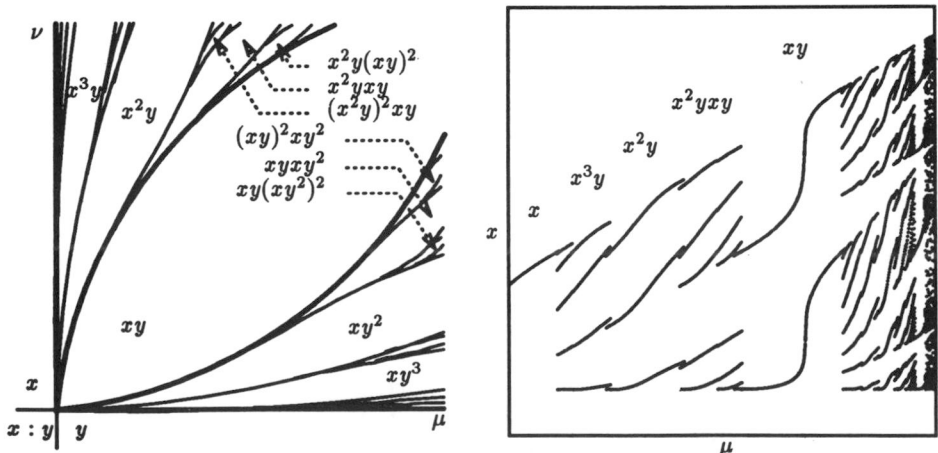

Figure 24.6   The parameter plane unfolding and a bifurcation diagram: case 1.

---

for which the rotation number is irrational correspond to the existence of a non-periodic attractor about which the unstable manifold winds irrationally (Zaks 1993).

We now consider the impact of introducing τ_x (even) positive half-twists along the x-branch. Corollary 1 is no longer applicable; however, we note that in the isotopy moves from the proof of Proposition 1, the x-branch may be fixed. As such, we may let τ_x be non-zero and perform the same moves without any interference on the part of the τ_x half-twists. In this case, since τ_x is even (i.e. there are $\frac{1}{2}\tau_x$ full twists) we then get a $(p, q + \frac{1}{2}p\tau_x)$ torus knot. Similarly, a related set of isotopy moves may be performed which will present the (p, q) torus knot as a braid on q strands, this time leaving the right y-branch fixed. As before, if $\tau_x = 0$, we may insert τ_y (even) twists to obtain a $(p + \frac{1}{2}q\tau_y, q)$ torus knot.

We cannot, however, obtain a torus knot when τ_x and τ_y are simultaneously non-zero and even: no set of isotopy moves exists which leaves both branches fixed. While we are unable to identify the knot types of orbits for arbitrary τ_x, τ_y in terms of well-known families, we can arrange the knot into a braid on p strands. Let it be an exercise for the reader to isotope the appropriate subtemplate to the braided form presented in Figure 24.7.

This presentation defines an element γ of the p-strand *braid group* B_p (Birman 1975): that is, the orbit is presented so that it travels monotonically around a *braid axis* (coming out of the page above Figure 24.7) p times. The generators of B_p, $\{\sigma_i : i = 1 \ldots p - 1\}$, each represent the crossing of the ith strand over the $(i + 1)$th: in this way, we can express the periodic orbit in terms of the σ_i and hence represent symbolically the crossing information. Though they are algebraic entities, the geometric interpretations of the braid groups permit numerous applications to the study of periodic orbits (Boyland 1984; McRobie and Thompson 1993a). We use the following notation: let Δ_k^m denote m half-twists on the first k strands, and let χ_k^m represent the crossing of the first m strands over the next k strands (see Figure 24.8).

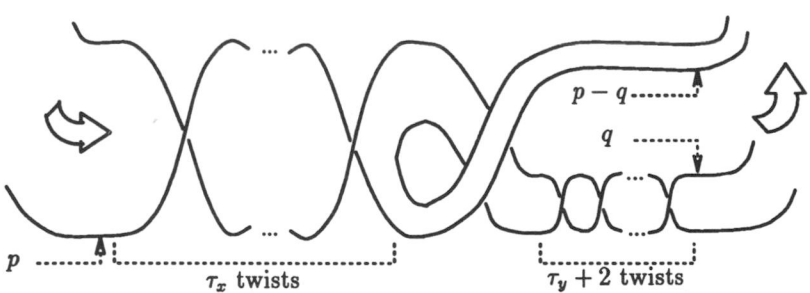

Figure 24.7 Braided subtemplate.

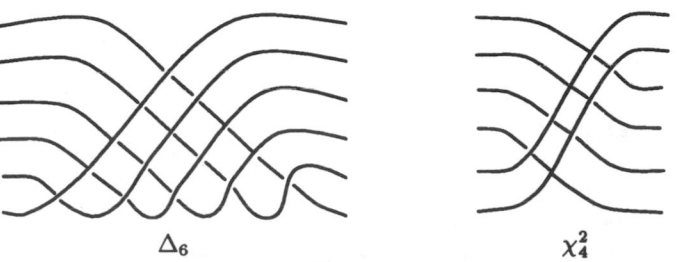

$\Delta_6 \qquad\qquad \chi_4^2$

Figure 24.8 Twists and crossovers.

Then, from Figure 24.7, we may write γ as

$$\gamma = \Delta_p^{\tau_x} \chi_q^{p-q} \Delta_q^{\tau_y+2}. \tag{5}$$

Now, noting that

$$\Delta_k^1 = \sum_{i=1}^{k-1} \left(\prod_{j=1}^{k-i} \sigma_j \right), \quad \chi_k^m = \prod_{i=1}^{m} \left(\sum_{j=1}^{k} \sigma_{m-i+j} \right), \tag{6}$$

we may explicitly write out the element of B_p defined by the periodic orbit γ for arbitrary τ_x, τ_y. From the braid presentation in Figure 24.7, it may be shown that for $\tau_x = \tau_y$ positive and even, the orbit lies on $T^2 \# T^2$, the standardly embedded surface of genus two.

In addition, when τ_x, τ_y are positive and even this presentation of the periodic orbit as a positive braid on p strands permits computation of the genus g of the knot by the formula (Birman and Williams 1983a; Ghrist and Holmes 1993):

$$g = \frac{c - p + 1}{2}, \tag{7}$$

where c is the number of crossings in the braid presentation. This number may be directly counted:

$$c(\Delta_k^1) = \sum_{i=1}^{k=1} i = \frac{1}{2} k(k-1), \quad c(\chi_k^m) = mk. \tag{8}$$

This implies

$$c(\gamma) = c(\Delta_p^{\tau_x}) + c(\chi_q^{p-q}) + c(\Delta_q^{\tau_y+2}) \tag{9}$$

$$= \tfrac{1}{2}(\tau_x p(p-1)) + (p-q)q + \tfrac{1}{2}((\tau_y+2)q(q-1)). \tag{10}$$

And thus, the genus of an arbitrary periodic orbit is

$$g = \frac{1}{2} \left(\frac{\tau_x}{2} p(p-1) + \frac{\tau_y}{2} q(q-1) + (p-1)(q-1) \right). \tag{11}$$

Note that when $\tau_x = \tau_y = 0$, this reduces to $g = \tfrac{1}{2}(p-1)(q-1)$, the genus of the (p, q) torus knot (Rolfsen 1977), as expected.

The braid presentation of Figure 24.7 also yields another invariant: the *braid index*, or, the minimal number of strands on which a knot may be presented as a braid. Again, assuming $p > q$ and τ_x, τ_y positive, if $\tau_x \geqslant 2$ it follows by a theorem of Franks and Williams (1987) that the braid index is precisely p (see Ghrist and Holmes (1993) and Holmes and Williams (1985) for applications of the braid index and genus in distinguishing periodic orbits).

Case 2: singly-twisted $(a > 0 > b)$

In this case where global (odd) twisting occurs along one branch of the template, the existence and uniqueness results available from rotation map theory are no longer available; however, we may still apply Theorem 2 to limit the possible behaviour. Assume without loss that the y-branch is twisted (i.e. $a > 0 > b$ in (4)). The other case is equivalent via symmetry. The class of maps (4) are difficult to analyse in general, but appealing to the template semi-flow will provide a geometric argument for the following.

Proposition 3   *If the y-branch of the butterfly template has a half-twist (case 2) then all periodic orbits appearing on it must have signature* x *or* $x^k y (k \geqslant 0)$. *The same holds reversing* x *and* y.

Proof   We begin by showing that no two consecutive y's can occur. Although this can be proved directly from the nature of the one-dimensional map, the following 'template' argument will set the tone for subsequent steps. Assume the existence of a closed orbit containing a pair of points $y_1 \neq y_2$ on the 'right half' of the branchline (I_y) which are consecutive with respect to the flow. Flow y_1 forwards on the template, travelling once around the y-branch and connecting to y_2. Now, flow y_2 forwards until it intersects the branchline again. Since the semi-flow is contracting and orientation-reversing about the y-branch, y_2 must flow again around the y-branch to a point $y_3 \in I_y$ which lies *between* y_1 and y_2: in this interval the orbit is now forever trapped (iterate the preceding argument), implying that the orbit was not periodic as assumed.

Now, we show that there can be only one y in the entire signature. This time, assume the existence of a closed orbit containing $y_1, y_2 \in I_y$ such that $y_1 < y_2$ with y_2 the first intersection of I_y with the image of y_1 under the semi-flow. As before, flow y_1 until it reaches y_2; then continue flowing. Eventually, perhaps after traversing the x-branch numerous times, the image of y_2 will intersect I_y at a point y_3 (see Figure 24.9). Since the y-branch reverses order, $y_3 < y_2$. In addition, since the semi-flow is contracting, $y_3 > y_1$: i.e. $y_3 \in (y_1, y_2)$ as before. Iterating the argument gives the next intersection point $y_4 \in (y_3, y_2)$, etc. The orbit is successively trapped and is consequently non-periodic.   □

The argument associated with Figure 24.9 implies the following.

Corollary 2   *All orbits in the unfolding of a singly twisted butterfly are either periodic, asymptotically periodic, or homoclinic: there is no irrational winding as is possible in the untwisted case.*

Proof   As soon as an orbit itinerary contains two distinct y's, Figure 24.9 implies that the orbit is 'trapped' and the itinerary is henceforth periodic. Thus, all itineraries are either finite (homoclinic orbit) or eventually periodic.   □

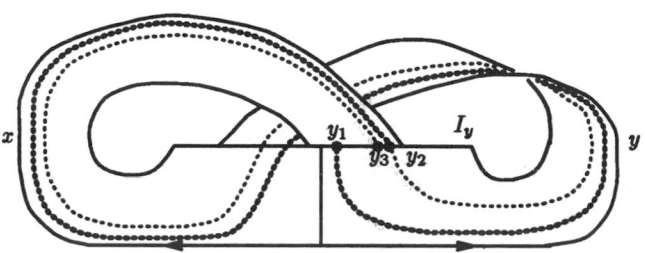

Figure 24.9 y_3 is 'trapped' within (y_1, y_2).

---

*Since completing this work we have discovered the note by Turaev and Šilnikov (1987), announcing results on unfoldings similar to and in agreement with Proposition 3, 4 and 5 of this work.

Knowledge of the possible signatures from Proposition 3 then determines the possible knot types. Although Figure 24.9 is drawn for the case $\tau_x = 0$, Proposition 3 and Corollary 2 are true for any evenly twisted x-branch, embedded in any way (even knotted). The following corollary, however, is specific to $\tau_x = 0$.

Corollary 3 *Any periodic orbit appearing in the unfolding of a singly twisted butterfly saddle is an unknot.*

Turning now to the unfolding of this bifurcation, we encounter a situation quite different from that of the untwisted case. In this singly twisted case we cannot relate the first-return maps to rotation maps in order to prove the uniqueness of periodic orbits: in fact coexistence does occur.

Proposition 4 *The unfolding of the singly twisted butterfly system is as given in Figure 24.10*. In particular, there exist open regions in the first quadrant of the parameter plane for which the periodic orbits $x^k y$ and $x^{k+1} y$ coexist, for each $k \geqslant 0$.*

As in Figure 24.6, Figure 24.10 also contains numerically computed stable orbits[†] along a slice of constant $\nu > 0$. *Exercise*: what is the linking number of the coexisting orbits?

The proof of Proposition 4 is based on the examination of homoclinic connections and their subsequent breakings. Given a homoclinic connection in the flow, one may pass to the template and look at the corresponding homoclinic connection there. [‡]Upon

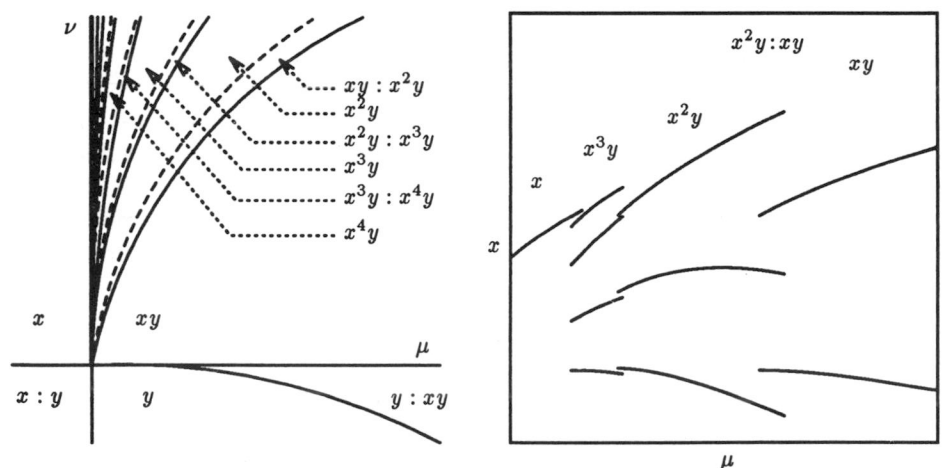

Figure 24.10   The parameter plane unfolding and a bifurcation diagram: case 2.

---

[*]Cf. Figure 3 of Turaev and Šilnikov (1987), where there is a mislabelling in the upper right corner.
[†]Note the Farey neighbouring orbits which coexist in the overlapping regions of the tongues.
[‡]The astute reader will note that a homoclinic connection is sent to a point by the template-collapsing map; thus, the corresponding homoclinic connection is technically not defined. We circumvent this technicality by considering the homoclinic orbit as the limit of a sequence of periodic orbits of increasing period, each of which has a well-defined template counterpart which approaches a homoclinic connection for the semi-flow.

breaking a connection by (say) fixing v and perturbing μ, the ω-limit sets of the two branches of the unstable manifold $W^u(p)$ are determined by the restricted topology of the branched two-manifold. Within the template, it is observed that the stable manifold $W^s(p)$ 'surrounds' the ω-limit set of one branch of $W^u(p)$ on one side of the bifurcation. Hence, the ω-limit set of the second branch of $W^u(p)$ cannot coincide with that of the first: i.e. there is a coexistence of periodic orbits on one side of the bifurcation. Knowing the signatures of the possible orbits from Proposition 3, the Farey neighbour property of Theorem 2 then implies that for $x^k y$ and $x^j y$ to coexist, $|k - j| = 1$. Without plunging into the details, we remark that these facts completely determine the behaviour associated with single homoclinic connections within this system. Having done this, the unfolding diagram is a simple matter of placing the bifurcations in the unique order which fills in the gaps in Figure 24.3 of Proposition 2. $\square$

Considering now more complicated embeddings of this template, let there be τ_x, τ_y half-twists in the x- and y-branches respectively, with τ_x even, τ_y odd. Since all periodic orbits (besides x) are of the form $x^k y$, there is a unique y-strand and thus τ_y does not influence the knot type. So, taking the k strands wrapped about the untwisted x-branch and inserting $\frac{1}{2}\tau_x$ full twists yields a braid on k strands which naturally fits onto a torus: these are $(k, \frac{1}{2}k\tau_x + 1)$ torus knots. It is worth noting that one cannot obtain arbitrary types of torus knot, as is possible in case 1. For example, the torus knots of types $(3, 5), (4, 7), (5, 7), (3, 8), \ldots$ are all incompatible with the above formula, and therefore non-existent, irrespective of global twisting.

Case 3: doubly twisted $(a, b < 0)$

This is the simplest case. Recall, the first step in the proof of Proposition 3: when the y-branch is twisted, a y must be followed by an x. This applies equally well to the doubly twisted case, for both x- and y-branches. Thus, there are no orbits having signatures containing x^2 or y^2.

Corollary 4 *Any periodic orbit appearing in the unfolding of a double twisted butterfly saddle has signature x, y, or xy: all of which are unknots.*

As this map may be interpreted as a monotone injective map of the circle with a single discontinuity, we may again prove that there is a unique periodic orbit for μ, $v > 0$. The corresponding unfolding is not difficult to determine, nor is it as interesting as the previous two cases. The simplicity of these periodic orbits prevents any knots from forming, even under arbitrary amounts of additional (odd) twisting in the branches.

24.3.2 Classification of the figure-of-eight

The associated figure-of-eight template is not branched, but is a standard compact two-manifold whose form mirrors its namesake: there are two ribbons which meet together in a neighbourhood of the fixed point. Along each branch of the template, there may be an arbitrary number of half-twists in the ribbons, depending upon the global twisting in the flow about $W^u(p)$. This yields the template given in Figure 24.11. Let τ_x, $\tau_y \in \mathbb{Z}$ be the number of half-twists in the template x- and y-branches. As in the butterfly configuration,

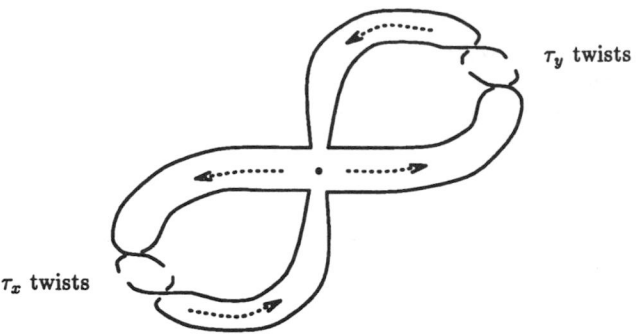

Figure 24.11 The figure-of-eight template.

the unfolding of the gluing bifurcation for this case will break down into untwisted, singly twisted and doubly twisted cases, depending on τ_x and τ_y, each mod 2. Imitiating the proof of Proposition 3, though, one can show:

Proposition 5 *In a figure-of-eight configuration, any periodic orbit resulting from a gluing bifurcation must have signature xy, $x(yx)^k$, or $y(xy)^k$ for some $k \geqslant 0$.*

In the untwisted and singly twisted cases, only the simplest orbits ($x^x y$, xy^k; $k \leqslant 2$) can appear. The unfoldings are likewise simple and left to the reader (see Baesens *et al.* (1991, § 4.5.2) for help in the untwisted case). In the doubly twisted case, one can use an argument similar to that of Proposition 4 to obtain the bifurcation diagram given as Figure 24.12. In the more general cases where τ_x, τ_y are arbitrary, any periodic orbits in the untwisted or singly twisted configurations are either unknots or simple two-strand torus knots. For the doubly twisted case as pictured in Figure 24.12, the reader may verify that, for $k > 0$, the

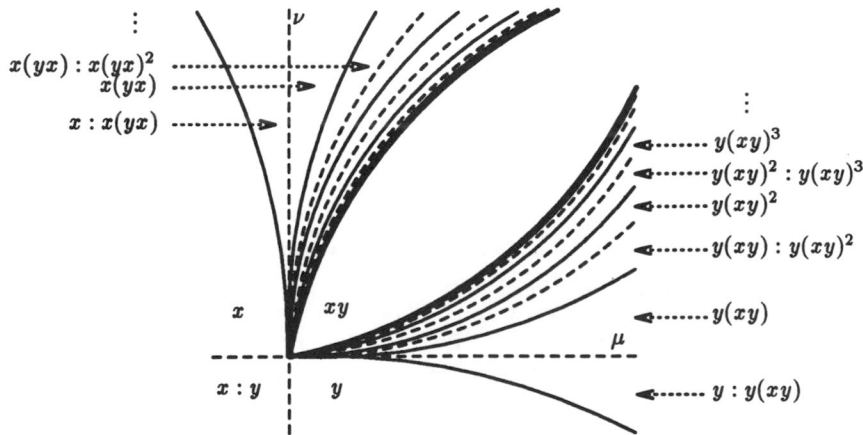

Figure 24.12 The parameter plane unfolding: doubly twisted figure of eight.

orbit $y(xy)^k$ has braid index k if $\tau_y = 1$ (resp. $k + 1$ if $\tau_y \geqslant 3$) and genus

$$g = \tfrac{1}{4}k[(\tau_x + \tau_y)k - \tau_x + \tau_y - 2]. \tag{12}$$

24.4   CONCLUSION

This exploration of knotting and linking phenomena present in the gluing bifurcation is the first step of the general strategy mentioned in Section 24.1. A thorough understanding of the topology and unfoldings associated with gluing bifurcations will then permit connections to more complex systems via parametrized families. The subsequent application of topological bifurcation invariants (knot types, linking information) may then lend insight into the bifurcation sequences giving rise to highly complicated systems. For example, if the eigenvalue ratio δ is regarded as a third bifurcation parameter in the unfolding, then as δ decreases through one and the saddle point's unstable eigenvalue begins to dominate, the butterfly template acquires the expansive property of the 'standard' Lorenz system and infinitely many periodic orbits and knot types appear. From classifications such as Corollary 1, we can say that, among these, only certain torus knots and unknots can be followed from $\delta > 1$.

We stress that, in establishing global genealogies of orbits in rather general flows, we cannot rely on special braid structures which exist in forced oscillators, for example, but must allow for gross distortions. Thus, the symbolic descriptions in terms of the letters $\{x, y\}$, which distinguish orbits as long as they remain within the small neighbourhood W, are no longer invariants. Here, the knotting and linking data is crucial: to this end, the template construction has been, and continues to be, a useful tool.

In addition to this overt goal, it is worthwhile to examine the gluing bifurcation for its own sake. As the periodic orbits associated with the bifurcations are attracting, they may be readily observed in experimental data. Indeed, gluing bifurcations have been identified in various physical models (e.g. Meron and Procaccia 1987; Baesens *et al.* 1991).

For the gluing bifurcations associated with a real saddle point, we have noted the predominance of unknots and torus knots among the orbits: this is particularly encouraging as such orbits then have numerous well-understood topological properties. We again stress the impact of global twisting upon the ensuing knot types and unfoldings: such is the flavour of global bifurcation theory.

ACKNOWLEDGEMENTS

This work is supported by an NSF Graduate Research Fellowship (RG) and by AFOSR grant #91-0329 (PH). The authors would also like to thank F. J. Wicklin and Mike Sullivan for their useful comments.

REFERENCES

Afraimovich, V., Bykov, V. and Šilnikov, L. (1982) On structurally unstable attracting limit sets of Lorenz attractor type, *Trans. Moscow Math. Soc.*, **44**, 153–216.

Baesens, C., Guckenheimer, J., Kim S. and MacKay, R. S. (1991) Three coupled oscillators: modelocking, global bifurcations and toroidal chaos, *Physica*, **49D**, 387–475.

Birman, J. S. (1975) *Braids, Links, and Mapping Class Groups*, Princeton, NJ, Princeton University Press.

Birman, J. and Williams, R. F. (1983a) Knotted periodic orbits in dynamical systems—I: Lorenz's equations, *Topology*, **22** (1), 47–82.

Birman, J. and Williams, R. F. (1983b) Knotted periodic orbits in dynamical systems—II: Knot holders for fibered knots, *Cont. Math.*, **20**, 1–60.

Boyland, P. (1984) Braid types and a topological method for proving positive entropy, Preprint.

Coullet, P., Glendinning P. and Tresser, C. (1984) Une nouvelle bifurcation de codimension 2: le collage de cycles, *C. R. Acad. Sc. Paris*, **299**, 253–256.

Franks, J. and Williams, R. F. (1987) Braids and the Jones polynomial, *Trans. Am. Math. Soc.*, **303** (1), 97–108.

Gambaudo, J. and Tresser, C. (1985) Dynamique régulière ou chaotique: applications du cercle ou de l'intervalle ayant une discontinuité, *C. R. Acad. Sc. Paris*, **300**, 311–313.

Gambaudo, J. and Tresser, C. (1988) On the dynamics of quasi-contractions, *Bol. Soc. Brasil Math.*, **19**, 61–114.

Gambaudo, J., Glendinning, P. and Tresser, C. (1984) Collage de cycles et suites de Farey, *C. R. Acad. Sc. Paris*, **299**, 711–714.

Gambaudo, J., Glendinning, P. and Tresser, C. (1988) The gluing bifurcation I: symbolic dynamics of the closed curves, *Nonlinearity*, **1**, 203–214.

Gambaudo, J., Procaccia, I. Thomae, S. and Tresser, C. (1988) New universal scenarios for the onset of chaos in Lorenz-like flows, *Phys. Rev. Lett.*, **57**, 925–928.

Ghrist, R. and Holmes P. (1993) Knots and orbit genealogies in three dimensional flows. In: *Bifurcations and Periodic Orbits of Vector Fields*, NATO ASI series Vol. C408, Dordrecht, Kluwer, pp. 185–239.

Glendinning P. (1988) Global bifurcation in flows. In: *New Directions in Dynamical Systems*, (T. Bedford and J. Swift eds.), London Math. Society, Cambridge University Press, pp. 120–149.

Guckenheimer, J. and Holmes, P. J. (1988) *Nonlinear Oscillations, Dynamical Systems, and Bifurcations of Vector Fields*, New York, Springer-Verlag.

Guckenheimer, J. and Williams, R. F. (1979) Structural stability of Lorenz attractors, *Inst. Hautes Études Sci. Publ. Math.*, **50**, 59–72.

Holmes, P. J. (1986) Knotted periodic orbits in suspensions of Smale's horseshoe: period multiplying and cabled knots, *Physica*, **21D**, 7–41.

Holmes, P. J. (1987) Knotted periodic orbits in suspensions of annulus maps, *Proc. Roy. Soc. London A*, **411**, 351–378.

Holmes, P. J. (1988) Knots and orbit genealogies in nonlinear oscillators. In: *New Directions in Dynamical Systems*, T. Bedford and J. Swift (eds.), London Math. Society, Cambridge University Press, pp. 151–191.

Holmes, P. J. (1989) Knotted periodic orbits in suspensions of Smale's horseshoe: extended families and bifurcation sequences, *Physica*, **40D**, 42–64.

Holmes, P. J. and Williams, R. F. (1985) Knotted periodic orbits in suspensions of Smale's horseshoe: torus knots and bifurcation sequences, *Archive for Rational Mech. and Anal.*, **90** (2), 115–193.

Lorenz, E. N. (1963) Deterministic non-periodic flow, *J. Atmospheric Sci.*, **20**, 130–141.

McRobie, F. A. and Thompson, J. M. T. (1993a) Braids and knots in driven oscillators. *Int. J. Bifurcation and Chaos*, **3**, 1343–1361.

McRobie, F. A. and Thompson, J. M. T. (1993b) Knot types and bifurcation sequences of homoclinic and transient orbits of a single-degree-of-freedom driven oscillator. Submitted to *Dynamics and Stability of Systems*, in press.

Meron, E. and Procaccia, I. (1987) Gluing bifurcations in critical flows: the route to chaos in parametrically excited surface waves, *Phys. Rev. A*, **35** (9), 4008–4011.

Mindlin, G. B., Solari, H. S., Natiello, M. A., Gilmore, R. and Hou X. J. (1991) Topological analysis of chaotic time series data from the Belousov-Zhabotinskii reaction, *J. Nonlinear Sci.*, **1**, 147–173.

Morgan, J. (1978) Nonsingular Morse-Smale flows on 3-dimensional manifolds, *Topology*, **18**, 41–54.

Procaccia, I., Thomae, S. and Tresser, C. (1987) First-return maps as a unified renormalization scheme for dynamical systems, *Phys. Rev. A*, **35** (4), 1884–1900.

Rolfsen, D. (1977) *Knots and Links*, Berkely, Calif, Publish or Perish.

Turaev, D. and Šilnikov (1987) On bifurcations of homoclinic 'figure eight' for a saddle with a negative saddle value, *Soviet Math. Dokl.*, **34**, 397–401.

Wada, M. (1989) Closed orbits of nonsingular Morse-Smale flows flows on S^3, *J. Math. Soc. Japan*, **41** (3), 405–413.

Zaks, M. (1993) Scaling properties and renormalization invariants for the 'homoclinic quasiperiodicity', *Physica*, **62D**, 300–316.

R. Ghrist, *Departments of Theoretical and Applied Mechanics and Mathematics and Center for Applied Mathematics, Cornell University, Ithaca, NY 14853, USA.*

P. Holmes, *Program in Applied and Computational Mathematics, Princeton University, Princeton, NJ 08544, USA.*

25 DRIVEN OSCILLATORS, KNOTS, BRAIDS AND NIELSEN– THURSTON THEORY

F. A. McRobie and J. M. T. Thompson

This chapter describes the application of knot, braid and Nielsen–Thurston theory to the orbits of single-degree-of-freedom driven oscillators. By interpreting the (x, t) graphs as braid diagrams much information on bifurcation structure can be obtained.

25.1 INTRODUCTION

Some topological methods applicable to nonlinear oscillators are described. The shortcoming is that these really only work with a three-dimensional (3D) phase space, and extension to higher dimensions is difficult to imagine. Nevertheless, the study of simple 1D maps has given insight into nonlinear phenomena and this chapter describes an extension of such analysis to 2D maps. An aim is to address the question: given a single-degree-of-freedom oscillator exhibiting a rich bifurcational structure, how far can the dynamics be understood? We consider oscillators of the form

$$\ddot{x} = g(x, \dot{x}, t) \tag{1}$$

where g is periodic in t. Familiar examples are the forced Duffing oscillators. We assume that x is an ordinary space coordinate, $-\infty < x < \infty$. (Any periodicity in space, or any other symmetry, will provide even more constraints on possibilities.) Almost all the mathematics discussed here can be found in the literature, and we generally refer to introductory texts where proper references to original sources may be found. Amongst those on whose work we draw are Boyland, Franks, Handel, Fried, Los, White, Gambaudo, van Strien, Tresser, Hall, Holmes, Birman, Williams, Gilmore and Tufillaro, along with more obvious earlier pioneers. The reference that we have drawn on most heavily is Boyland (1989), lecture notes from a seminar at Warwick University.

Knots and braids are probably familiar to most readers, but Nielsen–Thurston theory less

Nonlinearity and Chaos in Engineering Dynamics
Edited by J. M. T. Thompson and S. R. Bishop, © 1994 John Wiley & Sons Ltd

so. Loosely, it could be described as Thurston's theory of 2D maps incorporating Nielsen's theory of fixed point classes.

25.2   BACKGROUND THEORY

25.2.1   Braids, positive braids and (x, t) graphs

Because of the periodicity in time, the phase space of the oscillator may be written as either $\mathbb{R}^2 \times S^1$ or $\mathbb{R}^2 \times I$ (see Figure 25.1). In the latter case, the interval I is one period of applied forcing, and in Figure 25.2 we show two period-1 orbits. Their (x, t) graphs are a projection from the 3D phase space $(x, \dot{x}, t)$ down the velocity axis. Wherever the graphs cross, the segment with the greater slope dx/dt has the greater velocity $\dot{x}$, and thus passes *above* the other. The nature of all crossings is known, thus the (x, t) graphs can be interpreted as a *braid diagram*. Moreover, since all crossings are of the form 'left over right', the result is a *positive braid*.

In recent years, Poincaré sampling has proved to be a powerful method of analysis for such oscillators, revealing much about the structure of chaotic attractors and the convoluted basins of attraction with possibly fractal boundaries. Unfortunately this has led to some neglect of 'what happens between the planes'. This has been exacerbated by a propensity for projecting orbits onto $(x, \dot{x})$ planes. Unfortunately, this is not a generic projection, and leads to neither a braid nor a knot diagram (see Figure 25.2a).

The fact that all braids are positive is a feature of oscillator studies not found in standard Nielsen–Thurston theory. It follows that there is information in the (x, t) graphs which cannot be deduced from knowledge of the Poincaré map alone.

25.2.2   Knot theory

As control parameters are varied, the orbits of an oscillator evolve. Shapes change, but various topological features are invariant. For example, two orbits may be linked. To pull

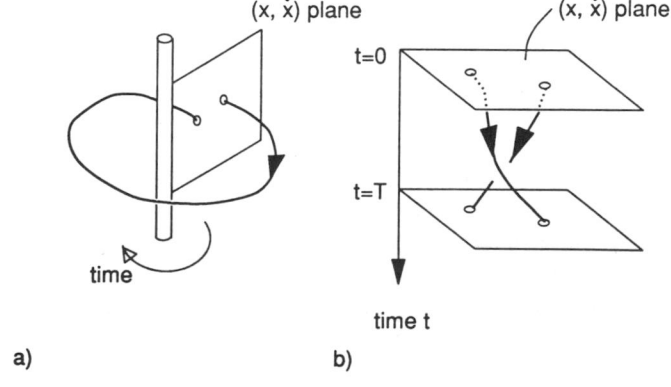

Figure 25.1   The phase space of a driven single-degree-of-freedom oscillator may be written variously as (a) $\mathbb{R}^2 \times S^1$ or (b) $\mathbb{R}^2 \times I$. Identifying planes at $t = 0$ and $t = T$ in the latter leads to the same space as the former.

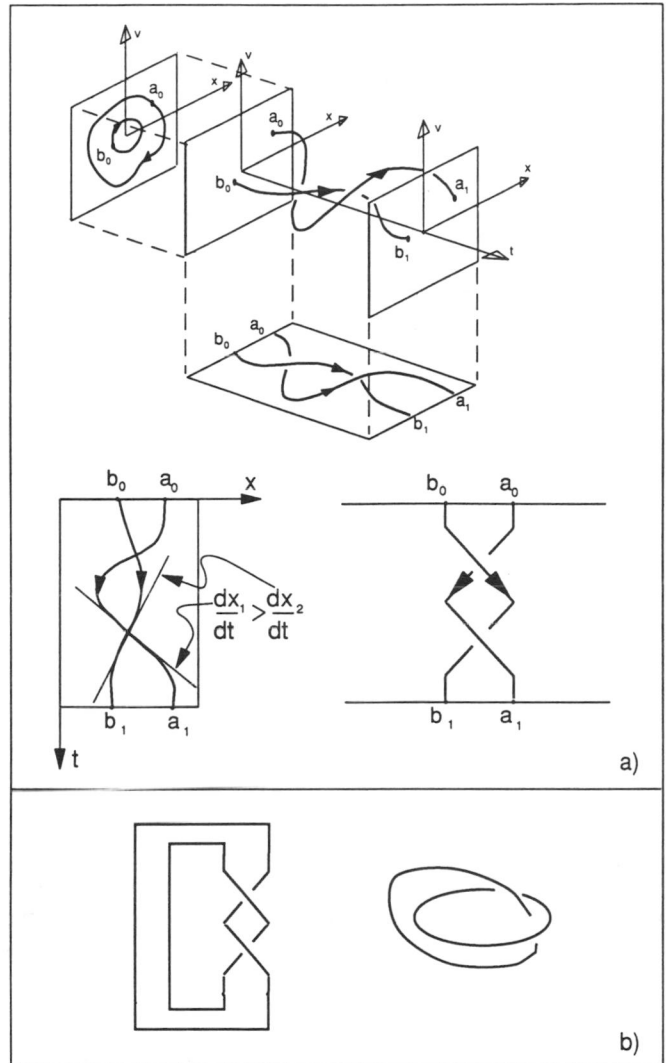

Figure 25.2 (a) Projecting to period-1 orbits from the phase space $(x, \dot{x}, t)$ onto the (x, t) and $(x, \dot{x} = v)$ planes. The former gives the braid diagram of a positive braid. Note that any attempt to interpret the $(x, \dot{x})$ projection as a knot diagram would conclude that the two orbits are not linked. However, it is shown in (b) by closing the braid that these two orbits are linked in $\mathbb{R}^2 \times S^1$.

them apart by changing parameters would be impossible, since this would require an intermediate condition where the orbits intersect, so contravening the uniqueness of solutions of the differential equation. Similarly knotted orbits cannot be unknotted by changing control parameters.

Definition: Two knots A and B, are *ambient isotopic* if there is a homeomorphism of $\mathbb{R}^3$ isotopic to the identity taking the knot A to the knot B. (Note: all homeomorphisms in this paper are assumed to be orientation-preserving.)

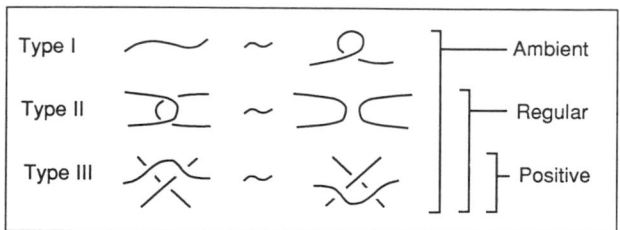

Figure 25.3   The Reidemeister moves.

Rather than working in 3D, it can be easier to project the knots onto some plane. This gives a *knot diagram*, with over- and under-crossings indicated. It is vaguely similar to dropping a knotted loop of string onto a table. Knot diagrams may be analysed using the *Reidemeister moves* (Figure 25.3). That each move can be performed on a string lying on a table is rather obvious. Their power lies in the theorem which states that if (and only if) one knot diagram can be converted into another by any sequence of Reidemeister moves then, back in three dimensions, the two knots are ambient isotopic. If, however, it is possible to construct a sequence involving only Type II and III moves, the two knots are said to be *regular isotopic*, and if only Type III moves are necessary, *positive isotopic*.

Ambient isotopy is a good general concept for knot theory in 3D. Regular isotopy is useful in physical applications when one of the dimensions is time, such that a Type I move is prohibited because it would involve some section of an orbit moving backwards in time. Positive isotopy is useful for oscillator (x, t) graphs, since Type II moves involve a negative crossing and so are not permitted. (Kauffman (1991) is a good introduction to modern knot theory and Birman (1974) the classic text on braids.)

A torus knot can be drawn on a torus. An iterated torus knot may be obtained by first creating a torus knot and then replacing the 'string' by a thin tube and drawing a knot on the surface of this thinner tube. Clearly the construction can be iterated further. In $\mathbb{R}^2 \times S^1$, both concepts are most useful in regular isotopy, when the original torus passes around the axis removed from $\mathbb{R}^3$ (Figure 25.4). Many knots can be constructed in $\mathbb{R}^2 \times S^1$ which (in

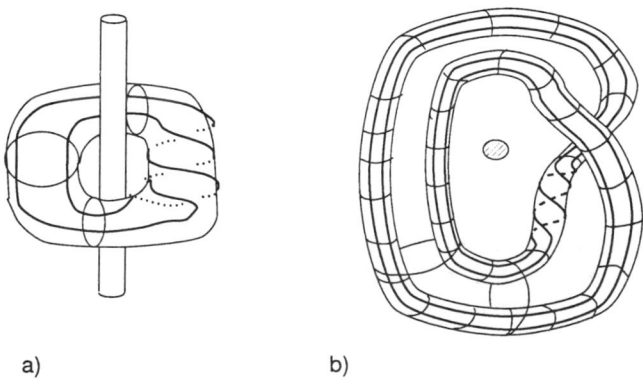

a)   b)

Figure 25.4   A torus knot (a) and an iterated torus knot (b). In both cases these knots are *regular* isotopic to torus and iterated torus knots (respectively) in $\mathbb{R}^2 \times S^1$.

$\mathbb{R}^3$) are ambient (but not regular) isotopic to a torus knot. However, the ambientization involves Type I moves pulling across the removed axis, and so such deformations are not physically possible for oscillator orbits.

25.2.3   Nielsen–Thurston theory

A basic object of study here is the punctured disc. (Studying a disc is tantamount to studying the plane, and discs are easier to draw.) Given a period-n orbit p of a map f of the disc to itself, the removal of the n points of p leaves the punctured disc $D^2 - p$. The aim now is to study the behaviour of the map everywhere else (i.e. on the complement of p) to see what other orbits may exist there.

To study $D^2 - p$, homotopy theory provides some basic tools such as the *fundamental group* of the space: a free group on n generators, which may be obtained by retracting $D^2 - p$ onto a 1-dimensional skeleton, as sketched in Figure 25.5a. The map f of $D^2 - p$ to itself induces a homomorphism of the fundamental group $\pi_1(D^2 - p, *)$ to itself (Figure 25.5b). Nielsen–Thurston theory applies group theory, fixed-point theorems, etc., to deduce facts about the behaviour out in $D^2 - p$. By this approach, *global* information is deduced from *local* knowledge.

How can one describe the orbits of a 2D map using knots and braids when orbits are point-like entities and there is no intervening flow? The rigorous approach involves the use of an isomorphism between the *mapping class group* of $D^2 - p$ and the Artin braid group $[B_n/Z_n]$. (The set of braids on n strings has a natural group structure, as does the set of homeomorphisms of $D^2 - p$ (up to isotopy). These two structures are in 1-to-1 correspondence; thus, given a map of $D^2 - p$ one may ascribe to p the corresponding braid (up to equivalence) from the group B_n. The symbol $/Z_n$ means that full twists are factored out, since these have no meaning for a map.) A simpler approach is to just create some suspension of f and then take some projection to obtain a braid diagram.

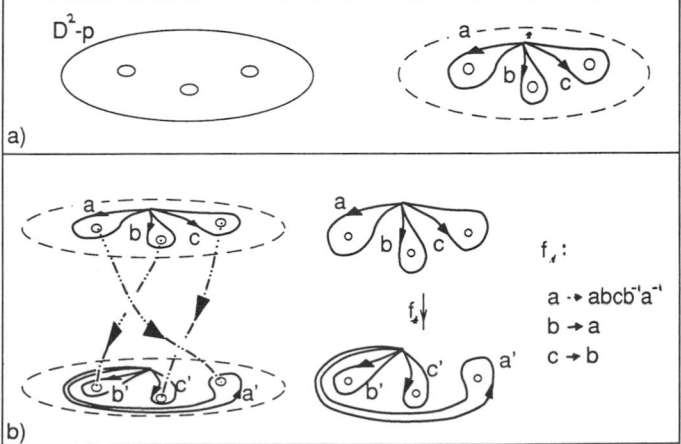

Figure 25.5   (a) The disc punctured by removing a period-3 orbit, together with a skeleton of generators for its fundamental group. (b) How a map f induces a homomorphism f_*.

One of the cornerstones of Nielsen–Thurston theory is Thurston's classification of mapping classes. For brevity, we crudely paraphrase this here. By this classification, if an orbit is

(i) regular isotopic to a torus knot then it is *finite order*;
(ii) regular isotopic to an iterated torus knot then it is *reducible*;
(iii) neither of the above, then it is *pseudo-Anosov*.

Essentially the first two correspond to simple dynamics, and the third to complexity.

25.2.4   Conjugacy and braid-type

Two braids A and B are *conjugate* if there exists a third braid γ such that $B = \gamma^{-1}A\gamma$. An example is given in Figure 25.6. The *braid-type* of a braid A is the set of all braids conjugate to A. These notions provide a convenient method of comparing orbits which factors out the dependence on the choice of suspension and projection (Figure 25.6b). For oscillators though, there is nothing arbitrary about the suspension or projection. The suspension is the flow and the projection gives (x, t) graphs. Conjugation with an arbitrary braid is not appropriate.

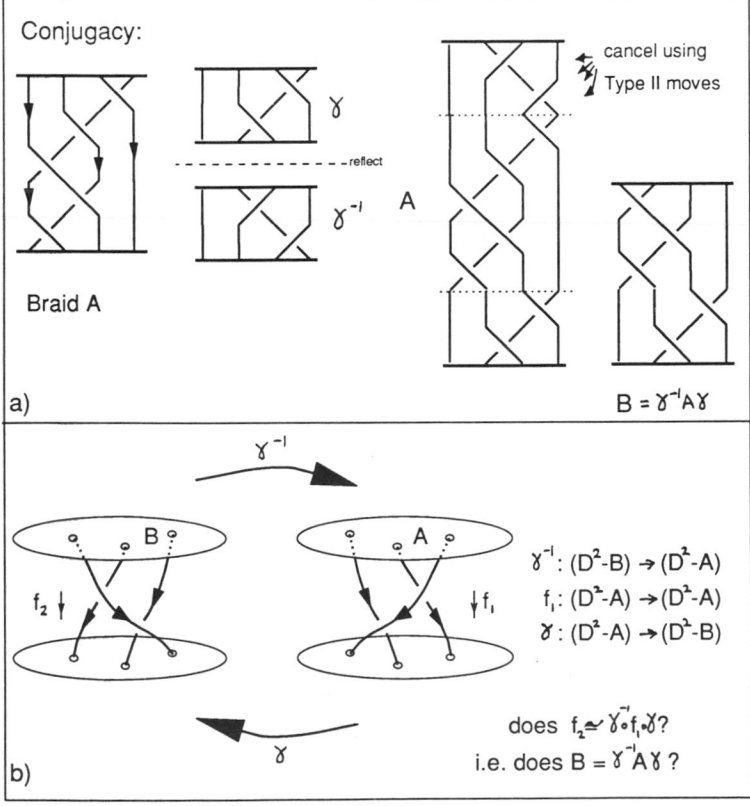

Figure 25.6   (a) An example of braid conjugacy. The braid A is conjugate to the braid B. (b) Comparing different orbits of different maps using conjugacy.

Given the arbitrariness of the phase of Poincaré sampling, crossings can move from the bottom of a braid diagram around to the top and vice versa. This gives a slightly weaker notion of conjugacy, namely *positive cyclic equivalence* (see McRobie and Thompson 1993a).

25.2.5   A 2D analogue of the Sarkovskii sequence

The Sarkovskii sequence for 1D maps is an ordering of the integers such that if a (continuous) 1D map contains an orbit of period n then there coexist orbits of all periods to the left of n in the sequence. The last integer is 3 and thus (loosely) 'period 3 implies chaos'. This ordering can be refined if one considers, rather than just the period, the way in which a periodic orbit permutes its points. In 2D, the analogue of comparing permutations is comparing braid-types. Recent work by Gambaudo, van Strien, Tresser, Boyland and others has established that for 2D maps there is a partial ordering of braid-types. Orbits of some braid-types imply the coexistence of orbits with other braid-types. The analogous dictum is 'a pseudo-Anosov braid-type implies positive topological entropy'.

25.2.6   Bifurcation sequences

Closely associated with the question 'given an orbit of one braid-type, which others must coexist?' is the problem 'given two orbits, can they collide at a bifurcation?'. In Figure 25.7 we summarize the interrelationship of knot, braid and Nielsen–Thurston theory applied to

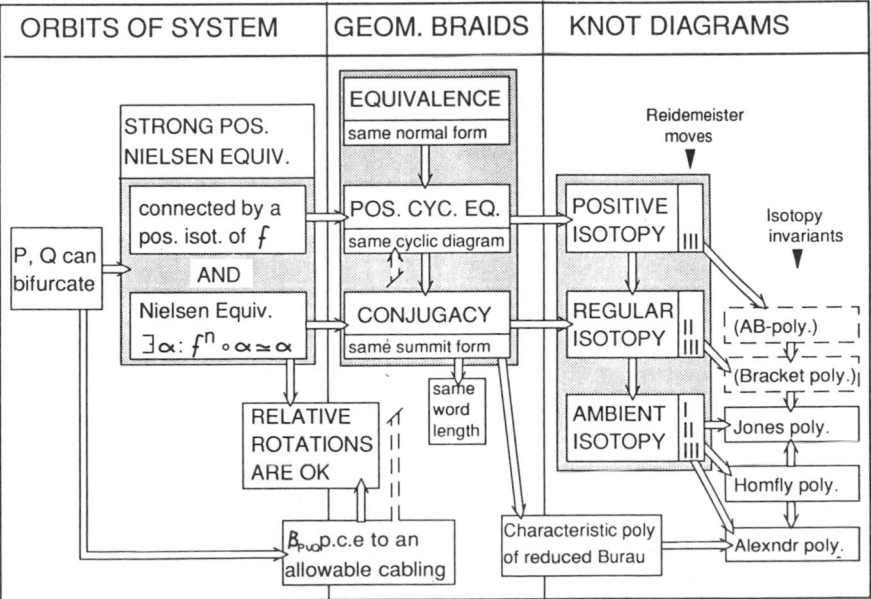

Figure 25.7   Overview of methods applicable to decide if two orbits P and Q can collide at a saddle–node bifurcation.

this problem in the context of a simple saddle–node bifurcation. Here we consider the positive braids of oscillator orbits, and some of the notions additional to a standard Nielsen–Thurston approach are included. One obstacle to progress in the analysis of horseshoe formation is the need to solve the *conjugacy problem* in the braid group. Although finite algorithms exist (such as Garside's summit forms), they can take impractically long times. For positive braids, though, it is only necessary to solve the *positive cyclic equivalence problem* in the positive braid semigroup. Unfortunately, this too can take excessive time. It is thus useful to have rapid methods (such as the knot polynomials) which can decide if two braids are not conjugate. There are a variety of polynomials available, and we concentrate here on the classical polynomial of knot theory – the Alexander polynomial. There are two reasons for this. One is that some of the more recent polynomials also have long computation times, and so do not provide much of a short cut. The other is that the relationship of the Alexander polynomial (through the reduced Burau representation) with Nielsen–Thurston theory is rather transparent in this case. Loosely speaking, the further to the lower right one goes in Figure 25.7, the greater the amount of information that is being thrown away. At the left-hand side are Nielsen–Thurston notions, based on the full structure of the fundamental group. As explained in Boyland (1989), to obtain the reduced Burau representation, one moves from homotopy theory of the fundamental group to a homology group of a non-universal cover of the skeleton of generators, thus obtaining a sort of 'abelianized Nielsen–Thurston theory'.

25.2.7   The reduced Burau representation

Given a braid on n strings, each elementary crossing σ_i may be represented by a square matrix $S_i(t)$ of dimension $n-1$ (see Figure 25.8). Multiplying the matrices in the order of the word gives the *reduced Burau matrix* $S(t)$, with characteristic polynomial $C(\lambda, t) = \det | S(t) - \lambda I |$ where I is the identity. The Alexander polynomial $\Delta(t)$ is given by $\Delta(t) = C(1, t)/p(t)$ where $p(t) = 1 + t + t^2 + \cdots + t^{n-1}$. The Alexander polynomial is an invariant of ambient isotopy, and the characteristic polynomial is an invariant of braid conjugacy. Neither are complete invariants. (Note: t is not time here.)

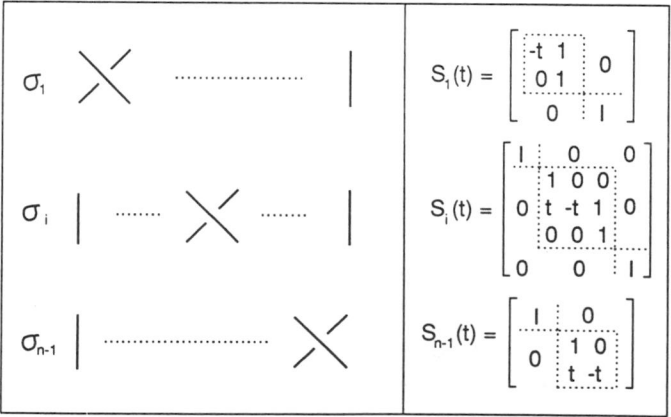

Figure 25.8   The reduced Burau representation.

25.2.8   Minimal periodic-orbit structure and minimal topological entropy

As stated above, orbits of some braid-types imply orbits of others. Thus given one orbit, what other orbits must coexist? Essentially one is seeking a map which contains an orbit of the given braid-type, but which contains no 'unnecessary' orbits. Bestvina and Handel (1991) provide the full solution to this problem. However, the method requires a certain mathematical ability to appreciate, and is not particularly suited to practical purposes. (That said, there is a program 'foldtool' (White 1993) which implements this and is easy to use.) Again, though, practical short cuts can be useful.

Hall (1991) provides a simpler algorithm which, given a braid, constructs a simple map containing such an orbit, and checks whether the map has minimal structure ('No Bogus Transitions'). If so, the problem is solved in that instance.

An alternative approach is to use the abelianized Nielsen–Thurston theory. We consider upper- and lower-bound methods below. Although we shall focus here on entropy, both the full theory and its simplifications can give detailed information about the minimal periodic-orbit structure implied at any period (see Boyland 1989). The topological entropy $h(\beta)$ of an orbit β is defined as the least topological entropy of any map (of D^2 to D^2, here) containing an orbit of the same braid-type.

25.2.9   Upper-bound approach – line diagrams and transition matrices

The *line diagram* of a geometric braid is constructed by drawing lines across the top of the braid, and passing these lines down the braid to obtain the line diagram at the bottom (see Figure 25.9). By thickening up the lines to 'boxes', one constructs a simple map of the disc to itself. The overlap of the images of the boxes at the bottom of the braid with their original locations at the top gives a *transition matrix*. A standard symbolic-dynamics interpretation

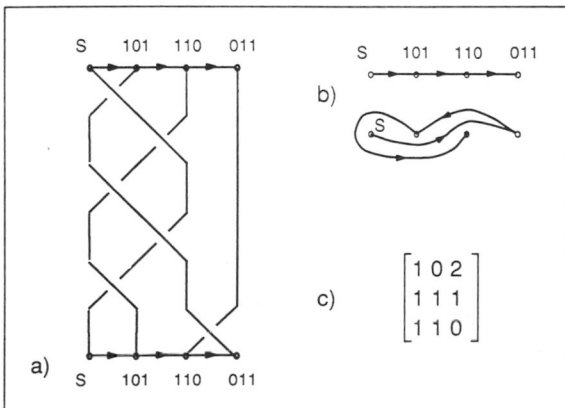

Figure 25.9   Line diagrams and transition matrices. Given a braid such as that shown at (a), the line diagram is obtained by connecting points at the top of the braid and passing these line segments down the braid. This gives the line diagram shown at (b). The overlap of the segments in the line diagram with their original locations gives the transition matrix shown at (c).

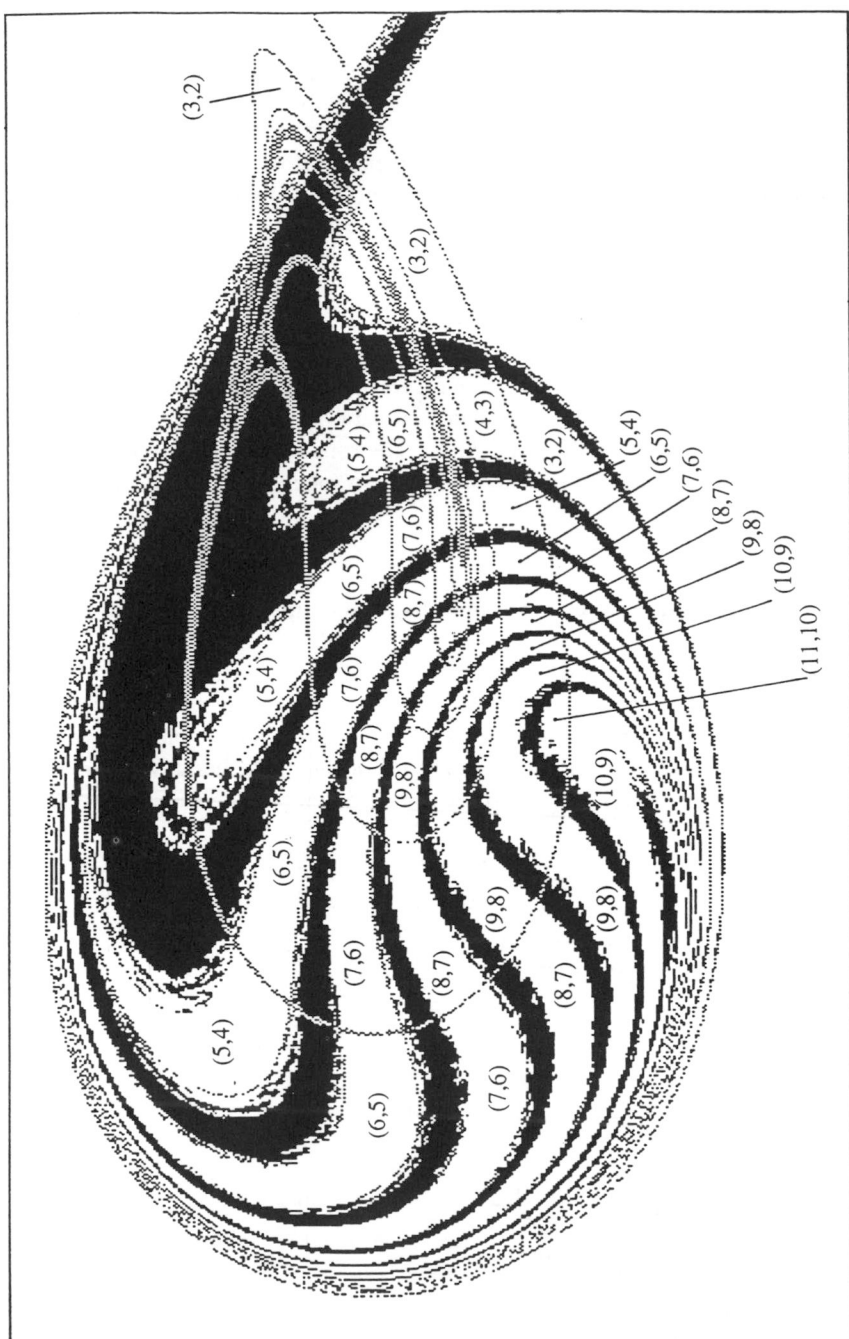

Figure 25.10 The 'ambient isotopy knot-types' of arriving/escaping transients in the cubic-well escape equation during basin erosion. The dominant areas of erosion are filled with transient orbits which ambientize to $(n+1, n)$ torus knots.

of this matrix gives the periodic-orbit structure. However, this may not be minimal. It follows that $h(\beta) \leqslant \log(\lambda_m)$ where λ_m is the spectral radius of the transition matrix. (See Figure 25.7).

25.2.10 Lower-bound approach – the reduced Burau representation

There is a theorem of Fried (see Boyland 1989) which implies that the topological entropy $h(\beta)$ of an orbit β is greater than or equal to the logarithm of the spectral radius of its reduced Burau matrix $S(t)$, for all $|t| = 1$. (This works for other representations, not just the reduced Burau.) Thus, given an (x, t) graph, write down the braid-word, multiply together the elementary reduced Burau matrices, substitute $t = e^{i\theta}$, (note: t here is not time) and let θ vary from $-\pi$ to $+\pi$. The topological entropy is then greater than or equal to the logarithm of the largest eigenvalue modulus encountered. In practice it often happens that this lower-bound approach arrives at the exact answer as given by 'foldtool'.

25.3   APPLICATION TO DRIVEN OSCILLATORS

It is well known that, for a 2D map, a transverse homoclinic orbit implies a Smale horseshoe *at some iterate* (n say) of the map. In practice it is most useful if a specific horseshoe can be identified at a specific iterate, preferably with n small (such as $n = 1$). Symbolic dynamics tecniques can then be used to annotate an infinite number of periodic solutions. Using the methods described above these may be partitioned into positive cyclic equivalence classes, and saddle-node bifurcations restricted to pairs of orbits from the same class. Nielsen equivalence, or simpler, relative rotation methods, may then be used to deduce bifurcational precedence structure (see McRobie and Thompson 1993a; McRobie 1992). In the latter reference, many details of the low-period subharmonic structure of the escape equation were deduced from a few hypotheses about creating a three-striped Smale horseshoe. These low-period orbits comprise most of the dominant features of the subharmonic periodic-orbit structure. In many engineering systems, chaos is a comparatively marginal phenomenon. Nonlinear motion, though, is prevalent. We would contend that the methods reviewed here provide, for single-degree-of-freedom driven oscillators at least, the machinery with which one can gain better understanding of the organization of nonlinear motions. An obvious extension to the methods would include an analysis of transient motion. Handel has already produced results in this direction applying Nielsen–Thurston theory to homoclinic trajectories. Less rigorous results based on 'knot-types of transient orbits' are presented in McRobie and Thompson (1993b). In Figure 25.10 we present a cell map showing the knot-types of transient orbits encountered during basin erosion in the escape equation. Clearly these are not 'fine structure' but dominant features of the phase space in this parameter regime.

ACKNOWLEDGEMENTS

F.A.M. acknowledges the support of The Royal Society, and J.M.T.T. the receipt of a Senior SERC Fellowship. F.A.M. is grateful for conversations with Nick Tufillaro and Toby Hall.

REFERENCES

Bestvina, M. and Handel, M. (1991) Train tracks for surface homeomorphisms, Preprint.

Birman, J. S. (1974) *Braids, Links and Mapping Class Groups*, Annals of Math. Study, no. 82, Princeton, NJ, Princeton University Press.

Birman, J. S. and Williams, R. F. (1983) Knotted periodic orbits in dynamical systems – 1. Lorenz's equations, *Topology*, **22**(1), 47–82.

Boyland, P. L. (1989) Braid-types of periodic orbits of surface automorphisms. In: *Notes on Dynamics of Surface Homeomorphisms* (R. Mackay, ed.), Lecture Notes, Maths Institute, Warwick University.

Hall, T. D. (1991) Periodicity in Chaos:, the Dynamics of Surface Automorphisms, PhD thesis, University of Cambridge.

Holmes, P. J. (1986) Knotted periodic orbits in suspensions of Smale's horseshoe: period-multiplying and cabled knots, *Physica*, **21D**, 7–41.

Holmes, P. J. and Whitley, D. (1984) Bifurcations of one- and two-dimensional maps, *Phil. Trans. Roy. Soc. Lond. A*, **311**, 43–102.

Holmes, P. J. and Williams, R. F. (1985) Knotted periodic orbits in suspensions of Smale's horseshoe: torus knots and bifurcation sequences, *Arch. Rat. Mech. Ann.*, **90**, 115–194.

Kauffman, L. H. (1991) *Knots and Physics*, Singapore, World Scientific.

McRobie, F. A. (1992) Bifurcational precedences in the braids of periodic orbits of spiral 3-shoes in driven oscillators, *Proc. R. Soc. Lond. A*, **438**, 545–569.

McRobie, F. A. and Thompson, J. M. T. (1993a) Braids and knots in driven oscillators, *Int. J. Bifurcation and Chaos*, **3**, 1343–1361.

McRobie, F. A. and Thompson, J. M. T. (1993b) Knot-types and bifurcation sequences of homoclinic and transient orbits of a single-degree-of-freedom driven oscillator, *Dyn. and Stabil. Systems*, to appear.

Solari, H. G. and Gilmore, R. (1988) Relative rotation rates for driven dynamical systems, *Phys. Rev. A*, **37**, 3096–3109.

Tufillaro, N. B., Abbott, T. and Reilly, J. (1992) *An Experimental Approach to Nonlinear Dynamics and Chaos*, New York, Addison-Wesley.

White, T. (1993) 'foldtool' (An implementation of the Bestvina-Handel algorithm)(tadpole@ ucrmath. ucr. edu.)

F. A. McRobie and J. M. T. Thompson, *Centre for Nonlinear Dynamics and Its Applications, University College London, Gower Street, London WC1E 6BT, UK*

26 A NEW TECHNIQUE FOR ESTIMATING THE DYNAMICS IN THE NOISE-REDUCTION PROBLEM

M. E. Davies and J. Stark

In this chapter we consider the problem of estimating the dynamics from a time series that has been embedded in a reconstructed space. Previous work on noise reduction has relied on initially optimizing the dynamics by an ordinary least-squares fit. However, this is not directly relevant to the noise-reduction problem. Here we incorporate the optimization of the dynamics into the noise-reduction process. We also include numerical evidence that this improves the stability of the noise-reduction algorithm.

26.1 INTRODUCTION

Over the last decade a variety of new techniques for the treatment of chaotic time series have been developed. Initially these concentrated on the characterization of chaotic time series using invariants such as fractal dimensions or Lyapunov exponents. Later, attention focused on the possibility of predicting the future short-term behaviour of such time series and this in turn has led to algorithms for noise reduction in time series having a chaotic component.

A varity of different objectives and assumptions have been chosen for the construction of noise-reduction procedures, for example Kostelich and Yorke (1990), Farmer and Sidorowich (1991) and Stark and Arumugam (1992). However, in this chapter we will restrict ourselves to some recent developments in the field of filtering time series based solely on reducing the dynamic noise as much as possible (we will not deal with the explicit enforcement that the filtered time series shadows the original data). The original noise-reduction algorithm for this problem, proposed by Hammel (1990), was based on the geometric concepts of shadowing

(Bowen 1970) as a means of solving the zero-finding noise-reduction problem in a stable manner. However, this algorithm was unstable in the vicinity of homoclinic tangencies. These problems can be dealt with systematically by replacing the zero-finding problem with the closely related minimization task. Initially this was done using a gradient-descent algorithm in Davies (1993a) and this was shown to simplify and give a systematic framework for a variety of other *ad hoc* schemes: Cawley and Hsu (1992), Sauer (1992) and Schreiber and Grassberger (1991). Finally further improvements were obtained by using a Levenberg–Marquardt minimization method in Davies (1993b). This provides the stability of the gradient-descent algorithm while retaining some of the speed advantages of Newton-based methods.

However, in all the previous work on noise reduction, the relation between the function estimation and the noise-reduction process has not been addressed. It is this point that we are concerned with here. Regarding the problem of noise reduction as a root-finding or minimization task can only really be justified when the underlying dynamical system is known *a priori*. In the case where the dynamics is unknown a method for approximating the mapping function (and its derivatives) is also necessary. However, unless we intend to fix the mapping function at the start of the noise reduction, this interpretation is no longer obvious.

Most of the algorithms that have tackled the problem of noise reduction for data from an unknown dynamical system (Farmer and Sidorowich 1991; Cawley and Hsu 1992; Schreiber and Grassberger 1991; Sauer 1992) have all used the *ad hoc* two-step, approach of alternating between the estimation of the dynamics and applying one iterate of their particular noise-reduction scheme. This creates a variety of questions. How much noise reduction should be done between each new approximation of the dynamics? Does re-fitting the dynamics improve the noise reduction at all? Is there not a more systematic approach altogether? In this chapter we investigate these problems and the relationship between the estimation of the dynamics and the noise-reduction procedure. Finally we aim to go some way to producing a more systematic approach such that the mapping-function approximation can be absorbed more naturally into an extended minimization scheme.

The remainder of this chapter is set out as follows. In the next section we briefly review the work done in Davies (1993b) on applying Levenberg–Marquardt minimization to the noise-reduction problem. We then investigate the relationship between the estimation of the dynamics and the noise-reduction process and discuss the two-step approach in this context. It is then possible to consider a more systematic approach that ·adjusts the estimation for the dynamics in conjunction with filtering the data. Finally we compare these two approaches and give a practical example of the extended minimization scheme to the Lorenz attractor.

26.2   A REVIEW OF THE LEVENBERG–MARQUARDT NOISE-REDUCTION PROCESS

Given a time series that originated from an experimental dynamical system, before any noise reduction can be done it is first necessary to embed the data in a reconstructed phase space and to approximate the resulting dynamics. This introduces approximation errors into the model:

$$\varepsilon_i = x_i - F(x_{i-1}, x_{i-2}, \ldots, x_{i-d}) \tag{1}$$

where ε_i is the dynamic error for the ith point, x_i, in the time series, $F(\,)$ is the approximating function for the dynamical system and d is the embedding dimension. It is not possible to separate this noise from the deterministic part of the data. However, an approximate solution can be found if the dynamics are linearized in the neighbourhood of the trajectory and a deterministic orbit is assumed to lie somewhere within this neighbourhood. Linearizing the above equation around a perturbation to the trajectory and setting the error of the perturbed system to zero we can write:

$$\Delta x_i - J_{i-1}\Delta x_{i-1} = -\varepsilon_i. \tag{2}$$

This can be written in matrix form in the following way:

$$D(x_{new} - x_{cur}) = -\varepsilon,$$

$$D = \begin{pmatrix} 1 & -J_1 & & & \\ & 1 & -J_2 & & \\ & & & \cdots & \\ & & & 1 & -J_n \end{pmatrix}. \tag{3}$$

Here D is an $(n-d) \times n$ matrix, and hence x_{new} cannot be determined by simply inverting D. Indeed equation (3) does not have a unique solution. To solve this equation we thus need to add additional constraints. One method of applying a reasonable set of additional constraints is to constrain the ends in the stable and unstable subspaces (Hammel 1990).

A simpler approach, and one that is not confined to the assumption that the deterministic orbit and the noisy orbit are both approximately within the same linear neighbourhood, is to calculate an orbit within this neighbourhood that merely has *less* noise than the original data. The process is therefore one of minimizing a function of the dynamic errors with respect to adjustments in the data. Using a Euclidean norm the total error can be written as a cost function H:

$$H = \sum_{i=d+1}^{n} \varepsilon_i^2. \tag{4}$$

The problem can then be stated as one of moving down the cost function until a trajectory that minimizes H is found. In fact, there will be a d-dimensional set of trajectories that will minimize H, so the aim is to arrive at one that is close to the original data. This can be achieved by a sensible choice of minimization algorithm.

Although it can be shown that the gradient-descent approach avoids the instabilities that are caused due to ill-conditioning, this is at the cost of speed. A Newton-based algorithm (Hammel's method is equally applicable to the minimization problem) would obviously be preferable. Ideally we would like to retain this quadratic convergence while achieving the stability of the gradient-descent approach. This is, in fact, possible, to a certain extent, by using a Levenberg–Marquardt method (see for example Davies 1993b).

Heuristically the idea is to produce an algorithm that can smoothly interpolate between the gradient-descent algorithm for equation (4) and the Newton algorithm for the same equation. The obvious approach is to try

$$x_{new} = x_{cur} - (D^T D + \delta I)^{-1} \nabla H(x) \tag{5}$$

where I is the identity matrix and δ is an arbitrary weight that defines whether the algorithm behaves more as a gradient-descent scheme or more as a Newton method. If δ is zero then equation (5) is just an ordinary Newton step. However, when δ becomes large then $(D^TD + \delta I)^{-1}$ tends to $(1/\delta)I$ and the equation converges towards the gradient-descent algorithm.

The most important benefit of this method is its effect on the singularities of D. Consider the singular-value decomposition of $D = USV^T$. Then the matrix $M = (D^TD + \delta I)$ can be written as

$$M = V(S^2 + \delta I)V^T. \qquad (6)$$

Hence, even though D may be singular, the singular values of M are bounded from below by δ. Indeed, a further impressive feature of this algorithm is that if the non-zero singular values are bounded away from zero then, as $\delta \to 0$, the solution for x from equation (5) will tend to the solution of the rank-deficient minimization problem by the Moore–Penrose pseudo-inverse which is, in some sense, optimal. In practice, however, δ cannot be allowed to become too small, otherwise implementation of the algorithm will lead to numerical instabilities.

Since M can be forced to be well-conditioned this allows the use of special matrix techniques to exploit the matrix's banded structure. Here we inverted M by applying Cholesky decomposition to M and then solving by forward and backward substitution.

26.3   AN *AD HOC* APPROACH TO FUNCTION APPROXIMATION

We consider the problem of noise reduction in terms of the minimization of the cost function H, as above. This is the same cost function that is minimized when we perform an ordinary least-squares approximation of the function F. The only difference between the two cases is that the minimization is with respect to different parameters. It is possible to consider the two processes simultaneously by treating $H = H(x, p)$ as a function both of the trajectory x and the parameters p. That is a function of both the trajectory, x, and the parameter family, p, of the function approximation.

When the problem is posed in terms of $H(x, p)$, alternating between function estimation and noise reduction can be seen as minimizing H by means of zigzagging down the cost function. We can now consider how such a method effects the noise-reduction process. Obviously, as long as the two steps are individually stable, the combination should be stable. However, the additional parameters, p will result in the minimum of the cost function being d_p dimensions larger than that for just minimizing $H(x)$. We therefore want to construct a method that, to some extent chooses the point in the minimum of H that is closest to the original data set. This we do in the next section.

26.4   A MORE THEORETICAL APPROACH

If we have a set of mapping functions identified by a vector p in some parameter space and a starting point in this space of maps p^*, then the first-order Taylor expansion of the extended error function $H(x, p)$ can be written in terms of perturbations, Δp, from p^*, as well as the trajectory adjustments, Δx:

$$e = \varepsilon^* - D\Delta x + P\Delta p \qquad (7)$$

where D is the same as the definition in equation (3) and P is the matrix of the partial derivatives of the mapping function around p^*, whose elements are $\partial f(x_i, p^*)/\partial p_j$. It is clear from this equation that the inclusion of the freedom to alter the mapping parameter p will introduce d_p additional singularities to the problem. For the moment let us ignore the singularities and construct the least-squares cost function. This results in having to solve the following equation:

$$
\begin{array}{|c|c|}\hline D^{\mathrm{T}}D & -D^{\mathrm{T}}P \\\hline -P^{\mathrm{T}}D & P^{\mathrm{T}}P \\\hline\end{array}
\begin{array}{|c|}\hline \Delta x \\\hline \Delta p \\\hline\end{array}
=
\begin{array}{|c|}\hline D^{\mathrm{T}} \\\hline -P^{\mathrm{T}} \\\hline\end{array}
\begin{array}{|c|}\hline \varepsilon \\\hline\end{array}
\tag{8}
$$

We can attempt to solve this by splitting up the problem into two simultaneous equations, the first of which is

$$
\Delta x = (D^{\mathrm{T}}D)^{-1}D^{\mathrm{T}}(\varepsilon - P\Delta p), \tag{9}
$$

which can then be substituted into the second equation to solve for Δp:

$$
P^{\mathrm{T}}P\Delta p - P^{\mathrm{T}}D(D^{\mathrm{T}}D)^{-1}D^{\mathrm{T}}(P\Delta p - \varepsilon) = P^{\mathrm{T}}\varepsilon. \tag{10}
$$

As we have already stated, these equations are rank-deficient and in the non-singular directions of D the above equation is trivial. Here, as before, we can solve this problem by adding an adjustment to the matrix on the left-hand side in equation (8) to make it full-rank and, hence, invertible.

Heuristically, and in contrast to the previous Levenberg–Marquardt adjustment, we do not with to restrict all the free parameters since the shadowing problem only requires that the trajectory adjustment, Δx, be kept small and inflicts no constraints on the size of Δp (obviously we still need to maintain the validity of the truncation of the Taylor expansion). Therefore we only add the Levenberg–Marquardt parameter to the $D^{\mathrm{T}}D$ term by replacing it by $D^{\mathrm{T}}D + vI$. Then equations (9) and (10) can be replaced by

$$
\Delta x = (D^{\mathrm{T}}D + vI)^{-1}D^{\mathrm{T}}(\varepsilon - P\Delta p) \tag{11}
$$

and

$$
P^{\mathrm{T}}(I - D(D^{\mathrm{T}}D + vI)^{-1}D^{\mathrm{T}})(P\Delta p - \varepsilon) = P^{\mathrm{T}}\varepsilon \tag{12}
$$

which is solvable for $v > 0$. This is an extended version of the noise-reduction algorithm presented in the first section and optimizes both the trajectory and the mapping function.

More rigorously, we can show that this approach has a theoretical basis, in that it provides the solution to the following constrained optimization problem:

minimize: $q(\Delta x, \Delta p) = (\varepsilon - D\Delta x + P\Delta p)^{\mathrm{T}}(\varepsilon - D\Delta x + P\Delta p)$

w.r.t. $\{\Delta x, \Delta p\}$

subject to: $\Delta x^{\mathrm{T}}\Delta x = h^2$

where h defines the step size that Δx is restricted to. We can now prove that this is equivalent to the heuristic method described above. First we need to introduce a Lagrangian multiplier v. We can write the Lagrangian function as

$$
\mathscr{L}(\Delta x, \Delta p, v) = (\varepsilon - D\Delta x + P\Delta p)^{\mathrm{T}}(\varepsilon - D\Delta x + P\Delta p) + v(\Delta x^{\mathrm{T}}\Delta x - h^2) \tag{13}
$$

A solution to the constrained optimization can now be found by solving the $\nabla\mathscr{L} = 0$ (see

Fletcher 1980). This can be solved by considering the following equations:

$$\nabla_{\Delta x}\mathscr{L} = -2D^{\mathrm{T}}(\varepsilon - D\Delta x + P\Delta p) + 2\nu\Delta x = 0, \tag{14}$$

$$\nabla_{\Delta p}\mathscr{L} = 2P^{\mathrm{T}}(\varepsilon - D\Delta x + P\Delta p) = 0, \tag{15}$$

$$\nabla_{\nu}\mathscr{L} = \Delta x^{\mathrm{T}}\Delta x - h^{2} = 0. \tag{16}$$

The first two equations, above, can be rearranged so that they are equivalent to equations (11) and (12). This merely leaves the third equation to be solved. However, the freedom to choose the value of h can be substituted for the freedom to choose ν since, for every positive value of ν, there will exist a solution for Δx and Δp and consequently a value of h that solves equation (16). Thus any choice of ν can be regarded as an implicit choice of h.

In a similar way to the solution of the usual Levenberg—Marquardt, this algorithm has some useful properties. As $\nu \to 0$, this restricted step method tends to the solution for the redundant problem (equation (8)) that minimizes the norm, $\|\Delta x\|$. Furthermore, as $\nu \to \infty$, the solution tends to a gradient-descent method constrained to the set of minimal least-squared estimates for p. This we have already stated is the two-step method described in the last section. We can see this by considering equation (12), remembering that, as $\nu \to \infty$, $(D^{\mathrm{T}}D + \nu I)^{-1} \sim O(1/\delta)$. This then becomes

$$P^{\mathrm{T}}P\Delta p = P^{\mathrm{T}}\varepsilon + O(1/\delta) \tag{17}$$

which is equivalent to the least-squared estimate for Δp and then Δx can be solved substituting this into equation (11). Here the term $(\varepsilon + P\Delta p)$ is the error for the new mapping function, defined by $p + \Delta p$.

Obviously this is not optimizing the nonlinear problem. For simple parameter estimation this was investigated in Davies (1992). However, this used a Newton-based algorithm and the nature of the constrained optimization required makes it impractical to use more stable algorithms (see Davies 1993c for further details). For this reason we advocate using the above method since it is easy to apply and can, at least in the linearized problem, be considered in some sense optimal.

26.5   A COMPARISON BETWEEN THE TWO APPROACHES

To evaluate what a good compromise might be, the Levenberg–Marquardt is an ideal algorithm since it allows us to consider the performance of a whole spectrum of methods from almost Newton to almost gradient-descent by varying the Levenberg–Marquardt parameter δ. The data for this comparison was a 1000-point trajectory taken from the Hénon map with parameters $a = 1.4$ and $b = 0.3$. The function approximation used was a set of Gaussian radial basis functions placed at the first ten points in the embedded data set, once the data had been rescaled to span $[-0.5, 0.5]$. The rescaling parameter in the Gaussian was set to 0.1. This function approximation was chosen since it produced a good fit, although no attempt was made to formally optimize it and the same function approximation was used in each case. Finally, the embedding dimension used was 2 and the delay used was 1.

In the test of the two-step method the mapping function was estimated using an ordinary least-squared estimate, as proposed by Broomhead and Lowe (1988) and was followed by a single step of the Levenberg–Marquardt algorithm described above. Both these stages were then repeated ten times, after which the distance E_{dist} between the cleaned trajectory and the original deterministic orbit was measured, where E_{dist} is defined as

$$E_{dist} = \sum_{i=1}^{n} (x - y)^2 \tag{18}$$

where x is the filtered trajectory and y is the original deterministic orbit. The sum of the dynamic error, E_{dyn}, was also measured:

$$E_{dyn} = \sum_{i=d+1}^{n} \varepsilon_i^2. \tag{19}$$

This was done for a variety of values of fixed δ until, as δ tended to zero the algorithm became unstable. The results of these tests are summarized in Figure 26.1. It is clear from these graphs that, in practice, some benefits can be gained by using the additional information available in the quadratic form (i.e. allowing δ to be less than 1). However, the algorithm loses stability as $\delta \to 0$ and a compromise is necessary to maximize the noise reduction. It should be noted that without the Levenberg–Marquardt algorithm this compromise would not be possible since, even when the dynamics were known, the gradient–descent method became unstable well before the step size got close to 1.0. Similarly it can be seen from the figure that a Newton-type approach would also be unstable.

Obviously this test is not conclusive since the results would probably be sensitive to the levels of noise present in the data as well as the dynamical system from which the data came. However, it will serve as a good benchmark against which the extended Levenberg– Marquardt can be compared. Furthermore, these results are significant in themselves since, if the type of function approximation is local (Farmer and Sidorowich 1987) there is no explicit parametrization for the function model and the extended method is not applicable.

We then repeated this test using the algorithm proposed above. Again E_{dist} and E_{dyn} were evaluated for a range of fixed values of δ and these are also given in Figure 26.1 for comparison. Initially when δ is large, as predicted above, the two methods appear synonymous. Then, as δ is reduced below 0.1, the extended Levenberg–Marquardt method performs better than the *ad hoc* approach of treating the function estimation and the noise reduction separately. Furthermore, this extended method can be seen to be stable for a far broader range of δ than could be achieved with the two-step method. This means that it can be used at far lower values of δ which allows us to exploit the improved performance associated with Newton-based algorithms. This can be seen from the plots of E_{dyn} where minimum error achieved by the two-step method was about 10^{-3} whereas the extended method at one point reduced the noise level down to about 10^{-17}.

Neither method reduces the measurement error below about 0.3. This is misleading and should not be judged as poor performance from the noise-reduction algorithms. We can see this by considering the distance between the cleaned orbit found for $\delta = 0.1$ with the extended method and the unfiltered data: $E_{dist} = 5.148$, in comparison with the distance from the initial deterministic orbit and the unfiltered data: $E_{dist} = 5.257$. It is clear that the algorithm has actually found a 'deterministic' orbit (although this orbit is not strictly deterministic we argue that we can ignore this since the dynamic error is only $\sim 10^{-5}$) that is *closer* to the noisy data than the initial orbit. This is therefore a statistical limitation of noise reduction since,

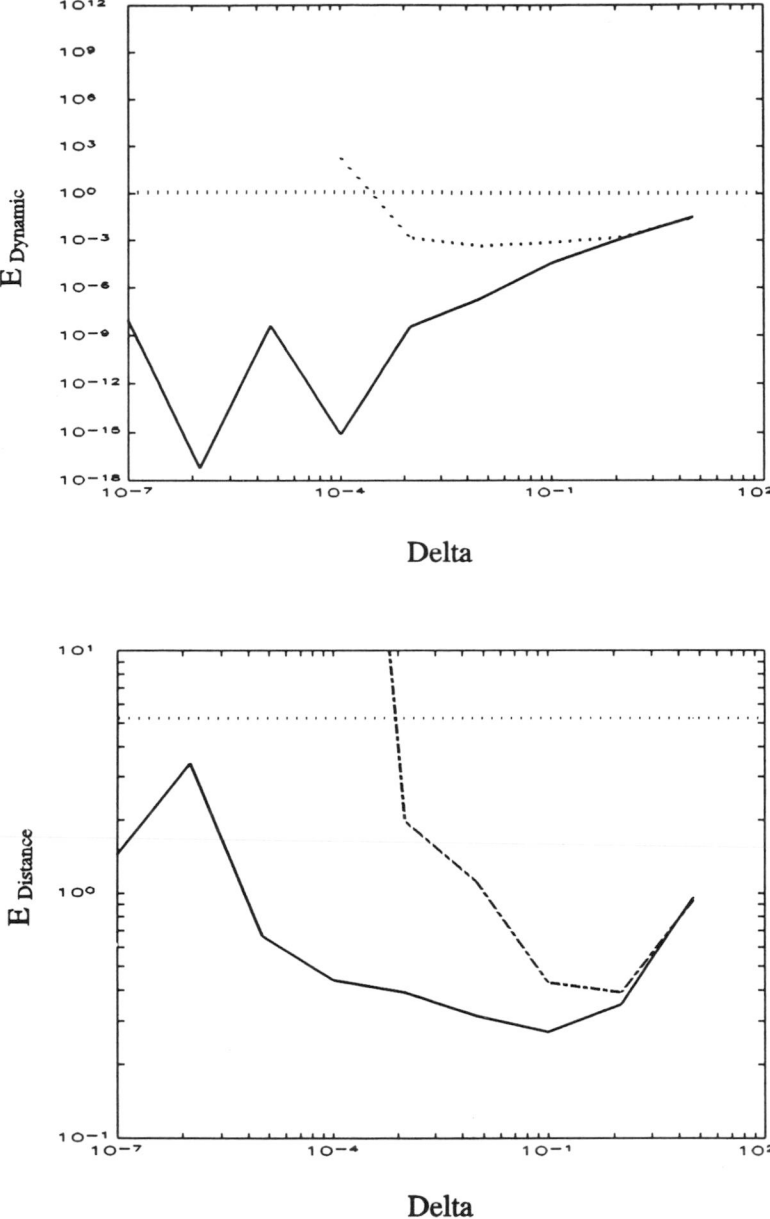

Figure 26.1 The top picture shows the dynamic error as a function of the Levenberg–Marquardt parameter, δ, for data from the Hénon map that has been filtered ten times with the two-step algorithm (dashed) and the extended algorithm (continuous). The bottom picture shows the measurement error for the same filtered data. The dotted line in both pictures is the level of error in the unfiltered data.

Figure 26.2 The top picture shows a 5000-point time series taken from the x coordinate of the Lorenz equations with parameter values $r = 28$, $b = 8/3$ and $\sigma = 10$. The bottom picture shows the projection of the reconstructed data using the first two singular directions.

without *a priori* knowledge, the unfiltered orbit would appear to be more likely to have come from the fiiltered data, as opposed to the original deterministic orbit.

26.6   A NUMERICAL EXAMPLE

Finally we apply the extended noise-reduction scheme to data from the standard chaotic attractor that occurs in the Lorenz equations at parameter values $r = 28$, $b = 8/3$ and $\sigma = 10$. The equations were integrated with a fixed-step-size Runge–Kutta algorithm with a step size

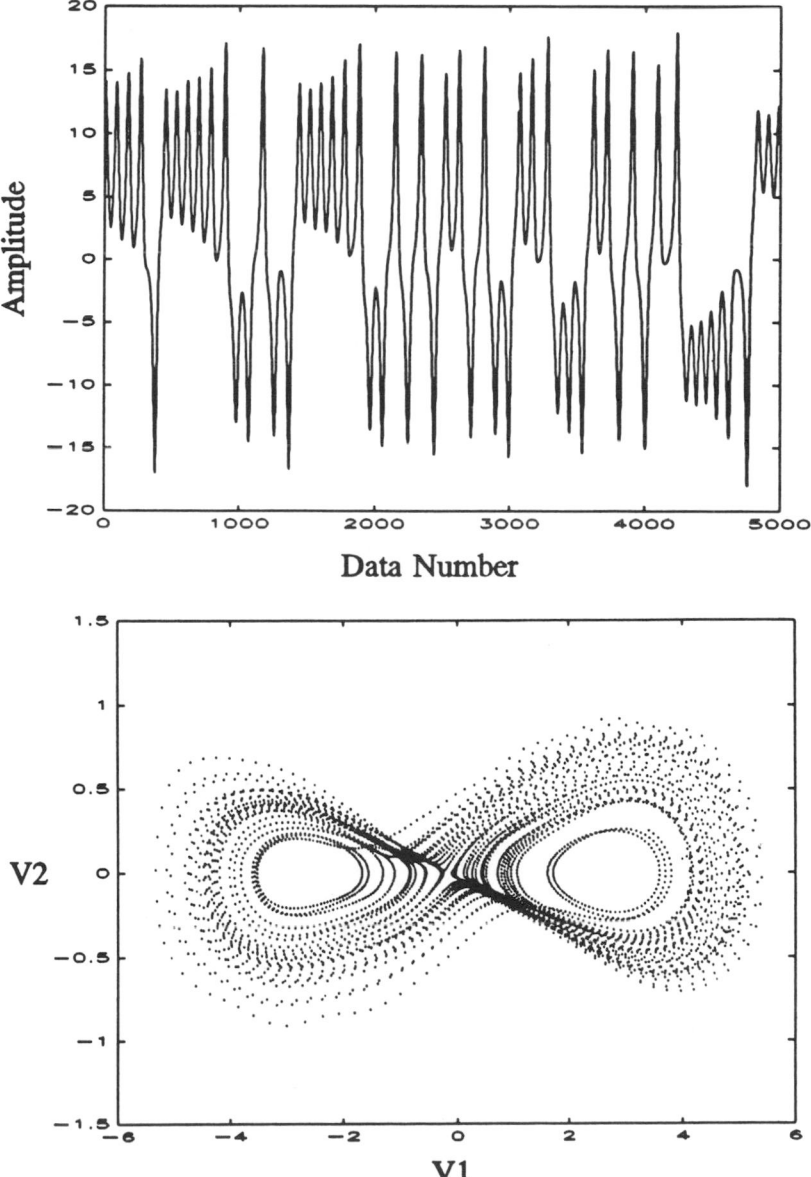

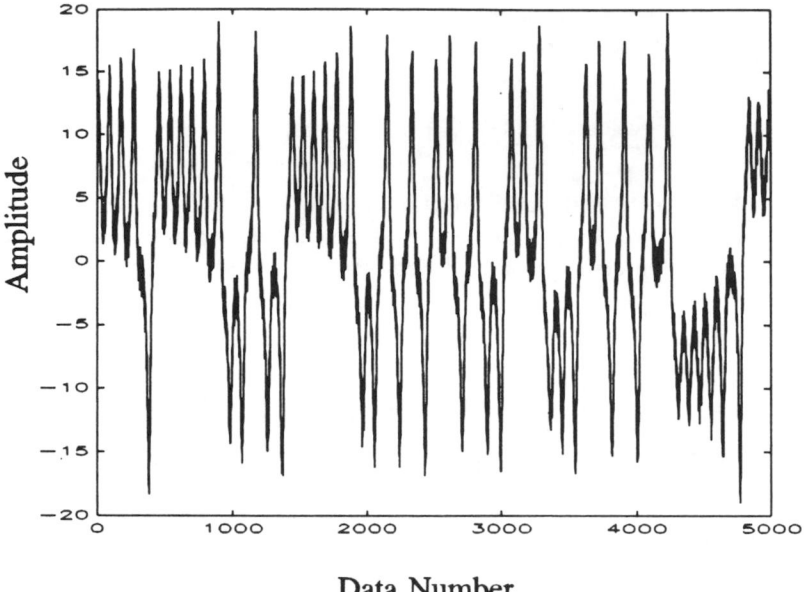

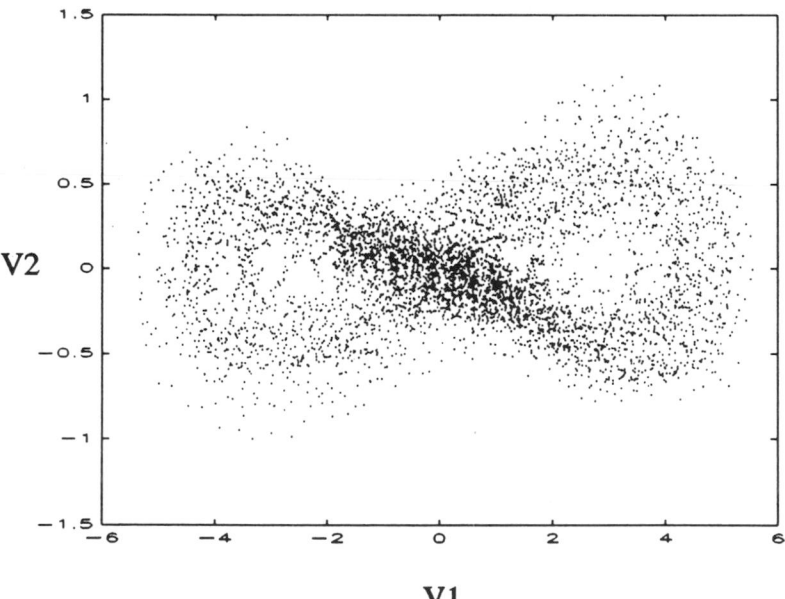

Figure 26.3   The top picture shows the data shown in Figure 27.2 with 10% noise added. The noise is i.i.d. with uniform distribution. The bottom picture shows the projection of the reconstructed data using the first two singular directions.

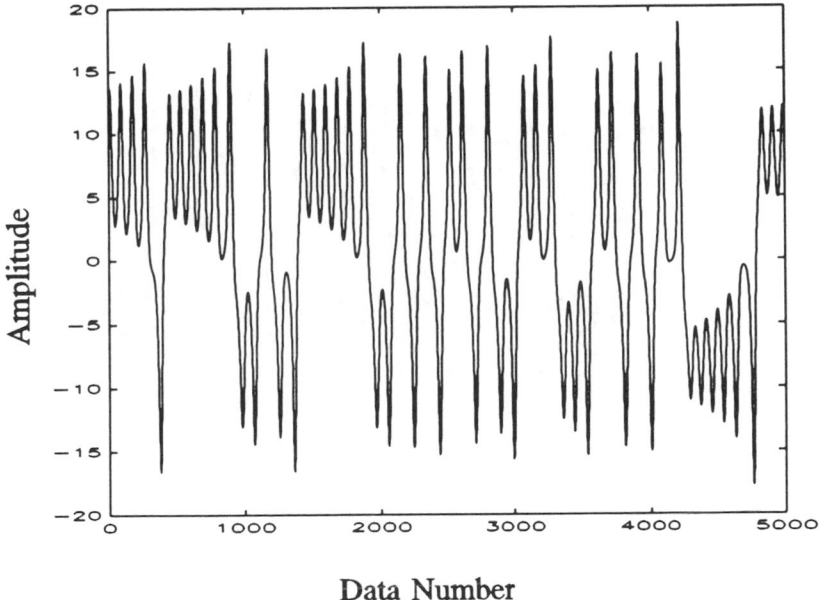

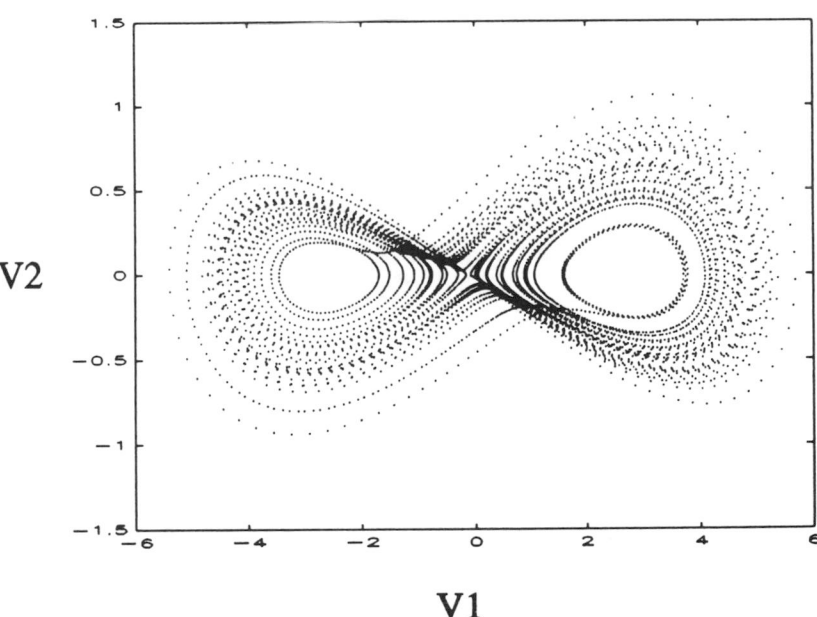

Figure 26.4 The top picture shows Lorenz data that was corrupted with 10% noise and then filtered ten times using the extended Levenberg–Marquardt algorithm. The bottom picture shows the same data reconstructed using the first two singular directions of the trajectory matrix.

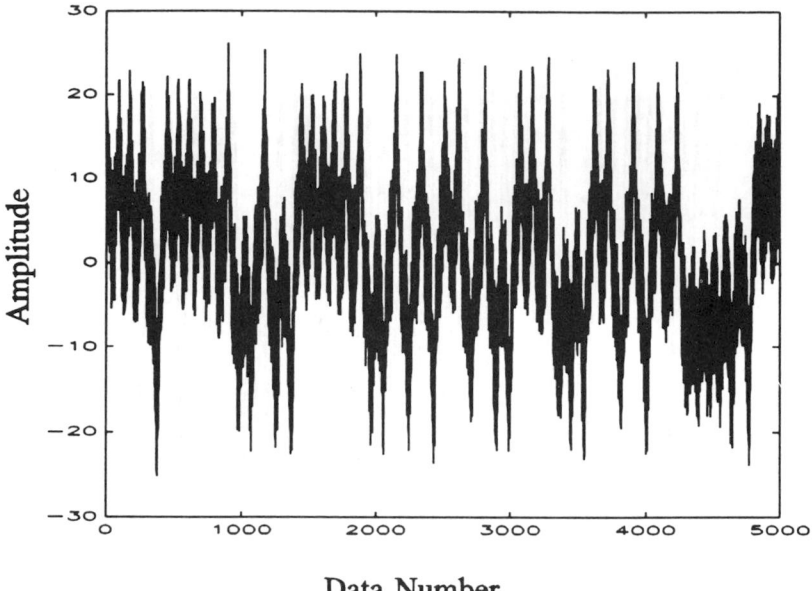

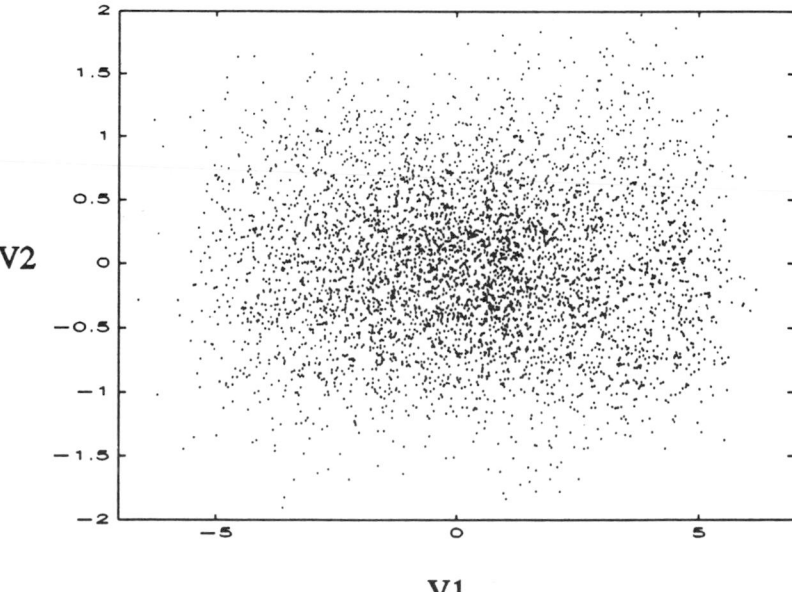

Figure 26.5   The top picture shows the data shown in Figure 26.2 with 50% noise added. The noise is i.i.d. with uniform distribution. The bottom picture shows the projection of the reconstructed data using the first two singular directions.

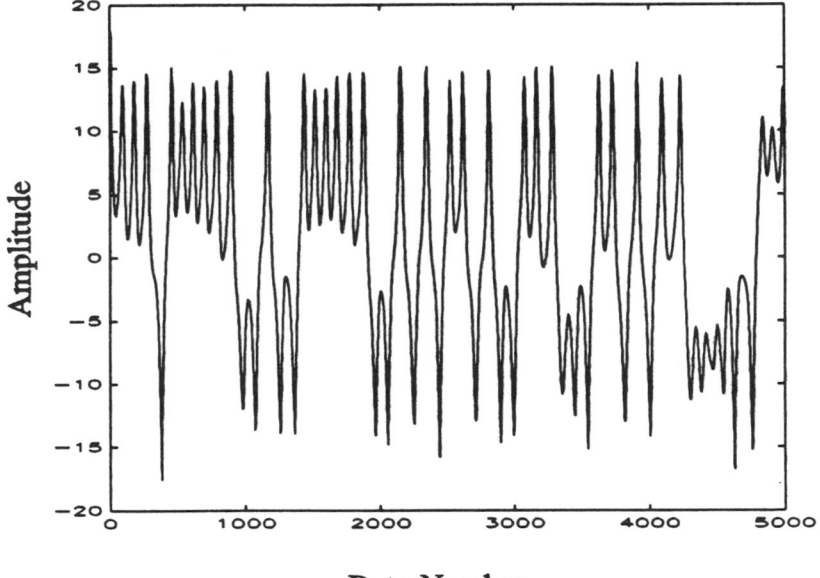

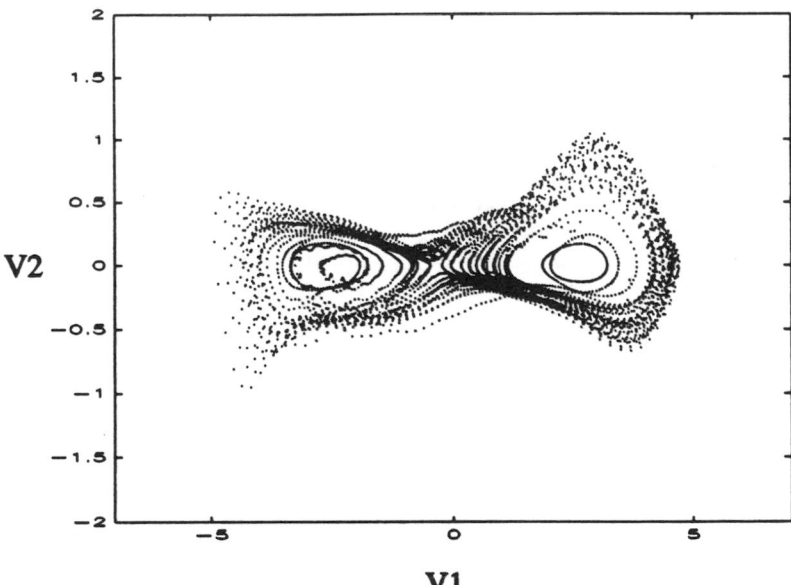

Figure 26.6 The top picture shows Lorenz data that was corrupted with 50% noise and then filtered ten times using the extended Levenberg–Marquardt algorithm. The bottom picture shows the same data reconstructed using the first two singular directions of the trajectory matrix.

of 0.01. Initially the equations were integrated for 1000 steps to allow the trajectory to settle down onto the attractor. We then took the next 5000 iterates as our data set. The time history of the data is shown in the top picture in Figure 26.2. The bottom picture shows the state-space reconstruction with the first two singular directions. In all the tests carried out here we used a three-dimensional singular system state space (see Broomhead and King 1987) in which to embed the data, and the delay window used was 10 iterates long.

Figure 26.3 shows the same data with some noise added. The noise was independent identically distributed (iid) with a uniform probability distribution. The noise level was set so that the total range of the noise was 10% of the total range of the data. It is clear from the figure that it is no longer possible to identify adjacent points in time from the embedded data.

We then applied our noise-reduction algorithm, using 100 Gaussian radial basis functions as our dynamics model. The centres were chosen as 100 evenly distributed points in the time series. Once the mapping function had initially been fitted we applied the noise-reduction process ten times. Figure 26.4 shows the results. The initial dynamic error prior to applying noise reduction was 4.88, and after noise reduction it had been reduced to 2.18×10^{-4}. This can be seen in the time-history plot where the data looks very similar to the original deterministic trajectory. Also, looking at the state space (bottom) we can see that it is now possible to identify adjacent points in time from the embedded data (i.e. determine the direction of the flow). The one noticeable difference between the embedded data in Figures 26.3 and 26.4 is that the outermost points have been contracted inwards due to the noise reduction. This can also be seen in the time history at around 4500 where the peak in the filtered data has been reduced. However, this is not surprising since these points are at the edge of the attractor and therefore there will be less points in their neighbourhood to define the flow accurately. Finally the original measurement error between the noisy time series and the deterministic orbit was 5.48×10^3. After applying noise reduction the error was reduced to 3.25×10^2.

In a second test we performed we added 50% noise (in the same sense as above) to the original deterministic orbit. In this case a great deal of the detail of the data has been lost (Figure 26.5). Although there is still evidence in the time series of the orbit switching between two states, the mechanism by which this happens has been lost. Furthermore, when the data is plotted in the reconstructed space (bottom picture) none of the structure is evident to the eye.

The filtered data for this test is shown in Figure 26.6. The state space shows that the attractor has been severely distorted in comparison with the bottom picture in Figure 26.2. However, the basic structure of the attractor has still been captured. The flow direction is still discernible for most points in the embedded space and the two foci of the attractor are clearly evident. The dynamic error, in this case, was reduced from 60.4 to 1.64×10^{-3}. Similarly, the measurement noise was reduced from 1.37×10^5 to 7.23×10^3, which is a similar performance to the results of the previous test. Thus, even when the noise level is high, the algorithm does an impressive job of filtering the data towards a deterministic orbit.

26.7   CONCLUSION

In this chapter a new approach to approximating the dynamics for the noise-reduction problem has been presented. Since the approximation of the mapping function can often be the limiting factor in applying noise reduction to experimental data it is important to concentrate on this aspect of the noise-reduction process. In the past, the general approach has been to fit a model to the data and then adjust the data to make it more consistent with that model. The consequence of doing this is that, since the approximation to the data is not likely to be the same as the fit that would have resulted from approximating the function to the clean data, any adjustment towards a trajectory on this model will throw away some information about the clean orbit. The worst methods from this point of view are therefore the Newton-based methods since they aim to arrive at a deterministic trajectory on the first iterate, where the deterministic trajectory is defined by the approximation function to the noisy data. Thus the next function approximation is going to be dominated by the fact that the new orbit is quite a good fit to the last approximation. On the other hand, improving the orbit only a small bit at a time allows the approximation to be optimized more while the noise reduction is being done, which produces a trade-off between speed and accuracy. The alternative approach presented here allows us to exploit some of the speed advantages associated with Newton-based algorithms by naturally merging the function approximation and noise reduction into the same optimization procedure.

Finally, it is interesting to interpret the noise reduction as fitting a deterministic model with measurement noise. As has already been stated above, the standard function approximation does not provide a purely deterministic description of the data since the errors manifest themselves in a dynamic form. Whereas there is a great deal of disagreement over whether estimates of invariant measures, such as fractal dimension, are good at answering the question: 'Is this data chaos or is this noise?', having a true orbit associated with the data enables a slightly weaker question to be answered: 'What is the possibility that this data came from a completely deterministic system?'. Here the level of measurement noise compared with the size of the deterministic orbit that the noise-reduction algorithm produces can be used as a measure of the likelihood that this model is a good one.

REFERENCES

Bowen, R. (1970) Markov partitions for Axiom A diffeomorphisms, *Amer. J. Math.*, **92**, 725–747.

Broomhead, D. S. and King, G. P. (1987) Extracting qualitative dynamics from experimental data, *Physica*, **20D**, 217–236.

Broomhead, D. S. and Lowe, D. (1988) Multivariate functional interpolation and adaptive networks, *Complex Systems*, **2**, 321–355.

Cawley, R. and Hsu, G.-H. (1992) SNR performance of a noise reduction algorithm applied to coarsely sampled chaotic data, *Phys. Lett.*, **166A**, 18–196.

Davies, M. (1992) An iterated function approximation in shadowing time series, *Phys. Lett.*, **169A**, 251–258.

Davies, M. (1993a) Noise reduction by gradient descent, *J. of Bifurcation and Chaos*, **3**, 133–119.

Davies, M. (1993b) Noise reduction schemes for chaotic time series, to appear in *Physica D*.

Davies, M. (1993c) Noise reduction in nonlinear time series analysis, *PhD Thesis*, University of London.

Farmer, J. D. and Sidorowich, J. (1987) Predicting chaotic time series, *Phys. Rev. Lett.*, **59** (8), 845–848.

Farmer, J. D. and Sidorowich, J. J. (1991) Optimal shadowing and noise reduction, *Physica*, **47D**, 373–392.

Fletcher, R. (1980) *Practical Methods of Optimisation, Vol 1. Unconstrained Optimisation*, Chichester, Wiley.

Hammel, S. M. (1990) A noise reduction method for chaotic systems, *Phys. Lett.*, **148A**, (8,9), 421–428.

Kostelich, E. J. and Yorke, J. A. (1990) Noise reduction: finding the simplest dynamical system consistent with the data, *Physica*, **41D**, 183–196.

Sauer, T. (1992) A noise reduction method for signals from nonlinear systems, *Physica*, **58D**, 193–202.

Schreiber, T. and Grassberger, P. (1991) A simple noise reduction method for real data, *Phys. Lett.*, **160A** (5), 411–418.

Stark, J. and Arumugam, B. (1992) Extracting slowly varying signals from a chaotic background, *Int. J. of Bifurcation and Chaos*, **2** (2), 413–420.

M. E. Davies and J. Stark, *Centre for Nonlinear Dynamics and Its Applications, University College London, Gower Street, London WC1E 6BT, UK*

27 STABILITY NUMBERS FOR NONLINEAR SYSTEMS

W. Schiehlen

The definition of Lyapunov stability is used for the introduction of stability numbers for nonlinear systems. A norm of the initial conditions $\| x_0 \|$ and a norm of the system response $\| x(t) \|$ are related to each other, resulting in stability numbers depending on the initial conditions. This concept presents some information on the global behaviour of the system including all types of solutions from limit cycles to strange attractors. The stability numbers characterize that part of the state space in which the motion occurs.

27.1   INTRODUCTION

The stability of motion in the sense of Lyapunov characterizes the qualitative behaviour of the equilibrium position of linear and nonlinear mechanical systems (Lyapunov 1907, 1966). The information on the stability of a system is the most fundamental one from both a theoretical and an engineering point of view. Unstable systems are not acceptable for engineering applications. Therefore, stability is a necessary requirement. For a well-designed engineering system the dynamical performance has to be evaluated, too. One approach is to provide quantitative stability information on the global dynamical behaviour of the system under consideration.

For linear systems the degree of stability or the degree of damping provides useful information on the dynamical behaviour of the system. These degrees are based on the eigenvalue distribution of the system considered (see e.g. Müller and Schiehlen 1985). Furthermore, the absolute-value criterion (absolute error − AE) is used to characterize the maximum amplitudes of some or all state variables. The absolute-value criterion depends on the initial conditions and the chosen state variables.

For nonlinear systems eigenvalues do not exist. Therefore, an extension of the degree of stability or the degree of damping is not possible. However, the absolute-value criterion may also be used for nonlinear systems. The stability numbers defined consider especially the influence of the initial conditions and include all state variables.

Nonlinearity and Chaos in Engineering Dynamics
Edited by J. M. T. Thompson and S. R. Bishop,   © 1994 John Wiley & Sons Ltd

The chapter is organized as follows. The equations of motion of nonlinear multi-body systems are represented in the canonical form or state space form, respectively. The stability definition of Lyapunov is extended to stability numbers of nonlinear dynamical systems. The application of this approach is demonstrated for the single and double pendulums as well as for the Van der Pol equation.

27.2   EQUATIONS OF MOTION

Multi-body systems are mechanical systems consisting of rigid bodies, constraint elements like bearings and joints, and coupling elements like springs, dampers or controlled actuators. For holonomic constraints and proportional-differential forces the equations of the motion read as

$$M(y, t)\ddot{y} + k(y, \dot{y}, t) = q(y, \dot{y}, t), \tag{1}$$

where y is the $f \times 1$ vector of generalized coordinates, $M(y, t)$ is the symmetric $f \times f$ inertia matrix, and $k(y, \dot{y}, t)$ and $q(y, \dot{y}, t)$ represent $f \times 1$ vectors of generalized Coriolis and applied forces, respectively.

In the more general case, non-holonomic constraints and proportional-integral forces result in an extended set of first-order equations of motion:

$$M(y, z, t)\dot{z} + k(y, z, t) = q(y, z, w, t),$$
$$\dot{y} = \dot{y}(y, z, t), \quad \dot{w} = w(y, z, t), \tag{2}$$

where z is a $q \times 1$ vector of generalized velocities and w is a $p \times 1$ vector representing the eigendynamics of the coupling elements and, in particular, the dynamical behaviour of the actuators. The equations (2) are in the literature sometimes known as Kane's equations. For more details see Schiehlen (1984, 1986).

In addition to the mechanical representation, (1) and (2), of a multi-body system, there exists also the possibility of using the more general representation of dynamical systems' in the state space, i.e.

$$\dot{x} = f(x, t), \tag{3}$$

where x means the $n \times 1$ state vector composed of generalized coordinates and velocities, and t is the time. In autonomous systems the $n \times 1$ vector function f does not depend on time.

The equations of motion presented may be automatically generated by the formalism NEWEUL described in Schiehlen (1990). NEWEUL is a software package for the dynamic analysis of mechanical systems with the multi-body system method. It deals with the computation of the symbolic equations of motion.

27.3   STABILITY NUMBERS

The dynamical equations of multi-body systems describing autonomous nonlinear oscillations are represented in a canonical form as

$$\dot{x} = f(x), \quad x(t_0) = x_0, \tag{4}$$

following on from (3). Here x is the $n \times 1$ state vector and f is an $n \times 1$ vector function not explicitly depending on time t. At initial time t_0 the initial state x_0 is given. It is assumed that $f(0, t) = 0$ represents an equilibrium position $x = 0$. Due to the nonlinearity of the system, there may exist additional equilibrium positions $x = x^*$.

The stability in the sense of Lyapunov characterizes the qualitative behaviour of the equilibrium position $x = 0$ of the dynamical system (4). For the stability definition the absolute-value norm of a vector is used.

The time-variant norm of the $n \times 1$ state vector $x(t) = [x_1(t), \ldots, x_n(t)]^T, t \in [t_0, \infty)$ is defined by

$$\|x(t)\| := \max_{1 \leqslant i \leqslant n} |x_i(t)|. \tag{5}$$

The time-interval norm reads as

$$\|x(t)\|_T := \max_{t \in [t_0, T]} \|x(t)\| \tag{6}$$

where time T may approach infinity, $T \to \infty$. These definitions are also valid for matrices.

The dynamical system (4) is called *stable* (in the sense of Lyapunov) if for every positive $\varepsilon > 0$ there exists a positive number $\delta = \delta(\varepsilon) > 0$ such that for all initial conditions bounded by

$$\|x_0\| < \delta = \delta(\varepsilon) \tag{7}$$

the corresponding trajectories $x(t)$ remain bounded for all t:

$$\|x(t)\| < \varepsilon. \tag{8}$$

The dynamical system (4) is asymptotically stable if it is stable and for all bounded initial conditions (7) the corresponding trajectory tends to zero:

$$\lim_{t \to \infty} \|x(t)\| = 0. \tag{9}$$

If the dynamical system (4) is not stable it is said be *unstable*.

There is a large literature on stability problems, and quite a number of textbooks, e.g. Hahn (1967) and Müller (1977). However, the stability analysis provides only a qualitative answer. For engineering applications some quantitative global information on the dynamical behaviour is of interest.

Based on the stability definitions in the sense of Lyapunov the following stability numbers are defined, see Hu (1992) and Schiehlen (1993).

The local stability number $S1$ given by

$$S1(x_0, t_0) = \begin{cases} \dfrac{\|x_0\|}{\|x(t)\|_\infty} & \text{for } x_0 \neq 0, \\ 1 & \text{for } x_0 = 0, \end{cases} \tag{10}$$

characterizes the ratio between a given initial state x_0 and the corresponding maximal displacement of the trajectory. The number $S1$ depends on x_0 and t_0.

The global stability number $S2$ defined by

$$S2(r, t_0) := \min_{x_0 \in \{x : \|x\| = r\}} S1(x_0, t_0) \tag{11}$$

is defined for a subspace of the initial-conditions state space. The number $S2$ characterizes the maximal displacement of all trajectories starting out of the initial-conditions subspace which is by definition a hypercube with respect to the equilibrium point $x = 0$. By definition it yields

$$0 \leqslant S1 \leqslant 1, \quad 0 \leqslant S2 \leqslant 1. \tag{12}$$

In a numerical analysis the integration interval is limited. Then, the numbers (10) and (11) have to be replaced by

$$S1_T(x, t_0) := \frac{\|x_0\|}{\|x(t)\|_T}, \tag{13}$$

$$S2_T(r, t_0) := \min_{x_0 \in \{x : \|x\| = r\}} S1_T(x_0, t_0). \tag{14}$$

For autonomous systems the initial time can be chosen as $t_0 = 0$ without loss of generality.

There is s direct relation between the stability and the above-defined stability numbers. For an unstable system one may choose a series of initial conditions satisfying $\|x_{0_1}\| > \|x_{0_2}\| > \cdots > \|x_{0_n}\|$ Then it yields

$$\lim_{n \to \infty} \|x_{0_n}\| = 0 \quad \text{and} \quad \lim_{n \to \infty} S1(x_{0_n}, t_0) = 0. \tag{15}$$

On the other hand, if the stability number $S1$ is limited, then

$$\|x(t)\|_\infty \leqslant s\|x_0\|, \text{ where } s \text{ is some positive number} \tag{16}$$

Usually the components of the state vector, $x_1(t), \ldots, x_n(t)$, have different units. For the application of the stability numbers it is necessary that all components have the same unit. This can be achieved by standardizing operations.

The stability numbers defined characterize dynamical systems of arbitrary dimension geometrically by a scalar number. Therefore, they are especially well suited for multi-body system analysis. More sophisticated geometric methods of nonlinear dynamics, like Poincaré mapping or cell mapping respectively, are usually restricted to two or at most three dimensions.

The above-defined stability numbers may be applied to linear systems, too (see Hu 1992).

27.4   APPLICATIONS TO ENGINEERING DYNAMICS

27.4.1   Single pendulum

As a first example, the single pendulum will be used, since, in addition to the numerical solution, an analytical solution is available. The equation of motion reads without units as

$$\varphi'' + \sin \varphi = 0. \tag{17}$$

After some calculation, the following results are obtained:
for $\varphi_0'^2 \leqslant 2(1 + \cos\varphi_0)$,

$$S1(x_0) = \frac{\max(|\varphi_0|, |\varphi_0'|)}{\max\{\sqrt{\varphi_0'^2 + 2(1 - \cos\varphi_0)}, \arccos(\cos\varphi_0 - \varphi_0'^2/2)\}},$$ (18)

for $\varphi_0'^2 > 2(1 + \cos\varphi_0)$,

$$S1(x_0) = 0,$$ (19)

for $0 < r \leqslant r^*$,

$$S2(r) = \frac{r}{\arccos(\cos r - r^2/2)},$$ (20)

for $r > r^*$

$$S2(r) = 0,$$ (21)

where $r^* = 1.478$ follows from the equation $r^2 = 2(1 + \cos r)$. Figure 27.1, presenting the stability number S2, shows clearly the instability of the equilibrium position $x = 0$ for a sufficiently large initial angular velocity. The numerical simulation yields the same results as shown in the figure.

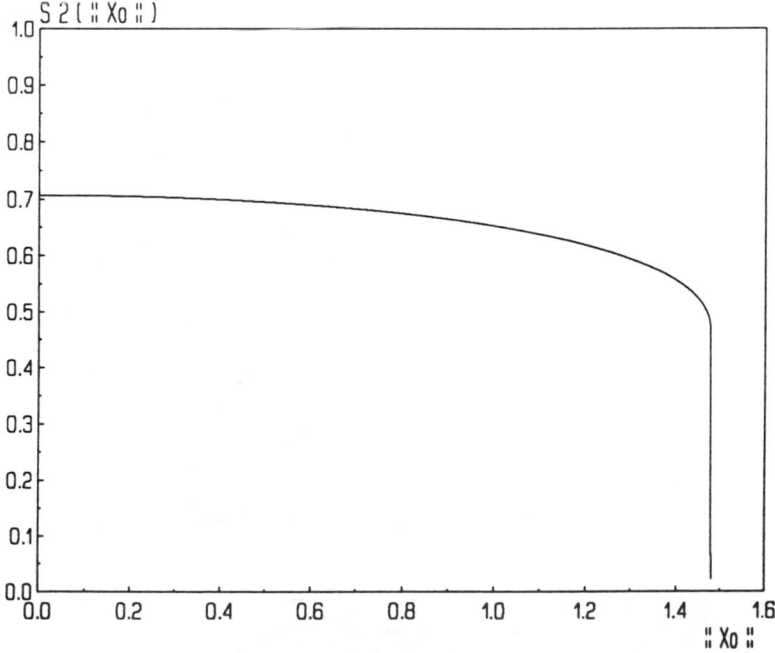

Figure 27.1 Stability number S2 of the single pendulum.

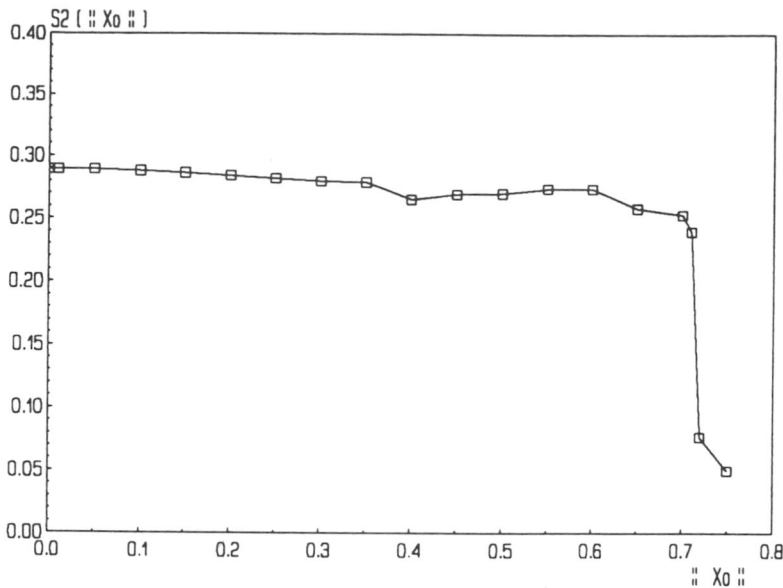

Figure 27.2 Stability number $S2$ of the double pendulum for $T = 100$.

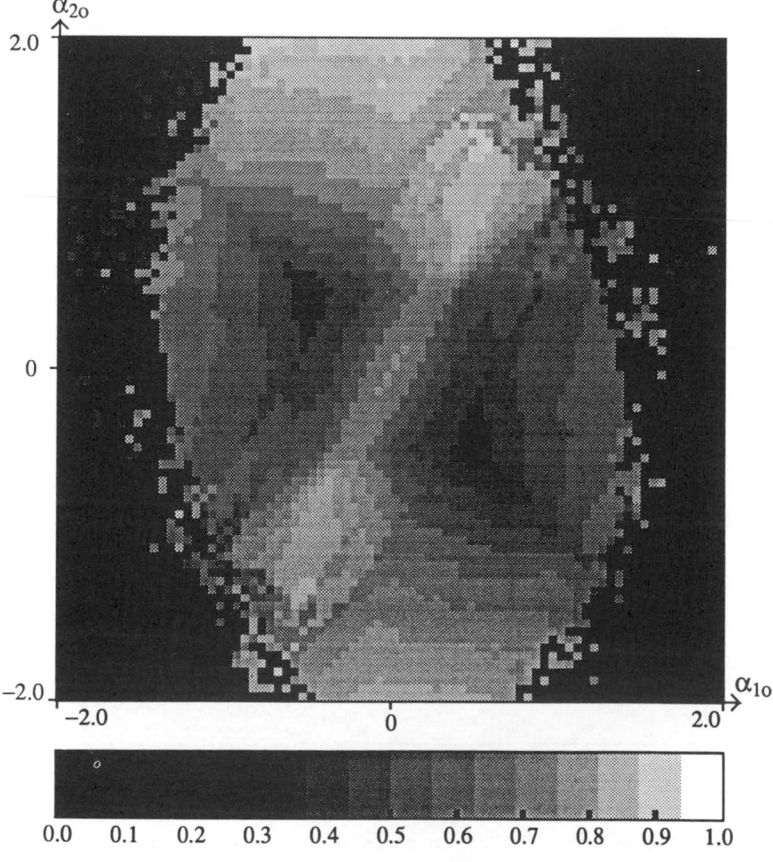

Figure 27.3 Stability number $S1_T$ of the double pendulum for $T = 100$, $\alpha_1' = \alpha_2' = 0.5$.

27.4.2   The double pendulum

The equations of the double pendulum in the standardized form result in

$$
\begin{bmatrix} (1+\lambda_1)\lambda_2^2 & \lambda_2\cos(\alpha_1-\alpha_2) \\ \lambda_2\cos(\alpha_1-\alpha_2) & 1 \end{bmatrix}\begin{bmatrix} \alpha_1'' \\ \alpha_2'' \end{bmatrix} + \begin{bmatrix} \sin(\alpha_1-\alpha_2)\lambda_2^2\alpha_2'^2 \\ -\sin(\alpha_1-\alpha_2)\lambda_1^2\alpha_1'^2 \end{bmatrix} = \begin{bmatrix} (1+\lambda_1)\lambda_2\sin\alpha_1 \\ -\sin\alpha_2 \end{bmatrix},
$$

$$(22)$$

and may be rewritten in state-space representation, too. Then, the state vector reads as $x = [\alpha_1, \alpha_2, \alpha_1', \alpha_2']^T$. A thorough numerical analysis was performed by Hu (1992). For the graphical representation the software for cell mapping methods developed by Schaub (1990) was extensively used.

The first step of the analysis requires the integration of the equations of motion. Then, by variation of all initial conditions, the stability number $S2$ is obtained (see, Figure 27.2). A comparison between Figures 27.1 and 27.2 shows that the double pendulum is much more sensitive than the single pendulum to initial disturbances in the displacement.

It is interesting to analyse the double pendulum also for larger initial displacement (see Figure 27.3). It turns out that there are two clearly separated regions. The left- and right-hand dark regions represent chaotic behaviour. There is a very high sensitivity to the initial conditions. From this point of view, the boundary $\|x_0\| = 0.72$ in Figure 27.2 is due to chaotic behaviour and not to simple instability. But, from an engineering point of view, neither chaos nor instability is acceptable.

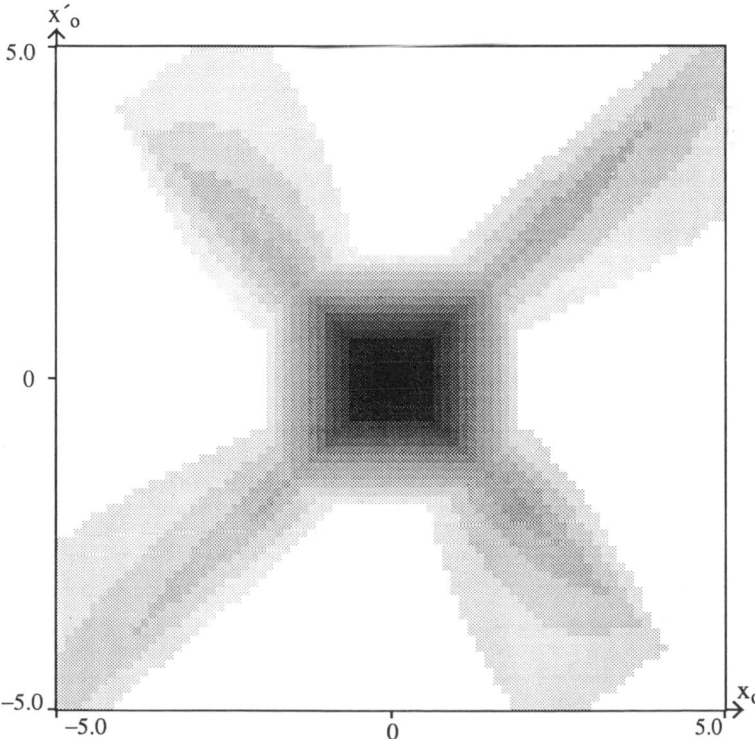

Figure 27.4   Stability number $S1_T$ of the van der Pol equation for $\mu = 0.1$ and $T = 200$.

27.4.3 Van der Pol equation

The stability numbers are also useful for nonlinear dissipative systems, an example of which is the Van der Pol equation:

$$\begin{bmatrix} x' \\ x'' \end{bmatrix} = \begin{bmatrix} x' \\ -\mu(x^2 - 1) - x \end{bmatrix}.$$

For $\mu > 0$, the equilibrium position $x = [x, x']^T = 0$ is unstable; on the other hand, there is a stable limit cycle. The stability number $S1_T$ shows clearly both phenomena, the instability in the origin of the state plane and the limit cycle (Figure 27.4). The grey chart is the same as in Figure 27.3.

27.5   CONCLUSION

Multi-body systems result in highly nonlinear equations of motion typical of engineering dynamics. In engineering applications only bounded motions are acceptable. The approach of stability numbers allows the systematic computation of a basin of bounded motions for systems of arbitrary dimension. Within that basin, the sensitivity of the systems considered with respect to initial conditions or parameters, respectively, may be investigated by other methods in more detail. Further, it has to be pointed out that the stability numbers are not related to the frequency of the system response, they consider the absolute value of all state variables often important in the engineering design of mechanical systems.

REFERENCECES

Hahn, W. (1967) *Stability of Motion*, Berlin, Springer.

Hu, B. (1992) Stabilitätsmaß nichtlinearer Systeme, Studienarbeit STUD-93, Institut B für Mechanik, Stuttgart.

Lyapunov, A. M. (1907) Probléme générale de la stabilité de mouvement, *Ann. Fac. Sci. Toulouse*, **9**, 203–474. (French translation of the 1893 published original work in Russian.)

Lyapunov, A. M. (1966) *Stability of Motion*, London, Academic Press.

Müller, P. C. (1977) *Stabilität und Matrizen*, Berlin, Springer.

Müller, P. C. and Schiehlen, W. (1985) *Linear Vibration*, Dordrecht, Kluwer.

Schaub, S. (1990) Interpolationsverfahren für Zellabbildungsmethoden, Diplomarbeit DIPL-30, Institut B für Mechanik, Stuttgart.

Schiehlen, W. (1984) Computer generation of equations of motion In: *Computer Aided Analysis of Optimization of Mechanical System Dynamics*, E. J. Haug (ed.), Berlin, Springer, pp. 183–216.

Schiehlen, W. (1986) *Technische Dynamik*, Stuttgart, Teubner.

Schiehlen, W. (ed.) (1990) *Multibody Systems Handbook*, Berlin, Springer.

Schiehlen, W. (1993) Nonlinear Oscillations in Multibody Systems – Modeling and Stability Assessment. In: Proc. 1st European Nonlinear Oscillations Conference, (E. Kreuzer and G. Schmidt, eds), Berlin, Akademie-Verlag, pp. 85–106.

W. Schiehlen, *Institut B für Mechanik, Universitat Stuttgart, D-70550 Stuttgart, Germany.*

28 ON ANALYTICAL ESTIMATES OF SAFE IMPULSIVE VELOCITY IN THE DRIVEN ESCAPE OSCILLATOR

W. Szemplińska-Stupnicka and J. Rudowski

Using the escape of a driven oscillator from a cubic potential well as an archetypal example, the problem of analytical approximate prediction of the system behaviour in a noisy environment is explored. Attention is focused on the criterion of the safe perturbation of a steady-state attractor, which employs the concept of safe impulsive velocity. The approximate studies of the stability of a periodic attractor are extended to consideration of finite disturbances and transient motion in the complete, nonlinear variational equation, and simple closed-form criteria are obtained. Comparison with computer simulations reveal that the estimates might be accepted and useful in analysis of engineering systems.

28.1   INTRODUCTION

Simple models of driven oscillators are widely used in engineering dynamics, in particular, when strongly nonlinear and chaotic phenomena are the point of interest. Within the class of nonlinear oscillatory systems, those with softening and escape characteristic deserve special attention and have been the subject of recent extensive studies. In this chapter we focus on the question of how to estimate and predict the escape phenomenon in the system parameter space and to interpret a mechanism of transition to the unbounded solution in relation to approximate solution of the periodic in-well attractor. Approximate, analytical estimates of the system parameters critical values and localisation of the regions, where escape can be expected, might be relevant to engineering design and would reduce considerably final computational efforts.

As an illustrative example the nonlinear escape oscillator with quadratic nonlinearity is considered. The governing equation of motion is written in the standard form:

$$\ddot{x} + h\dot{x} + x - x^2 = -F \sin \omega t, \qquad (1)$$

Nonlinearity and Chaos in Engineering Dynamics
Edited by J. M. T. Thompson and S. R. Bishop,   © 1994 John Wiley & Sons Ltd

where h is the damping coefficient, assumed to be small, F denotes forcing parameter and ω is driving frequency. The system is relevant to many problems of physics and engineering and has received wide attention in recent literature (see Soliman and Thompson 1989, 1991, Szemplińska-Stupnicka 1992, and the references therein). In a series of pape-s by Thompson and his co-workers, the problem of escape from the potential well, $x = \mathcal{J}$, was thoroughly studied by the use of a topological approach and numerical computations of invariant manifolds related to the unstable hilltop saddle point. In particular, the basins of attrraction of the bounded and unbounded (escape) solution, the fractal character of the basin boundaries, and various engineering integrity measures related to the problem of quantifying the safe and the dangerous (leading to escape) system parameters were points of interest (Soliman and Thompson 1989, and references therein).

The topological approach based on massive numerical computations, which result in diagrams of fixed points, stable and unstable manifolds and fractal basin boundaries for prescribed system parameters, do not enable us to draw *a priori* conclusions on what happens under a change of system parameters. Moreover, an engineering interpretation of the diagrams is by no means an easy question. In contrast, the approximate analytical methods lead to closed form formulae and straightforward interpretation. The methods such as perturbation techniques, average and harmonic balance methods, and investigation of local stability of periodic solutions by linear variational equations, were widely and successfully used in engineering dynamics and served as a source of information on the regular, periodic or almost-periodic oscillations. It is known, however, that application of the approximate methods is, from the theoretical point of view, confined to the so-called weakly nonlinear systems, although no useful measures which might put a boundary between 'weakly' and 'strongly' nonlinear systems are available. Therefore, in seeking an answer to the question of the range of applicability, one may turn to the results of numerical experiments. It turns out that many computer simulations reveal that the low-order approximate solutions are relevant also to analysis in the system parameter regions where the system is by no means 'weakly nonlinear, i.e. even close to the boundary of irregular, chaotic or escape phenomena. This observation was exploited by the present authors, and in a series of papers the approximate theory was adapted to the problem of the bifurcations of periodic attractors, which leads to chaotic or escape motion (Szemplińska-Stupnicka 1990, 1992; Szemplińska-Stupnicka and Rudowski 1993).

In this chapter some efforts are made towards application of the analytical approximate methods to investigation of transient behaviour of the system (1) under impulsive loading, and to determination of the safe impulsive velocity. The concept of safe impulsive velocity

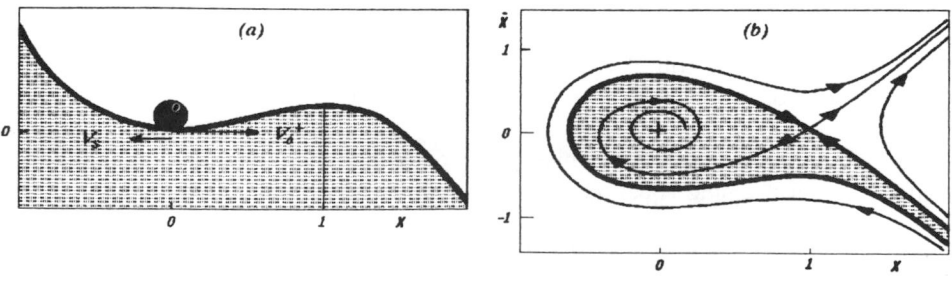

Figure 28.1 A model of equation (1) – a ball in a cubic potential well. (b) Basins of attraction of undriven system (1) with $F = 0$.

was introduced by Soliman and Thompson in a thoughtful study of various measures of engineering integrity of a periodic attractor – measures which quantify the danger of escape from the potential well. (Soliman and Thompson 1989; Thompson and Soliman 1990). It is assumed that the oscillator (1) exhibits forced, steady-state oscillations and, due to an impulsive loading, experiences a sudden change of velocity at $t = 0$. This induces free, transient motion,

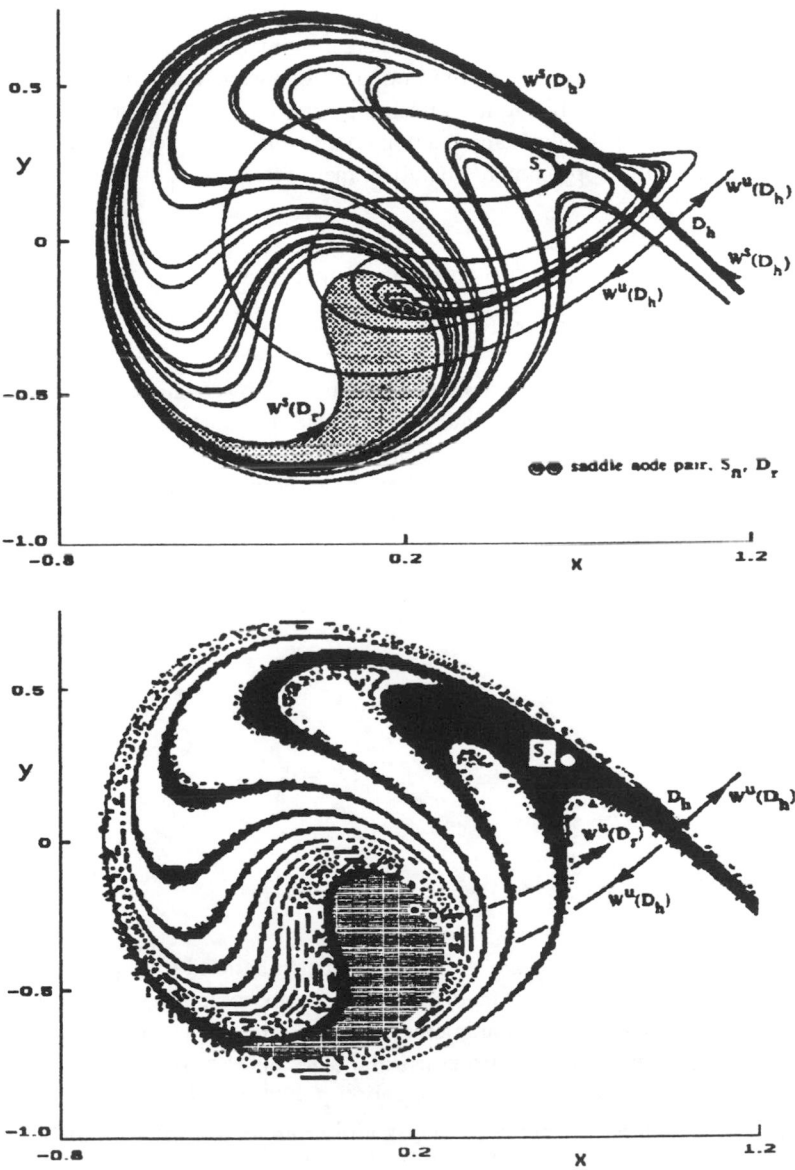

Figure 28.2   (a) Invariant manifolds and fixed points and (b) fractal basin boundaries at large F: basins of attraction of resonant attractor (black area), non-resonant attractor (grey area), and attractor in infinity (white area).

which might cause escape from the potential well. The largest value of the impulsive velocity that can be sustained without failure is defined as *safe impulsive velocity*.

For illustration a simple physical model of equation (1) is depicted in Figure 28.1a, and phase portrait and basins of attraction of bounded and escape solutions of autonomous system in Figure 28.1b. Next, Figure 28.2 aims to illustrate the complexity of invariant manifolds and basins of attraction in the complete system (1) under excitation higher than the Melnikov estimate of homoclinic tangling. The Poincaré sections presented, reproduced from Soliman and Thompson (1991) and Thompson and Soliman (1991), correspond to the driving phase $\phi = 0$, $t_0 = 0$ in equation (1). The safe impulsive velocity appears here as a distance from the resonant attractor to the boundary of escape basin in the $\dot{x} \equiv y$ direction. We see that the quantity is only slightly affected by the erosion of fractal basin boundaries. This observation signals that the escape phenomenon, which is caused by impulsive loading, might be captured by the approximate analytical methods.

28.2   ESCAPE DUE TO INFINITESIMAL AND LARGE DISTURBANCE

First we outline how the approximate analysis enable us to draw conclusions on local instability and the system parameters region where the stable periodic attractor does not exist (Szemplińska-Stupnicka 1992; Szemplińska-Stupnicka and Rudowski 1993.)

In the region of principal resonance the lowest-order approximate periodic solution which accounts for the effects of the quadratic term in equation (1) is assumed to be

$$x_s(t) = a_0 + a_1 \sin(\omega t + \varphi) + a_2 \cos(2\omega t + \varphi), \tag{2}$$

where

$$a_0 = \tfrac{1}{2}a_1^2, \quad a_2 = \tfrac{1}{6}a_1^2,$$

$$a_1 = \frac{F}{\sqrt{(\Omega^2(a_1) - \omega^2) - h^2\omega^2}}, \quad \Omega^2 = 1 - \kappa a^2, \quad \kappa = \tfrac{5}{6}, \tag{3}$$

$$\tan\varphi = \frac{-\omega h}{\Omega^2(a_1) - \omega^2}.$$

The solution can be determined by any of the classic perturbation techniques (e.g. asymptotic method, multiple scale method) provided that the zero approximate solution is assumed as $x^{(0)}(t) = a_1 \sin(\omega t + \varphi)$ and a small parameter is introduced into equation (1). Equations (3) enable us to calculate the resonance curves $a_1(\omega)$. Figure 28.3 shows that the curves possess the top resonant amplitude and two fold bifurcation points A and B until F reaches a critical value F_1 ($F_1 \approx h/\sqrt{4\kappa}$ at low damping). The two folds are associated with jump phenomena and hysteresis behaviour and they define limits of the first-order instability. It turns out that for $F > F_1$, when point B disappears, the resonant (upper) branch of the solution experiences another form of instability, namely the instability which manifests itself by period-doubling. To examine the period-doubling instability we add an infinitesimal disturbance δx to $x_s(t)$

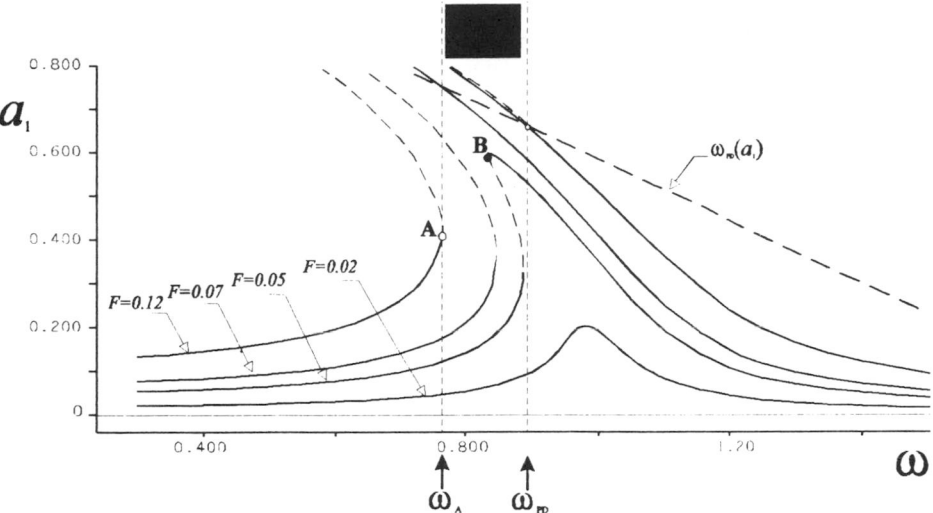

Figure 28.3 Resonance curves and unstable region $h = 0.1$ ——— and —·—·—·— period-doubling instability, ■-escape region at $F = 0.12$ (computer simulation results).

and consider the linear variational equation

$$\delta\ddot{x} + h\delta\dot{x} + \delta x[\lambda_0(a_1) + \lambda_{1S}(a_1)\sin(\omega t + \varphi) + \lambda_{2C}(a_1)\cos(2\omega t + 2\varphi)] = 0. \qquad (4)$$

Then, Floquet theory combined with an approximate technique allows us to seek an approximate solution for the period-double instability in the form

$$\delta x(t) = e^{\varepsilon t}\left[b_1\sin\left(\frac{\omega}{2}t + \delta_1\right) + b_3\sin\left(\frac{3}{2}\omega t + \delta_3\right)\right], \qquad (5)$$

where ε is real and positive, so that δx grows exponential with time. Now the instability of the resonant upper branch of the resonance curve is defined by the period-doubling stability limit $\omega_{PD}(a_1)$ and that of the non-resonant branch by the locus of point A, saddle–node bifurcation (Figure 29.3). Computer simulations reveal that, when $\omega_A < \omega_{PD}$, the first period-doubling is followed by a cascade of further bifurcations and that finally the stable periodic attractor ceases to exist, and the system tends to an 'attractor in infinity'. But we also learn from numerical experiments and from Feigenbaum theory that all the complex bifurcations occur in a very narrow band of the bifurcation parameter (here, the driving-frequency band), so that in the approximate analysis we assume the first period-doubling bifurcation as that which defines the boundary of existence of the stable periodic attractor. It follows that within the two stability limits at $\omega_A < \omega < \omega_{PD}$, the escape solution is expected (see Figure 28.3).

To consider the effects of large initial disturbance, we include a nonlinear term into equation (4). The linear variational equation is, therefore, transformed into

$$\ddot{z} + h\dot{z} + z(\lambda_0 + \lambda_{1S}\sin(\omega t + \varphi) + \lambda_{2C}\cos(2\omega t + 2\varphi)) - z^2 = 0, \qquad (6)$$

where z stands for the large disturbance of the steady-state solution (2).

To attack the problem of the safe (or critical) initial velocity, i.e. the maximal disturbance $\dot{z}(0)$ which does not result in unbounded solution, it is useful to consider first an autonomous system by setting $a_1 = 0$ in equation (6). This yields

$$\ddot{z} + h\dot{z} + z - z^2 = 0. \tag{7}$$

Thus the variational equation (6) is reduced to the original system (1) with $F = 0$ and the approximate solution takes the form (see equation (2)):

$$z(t) \equiv x(t)_{F=0} = B_0 + B_1 \sin(\Omega t + \theta) + B_2 \cos(2\Omega t + 2\theta),$$

$$B_0 = \tfrac{1}{2}B_1^2, \quad B_2 = \tfrac{1}{6}B_1^2, \quad \frac{dB_1}{dt} = \frac{-B_1 h}{2}, \quad \Omega^2(B_1) = 1 - \tfrac{5}{6}B_1^2. \tag{8}$$

where B_1 and θ denote arbitrary constants depending on initial conditions. The solution is, of course, legitimate for low values of the amplitude only, and the question of whether it can be used for large initial disturbances is, for the time being, an open one.

In the chapter we seek the answer in computer simulation results. We find out that the system response is close to that defined by equation (8) at initial disturbance that high, that the maximum displacement is very close to the hilltop critical value at $z = 1$ (Figure 28.4). Therefore, our analytical estimates of the critical impulsive velocity in autonomous system (7) is based on the approximate solution (8). The critical conditions are assumed to be:

$$z(0) = 0, \tag{9a}$$

$$\dot{z}(t_1) = 0, \tag{9b}$$

$$z(t_1) = 1, \tag{9c}$$

where t_1 denotes time for which $z(t)$ reaches the hilltop value with zero velocity. Conditions (9) at $h = 0$ yield a set of equations for B_1, θ and t_1:

$$\tfrac{1}{2}B_1^2 + B_1 \sin\theta + \tfrac{1}{6}B_1^2 \cos 2\theta = 0, \tag{10a}$$

$$B_1\Omega \cos(\Omega t_1 + \theta) - \tfrac{1}{3}B_1^2\Omega \sin(2\Omega t_1 + 2\theta) = 0, \tag{10b}$$

$$\tfrac{1}{2}B_1^2 + B_1 \sin(\Omega t_1 + \theta) + \tfrac{1}{6}B_1^2 \cos(2\Omega t_1 + 2\theta) = 1. \tag{10c}$$

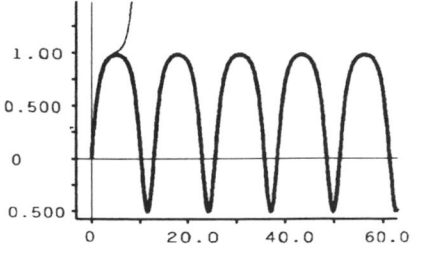
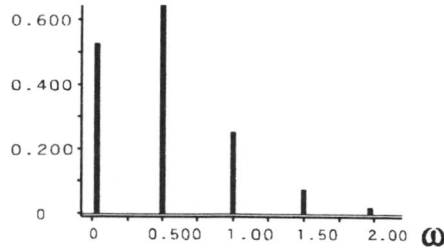

Figure 28.4　Time history and frequency spectrum in autonomous, undamped system with initial impulsive velocity 0.577; escape occurs at $V^+ = 0.578$.

Eventually, the sought critical value of the initial velocity is obtained as:

$$\dot{z}(0) \equiv V_0 = B_1\Omega(B_1)\cos\theta - \tfrac{1}{3}B_1^2\Omega(B_1)\sin 2\theta; \tag{11}$$

where B_1, θ satisfy equations (10). The critical velocity thus defined is equal 0.60 and the value appears to be very close to that obtained by computer simulations (Figures 28.4 and 28.6).

Next we proceed to calculation of the critical, escape initial velocity of the system (1) with $F \neq 0$, and we begin with low values of amplitudes a_1 of the steady-state oscillation, which corresponds to the non-resonant branch of the solution and low values of F (see equations (3) and Figure 28.3). Now the impulsive velocity (positive impact) is applied at the instant $t = 0$, when the system exhibits steady-state motion with velocity denoted as V_S, and, in the non-resonant case, the two components are almost out of phase (Figure 28.1). A simple physical reasoning makes us believe that now the system requires a larger impulse to leave the potential well, and that the critical impulsive velocity is close to

$$V_F^+ = V_0^+ + |V_S(0)|, \tag{12}$$

where V_0^+ stands for the escape velocity with $F = 0$, (assumed to be positive), and V_F^+ for the corresponding quantity for the system (1) at $F \neq 0$. With the initial velocity due to steady-state oscillations assumed as $V_S(0) \approx a_1\omega\cos\varphi$, equation (12) yields

$$V_F^+ = 0.60 + |a_1\omega\cos\varphi|. \tag{13}$$

Results of the simple calculations are plotted in the $V_F^+ - F$ plane for $\omega = 0.85$ (Figure 28.6). Computer simulation values of the critical impulsive velocity are based on Soliman and Thompson (1989) and were partly verified by the present authors. The line marked by squares denotes the critical V_F^+ for which escape occurs (i.e. the system reaches $x = 20$) within one cycle of forcing frequency. The curve marked by triangles corresponds to that for which the system leaves the potential well after 16 forcing cycles. Where the two values are different the region between them is shown shaded.

At this point we should remember that our approximate theoretical analysis corresponds to the former case, i.e. to escape occurring within the first cycle. This phenomenon is illustrated in Figure 28.5a, where the time history of the perturbed motion and its two components, steady state $x_s(t)$ and perturbation $z(t)$, are plotted. The initial velocity added to the steady-state oscillation is very close to the escape value. We see that the displacement takes a maximum value after about a quarter of the cycle and is about equal to 1. The figure illustrates also the opposite phase of the steady-state and disturbance oscillations, so that it becomes clear why the forced system requires a larger value of impulsive velocity to reach the hilltop than the autonomous one.

An approximate treatment of escape phenomena in the non-autonomous system exhibiting a steady state with large, resonant amplitude requires special attention. In this chapter we begin with an assumption that the disturbed solution is a superposition of the forced, steady states $x_s(t)$ and free oscillations of the homogeneous undamped system (equations (2) and (8)):

$$x(t) = x_s(t) + z(t) = \tfrac{1}{2}a_1^2 + a_1\sin(\omega t + \varphi) + \tfrac{1}{6}a_1^2\cos(2\omega t + 2\varphi) + \tfrac{1}{2}B_1^2 + B_1\sin(\Omega t + \theta)$$
$$+ \tfrac{1}{6}B_1^2\cos(2\Omega t + 2\theta). \tag{14}$$

Thus we ignore, for the time being, coupling effects that arise in the nonlinear system and

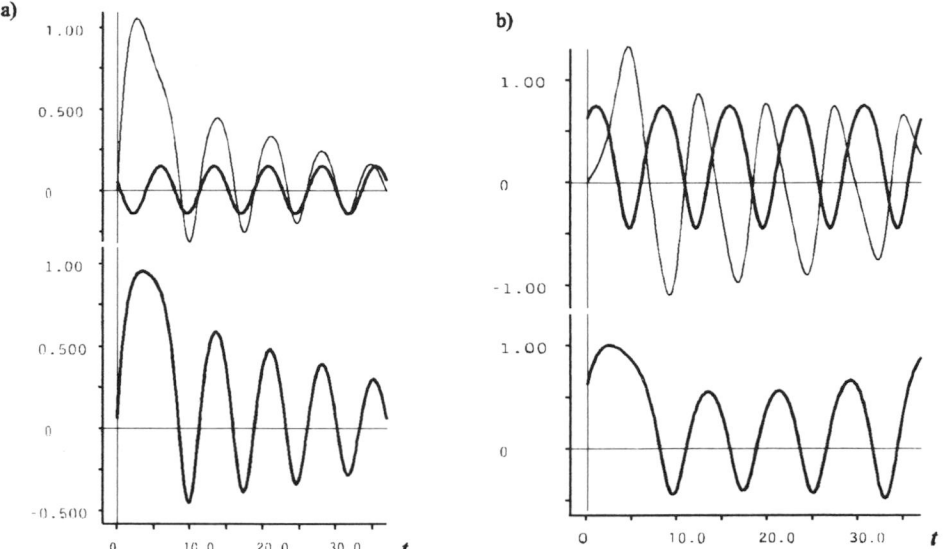

Figure 28.5   Time history of the steady-state component $x_s(t)$ (thick line), the perturbation $z(t)$ and (bottom) total response $x(t)$: (a) $F = 0.04$, $V_0^+ = 0.78$; (b) $F = 0.08$, $V_0^+ = 0.14$.

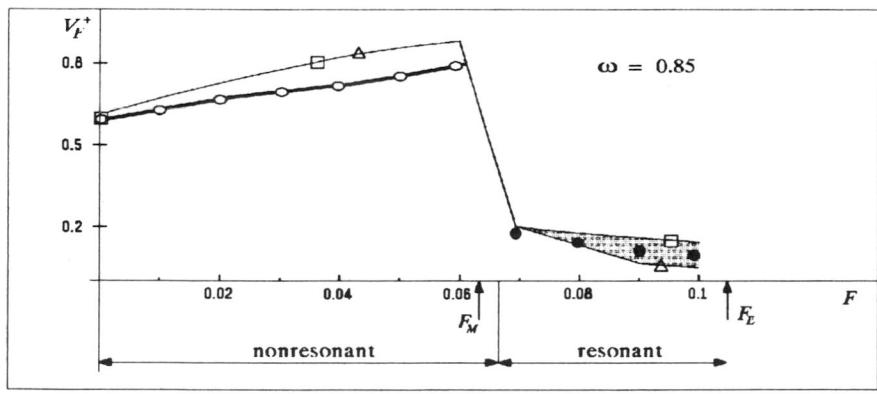

Figure 28.6   Critical impulsive velocity (positive impulse) plotted against F at $\omega = 0.85$, $h = 0.1$: simulation results, --◯--◯--◯-- theoretical results equation (13), --●--●--●-- equations (11), (15).

set the critical escape conditions as:

$$z(0) = 0, \tag{15a}$$

$$\dot{x}(t_1) \equiv \dot{x}_S(t_1) + \dot{z}(t_1) = 0, \tag{15b}$$

$$x(t_1) \equiv x_S(t_1) + z(t_1) = 1. \tag{15c}$$

The conditions lead to a set of equations for the unknowns B_1, θ and t_1, and finally for the critical velocity $\dot{z}(0) \equiv V_F^+$ analogous to equations (10) and (11).

In turns out that the crude approximate assumptions and the theoretical analysis reduced to the simple, algebraic calculations to give us results close to those obtained by high-accuracy computational techniques (Figure 28.6).

28.3   DISCUSSION

The case $\omega = 0.85$ considered is the most celebrated in recent literature (e.g. Soliman and Thompson 1989, 1991; Thompson and Soliman 1990). It is, indeed, of considerable interest, because it is close to the condition of optimal escape (under minimum F). Moreover, within the range of the forcing term. F from 0 to the escape value F_E, the system jumps from non-resonant to resonant motion at the saddle–node bifurcation point at $F \approx 0.07$, and, the value of F which corresponds to the jump phenomenon is very close to F_M, i.e. to the critical value defined by the Melnikov criterion for homoclinic tangling. As is known for $F > F_M$, basin boundaries of periodic and 'attractor in infinity' become fractal and this evidently brings a drop in the area of initial conditions, which lead to a bounded attractor (see Figure 28.2 and Soliman and Thompson 1989). Therefore, the sudden fall of the integrity measures which are based on areas of the corresponding basins of attraction, are evidently related to the Melnikov criterion. The sudden jump in the magnitude of critical (or safe) impulsive velocity close to $F \approx 0.07$, is however, a point to be discussed. Firstly, one can notice that the theoretical calculations have no relation to Melnikov criteria or with the fractal nature of basin boundaries. The sudden fall of the theoretical critical impulsive velocity close to $\omega = 0.7$ is caused exclusively by the jump of the forced oscillations amplitude and phase angle at the saddle–node bifurcation (point A on the resonant curve $a_1(\omega)$ (Figure 28.3). Secondly, the computer simulation results do not reveal dramatic, jump-like fall of V_F^+ close to the Melnikov criterion for other values of the driving frequency, the values at which steady-state oscillations do not pass the saddle–node bifurcation (see Figure 10 in Soliman and Thompson 1989).

The two observations lead us to the conclusion that the sudden discontinuous fall of $V_F^+(F)$ close to $F = 0.07$ at $\omega = 0.85$ is related to the discontinuous jump of the system steady-state oscillations from non-resonant to resonant branch at the saddle–node bifurcation point, rather than to passing the Melnikov critical value F_M.

The next point in the discussion is a comparison of the safe impulsive velocity criterion to another criterion, which is also based on the concept of safe initial disturbance. Here we refer to the concept of the safe-basin locus in parameter space corresponding to motions starting with zero displacement and zero velocity, $x(0) = \dot{x}(0) = 0$. This was developed by Thompson and Soliman, and we refer in particular to the computer simulation results in Thompson and Soliman (1990). The results are presented in the control space F–ω by showing regions characterized by different levels of the danger of escape (Figure 28.7). If the parameters F and ω are within the region denoted in white, the system escapes in less than 4 forcing cycles, in the grey area the system escapes between 4 and 16 forcing cycles, and in the black area there is no escape after 16 cycles. To find a relation between the two criteria: the safe impulsive velocity $V_F^+(F)$ and the safe-basin locus in the F–ω plane, we should first notice that the former involves initial conditions, which are different for different (F, ω) points, and the latter is based on constant, zero initial conditions. To compare the two concepts one might, for instance, plot curves of equal-value V_F^+ in the F–ω plane. Here we apply an *ad hoc* criterion of $\frac{1}{3}V_0^+$, that is, we single out the values of F and ω for which V_F^+ is equal to $\frac{1}{3}$ of the safe impulsive velocity for autonomous system V_0^+. At $\omega = 0.85$ the condition is satisfied

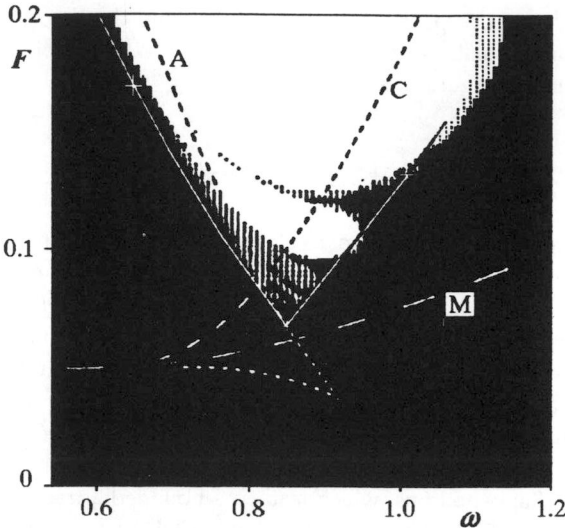

Figure 28.7 The safe basins locus in the $F–\omega$ control plane for starts from $x(0) = \dot{x}(0) = 0$ and $\frac{1}{3}V_0^+$ criterion (solid white line).

for $F > 0.07$ (after the saddle–node bifurcation point). The corresponding values at $\omega = 0.65$ and $\omega = 1.0$ are taken from computer simulation results in Soliman and Thompson (1989). It turns out that the three points defined by the $\frac{1}{3}V_0^+$ criterion lie very close to the boundary of the grey and white areas in the $F–\omega$ plane (white crosses in Figure 28.7). When the three points are connected by two segments of straight lines, they seem to form an envelope of the fractal boundary of the transient escape region.

28.4   CONCLUDING REMARKS

Analytical methods of the approximate theory of nonlinear oscillations have played a significant role in engineering applications and were commonly accepted as long as system considered might be assumed to be weakly nonlinear. Our study suggests that the macroscopic level of consideration of approximate analytical methods and simple physical arguments leads to a deeper understanding and to estimation of a structure for strongly nonlinear phenomena as well. This is of particular importance in the age of fast computers and automatic algorithms, because a clear physical interpretation of a huge amount of numerical results becomes an essential point in the theoretical analysis of engineering systems

REFERENCES

Soliman, M. S. and Thompson, J. M. T. (1989) Integrity measures quantifying the erosion of smooth and fractal basin of attraction, *J. Sound Vib.*, **135**, 453–475.

Soliman, M. S. and Thompson, J. M. T. (1991) Basin organisation prior to a tangled saddle-node bifurcation, *Int. J. Bif. and Chaos.*, **1**, 107–118.

Szemplińska-Stupnicka, W. (1990) *The Behaviour of Nonlinear Vibrating System*, Dodrecht, Kluwer.
Szemplińska-Stupnicka, W. (1992) Cross-well chaos and escape phenomena in driven oscillators, *Nonlinear Dynamics*, **3**, 225–243.
Szemplińska-Stupnicka, W. and Rudowski, J. (1993), Bifurcations phenomena in a nonlinear oscillator: approximate analytical studies versus computer simulation results, *Physica D*, **66**, 368–380.
Thompson, J. M. T. and Soliman, M. S. (1990) Fractal control boundaries of driven oscillators and their relevance to safe engineering design, *Proc. R. Soc. London A*, **428**, 1–13.
Thompson, J. M. T. and Soliman, M. S. (1991) Indeterminate jumps to resonance from a tangled saddle-node bifurcation, *Proc. R. Soc. London A*, **432**, 101–111.

W. Szemplińska-Stupnicka and J. Rudowski, *Institute of Fundamental Technological Research, Swietokzyska 21, Warsaw, Poland*

29 BIFURCATION AND CHAOS OF COUPLED SYSTEMS BY FAST INCREMENTAL HARMONIC BALANCING

S. Narayanan and P. Sekar

A numeric–analytic method of finding periodic solutions of a general nonlinear dynamical system is developed using the fast Fourier transform (FFT) algorithm. By incorporating the stability analysis, bifurcation points and type of bifurcation can be obtained. Systems with time-delay terms also can be treated with this method. A generalized Fourier method has been proposed for multiple excitation systems.

29.1   INTRODUCTION

A fast incremental harmonic balancing (FIHB) method of obtaining steady-state periodic solutions of multi-degree-of-freedom nonlinear dynamical systems is proposed. The FFT technique is used to advantage in conjunction with harmonic balancing, making the method general and capable of treating analytical and/or non-analytical nonlinearities (e.g. piecewise linear/nonlinear characteristics) and time-delay terms in the coupled set of nonlinear ordinary differential equations. Apart from obtaining the fundamental periodic solution it is shown that the method can also determine various orders of subharmonic, supersubharmonic, superharmonic solutions efficiently and accurately. It can also be applied to analyse the behaviour of both autonomous and non-autonomous systems. The stability of the periodic solutions is investigated by constructing the monodromy matrix and determining the eigenvalues of the same. The bifurcation behaviour is studied by observing the movement of the eigenvalues in the unit ball. The routes to chaos are analysed from the nature of the bifurcations.

The incremental harmonic balance (IHB) method is a general multiharmonic balance procedure to obtain the steady-state periodic solutions of nonlinear systems and uses the Newton–Raphson

Nonlinearity and Chaos in Engineering Dynamics
Edited by J. M. T. Thompson and S. R. Bishop,   © 1994 John Wiley & Sons Ltd

method for the solution of the resulting nonlinear algebraic equations in the harmonic coefficients (Lau *et al.* 1982). It requires modification to treat non-analytical nonlinearities and does not estimate the error involved in the approximate solution due to truncation of higher harmonics (Wong *et al.* 1991). In the proposed FIHB method FFT is extensively used to evaluate the integrals arising out of the Galerkin process and the error due to the approximate Fourier expansion of the solution within the number of harmonics assumed in the solution as well as the error due to the truncation of higher harmonics can be evaluated (Ling and Wu 1987). The response curves and the bifurcation diagrams are obtained by an iterative and augmentation process (Lau *et al.* 1982).

29.2 FAST INCREMENTAL HARMONIC BALANCING, FOURIER–GALERKIN–NEWTON ALGORITHM

We consider the multi-degree-of-freedom system with l degree of freedom in non-dimensional form:

$$\omega^2 M\ddot{x} + \omega C\dot{x} + Kx + g(\ddot{x}, \dot{x}, x, \tau) - f(\tau) = \Phi(\tau) = 0 \qquad (1)$$

where $\omega t = \tau$, and M, C, K are the mass, damping and stiffness matrices, x the response vector, ω the frequency of excitation, and g and f are respectively the l-dimensional nonlinear and forcing vector functions.

Substituting $x = x_0 + \Delta x; \omega = \omega_0 + \Delta\omega$ in equation (1), we get

$$\Phi_0 + \left[\frac{\partial\Phi}{\partial\ddot{x}} \frac{\partial\Phi}{\partial\dot{x}} \frac{\partial\Phi}{\partial x} \frac{\partial\Phi}{\partial\omega} \right]_0 [\Delta\ddot{x} \, \Delta\dot{x} \, \Delta x \, \Delta\omega]^T = \varepsilon(\tau). \qquad (2)$$

Expanding x in a Fourier series we obtain

$$x = a_0 + \sum_{j=1}^{M} (a_{2j-1} \cos j\tau + a_{2j} \sin j\tau); \quad \Delta x = \Delta a_0 + \sum_{j=1}^{M} (\Delta a_{2j-1} \cos j\tau + \Delta a_{2j} \sin j\tau);$$

$$Y = \{a_0^T a_1^T a_2^T \dots a_{2M-1}^T a_{2M}^T\}^T; \quad \Delta Y = \{\Delta a_0^T \Delta a_1^T \Delta a_2^T \dots \Delta a_{2M-1}^T \Delta a_{2M}^T\}^T.$$

By applying the Galerkin method for equation (2) the error $\varepsilon(\tau)$ can be minimized with weighting function $(1 \cos \tau \sin \tau \dots \cos M\tau \sin M\tau)^T$. Hence, we get

$$[J^L]\{Y\} + \{G\} - \{F\} + [J^L + J^N]\{\Delta Y\} + \{H\}\Delta\omega = r \qquad (3)$$

where $[J^L], [J^N]$ are the Jacobian matrices for linear and nonlinear terms respectively.

$$\{G\} = \{G_0^T G_1^T G_2^T \dots G_{2M-1}^T G_{2M}^T\}^T; \quad \{F\} = \{F_0^T F_1^T F_2^T \dots F_{2M-1}^T F_{2M}^T\}^T;$$

$$\{H\} = \{H_0^T H_1^T H_2^T \dots H_{2M-1}^T H_{2M}^T\}^T;$$

$$G_0 + \sum_{j=1}^{M} (G_{2j-1} \cos j\tau + G_{2j} \sin j\tau) = FT[g(\ddot{x}, \dot{x}, x, \tau)];$$

$$F_0 + \sum_{j=1}^{M} (F_{2j-1} \cos j\tau + F_{2j} \sin j\tau) = FT[f(\tau)];$$

$$H_0 + \sum_{j=1}^{M} (H_{2j-1} \cos j\tau + H_{2j} \sin j\tau) = \text{FT}[2\omega_0 M \ddot{x}_0 + C\dot{x}_0];$$

where FT[..] denotes Fourier Transform;

$$[J^L] = \begin{bmatrix} K & 0 & 0 & & & \\ 0 & -\omega^2 M + K & \omega C & & [0] & \\ 0 & -\omega C & -\omega^2 M + K & & & \\ \rule{0pt}{0pt} & & & -j^2\omega^2 M + K & & j\omega C \\ & [0] & & & & \\ & & & -j\omega C & & -j^2\omega^2 M + K \end{bmatrix};$$

$$[J^N] = \begin{bmatrix} P_{00} & P_{01} & R_{01} & P_{0j} & R_{0j} \\ P_{10} & P_{11} & R_{11} & P_{1j} & R_{1j} \\ Q_{10} & Q_{11} & S_{11} & Q_{1j} & S_{1j} \\ \hline P_{k0} & P_{k1} & R_{k1} & P_{kj} & R_{kj} \\ Q_{k0} & Q_{k1} & S_{k1} & Q_{kj} & S_{kj} \end{bmatrix}.$$

Elements of $[J^N]$ are given by

$$P_{00} + \sum_{k=1}^{M} (P_{k0} \cos k\tau + Q_{k0} \sin k\tau) = \text{FT}\left\{\frac{\partial g}{\partial x}\right\};$$

$$P_{0j} + \sum_{k=1}^{M} (P_{kj} \cos k\tau + Q_{kj} \sin k\tau) = \text{FT}\left\{\frac{\partial g}{\partial \ddot{x}}(-j^2 \cos(j\theta_i)) + \frac{\partial g}{\partial \dot{x}}(-j \sin(j\theta_i)) + \frac{\partial g}{\partial x}(\cos(j\theta_i))\right\};$$

$$R_{0j} + \sum_{k=1}^{M} (R_{kj} \cos k\tau + S_{kj} \sin k\tau) = \text{FT}\left\{\frac{\partial g}{\partial \ddot{x}}(-j^2 \sin(j\theta_i)) + \frac{\partial g}{\partial \dot{x}}(j \cos(j\theta_i)) + \frac{\partial g}{\partial x}(\sin(j\theta_i))\right\};$$

$j = 1, M$; $i = 1, 4M$; $\theta_i = 2\pi(i-1)/4M$. r is the error function to be minimized.

The elements of $[J^N]$ can be evaluated using cosine and sine transforms of the derivatives of the nonlinear function g. By solving equation (3) for a particular set of parameters it is possible to obtain the periodic solutions of equation (1). The spectral components of the linear terms of equation (2) are obtained directly from the assumed harmonic coefficients while that of the nonlinear terms are obtained as follows. Using FFT the waveform of the assumed solution is obtained at $4M$ equal time intervals. Using this time waveform the nonlinear functions are evaluated at $4M$ equidistant points, and further using FFT the spectral components of the nonlinear functions can be evaluated. With this procedure $(2M + 1)l$ linear algebraic equations containing the incremental coefficients are obtained which are solved by applying iterative and augmentation processes alternatively (Lau *et al.* 1982).

The errors $(\varepsilon_1, \varepsilon_2)$ introduced into the solution are given by

$$\varepsilon_1 = \sqrt{r_0^2 + \sum_{j=1}^{M} (r_{2j-1}^2 + r_{2j}^2)}; \quad \varepsilon_2 = \sqrt{\sum_{j=M+1}^{2M} (G_{2j-1}^2 + G_{2j}^2)}. \qquad (4)$$

29.3   STABILITY ANALYSIS

Stability analysis is performed by using a multi-valued Floquet analysis. By perturbing the state variables one can get a variational equation as follows:

$$\{\dot{z}\} = [A]\{z\} \qquad (5)$$

where $\{z\} = \{\delta x \delta \dot{x}\}^T$ and

$$[A] = \begin{bmatrix} 0 & I \\ -M^{-1}K & -M^{-1}C \end{bmatrix} + \begin{bmatrix} 0 & 0 \\ -M^{-1}\left[\dfrac{\partial g}{\partial x}\right] & -M^{-1}\left[\dfrac{\partial g}{\partial \dot{x}}\right] \end{bmatrix}.$$

As the elements of $[A]$ are periodic with the period equal to the period of the equilibrium solution, the matrix exponential can be calculated by considering the equation as the differential equation with constant coefficients in a small interval of time. The monodromy matrix relates the state of the system at the start and end of one period (Hsu 1974). By observing the movement of the eigenvalues it is possible to predict bifurcation points and type of bifurcation.

29.4   SYSTEMS WITH TIME DELAY

The response of systems with time delay can be obtained by evaluating the Galerkin approximation of the time-delay terms as follows.

The response can be expressed in the form

$$x(t) = a_0 + \sum_{j=1}^{M} (a_{2j-1} \cos jt + a_{2j} \sin jt);$$

$$x(t - \tau) = a_0 + \sum_{j=1}^{M} (a_{2j-1} \cos j\tau - a_{2j} \sin j\tau) \cos jt + (a_{2j-1} \sin j\tau + a_{2j} \cos j\tau) \sin jt.$$

This can be expressed as

$$\{X(\tau)\} = [T]\{X\}$$

where

$$[T] = \left[\begin{array}{ccc|cc} 1 & 0 & 0 & & \\ 0 & \cos \tau & -\sin \tau & & [0] \\ 0 & \sin \tau & \cos \tau & & \\ \hline & & & \cos j\tau & -\sin j\tau \\ & [0] & & & \\ & & & \sin j\tau & \cos j\tau \end{array} \right].$$

By tracing the bifurcation diagram the regions where all the possible periodic solutions are unstable (i.e. chaos) can be obtained. With the transformation $m\omega t = n\tau$, the subharmonics of order n, supersubharmonics of order m/n, and superharmonics of order m can be obtained.

29.5   QUASI-PERIODIC SOLUTION

The quasi-periodic solutions arising due to multiple excitation with incommensurate frequencies are obtained by expanding the solution in a generalized Fourier series form in which the various frequencies are a linear combination of the basic frequencies.

Assume a solution in a generalized Fourier series form

$$\{x(t_i)\} = a_0 + \sum_{j=1}^{M} (a_{2j-1} \cos v_j t_i + a_{2j} \sin v_j t_i) = [\Gamma]\{X\} \tag{7}$$

where
$t_i = i\Delta t$; T_p = total sampling time; Δt = sampling rate; N = number of samples; and

$$[\Gamma] = \begin{bmatrix} 1 & 1 & 0 & \cdots & 1 & 0 \\ 1 & \cos v_1 t_1 & \sin v_1 t_1 & \cdots & \cos v_M t_1 & \sin v_M t_1 \\ 1 & \cos v_1 t_2 & \sin v_1 t_2 & \cdots & \cos v_M t_2 & \sin v_M t_2 \\ \vdots & \vdots & \vdots & \cdots & \vdots & \vdots \\ \vdots & \vdots & \vdots & \cdots & \vdots & \vdots \\ 1 & \cos v_1 t_N & \sin v_1 t_N & \cdots & \cos v_M t_N & \sin v_M t_N \end{bmatrix},$$

where $[\Gamma]$ is generally a rectangular $(N+1) \times (2M+1)$ matrix.

Hence, by applying the least squares technique, we get

$$\{X(\omega)\} = [[\Gamma^T \Gamma]^{-1} \Gamma^T]\{(x(t)\}. \tag{8}$$

With the above generalized Fourier transformation we can get the harmonic balancing equations (in state variables alone) of a quasi-periodically excited system which are given by

$$[J^L]\{Y\} + \{G\} - \{F\} + [J^L + J^N]\{\Delta Y\} = 0; \tag{9}$$

$$[J^L] = \left[\begin{array}{ccc|ccc} K & 0 & 0 & & & \\ 0 & -v_1^2 M + K & v_1 C & & [0] & \\ 0 & -v_1 C & -v_1^2 M + K & & & \\ \hline - & - & - & - & - & - \\ & & & -v_j^2 M + K & & v_j C \\ & [0] & & & & \\ & & & -v_j C & & -v_j^2 M + K \end{array} \right];$$

$$[J^N] = J_0 + J_1 + J_2;$$

$$J_0 = [[\Gamma^T \Gamma]^{-1} \Gamma^T] \text{diag} \left\{ \left[\frac{\partial g(t_0)}{\partial x} \frac{\partial g(t_1)}{\partial x} \cdots \frac{\partial g(t_N)}{\partial x} \right] \right\} [\Gamma_0];$$

$$J_1 = [[\Gamma^T \Gamma]^{-1} \Gamma^T] \text{diag} \left\{ \left[\frac{\partial g(t_0)}{\partial \dot{x}} \frac{\partial g(t_1)}{\partial \dot{x}} \cdots \frac{\partial g(t_N)}{\partial \dot{x}} \right] \right\} [\Gamma_1];$$

$$J_2 = [[\Gamma^T \Gamma]^{-1} \Gamma^T] \text{diag} \left\{ \left[\frac{\partial g(t_0)}{\partial \ddot{x}} \frac{\partial g(t_1)}{\partial \ddot{x}} \cdots \frac{\partial g(t_N)}{\partial \ddot{x}} \right] \right\} [\Gamma_2];$$

$[\Gamma] = [\Gamma_0];$

$$[\Gamma_1] = \begin{bmatrix} 0 & 0 & 1 & \cdots & 0 & 1 \\ 0 & -v_1 \sin v_1 t_1 & v_1 \cos v_1 t_1 & \cdots & -v_M \sin v_M t_1 & v_M \cos v_M t_1 \\ 0 & -v_1 \sin v_1 t_2 & v_1 \cos v_1 t_2 & \cdots & -v_M \sin v_M t_2 & v_M \cos v_M t_2 \\ \vdots & \vdots & \vdots & \cdots & \vdots & \vdots \\ \vdots & \vdots & \vdots & \cdots & \vdots & \vdots \\ 0 & -v_1 \sin v_1 t_N & v_1 \cos v_1 t_N & \cdots & -v_M \sin v_M t_N & v_M \cos v_M t_N \end{bmatrix};$$

$$[\Gamma_2] = \begin{bmatrix} 0 & -1 & 0 & \cdots & -1 & 0 \\ 0 & -v_1^2 \cos v_1 t_1 & -v_1^2 \sin v_1 t_1 & \cdots & -v_M^2 \cos v_M t_1 & -v_M^2 \sin v_M t_1 \\ 0 & -v_1^2 \cos v_1 t_2 & -v_1^2 \sin v_1 t_2 & \cdots & -v_M^2 \cos v_M t_2 & -v_M^2 \sin v_M t_2 \\ 0 & \vdots & \vdots & \cdots & \vdots & \vdots \\ \vdots & \vdots & \vdots & \cdots & \vdots & \vdots \\ 0 & -v_1^2 \cos v_1 t_N & -v_1^2 \sin v_1 t_N & \cdots & -v_M^2 \cos v_M t_N & -v_M^2 \sin v_M t_N \end{bmatrix}.$$

By solving the linear set of equations (9) an accurate solution can be obtained. The incrementation in the non-state variable can also be performed as explained in Section 29.2.

29.6   EXAMPLES

29.6.1   Single-degree-of-freedom model with analytical and non-analytical nonlinearities

A single-degree-of-freedom system with piecewise linear stiffness under harmonic and flow excitations is considered (Narayanan and Sekar in Press):

$$\ddot{x} + \zeta_0 \dot{x} + \zeta_3 \dot{x}^3 + g(x) = \cos(\Omega \tau), \tag{10}$$

where $g(x) = x$ when $x \leqslant \delta$,

$$= \sigma^2 x + (1 - \sigma^2)\delta \quad \text{otherwise.}$$

$$\zeta_0 = \zeta - 0.5\beta B_1 U, \quad \zeta_3 = -0.5\beta B_3 / U,$$

$$\zeta = 0.05, \beta = 0.1, \sigma = 5.0, \delta = 0.0, \Omega = 1.0, B_1 = 2.7, B_3 = -31.0.$$

Equation (10) has been solved by the method proposed in Section 29.2 for the following system parameters. The dynamics of the system is studied by considering U as the bifurcation parameter which represents the flow velocity of a square prism. Periodic response curves for period-1($U = 1.5$) and period-2 ($U = 2.5$) orbits have been obtained (Figure 29.1). As the flow velocity increases, period-doubling bifurcations occur and the system exhibits chaotic behaviour when U reaches 4.0.

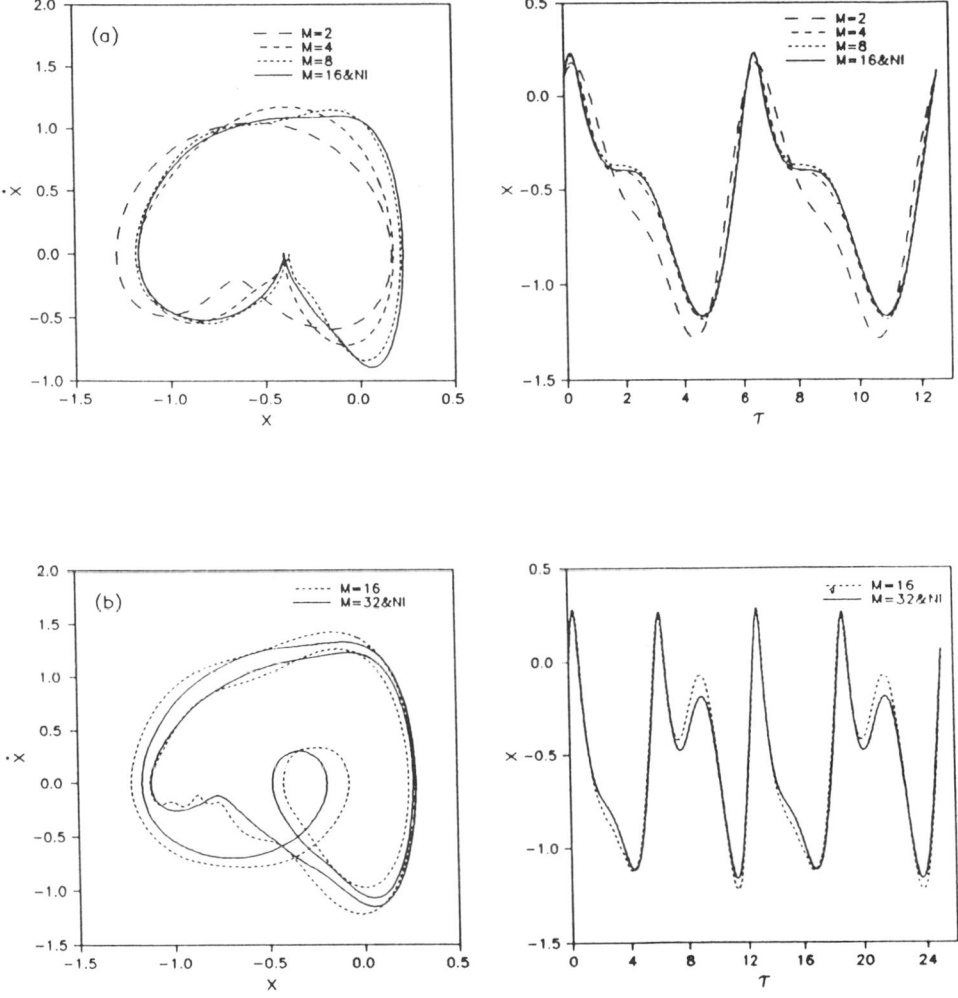

Figure 29.1 Effect of number of harmonics (M): (a) period-1 orbit ($U = 1.5$), (b) period-2 orbit ($U = 2.5$).

29.6.2 Van der Pol–Duffing oscillator with time delay

Consider a van der Pol–Duffing oscillator with time-delay term (Tsuda *et al.* 1992):

$$\ddot{x} - 0.1(1 - x^2)\dot{x} + x + 3x^3 + x(t - \tau) = 5\cos(\omega t), \tag{11}$$

$$\tau = 0.1.$$

The periodic responses are obtained by FIHB and the phase plane and time history are given in Figure 29.2 for $\omega = 1.3$ and 2.3. This system undergoes a series of period-doubling bifurcations which results in chaos when $\omega = 1.22$.

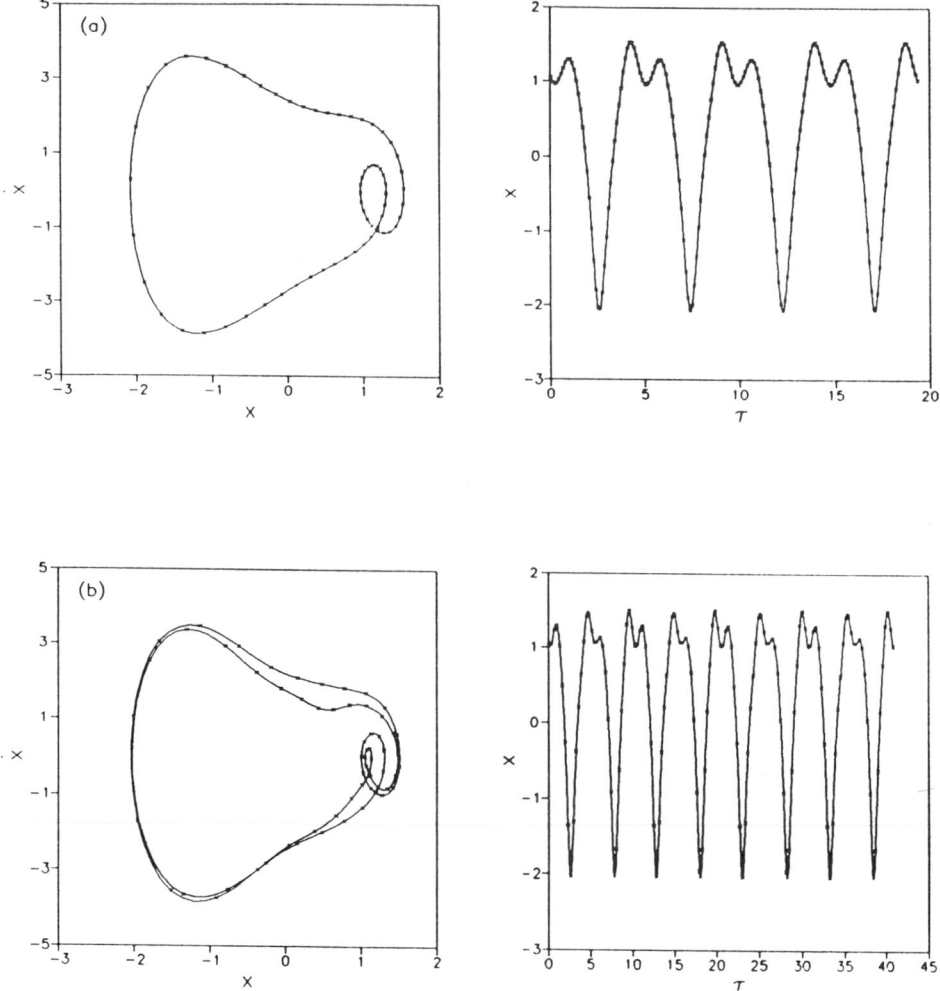

Figure 29.2   Phase plane and time history: (a) period $1(\omega = 1.3)$, (b) period 2 $(\omega = 1.23)$.

29.6.3   Duffing oscillator with multiple excitations

Consider a Duffing oscillator with two incommensurate frequencies (Chua and Ushida 1981):

$$\ddot{x} + 0.1\dot{x} + x + x^3 = 0.3\cos(t) + 1.5\cos(0.115t). \tag{12}$$

Response curves are obtained by using the generalized Fourier series method explained in Section 29.5 and the time history and Fourier transform of quasi-periodic solution is given in Figures 29.3.

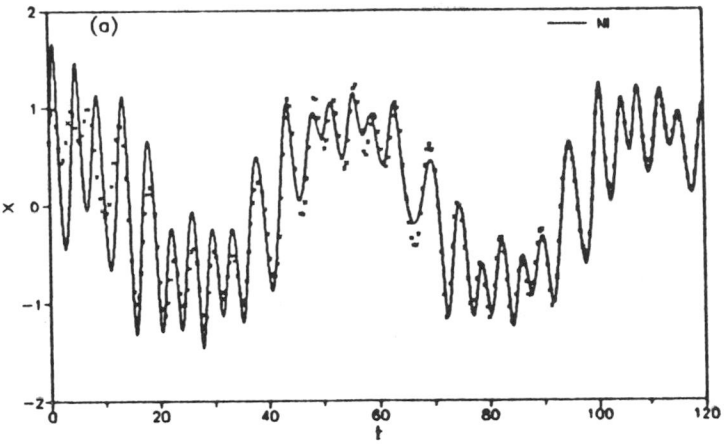

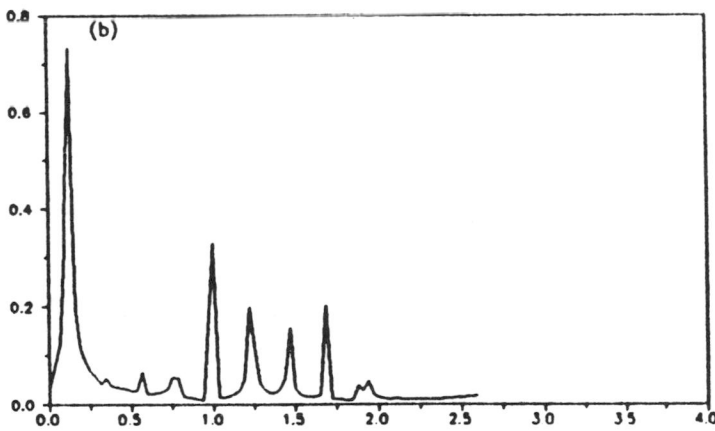

Figure 29.3(a) Time history, (b) Fourier transform.

29.6.4 Three-degree-of-freedom model with piecewise linear stiffness

Consider a three-degree-of-freedom model representing the dynamics of a gear-bearing system (Kahraman and Singh 1991)

$$
\begin{bmatrix} 1 & 0 & 0 \\ 0 & 1 & 0 \\ -1 & 1 & 1 \end{bmatrix} \begin{Bmatrix} \ddot{x}_1 \\ \ddot{x}_2 \\ \ddot{x}_3 \end{Bmatrix} + \begin{bmatrix} c_{11} & 0 & c_{13} \\ 0 & c_{22} & -c_{23} \\ 0 & 0 & c_{33} \end{bmatrix} \begin{Bmatrix} \dot{x}_1 \\ \dot{x}_2 \\ \dot{x}_3 \end{Bmatrix} + \begin{bmatrix} k_{11} & 0 & k_{13} \\ 0 & k_{22} & -k_{23} \\ 0 & 0 & k_{33} \end{bmatrix} \begin{Bmatrix} x_1 \\ x_2 \\ g_3(x_3) \end{Bmatrix}
$$

$$
= \begin{Bmatrix} 0 \\ 0 \\ f_m + f_a \omega^2 \sin(\omega t) \end{Bmatrix} \tag{13}
$$

with

$$
\begin{aligned}
g_3(x_3) &= 0 && \text{if } -1 \leqslant x_3 \leqslant 1, \\
&= x_3 - 1 && \text{if } x_3 > 1, \\
&= x_3 + 1 && \text{if } x_3 < 1.
\end{aligned}
$$

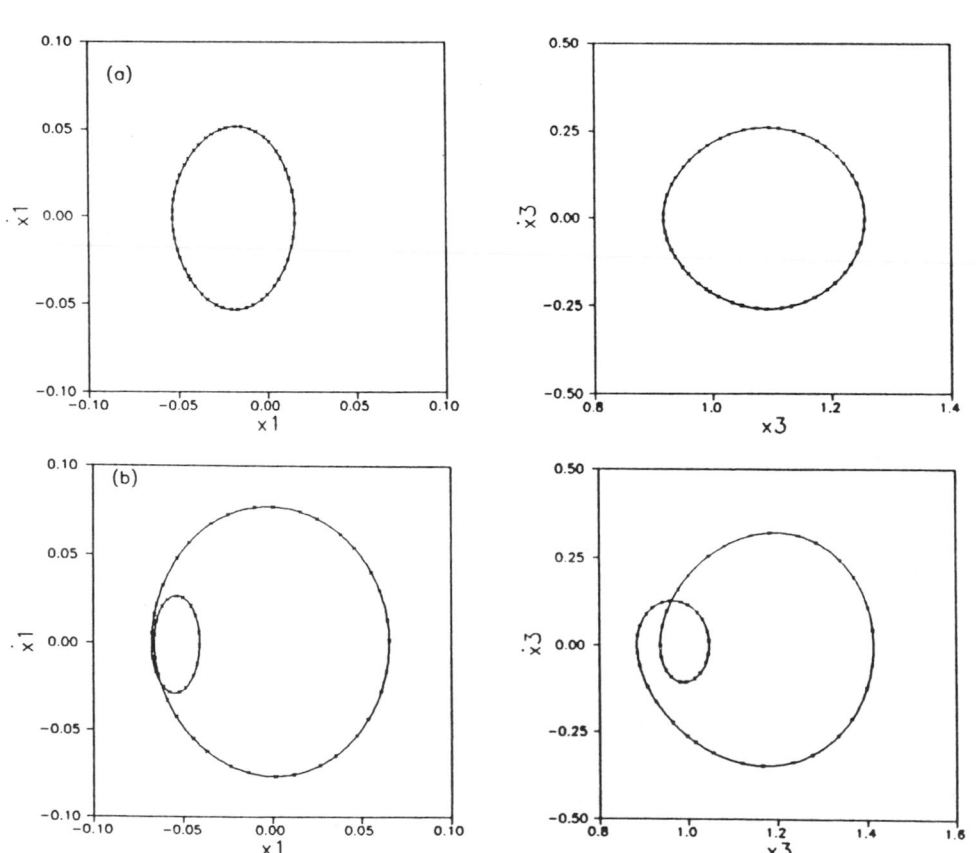

Figure 29.4   Phase-plane plots: (a) period-1 ($\omega = 1.5$), (b) period-2 ($\omega = 1.48$).

Periodic response curves are obtained for a range of excitation frequency (ω) and phase-plane plots for x_1 and x_3 are given in Figure 29.4.

29.7 CONCLUSIONS

A numerical-analytical method has been developed and numerous problems have been solved. By assuming a sufficient number of harmonics in the solution it is possible to get periodic solutions and their stability so as to enable us to perform bifurcation analysis.

REFERENCES

Chua, L. O. and Ushida, A. (1981) Algorithms for computing almost periodic steady state response of nonlinear systems to mutiple input frequency, *IEEE Transactions on Circuits and Systems.*, **CAS-28** (10).

Hsu, C. S. (1974) On approximating a general periodic system, *Journal of Mathematical Analysis and Applications*, **45**, 234–251.

Kahraman, A. and Singh R. (1991) Interactions between time-varying mesh stiffness and clearance non-linearities in geared system, *Journal of Sound and Vibration*, 146(1), 135–156.

Lau, S. L., Cheung, Y. K. and Wu, S. Y. (1982) A variable parameter incrementation method for dynamic instability of linear and nonlinear elastic systems, *Journal of Applied Mechanics*, **49**, 849–853.

Ling, F. H. and Wu, X. X. (1987) Fast Galerkin method and its application to determine periodic solutions of nonlinear oscillators, *International Journal of Nonlinear Mechanics*, **22**(2), 89–98.

Narayanan, S. and Sekar, P. (in press) Periodic and chaotic responses of a sdf system with piece-wise linear stiffness subjected to combined harmonic and flow induced excitations, *Journal of Sound and Vibration.*

Tsuda, Y., Tamura, H., Sueoka, A. and Fujii, T. (1992) Chaotic behavoir of a nonlinear vibrating system with a retarded argument, *JSME International Journal Series-III*, **35**(3), 259–267.

Wong, C. W., Zhang, W. S. and Lau, S. L. (1991) Periodic forced vibration of unsymmetrical piecewise linear systems by incremental harmonic balance method, *Journal of Sound and Vibration*, **149**(1), 91–105.

S. Narayanan and P. Sekar, *Machine Dynamics Laboratory, Department of Applied Mechanics, Indian Institute of Technology, Madras 600 036, India.*

PART VII
Uses of Chaos

With the discovery of chaos it was originally thought that the complicated, non-periodic motions from a chaotic system could be used to understand turbulence in fluid flow. It seems now that chaos does indeed improve our understanding of weak turbulence and the way in which turbulence is approached; but fully developed turbulent fluids form a much more complex structure than a simple low-dimensional system (see Ruelle and Takens 1971 and Mullin 1993). To date most of the research into chaotic dynamics has been aimed at either predicting its occurrence or understanding the behaviour once a system is chaotic. In engineering for the most part, chaos is seen as an undesirable behaviour and efforts are directed to try and adapt the system so as to avoid the chaotic regimes. New ideas are, however, now being formulated that utilize the powerful properties of chaotic attractors.

TIME SERIES AND CONTROL

In many experimental systems we have little if any knowledge of the governing equations but yet we are able to easily record data. It is not unusual for physical experiments to produce a response that appears to be random. However, from the recorded measurements we can now use the methods of time series analysis to reconstruct the attracting solution in the appropriate phase space, provided that the system has some underlying low-dimensional behaviour; see the chapter by *Abarbanel*. Using such methods of analysis, together with topological arguments, we can identify desirable characteristics that may form part of the attractor. The technique proposed by Grebogi and his colleagues at Maryland (also see Shinbrot *et al.* 1993) aims to identify one of these orbits which has a desired periodic solution (an unstable solution embedded within the chaotic attractor). Since a trajectory wanders through the chaotic attractor it will eventually land arbitrarily close to this unstable orbit. Their method then controls the system and proceeds to stabilize the solution so that this desired state is maintained. This technique has been applied to a variety of disciplines and is reviewed for us here in the chapter by *Grebogi and Lai*.

FLUID MIXING

Perhaps the most visual use of chaos has been in the chaotic mixing of fluids. The work of Ottino (1989) shows that chaos is a most desirable attribute when trying to achieve uniform mixing of fluids. This improved mixing and the chaos-enhanced transport in cellular flows have been observed experimentally. It should be said, however, that this work relies upon knowledge of Hamiltonian systems and fluid dynamics that have not, for the most part, been considered in this book.

REFERENCES

Mullin, T. (1993) *The Nature of Chaos*, Oxford, Clarendon Press.

Ottino, J. M. (1989) *The Kinematics of Mixing: Stretching, Chaos and Transport* Cambridge, Cambridge University Press.

Ruelle, D. and Takens F. (1971) On the nature of turbulence, *Commun., Math. Phys.*, **20**, 167–192.

Shinbrot, T., Grebogi, C., Ott, E. and Yorke, J. A. (1993) Using small perturbations to control chaos, *Nature*, **363**, 3 June, 411–417.

30 ANALYSING AND UTILIZING TIME-SERIES OBSERVATIONS FROM CHAOTIC SYSTEMS

H. D. I. Abarbanel

30.1 INTRODUCTION

Studies of nonlinear systems have moved out of the arena of classifying what can happen with a given set of nonlinear differential equations or discrete time maps. We have moved on to the analysis of observations from such systems for purposes of learning about the systems themselves or exploiting, for engineering design and application, the properties deduced from those observations (Abarbanel *et al.* 1993). This chapter cannot possibly cover the full spectrum of such analyses of observations, and I will not try. The focus of the chapter will be on a few things connected with Lyapunov exponents, *local and global*, their use in *model building* and in determining a *Cramer–Rao-like bound* on the accuracy with which one may estimate a chaotic signal in the presence of contamination. In Table 30.1 is a comparison between the issues in linear and nonlinear signal processing or time-series analysis which shows the issues to be close in thrust and quite different in methodology. We shall address the bottom two entries in right-hand column.

Lyapunov exponents are not the only invariant quantities which characterize an attractor, but among all the quantities one can think of extracting from observations they tell us much more than a single number. Knowing *global* Lyapunov exponents gives us an indication about the *predictability* of the system, a sense of the *dimension* of the attractor, and knowledge of the number of active dynamical degrees of freedom involved in the observed system. Local exponents allow us to determine these items locally, and since short-term (local) prediction is much more of interest than infinite time prediction, to which the global exponents refer, they provide real insight as to how useful one can expect models to be.

Nonlinearity and Chaos in Engineering Dynamics
Edited by J. M. T. Thompson and S. R. Bishop,   © 1994 John Wiley & Sons Ltd

Table 30.1   Comparison of linear and nonlinear signal processing.

	Linear signal processing	Nonlinear signal prrocessing
Finding the signal	Noise reduction; detection Separate broad-band noise from narrow-band signal using different spectral characteristics. System known: make matched filter in frequency domain:	Noise reduction; detection Separate broad-band signal from broad-band noise using deterministic nature of signal. System known: make 'matched filter' in time-domain. Use dynamics or invariant distribution and Markov transitions probabilities.
Finding the space	Fourier transforms Use Fourier-space methods to turn differential equations or recursion relations into algebraic forms: $x(n)$ is observed; $x(f) = \Sigma x(n) \exp[i2\pi nf]$ is used.	Phase-space reconstruction Using time-lagged variables, form coordinates for the phase space in d dimensions: $y(n) = [x(n), x(n+T), \ldots, x(n+(d-1)T)]$ How are we to determine d and T? Mutual information; false nearest neighbours.
Classifying the signal	Sharp spectral peaks. Resonant Frequencies of the system. *Quantities independent of initial conditions.*	Invariants of orbits. Lyapunov exponents; various fractal dimensions; topological invariants; linking numbers of unstable periodic orbits. *Quantities independent of initial conditions.*
Making models, Predicting	$x(n+1) = \Sigma c_j x(n-j)$ Find parameters c_j consistent with invariant classifiers (spectral peaks).	$y(n) \rightarrow y(n+1)$ as time evolution. $y(n+1) = F[y(n), a_1, a_2, \ldots, a_p]$. Find parameters a_j consistent with invariant classifiers (Lyapunov exponents, dimensions). Find dynamical dimensions D_L from data; use local false nearest neighbours.

30.2   LYAPUNOV EXPONENTS

For review we cover the definition of Lyapunov exponents and restate the main mathematical properties of the exponents which follow from the Oseledec multiplicative ergodic theorem (Oseledec 1968; Ruelle 1979). We begin by imagining that we have an appropriate state space of observed d-dimensional vectors $y(n) = y(t_0 + n\tau_s)$ where observations begin at t_0 and are made every sampling time τ_s. We may not actually know the dynamics which evolves the system, but we know it is a d-dimensional vector field $F(x)$ which acts as

$$y(n+1) = F(y(n)). \tag{1}$$

$$\mathbf{D}F(\mathbf{y}(n)) = \frac{\partial F(\mathbf{x})}{\partial \mathbf{x}}\bigg|_{\mathbf{x}=\mathbf{y}(n)} \tag{2}$$

is the Jacobian matrix for the system. From the composition of L Jacobian matrices

$$\mathbf{D}F^L(\mathbf{y}(n)) = \mathbf{D}F(\mathbf{y}(n+L-1))\cdot\mathbf{D}F(\mathbf{y}(n+L-2))\cdots\mathbf{D}F(\mathbf{y}(n)), \tag{3}$$

we form the Oseledec matrix (Oseledec 1968):

$$\mathrm{OSL}(\mathbf{x}) = [\mathbf{D}F^L(\mathbf{x})^{\mathrm{T}}\cdot\mathbf{D}F^L(\mathbf{x})]^{\frac{1}{2L}} \tag{4}$$

about which Oseledec proved the *multiplicative ergodic theorem* stating, for our purposes, that for essentially all $\mathbf{x}$ within the basin of attraction of an attractor: (1) $\mathrm{OSL}(\mathbf{x})$ exists; (2) for $L \to \infty$ it becomes independent of $\mathbf{x}$, and (3) its eigenvalues and eigendirections exist with the latter spanning the d-dimensional state space. The eigenvalues $\exp[\lambda_a]; a = 1, 2, \ldots, d$ define the global Lyapunov exponents λ_a. These exponents are also unchanged under smooth changes of the coordinate system. These properties mean that the λ_a are characteristic of the source of the observation $\mathbf{y}(n)$, that is, the dynamics, and, can be computed in any coordinate system which is smoothly related to the 'true' coordinates of the dynamics. Since we rarely know these true coordinates, this fact about the λ_a is good news.

Local Lyapunov exponents $\lambda_a(\mathbf{x}, L)$ are the eigenvalues of the Oseledec matrix for finite L. The average local Lyapunov exponents

$$\bar{\lambda}_a(L) = \int \mathrm{d}^d x \rho(\mathbf{x}) \lambda_a(\mathbf{x}, L), \tag{5}$$

where $\rho(\mathbf{x})$ is the natural measure on the attractor

$$\rho(\mathbf{x}) = \frac{1}{N}\sum_{m=1}^{N}\delta^d(\mathbf{x} - \mathbf{y}(m)), \tag{6}$$

answer the question: on the average over the attractor how does a perturbation to an orbit grow in the L steps after it is imposed on an orbit? This is the question of direct interest to estimating the predictability of a dynamical system. One wants to know the weather in 5 days, not 10^5 days (274 years!). The $\bar{\lambda}_a(L)$ for large L behave as

$$\bar{\lambda}_a(L) = \lambda_a + \frac{c_a}{L^{1-q}} + \frac{c_a'}{L}, \tag{7}$$

where $0 < q \leqslant 0.5$, c_a and c_a' are constants, and the last term comes from the coordinate system dependence of the $\lambda_a(\mathbf{x}, L)$. The $\bar{\lambda}_a(L)$ are independent of initial conditions on the attractor because of the average taken with $\rho(\mathbf{x})$.

30.2.1 Lyapunov exponents from observed data

Experiments or field observations rarely, if ever, evaluate enough independent variables to fill the d-dimensional data vectors $\mathbf{y}(n) = [y_1(n), y_2(n), \ldots, y_d(n)]$ with *different* measured quantities (say, $y_1(n)$ is pressure, $y_2(n)$ is temperature, etc.). Typically one measures *one* dynamical variable which we call $x(n) = x(t_0 + nt_s)$. From this scalar variable we build up a phase space or state space from time-lagged versions of $x(n)$ (Eckmann and Ruelle 1985; Mañé 1981; Takens 1981):

$$y(n) = [x(n), x(n + T), x(n + 2T), \ldots, x(n + (d - 1)T)], \tag{8}$$

where T is an integer which tells what multiple of τ_s, $T\tau_s$, we use in lagging the components of $y(n)$.

Choosing the time lag T has been the subject of much discussion. We use average mutual information (Fraser and Swinney 1986). Choosing the appropriate d for the data vector comes from the consideration of false nearest neighbours. The basic idea is that unfolding the attractor from its projection on the observation axis $x(n)$ is completed when neighbouring points on the attractor are nearby because of the dynamics (true neighbours), not because of projection (false neighbours), from a higher-dimensional space (Kennel *et al.* 1992). Surveying whether the nearest neighbour of every point in dimension d remains a neighbour in dimension $d + 1$ tells us when false neighbours cease. This identifies the necessary dimension of the data vector $y(n)$ in the reconstructed state space. The usual criterion that the embedding dimension should be the integer d greater than $2d_A$, with d_A the fractal (box-counting) dimension, is sufficient. The dimension from the false-nearest-neighbour criterion is from the data and is necessary.

It $x(n)$ and the reconstructed d-dimensional vectors $y(n)$ are all that is observed, how are we to estimate the Jacobian matrices $DF(x)$ required to determine the Oseledec matrix? The basic idea of the answer is in Eckmann and Ruelle (1985) and Sano and Sawada (1985): construct maps from neighbourhood to neighbourhood utilizing phase-space information as well as temporal information to provide estimates of the local Jacobian. Represent the vector field $F(x)$ which evolves points in terms of some basis functions $\phi_k(x)$:

$$F(x) = \sum_{k=1}^{M} c(k)\phi_k(x). \tag{9}$$

Determine the coefficients $c(k)$ by a least-squares fit to the requirement that phase-space neighbours of $y(n)$ map into neighbours of $y(n + 1)$. $DF(y(n))$ is then

$$DF_{ab}(y(n)) = \sum_{k=1}^{M} c_a(k) \frac{\partial \phi_k(x)}{\partial x_b}\Big|_{x=y(n)}. \tag{10}$$

For basis functions one has used polynomials (Bryant *et al.* 1990; Brown *et al.* 1991), radial basis functions (Parlitz 1992), sigmoidal functions (Gencay and Dechert 1992), and probably others I am not aware of. One can also attempt a global fit to the data of this form (Brown 1992).

Each of these methods works. Each of these methods requires a substantial amount of computation, though with existing workstations the total wall-clock time elapsed is still short. Each of these methods is subject to uncertainties when the data is contaminated by 'noise'. The issue with contamination is that the product of local Jacobians entering the Oseledec matrix is highly ill-conditioned, so small errors in the determination of the product can lead to significant errors in the determination of the exponents.

30.3   MODELLING CHAOTIC SYSTEMS

While many goals can be achieved by the classification of chaotic systems through fractal dimensions or Lyapunov exponents, often one wishes for various purposes to make a detailed model of the state-space evolution of a dynamical system. Among these purposes could be design of a control system which would drive the chaotic dynamics into a regular regime or equally likely drive the chaotic system into a *more* chaotic regime, for example, to increase mixing in a chemical process. Another possible purpose of having a model would be for design of synthetic signals from chaotic systms to be used in communications applications or for the study of synchronizing chaotic systems for the same general purpose. Given data alone and limited or no knowledge of the underlying dynamics one can still make substantial progress by building models in the reconstructed state space. These models attempt to produce evolution rules $y(n) \rightarrow y(n + 1)$ in discrete time through the construction of a parametrized vector field $F(x, a)$ which satisfies

$$y(n + 1) = F(y(n), a), \tag{11}$$

as accurately as possible. The a would be typically chosen by a least-squares approximation to this equality. The map $x \rightarrow F(x, a)$ can either be local in phase space, that is, map a given neighbourhood to other neighbourhoods visited at later times, or it can be global, that is, provide an evolution rule applicable throughout phase space, or at least applicable throughout the basin of attraction which has been observed. In a sense this is a 'black-box' approach to modelling, and a physicist might be uncomfortable with that. Nonetheless, suppose one is dealing with a system with many or, in principle, an infinite number of degrees of freedom as with continuum mechanics or with fluids; then if one has an attractor of just a few dimensions one may conclude that among the many possible degrees of freedom only a few are active in the sense that the amplitudes of any others are quite unimportant to the observations. Since those few degrees of freedom are some complex, unknown projection from the very large underlying system, we can productively put aside the discovery of that projection and focus on extracting interesting answers from the behaviour observed in phase space.

30.3.1   Model dimension

An important issue in making models is the dimension of the model to be constructed. Using the idea of false nearest neighbours one can readily establish a *global* integer dimension d_E in which the attractor is unfolded. This global dimension may, however, be larger than the dimension of the dynamics d_L. To see this, consider a two-dimensional flow which evolves on the surface of a Möbius strip. The strip is unfolded only in three dimensions, and this is the embedding dimension which would be revealed by the false-nearest-neighbour method. How then are we to identify that the dynamics evolves in two dimensions?

A first idea is to evaluate the local Lyapunov exponents of the data forward in time and then backward along the same data stream (Eckmann and Ruelle 1985; Parlitz 1992; Abarbanel and Sushchik 1993). True exponents will reverse in time, while false exponents will do something else. Examining the local exponents rather than just the global exponents gives interesting additional information. In Figure 30.1 we show the local Lyapunov exponents for the Lorenz attractor in $d_E = 4$, which is one larger than required by global false nearest

Average Lyapunov Exponents

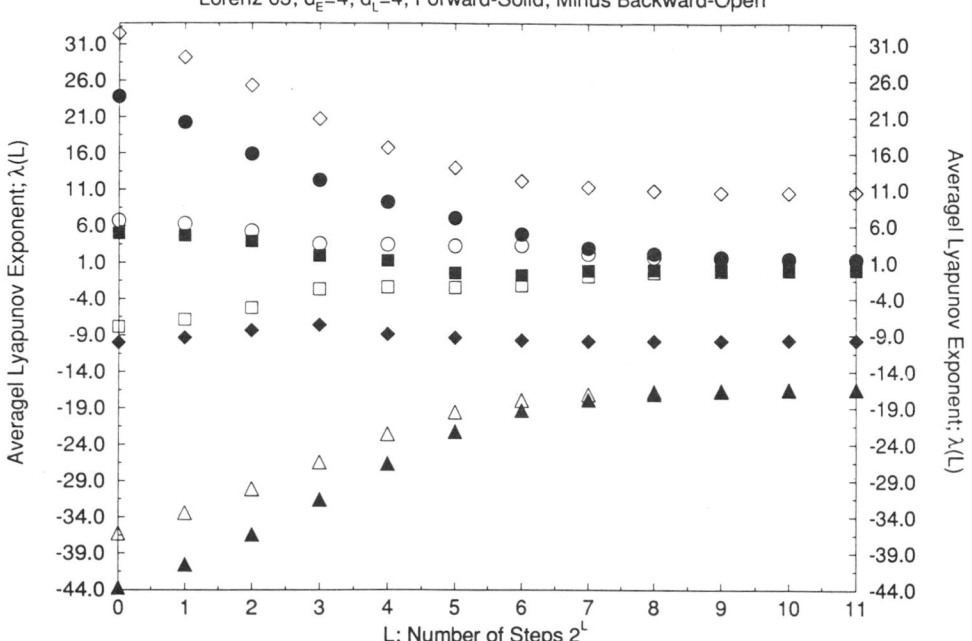

Figure 30.1 Average Lyapunov exponents for the Lorenz system. $d_E = 4$ and $d_L = 4$. Forward and Minus Backward exponents are plotted together; one exponent is false.

neighbours, and with $d_L = 3$. Clearly one of the exponents is false. This method is, unfortunately, not robust to the presence of noise. Small amounts of contamination, as indicated above, ruin one's ability to establish the local Lyapunov exponents accurately.

A more robust method is given in Abarbanel and Kennel (1993). Start in $d \geqslant d_E$ chosen by global false nearest neighbours, and ask in what d_L can one construct accurate maps which take neighbourhoods of phase-space points to neighbourhoods visited by these points in the future. False *local* neighbours are near each other because of projection from a higher dimension. Two false neighbours evolve into very different parts of state space because they are really not close to one another dynamically. Thus mixing a dynamical criterion with the geometrical criterion of neighbourliness we are able to identify a d_L which captures the dynamics even if the choice of coordinate system has required a larger d_E to globally unfold the attractor. By plotting the percentage of bad predictions versus d_L one finds where this becomes independent of d_L. When this d_L is also independent of the number of neighbours used, we have identified the correct local dimension. Doing this forward and backward in time gives further insight into d_L.

30.4   AN EXAMPLE FROM A CHAOTIC LASER SYSTEM

To illustrate the whole panoply of analysis tools we have described, we now consider data from a Nd:YAG laser located in the laboratory of Professor R. Roy at Georgia Institute of

Technology in Atlanta (Roy *et al.* 1992; Bracikowski and Roy 1991). This laser is pumped by an infrared source and lases at 1.06 *μ*m. In the laser cavity is a crystal of potassium titanyl phosphate (KTP) which is birefringent and doubles the frequency of the laser light to green. From the earliest experiments where the KTP crystal was present, experimenters found that there were irregular fluctuations in the intensity of the output beam with a broad spectrum centring around 100 kHz or so. Roy and his colleagues demonstrated only recently that these fluctuations are chaotic and arise from the interaction of the modes in the laser cavity through the medium of the birefringence in the KTP crystal. By rotating the KTP crystal they were able to 'dial up' regular, quasi-periodic, and various chaotic behaviours of the laser intensity. In addition Roy and his colleagues have been able to control these chaotic fluctuations and put the energy which was spread across a broad band back into a steady-intensity output of the laser. In this way they have been able to increase the power in the desired laser line by a factor of ten to fifteen!

We have analysed many chaotic intensity traces from Roy's Nd:YAG laser using the tools described. Our original intention was to characterize this system so we could use it for a demonstration of secure communication in the optical waveband. For our discussion here we use it as a class of experimental chaotic laser signals which yield nothing to Fourier analysis but clearly reveal their character to our nonlinear tools. In Figure 30.2 is the time trace of one of the laser signals. The sampling time was $\tau_s = 100$ ns. In Figure 30.3 is the power spectrum of this signal. The units along the frequency axis are 1.2 kHz. In Figure 30.4 is the average mutual information used to determine the time lag $T\tau_s$ used in reconstructing the

Laser Intensity

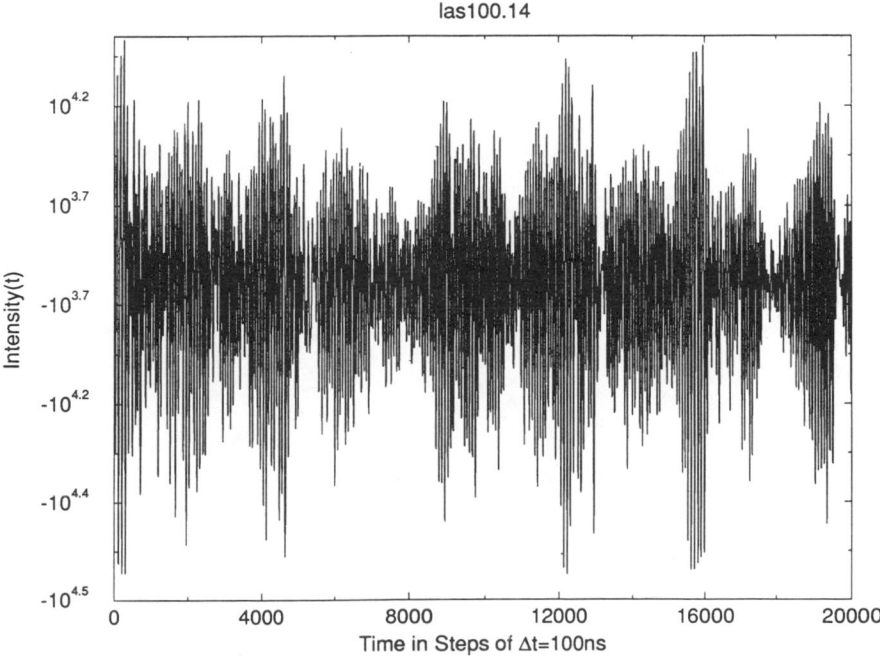

Figure 30.2 The chaotic intensity of the Nd:YAG laser from Roy's laboratory. $\tau_s = 100$ nanoseconds.

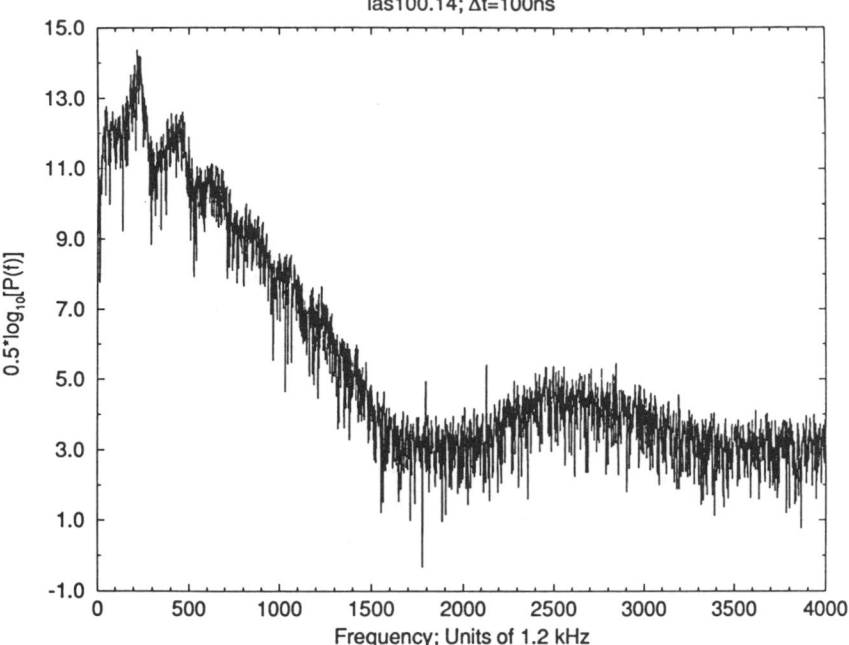

Figure 30.3 The power spectrum of the chaotic laser. The frequencies are in units of 1.2 kHz; the broad peak is around 250 kHz.

Average Mutual Information

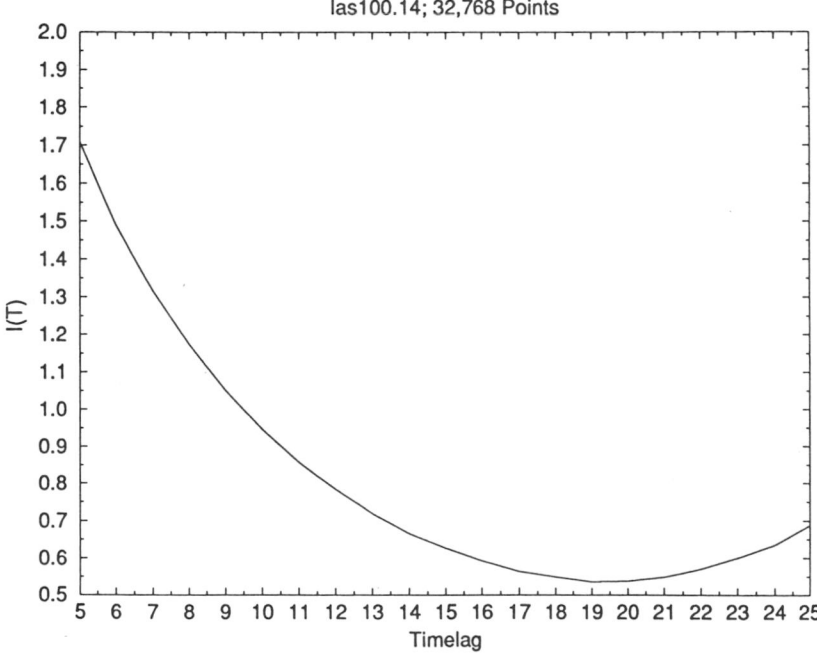

Figure 30.4 The average mutual information from the chaotic laser intensity in the region of its first minimum at $T = 19\tau_s$.

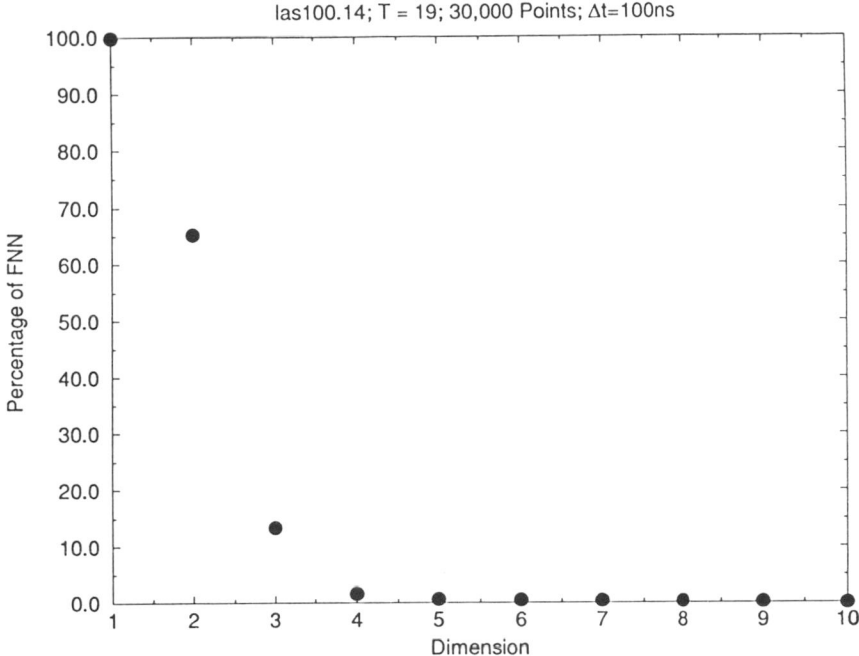

Figure 30.5   Global false nearest neighbours for the chaotic laser.

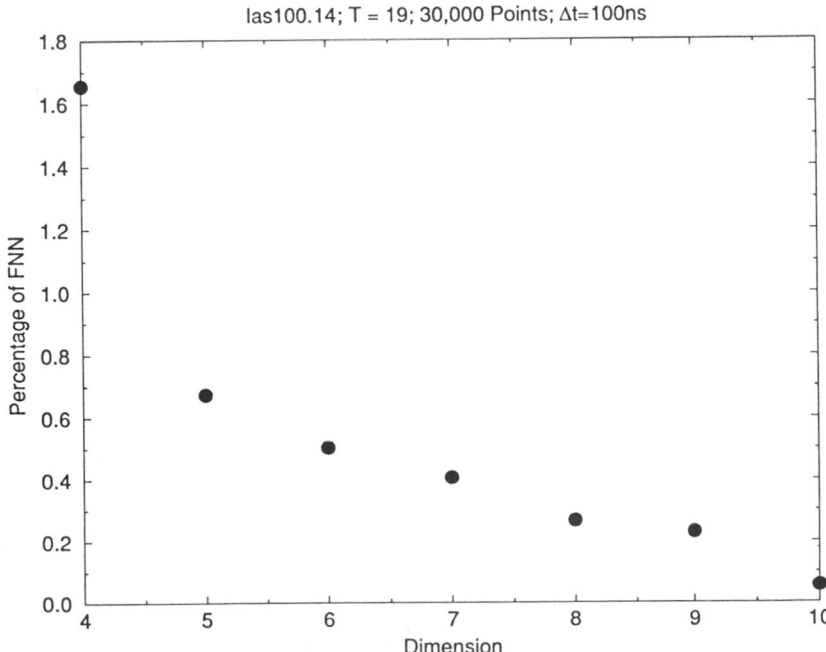

Figure 30.6   Global false nearest neighbours for the chaotic laser. This shows the approach to zero false neighbours for $d \geqslant 4$.

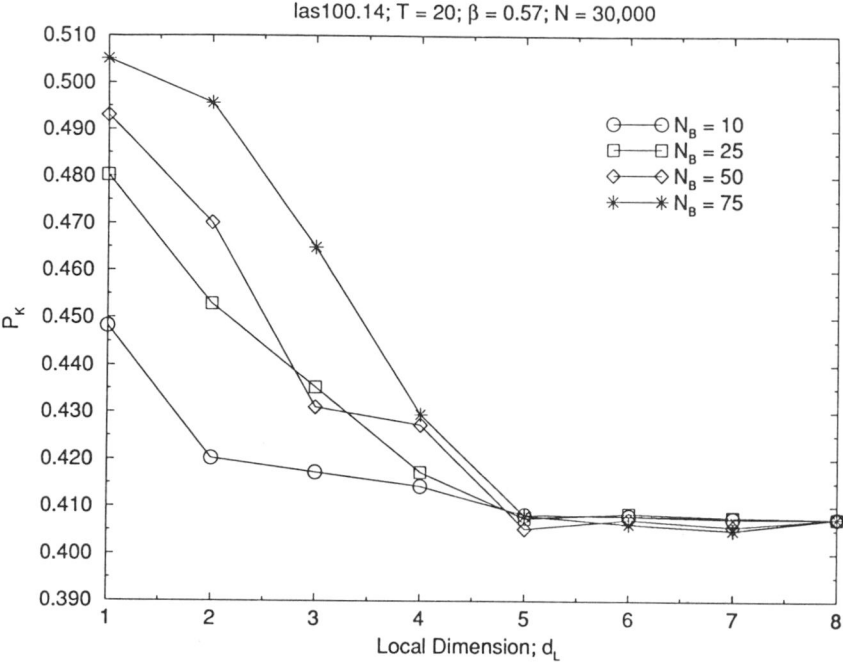

Figure 30.7 Local false nearest neighbours for the chaotic laser. The percentage of bad predictions becomes independent of d_L and N_B at $d_L = 5$.

Average Lyapunov Exponents

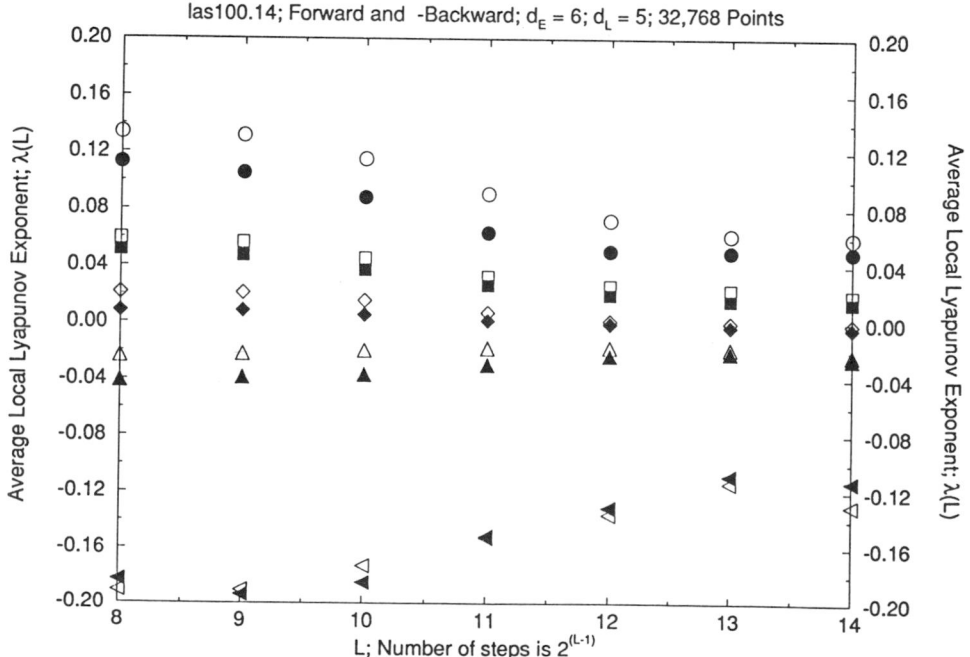

Figure 30.8 The average local Lyapunov exponents for the chaotic laser. These have been by the local false neighbour test.

phase space for the dynamics of this laser. There is a clear minimum at $T = 19$; $T\tau_s = 1.9\,\mu s$. Using this value for T we evaluated the percentage of false nearest neighbours using 30 000 data points from the chaotic intensity; this is shown in Figure 30.5. Clearly there is a low-dimensional system as the source of these chaotic intensity fluctuations. The dimension where the number of false nearest neighbours appears to reach zero is $d_E = 5$, but blowing up the regim of false negihbours near this dimension as in Figure 30.6 shows that we never quite reach zero. If we evaluate the number of bad predictions from neighbourhood to neighbourhood locally in the data varying both local dimension and number of neighbours working in a large global dimension, $d_E = 8$, where the number of false neighbours is likely governed by noise, we see in Figure 30.7 that the predictability saturates at $d_L = 5$. This indicates two things: (1) we can make models of the chaotic laser fluctuations in dimension 5 with confidence; and (2) the noise level in the data is about 0.5% as indicated by the residue of false neighbours above $d = 5$. Finally we examine the local Lyapunov exponents for this data in Figure 30.8. The value of the largest Lyapunov exponent is 500 kHz, indicating a predictability time for this laser of about $2\,\mu s$. Taking the set of Laypunov exponents we evaluate the Lyapunov dimension (Kaplan and Yorke 1979) of the system to be $D_L \approx 4.5$.

30.5   CRAMER–RAO BOUNDS

A classical problem in signal estimation is to take noisy or contaminated observations and estimate from them the true state of the system. There is a standard lower bound on how well one can do in this effort. It is called the Cramer–Rao bound (Van Trees 1968). If one has a dynamical system $x(n + 1) = F(x(n))$ producing a signal $x(n)$ but observes the noisy $z(n) = x(n) + v(n)$, how well can one estimate the values of $x(n)$? It turns out that in the case when $x(n)$ comes from a regular system, the lower bound on the quality of the estimate (the variance about the mean estimate) is the noise level divided by the number of observations. When the system is chaotic, this same bound is exponentially small in the number of observations (Richard and Abarbanel 1993) rather than going to zero as a power.

To be more precise, suppose we have made $N + M + 1$ observations of $z(n); n = -M,$ $-M + 1, \ldots, D, \ldots, N - 1, N$. Call the collection of observations Z_M^N. We wish to know aspects of the conditional probability density of the observations Z_M^N given $x(n)$, the actual state of the system; call this $p(Z_M^N)\,|x(n))$. From the observations we want to establish an estimate $x_E(n;\,Z_M^N)$ of the state $x(n)$ of the system. The dependence of this estimate on the observations is made explicit. We want to obtain a lower bound on the error covariance matrix

$$C(x_E(n;\,Z_M^N)) = \int dZ_M^N[(x_E(n;\,Z_M^N) - x(n))(x_E(n;\,Z_M^N) - x(n))^T\,|x(n)]p(Z_M^N\,|x(n)). \quad (12)$$

The Cramer–Rao lower bound uses the fact that the estimate we make is unbiased, namely

$$E[x_E(Z_M^N\,|x(n)) = x(n), \quad (13)$$

and then uses the Cauchy–Schwartz inequality to show that

$$C(x_E(n;\,Z_M^N)) \geqslant J^{-1}(x(n)), \quad (14)$$

where the *Fisher information matrix* is

$$J(x(n)) = E\left\{ \frac{\partial \log p(Z_M^N | x(n))}{\partial x(n)} \left[\frac{\partial \log p(Z_M^N | x(n))}{\partial x(n)} \right]^{\mathrm{T}} \Big| x(n) \right\}. \tag{15}$$

If we take the noise $v(n)$ to be i.i.d. with a distribution function p_v, then we can write the matrix J as

$$J(x(n)) = \sum_{k,l=-M}^{N} [DF^l(x(n))]^{\mathrm{T}} S(l,k) [DF^k(x(n))], \tag{16}$$

where

$$S(l,k)_{\alpha\beta} = \int \prod_{k,l=-M}^{N} \sigma_\alpha(z(l) - x(l)) \sigma_\gamma(z(k) - x(k)) p(Z_M^N | x(n)) dZ_M^N, \tag{17}$$

and

$$\sigma_\alpha(w) = -\frac{\partial \log p_v(w)}{\partial w_\alpha}. \tag{18}$$

This decomposition effectively decouples the dependence of J on the statistics of the noise, placing it in S, and the dynamics of the system.

We recognize the essential part of the Fisher information matrix as the Oseledec matrix which determines the local Lyapunov exponents. If the noise is Gaussian, white with a diagonal covariance matrix with entries v^2, the Fisher matrix is

$$J(x(n)) = \frac{1}{v^2} \sum_{k=-M}^{N} DF^k(x(n))^{\mathrm{T}} \cdot DF^k(x(n)). \tag{19}$$

We first consider an idealized case of DF where the entries in the Fisher matrix are precisely $e^{2\lambda a}$ which are the eigenvalues entering the Oseledec matrix. This is true for large index in the previous expression, but let us see what happens if we take it generally true. Then the Cramer–Rao bound reads

$$C(x_E(n; Z_M^N)) \geqslant v^2 \begin{bmatrix} S_1^{-1} & 0 & \cdots & 0 \\ 0 & S_2^{-1} & & 0 \\ \vdots & & \ddots & \vdots \\ 0 & 0 & \cdots & S_d^{-1} \end{bmatrix}, \tag{20}$$

with

$$S_a = e^{-2n\lambda a} \frac{e^{-2M\lambda a} - e^{(2N+2)\lambda a}}{1 - e^{2\lambda a}}. \tag{21}$$

Now imagine we wish to estimate $x(n)$ which is 'in the middle' of the data: $-M \ll n \ll N$. Then each entry in the inverse Fisher matrix corresponding to $\lambda_a \neq 0$ goes to zero as N and M become large. If $\lambda_a > 0$, then the entries behave as $\approx e^{-2N\lambda a}$, and if $\lambda_a < 0$, they behave as $e^{-2M|\lambda_a|}$. If an exponent is *precisely* zero, then the corresponding entry behaves as $1/(N+M+1)$. The latter case is the familiar one from linear dynamics where Lyapunov exponents are either zero or negative.

This argument can be generalized by considering the more general form of the Fisher matrix and working in a coordinate system where its eigenvectors are the basis. It is an orthogonal matrix, so this is always possible. The argument goes through as just stated heuristically with the interesting exception of the zero exponent. The 'zero' exponent is a feature of the inifinit limit of the Oseledec matrix. We saw earlier that all exponents approach their limit as a power of the number of steps away from some point on the atteractor. The exponent which corresponds to the global zero exponent approaches that limit as $1/(L^{1-q})$; $q > 0$, so the relevant element of the Fisher matrix behaves as e^{2Nq} as N becomes large (or with N replaced by M if the zero exponent is approached from above).

This means that every element of the inverse Fisher matrix approaches zero for a chaotic nonlinear system. So the ability to estimate any state $x(n)$ using information from the future and from the past is exponentially accurate regardless of the noise level or the statistics of the noise. In a linear system, the maximum accuracy of this kind of estimate is the noise level divided by the square root of the number of observations (this is the $\lambda_a \equiv 0$ case above). The actual rate of exponential accuracy is governed by the exponent closest to zero in the case of a map and the approach to the global zero exponent for a set of differential equations.

The bound does not tell us how to reach this accuracy in estimation, but the geometrical idea behind this estimate is essentially that a noise ball when carried into the future becomes exponentially thin along the stable directions and when carried into the past becomes exponentially thin along unstable directions. The overlap of these noise balls at the value $x(n)$ we wish to estimate is then determined by the overlap of these two thin sections of phase space.

ACKNOWLEDGEMENTS

I thank the members of INLS for numerous discussions on this subject; the comments and assistance of Reggie Brown, J.J. ('Sid') Sidorowich, Lev Tsimring and Matt Kennel were especially helpful. This work was supported in part by the US Department of Energy, Office of Basic Energy Sciences, Division of Engineering and Geosciences, under contract DE-FG03-90ER14138, and in part by the Army Research Office (Contract DAAL03-91-C-052), and by the Office of Naval Research (Contract N00014-91-C-0125), under sub-contract to the Lockheed/Sanders Corporation.

REFERENCES

Abarbanel, H. D. I., Brown, R., Sidorowich, J. J. and Tsimring, L. Sh. (1993) The analysis of observed chaotic data in physical systems, *Rev. Med. Phys.*, **65**, 1331–1392.

Abarbanel, H. D. I. and Kennel, M. B. (1993) Local false nearest neighbours and dynamical dimensions from observed chaotic data, *Phys. Rev. E*, **47**, 3057–3068.

Abarbanel, H. D. I. and Sushchik, M. M. (1993) True Lyapunov exponents and models of chaotic data, *Int. J. Bif. Chaos*, **3**, 543–550.

Bracikowski, C. and Roy, R. (1991) *Chaos*, **1**, 49–64.

Brown, R. (1992) *Phys. Rev. E*, to appear.

Brown R. Bryant, P. and Abarbanel, H. D. I. (1991) Computing the Lyapunov Spectrum of a dynamical system from observed time series, *Phys. Rev. A*, **43**, 2787–2806.

Bryant, P., Brown, R. and Abarbanel, H. D. I. (1990) Lyapunov exponents from observed time series, *Phys. Rev. Lett.*, **65**, 1523.

Eckmann, J.-P. and Ruelle, D. (1985) Ergodic theory of chaos and strange attractors, *Rev. Mod. Phys.*, **57**, 617.

Fraser, A. M. and Swinney, H. L. (1986) Independent coordinates for strange attractors, *Phys. Rev. A*, **33**, 1134–1140.

Fraser, A. M. (1989) Information and entropy in strange attractors, *IEEE Trans. Inf. Theory*, **IT-35**, 245–262.

Fraser, A. M. (1989) Reconstructing attractors from scalar time series: a compariosn of singular system and redundancy criteria, *Physica*, **34D**, 391–404.

Gencay, R. and Dechert, W. D. (1972) *Physica*, **59D**, 142–157.

Kaplan, J. L. and Yorke, J. A. (1979) Chaotic behaviour of multidimensional difference equations, In: *Functional Differential Equations and Approximations of Fixed Points*, (H.-O. Peitgen and H.-O. Walter, eds.), Lecture Notes in Maths., **730**, p. 204.

Kennel, M. B., Brown, R. and Abarbanel, H. D. I. (1992) Determining embedding dimension for phase space reconstruction using a geometrical method, *Phys. Rev. A*, **45**, 3403.

Mañé, N. (1981) In: *Dynamical Systems and Turbulence, Warwick, 1980*, (L. D. Rand and L. S. Young, eds.), Lecture Notes in Maths., 898, p. 230.

Oseledee, V. I. (1968) A multiplicative ergodic theorem. Lyapunov characteristic numbers for dynamical systems, *Trudy Mosk. Mat. Obsc.*, **19**, 197. (*Moscow Math. Soc.*, **19**, 197).

Parlitz, U. (1992) Identification of true and spurious Lyapunov exponents from time series, *Int. J. Bif. Chaos*, **2**, 155–165.

Ruhard, M. D. and Abarbanel, H. D. I. (1993) Estimating chaotic time series values Cramer–Rao bounds, *J. Nonlin. Sci.*, to appear.

Roy, R., Bracikowski, C. and James, G. E. (1992) In: *Proc. Int. Conf. on Quantum Optices*, (R. Ingura and G. S. Agarwal, eds.), New York, Plenum.

Ruelle, D. (1979) Ergodic theory of differentiable dynamical systems, *Publ. Math. Inst. Hautes Etudes Scientifiques*, No. 50, 27.

Sano, M. and Sawada, Y. (1985) *Phys. Rev. Lett.*, **55**, 1082.

Takens, F. (1981) In: *Dynamical Systems and Turbulence, Warwich 1980*, (D. Rand and L. S. Young, eds.), Lecture Notes in Maths., 898, p. 366.

Van Trees, H. L. (1968) *Detection, Estimation and Modulation Theory, Part 1: Detection, Estimation, and Linear Modulation Theory*, New York, Wiley.

H. D. I. Abarbanel, *Institute for Nonlinear Science, Department of Physics and Marine Physical Laboratory, Scripps Institute of Oceanography, University of California at San Diego, La Jolla, CA 92093-0402, USA*

31 CONTROL AND USE OF CHAOS

C. Grebogi and Y.-C. Lai

This chapter addresses the problem of controlling chaos by applying small perturbations to an accessible parameter of the system. The control of chaos is accomplished by first determining some of the unstable low-period periodic orbits or unstable steady states that are embedded in the chaotic attractor. We then examine these orbits and choose one which yields improved system performance. Finally, we apply small controls so as to stabilize the selected orbit or steady state

31.1 INTRODUCTION: A ONE-DIMENSIONAL EXAMPLE

Control of chaos using unstable periodic orbits embedded in a chaotic attractor was proposed in Ott *et al.* (1990). The basic idea is as follows. First one chooses an unstable periodic orbit embedded in the attractor, one then defines a small region around the desired periodic orbit. Due to ergodicity of trajectories on the chaotic attractor, a trajectory eventually falls into this small region. When this occurs, small judiciously chosen temporal parameter perturbations are applied to force the trajectory to approach the unstable periodic orbit. This method allows for the stabilization of different periodic orbits, depending on one's needs, for the same set of nominal values of the parameter. Hence, control of chaos gives flexibility because it allows the same system to have different kinds of behaviour.

The basic idea can be understood by considering a simple model system. We consider one of the best-understood chaotic systems, the simple one-dimensional logistic map:

$$x_{n+1} = f(x_n, \lambda) = \lambda x_n (1 - x_n), \tag{1}$$

where x is restricted to the unit interval $[0, 1]$, and λ is a control parameter. It is known that this map develops chaos via the period-doubling bifurcation route (Feigenbanm 1978). For $0 < \lambda < 1$, the asymptotic state of the map (or the attractor of the map) is $x = 0$; for $1 < \lambda < 3$, the attractor is a non-zero fixed point $x_F = 1 - 1/\lambda$; for $3 < \lambda < 1 + \sqrt{6}$, this fixed point is unstable and the attractor is a stable period-2 orbit. As λ is increased further, a sequence of period-doubling bifurcations occurs in which successive period-doubled orbits become stable. The period-doubling cascade accumulates at $\lambda = \lambda_\infty \approx 3.57$, after which chaos exists.

Nonlinearity and Chaos in Engineering Dynamics
Edited by J. M. T. Thompson and S. R. Bishop,   © 1994 John Wiley & Sons Ltd

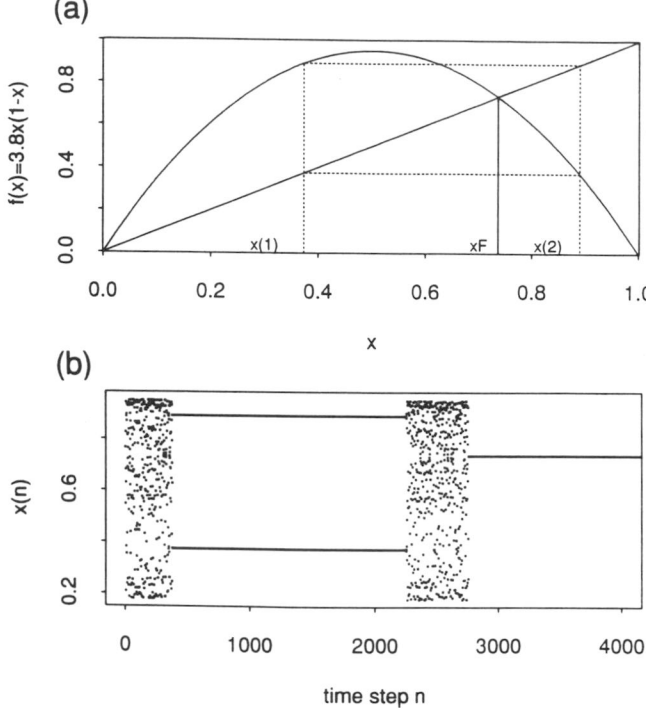

Figure 31.1 (a) The logistic map $x_{n+1} = f(x_n) = 3.8x_n(1 - x_n)$. An unstable fixed point and an unstable period-2 orbit are also shown. (b) Time series illustrating the control of the period-2 orbit and the fixed point in Figure 32.2a. The chaotic trajectory begins from $x_0 = 0.28$. At $n = 381$, the trajectory falls in an ϵ-neighbourhood of the period-2 orbit, after which the parameter control is turned on to stabilize the trajectory around the period-2 orbit. At $n = 2200$, the control is turned off. At $n = 2757$, the chaotic trajectory comes close to the fixed point and is controlled in subsequent iterates. We choose $\epsilon = 10^{-3}$. The maximum allowed parameter perturbation is $\delta = 5 \times 10^{-3}$.

Consider the case $\lambda = 3.8$ shown in Figure 31.1a for which the system is apparently chaotic. An important characteristic of a chaotic attractor is that there exists *an infinite number of unstable periodic orbits embedded within it*. For example, a fixed point $x_F \approx 0.7368$ and a period-2 orbit with components $x(1) \approx 0.3737$ and $x(2) \approx 0.8894$, where $x(1) = f(x(2))$ and $x(2) = f(x(1))$, are shown in Figure 31.1a.

Now suppose we want to avoid chaos at $\lambda = 3.8$. In particular, we want trajectories resulting from a randomly chosen initial condition x_0 to be as close as possible to the period-2 orbit shown in Figure 31.1a, assuming that this period-2 orbit gives the best system performance. Of course, in principle we can choose the desired asymptotic state of the map to be any of the infinite number of unstable periodic orbits. Suppose that the parameter λ can be finely tuned in a small range around the value $\lambda_0 = 3.8$, namely, we allow λ to vary in the range $[\lambda_0 - \delta, \lambda_0 + \delta]$, where $\delta \ll 1$. Due to the nature of the chaotic attractor, a trajectory that begins from an arbitrary value of x_0 will fall, with probability one, into the neighbourhood of the desired period-2 orbit at some later time. Because of the nature of chaos, the trajectory would diverge quickly from the period-2 orbit if we did not intervene. Our task is to program the variation of the control parameter so that the trajectory stays in the neighbourhood of the period-2 orbit as long as the control is present. In general, the small parameter perturbations will be time-dependent. We emphasize that it is important to apply only small parameter perturbations. If large parameter perturbations are allowed, then obviously we can eliminate chaos by varying λ from 3.8 to 2.0, for example. Such a large variation of λ is not interesting since it modifies the system so radically.

The logistic map in the neighbourhood of a periodic orbit can be approximated by a linear equation expanded around the period orbit. Denote the target period-m orbit to be controlled as $x(i)$, $i = 1, \ldots m$, where $x(i + 1) = f(x(i))$ and $x(m + 1) = x(1)$. Assume that at time n, the trajectory falls into the neighbourhood of component i of the period-m orbit. The linearized dynamics in the neighbourhood of component $i + 1$ is then

$$
\begin{aligned}
x_{n+1} - x(i + 1) &= \frac{\partial f(x, \lambda)}{\partial x}[x_n - x(i)] + \frac{\partial f(x, \lambda)}{\partial \lambda}\Delta\lambda_n \\
&= \lambda_0[1 - 2x(i)][x_n - x(i)] + x(i)[1 - x(i)]\Delta\lambda_n,
\end{aligned}
\tag{2}
$$

where the partial derivatives in (2) are evaluated at $x = x(i)$ and $\lambda = \lambda_0$. We require x_{n+1} to stay in the neighbourhood of $x(i + 1)$. Hence, we set $x_{n+1} - x(i + 1) = 0$, which gives the required parameter perturbation

$$
\Delta\lambda_n = \lambda_0 \frac{[2x(i) - 1][x_n - x(i)]}{x(i)[1 - x(i)]}.
\tag{3}
$$

Equation (3) holds only when the trajectory x_n enters a small neighbourhood of the period-m orbit, i.e. when $|x_n - x(i)| \ll 1$, and hence the required parameter perturbation $\Delta\lambda_n$ is small. Let the length of a small interval defining the neighbourhood around each component of the period-m orbit be 2ε. In general, the required maximum parameter perturbation δ is proportional to ε. Since ε can be chosen to be arbitrarily small, δ can also be made arbitrarily small. As we will see, the average transient time before a trajectory enters the neighbourhood of the target periodic orbit depends on ε (or δ). When the trajectory is outside the neighbourhood of the target periodic orbit, we do not apply any parameter perturbation, and the system evolves at its nominal parameter value λ_0. Hence we usually set $\Delta\lambda_n = 0$ when $\Delta\lambda_n > \delta$. Note that the parameter perturbation $\Delta\lambda_n$ depends on x_n and is time-dependent.

The above strategy for controlling the orbit is very flexible for stabilizing different periodic orbits at different times. Suppose we first stabilize a chaotic trajectory around the period-2 orbit shown in Figure 31.1a. Then we might wish to stabilize the fixed point in Figure 31.1a, assuming that the fixed point would correspond to a better system performance at a later time. To achieve this change of control, we simply turn off the parameter control with respect to the period-2 orbit. Without control, the trajectory will diverge from the period-2 orbit exponentially. We let the system evolve at the parameter value λ_0. Due to the nature of chaos, there comes a time when the chaotic trajectory enters a small neighbourhood of the fixed point. At this time we turn on a new set of parameter perturbations calculated with respect to the fixed point. The trajectory can then be stabilized around the fixed point.

Figure 31.1b shows an example where we first control the period-2 orbit and then the fixed point shown in Figure 31.1a. The initial condition is $x_0 = 0.28$. At time $n = 381$, the trajectory enters the neighbourhood of the component $x(1)$ of the period-2 orbit. For subsequent iterations, the parameter control calculated from (3) is used to stabilize the trajectory around the period-2 orbit. At time $n = 2200$, we choose to stabilize the trajectory around the fixed point, and hence we turn off the parameter perturbation. The trajectory quickly leaves the period-2 orbit and becomes chaotic. At time $n = 2757$, the trajectory falls into the neighbourhood of the fixed point. Parameter perturbations calculated with respect to the fixed point are then turned on to stabilize the trajectory around the fixed point.

In the presense of external noise, a controlled trajectory will occasionally be 'kicked' out of the neighbourhood of the periodic orbit. If this behaviour occurs, we turn off the parameter perturbation and let the system evolve by itself. With probability one, the chaotic trajectory will enter the neighbourhood of the target periodic orbit and can be controlled again. This situation is illustrated in Figure 31.2a where are control the period-2 orbit. The noise is modelled by an additive term in the logistic map of the form $\eta\sigma_n$, where σ_n is a Gaussian distributed random variable with zero mean and unit standard deviation, and η is the noise amplitude. The effect of the noise is to turn a controlled periodic trajectory into an intermittent one in which chaotic phases (uncontrolled trajectories) are interspersed with laminar phases

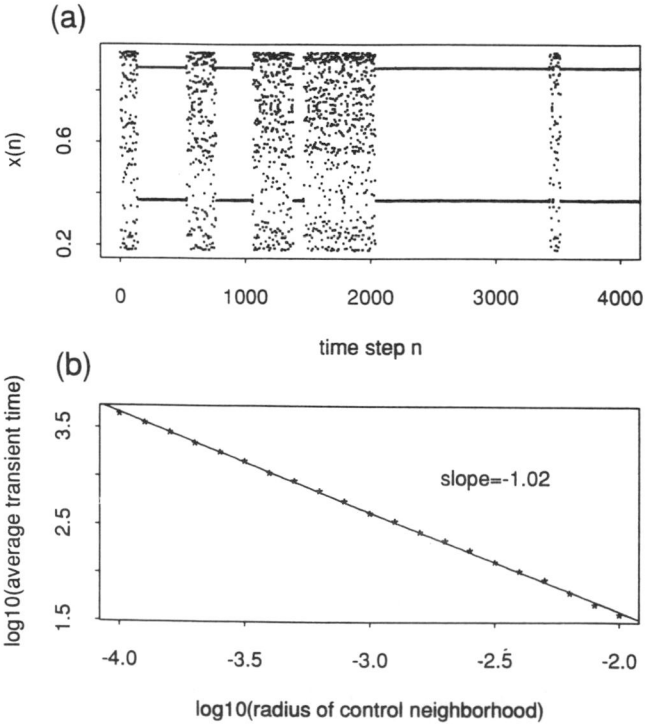

Figure 31.2   (a) The effect of additive noise modelled by $2.6 \times 10^{-4}\sigma_n$, where σ_n is a Gaussian random variable with zero mean and unit standard deviation. The noise can occasionally kick the controlled trajectory out of the neighbourhood of the periodic orbit. (b) Log-log plot of the average time to achieve control $\tau(\varepsilon)$ versus ε, the size of the controlling neighbourhood. Twenty values of ε are chosen on a logarithmic scale. For each ε, 2000 random initial conditions uniformly distributed in $[0, 1]$ are chosen to compute $\tau(\varepsilon)$. The scaling relation between $\tau(\varepsilon)$ and ε is well fitted by $\tau(\varepsilon) \sim \varepsilon^{-1}$.

(controlled periodic trajectories). It is easy to verify that the averaged length of the laminar phase increases as the noise amplitude decreases, and the length tends to infinity as $\eta \to 0$.

It is interesting to ask how many iterations are required on average for a chaotic trajectory originating from an arbitrarily chosen initial condition to enter the neighbourhood ε of the target periodic orbit. Clearly, the smaller the value of ε, the more iterations are required. In general, the average transient time $\tau(\varepsilon)$ before turning on control scales with ε as

$$\tau(\varepsilon) \sim \varepsilon^{-\gamma}, \tag{4}$$

where $\gamma > 0$ is a scaling exponent. For one-dimensional maps such as the logistic map, there usually exists a smooth probability density $\rho(x)$ for trajectory points on the attractor. The probability density ρ can be defined as the frequency that a chaotic trajectory visits a small neighbourhood of the point x on the attractor. In such a case, we have $\gamma = 1$, as can be seen from the following argument. The probability that a trajectory enters the neighbourhood of a particular component (component i) of the periodic orbit is given by

$$P(\varepsilon) = \int_{x(i)-\varepsilon}^{x(i)+\varepsilon} \rho x(i))\, \mathrm{d}x \approx 2\varepsilon \rho(x(i)). \tag{5}$$

Hence $\tau(\varepsilon) = 1/P(\varepsilon) \sim \varepsilon^{-1}$, and therefore $\gamma = 1$. This behaviour in illustrated in Figure 32.2b, where $\tau(\varepsilon)$ is plotted on a logarithmic scale for the case of stabilizing the period-2 orbit in Figure 32.1a. Twenty values of ε were chosen in the range $[10^{-4}, 10^{-2}]$. For each ε, we randomly choose 2000 initial conditions (with a uniform probability distribution) and calculate an average transient time. The linearly fitted slope of the line is approximately -1.02, indicating good agreement with the theoretical prediction of $\gamma = 1$. For higher-dimensional chaotic systems, the exponent γ can be related to the eigenvalues of the target periodic orbit (Ott *et al.* 1990).

A major advantage of the controlling chaos idea in Ott *et al.* (1990) is that it can be applied to experimental systems in which *a priori* knowledge of the system is usually not available. A time series found by measuring one of the system's dynamical variables in conjunction with the time-delay embedding method (Takens 1981), which transforms a scalar time series into a trajectory in phase space, is sufficient to determine the desired unstable periodic orbits to be controlled and the relevant quantities required to compute parameter perturbations (Ott *et al.* 1990). Another advantage of the method is its flexibility in choosing the desired periodic orbit to be controlled. The method has attracted growing interest for controlling dynamical systems. Theoretically, it has been extended to higher-dimensional dynamical systems (Romeiras *et al.* 1992; Auerbach *et al.* 1992). Hamiltonian systems (Lai *et al.* 1993a), control of transient chaos (Tél 1991) and chaotic scattering (Lai *et al.* 1993b), and the synchronization of chaotic systems (Lai and Grebogi 1993a). In Section 31.2 we will describe the method formulated for two-dimensional maps.

31.2   TWO-DIMENSIONAL ALGORITHM

The general algorithm for controlling chaos for two-dimensional maps or three-dimensional autonomous flows that can be reduced to two-dimensional maps on the Poincaré surface of section can be formulated in a similar way. (By autonomous flow we mean that the vector field does not contain an explicit time dependence.) Consider the two-dimensional map

$$X_{n+1} = F(X_n, p), \tag{6}$$

where $X_n \in \mathbb{R}^2$, F is a smooth function of its variables, and $p \in \mathbb{R}$ is an externally accessible control parameter. We restrict parameter perturbations to be small, i.e.

$$|p - p_0| < \delta, \tag{7}$$

where p_0 is some nominal parameter value, and $\delta \ll 1$ defines the range of parameter variation. We wish to program the parameter p so that a chaotic trajectory is stabilized when it enters an ε-neighbourhood of the target periodic orbit. Specifically, let the desired period-m orbit be $X(1, p_0) \to X(2, p_0) \to \cdots \to X(m, p_0) \to X(m + 1, p_0) = X(1, p_0)$. The linearized dynamics in the neighbourhood of component $i + 1$ of the period-m orbit is

$$X_{n+1} - X(i + 1, p_0) = A \cdot [X_n - X(i, p_0)] + B\Delta p_n, \tag{8}$$

where $\Delta p_n = p_n - p_0$, $\Delta p_n \leq \delta$, A is a 2×2 Jacobian matrix, and B is a two-dimensional column vector:

$$A = D_x F(X, p)|_{X=X(i), p=p_0}, \quad B = D_p F(X, p)|_{X=X(i)=p_0}. \tag{9}$$

To stabilize periodic orbits in two-dimensional maps, we take advantage of the geometric structure associated with unstable periodic orbits. Specifically, there exist a stable and an unstable direction at each component of an unstable periodic orbit. The stable (unstable) direction is a direction along which points approach (leave) the periodic orbit exponentially on the average. (For higher-dimensional maps, there may be several stable and unstable directions. The algorithm to control chaos in such cases is more complicated (Romeiras *et al.* 1992; Anerbach *et al.* 1992) and will not be discussed here.) The existence of both stable and unstable directions at each point of the trajectory can be seen as follows. Choose a small circle of radius ε around an orbit point $X(i)$. This circle can be written as $dx^2 + dy^2 = \varepsilon^2$ in the Cartesian coordinate system whose origin is at $X(i)$. The image of the circle under F^{-1} can be expressed as $A dx'^2 + B dx' dy' + C dy' = 1$, an equation for an ellipse in the Cartesian coordinate system whose origin is at $X(i-1)$. The coefficients A, B and C are functions of elements of the inverse Jacobian matrix at $X(i)$. This deformation from a circle to an ellipse means that the distance along the major axis of the ellipse at $X(i-1)$ contracts as a result of the map. Similarly, the image of a circle at $X(i-1)$ under F is typically an ellipse at $X(i)$, which means that the distance along the inverse image of the major axis of the ellipse at $X(i)$ expands under F. Thus the major axis of the ellipse at $X(i-1)$ and the inverse image of the major axis of the ellipse at $X(i)$ approximate the stable and unstable directions at $X(i-1)$. We note that typically the stable and unstable directions are not orthogonal to each other, and in rare situations they can be identical (non-hyperbolic dynamical systems) (Lai *et al.* in press).

To calculate these stable and unstable directions, we use an algorithm developed in Lai *et al.* (1993). This algorithm can be applied to cases where the period of the orbit is arbitrarily large. To find the stable direction at a point X, we first iterate this point forward N times under the map F and obtain the trajectory $F^{-1}(X), F^2(X), \ldots, F^N(X)$. Now imagine we place a circle of arbitrarily small radius ε at the point $F^N(X)$. If we iterate this circle backward once, the circle will become an ellipse at the point $F^{N-1}(X)$, with the major axis along the stable direction of the point $F^{N-1}(X)$. We continue iterating this ellipse backwards, while at the same time rescaling the ellipse's major axis to be order ε. When we iterate the ellipse back to the point X, the ellipse becomes very thin with its major axis along the stable direction at the point X, if N is sufficiently large. For a short period-m orbit, we choose $N = km$ with k an integer. In practice, instead of using a small circle, we take a unit vector at the point $F^N(X)$, since the Jacobian matrix of the inverse map F^{-1} rotates a vector in the tangent space of F towards the stable direction. Hence we iterate a unit vector backward to the point X by multiplying by the Jacobian matrix of the inverse map at each point on the already existing orbit. We rescale the vector after each multiplication to unit length. For sufficient large N, the unit vector so obtained at X is a good approximation to the stable direction at X.

Similarly, to find the unstable direction at point X, we first iterate X backward under the inverse map N times to obtain a backward orbit $F^{-j}(X)$ with $j = N, \dots, 1$. We then choose a unit vector at point $F^{-N}(X)$ and iterate this unit vector forward to the point X along the already existing orbit by multiplying by the Jacobian matrix of the map N times. (Recall that the Jacobian matrix of the forward map rotates a vector towards the unstable direction.) We rescale the vector to unit length at each step. The final vector at point X is a good approximation to the unstable direction at that point if N is sufficiently large.

The above method is efficient. For instance, the error between the calculated and real stable or unstable directions is on the order of 10^{-10} for chaotic trajectories in the Hénon map if $N = 20$ (Lai *et al.* 1993).

Let $e_{s,i}$ and $e_{u,i}$ be the stable and unstable directions at $X(i)$, and let $f_{s,i}$ and $f_{u,i}$ be the corresponding contravariant vectors that satisfy the conditions $f_{u,i} \cdot e_{u,i} = f_{s,i} \cdot e_{s,i} = 1$, and $f_{u,i} \cdot e_{s,i} = f_{s,i} \cdot e_{u,i} = 0$. To stabilize the orbit, we require that the next iteration of a trajectory point, after falling into a small neighbourhood about $X(i)$, along the stable direction at $X(i+1, p_0)$, i.e.

$$[X_{n+1} - X(i+1, p_0)] \cdot f_{u,i+1} = 0. \tag{10}$$

If we take the dot product of both side of (8) with $f_{u,i+1}$ and use (10), we obtain the following expression for the parameter perturbations:

$$\Delta p_n = \frac{\{A \cdot [X_n - X(i, p_0)]\} \cdot f_{u,i+1}}{-B \cdot f_{u,i+1}}. \tag{11}$$

The general algorithm for controlling chaos for two-dimensional maps can be summarized as follows:

1. Find the desired unstable periodic orbit to be stabilized.
2. Find a set of stable and unstable directions, e_s and e_u, at each component of the periodic orbit. The set of corresponding contravariant vectors f_s and f_u can be found by solving $e_s \cdot f_s = e_u \cdot f_u = 1$ and $e_s \cdot f_u = e_u \cdot f_s = 0$.
3. Randomly choose an initial condition and evolve the system at the parameter value p_0. When the trajectory enters the ε-neighbourhood of the target periodic orbit, calculate parameter perturbations at each time step according to (11).

The algorithm described above applies to two-dimensional invertible maps. In general, dynamical systems that can be described by a set of first-order autonomous differential equations are invertible, and the inverse system is obtained by letting $t \rightarrow -t$ in the original set of differential equations. Hence, the discrete map obtained on the Poincaré surface of section is also invertible. Most dynamical systems encountered in practice fall into this category. Non-invertible dynamical systems possess very distinct properties (Chossat and Golubitsky 1988; Chin *et al.* 1992–3). For instance, for two-dimensional non-invertible maps, a point on a chaotic attractor may not have a unique stable (unstable) direction. A method for determining all these stable and unstable directions is not known. If one or several such directions at the target unstable periodic orbit can be calculated, our method can in principle be applied to non-invertible systems by forcing a chaotic trajectory to fall in one of the stable directions of the periodic orbit.

31.3 DISCUSSION

In summary, we have presented an algorithm for converting chaos into periodic motion for one- and two-dimensional maps by using only small parameter perturbations. For two-dimensional maps, the geometrical structure associated with an unstable periodic orbit (namely, the stable and unstable directions) is utilized to effectively achieve control. In this way the original control method of Ott et al. (1990) can be extended to problems such as controlling chaos in Hamiltonian dynamical systems (Lai *et al.* 1993a), synchronizing chaotic systems (Lai and Grebogi 1993a), and converting transient chaos into sustained chaos (Lai and Grebogi 1994). In the following we briefly discuss these issues. Hamiltonian chaotic systems often arise in the study of conservative dynamical systems such as the gravitational three-body system. In such systems, the Jacobian matrices at some components of an unstable periodic orbit often possess complex-conjugate eigenvalues (Lai *et al.* 1993a) and, hence, eigenvectors are not even defined in the real plane for these components. It is important to note that while eigenvectors may not be defined, at each component of the unstable periodic orbit there exist both a stable and an unstable direction which are associated with the m-times iterated system, where m is the period of the unstable orbit. The method we presented in Section 32.2 captures the unstable nature of the entire periodic orbit, although each local Jocobian matrix may not be unstable. Thus, by exploring these stable and unstable directions, one can successfully apply the controlling-chaos algorithm at *each iteration* of the system to significantly reduce the effect of external noise (Lai *et al.* 1993a).

The use of stable and unstable directions can also be applied to synchronize chaotic trajectories (Lai and Grebogi). The key point is that each point of a chaotic trajectory possesses both stable and unstable directions. The idea is as follows. Consider two chaotic systems A and B. Assume that some parameter of one system (assume B) is externally adjustable and that some state variables of both systems A and B can be measured. Based on this measurement and our knowledge about the system (we can, for example, observe and learn the system beforehand), when it is determined that the state variables of A and B are close, we calculate a small parameter perturbation based on equation (11) and apply it to system B. Two systems can then be synchronized. Under the influence of external noise, there is a finite probability that the two already-synchronized trajectories may lose synchronization. However, with probability one (due to the ergodicity of chaotic trajectories), after a finite amount of transient time, the trajectories of A and B will get close and will be synchronized again. In this sense, the synchronization method is robust against small external noise. This idea has also been applied to a model of a chaotic laser system (Lai and Grebogi 1993b) and has been experimentally verified using an electronic circuit (Newell *et al.* 1994).

The formulation of Section 32.2 can also be applied to convert transient chaos into sustained chaos (Lai and Grebogi 1994) so as to restore sustained chaotic motion from catastrophic events such as boundary crises (Grebogi *et al.* 1982, 1983) in which a chaotic attractor is suddenly destroyed. It has been known that the origin of transient chaos is a non-attracting chaotic saddle in the phase space. The key observation is that on the chaotic saddle there exist dense chaotic orbits. By selecting one such non-attracting orbit as the *reference orbit*, which can be constructed by using the PIM-triple method (Nusse and Yorke 1989), we can make other trajectories stay in the neighbourhood of this reference orbit for as long as we wish by applying small, judiciously chosen temporal parameter perturbations. In this sense, non-attracting trajectories in the neighbourhood of the chaotic saddle are transformed into stable chaotic trajectories. This can indeed be achieved since there exist stable and unstable directions at each point of the reference orbit on the chaotic saddle. Hence, in principle, controlling a trajectory on the chaotic saddle is equivalent to stabilizing a long unstable periodic orbit.

Finally, we remark that the controlling-chaos idea may also help solve technologically important problems in many disciplines. In communications, it has been proposed that chaotic fluctuations can be put into use to send controlled, pre-planned signals (Hayes *et al.* 1993). The first controlling-chaos experiment was performed by Ditto *et al.* (1990) who successfully controlled a chaotically oscillating magnetic ribbon. Azevedo and Rezende (1991) controlled chaos in a spin-wave system. In fluid mechanics, it has been demonstrated in a simple configuration that chaotic convection can be controlled (Singer *et al.* 1991). In chemistry, researchers have developed mechanisms for controlling chaotic autocatalytic reactions (Pong *et al.* 1991; Petrov *et al.* 1993). In physiology, applications have been demonstrated for controlling chaos in the heart (Garfinkel *et al.* 1992). Chaotic lasers have been controlled (Roy *et al.* 1992; Gills *et al.* 1992; Bielawski *et al.* 1992a, b; Reyl *et al.* 1992), as has the chaotic diode circuit (Hunt 1992; Johnson and Hunt 1993). These and other achievements will certainly stimulate more applications of control of chaos in the future.

ACKNOWLEDGEMENT

This work was supported by DOE (Office of Scientific Computing, Office of Energy Research).

REFERENCES

Auerbach, D., Grebogi, C., Ott, E. and Yorke, J. A. (1992) *Phys. Rev. Lett.*, **69**, 3479.

Azevedo, A. and Rezende, S. M. (1991) *Phys. Rev. Lett.*, **66**, 1342.

Bielawski, S., Bouazaoui, M., Derozier, D. and Glorieux, P. (1992a) In: *Proc. Nonlinear Dynam. Optical Syst. Topical Mtg., Alpbach, June 22–26* Washington DC, Optical Society of America.

Bielawski, S., Bouazaoui, M., Derozier, D. and Glorieux, P. (1992b) Stabilization and characterization of unstable steady states in a laser, Preprint.

Chin, W., Kan, I. and Grebogi, C. (1992–3) *Random and Computational Dynamics*, **1**, 349.

Chossat, P. and Golubitsky, M. (1988) *Physica*, **32D**, 423.

Ditto, W. L., Rauseo, S. N. and Spano, M. L. (1990) *Phys. Rev. Lett.*, **65**, 3211.

Feigenbaum, M. J. (1978) *J. Stat. Phys.*, **19**, 25.

Garfinkel, A., Spano, M. L., Ditto, W. L. and Weiss, J. N. (1992) *Science*, **257**, 1230.

Gills, Z., Iwata, C., Roy, R., Schwartz, I. B. and Triandaf, I. (1992) *Phys. Rev. Lett.*, **69**, 3169.

Grebogi, C., Ott, E. and Yorke, J. A. (1982) *Phys. Rev. Lett.*, **48**, 1507.

Grebogi, C., Ott, E. and Yorke, J. A. (1983) *Physica*, **70**, 181.

Hayes, S., Grebogi, C. and Ott, E. (1993) *Phys. Rev. Lett.*, **70**, 3031.

Johnson, G. A. and Hunt, E. R. (1993) *Int. J. Bifurcation and Chaos*, **3**, 789.

Lai, Y. C. and Grebogi, C. (1993a) *Phys. Rev. E.*, **47**, 2357.

Lai, Y. C. and Grebogi, C. (1993b) In: *Chaos in Communications, San Diego '93*, SPIE Proceedings, Vol. 2038.

Lai, Y. C. and Grebogi, C. (1994) Converting transient chaos into sustained chaos by feedback control, *Phys. Rev. E*, **49**, 1094.

Lai, Y. C., Ding, M. and Grebogi, C. (1993a) *Phys. Rev. E.*, **47**, 86.

Lai, Y. C., Tél, T. and Grebogi, C. (1993b) *Phys. Rev. E.*, **48**, 709.

Lai, Y. C., Grebogi, C., Yorke, J. A. and Kan, I. (1993) *Nonlinearity*, **6**, 779–797.

Newell, T. C., Alsing, P. M., Gavrielides, A. and Kovanis, V. (1994) *Phys. Rev. Lett.* **72**, 1647.

Nusse, H. E. and Yorke, J. A. (1989) *Physica*, **36D**, 137.

Hunt, E. R. (1992) *Phys. Rev. Lett.*, **67**, 1953.

Peng, B. Petrov, V. and Showalter, K. (1991) *J. Phys. Chem.*, **95**, 4957.

Petrov, V., Gáspár, V., Masere, J. and Showalter, K. (1993) *Nature*, **361**, 240.

Reyl, C., Flepp. L., Badii, R. and Brun, E. (1992) Control of NMR-laser chaos in high-dimensional embedding space, Preprint.

Romeiras, F. J., Grebogi, C., Ott, E. and Dayawansa, W. (1992) *Physica*, **58D**, 165.

Roy, R., Murphy, Jr., T. W., Maier, T. D. and Gills, Z. (1992) *Phys. Rev. L ett.*, **68**, 1259.

Singer, J., Wang, Y.-Z. and Bau, H. H. (1991) *Phys. Rev. Lett.*, **66**, 1123.

Takens, F. (1981) In: *Lecture Notes in Mathematics*, No. 898, Berlin, Springer-Verlag.

Tél T. (1991) *J. Phys. A: Math. Gen.*, **24**, L1359.

C. Grebogi, *Laboratory for Plasma Research and Department of Mathematics and Institute for Physical Science and Technology, University of Maryland, College Park, MD 20782, USA.*

Y.-C. Lai, *Department of Physics and Astronomy, Department of Mathematics, and Kansas Institute for Theoretical and Computational Science, The University of Kansas, Lawrence, KS 66045, USA.*

Appendix I Short Reports of Current Research

This appendix contains a brief survey of additional current research with references to help the reader. The contributions were presented as posters at the IUTAM Symposium.

LIMIT CYCLES IN PERTURBED EULER EQUATIONS FOR ASYMMETRIC BODIES

V. T. Coppola

University of Michigan, UK

Traditional perturbatative approaches for investigating Euler's equations of rotational motion have focused on predicting the effects of small disturbing torques on either (i) motions which are nominally simple spins or (ii) nearly symmetric bodies. In these cases, Euler's equations reduce to a perturbed linear oscillator whose study may proceed using a wide assortment of standard perturbation techniques. The case of asymmetric bodies performing general rigid-body motion (including large coning/nutation) has been largely avoided because the strong nonlinearity of Euler's equations causes many of the standard perturbation techniques to fail. In spite of this, the method of averaging (Hale 1969; Sanders and Verhulst 1985) may still be applied because the solution for rigid-body motion involves periodic functions of time. The difficulty, however, is that the functions are elliptic rather than trigonometric.

Euler's equations for the rotational motion of a rigid body are

$$I\frac{d\omega}{dt} + \omega \times (I\omega) = \varepsilon M(\omega) \tag{1}$$

where ω is the angular velocity vector, I is the inertia matrix in principle axes (assumed asymmetric) about the body's centre of mass, $\varepsilon M(\omega)$ is the disturbing torque vector, and ε is a small scalar parameter. Both the angular momentum $L(\omega)$ and rotational energy $T(\omega)$ are constants of motion in the torque-free case. Moreover, the orientation of the body can be found by investigating the flow of (1) on the momentum sphere (Hughes 1986).

The perturbation approach seeks to determine the effects of small disturbing torques on the evolution of L, T and the orientation. The unperturbed system is explicitly solvable for ω using elliptic functions (Hughes 1986). The solution is expressed using two 'slow' variables

Nonlinearity and Chaos in Engineering Dynamics
Edited by J. M. T. Thompson and S. R. Bishop,   © 1994 John Wiley & Sons Ltd

and an angle, the 'fast' variable. The variational equations are then computed for the perturbed system and the method of averaging using elliptic functions is applied to first order. The result is the slow-flow equations whose evolution depends only upon the slow variables, the angle being averaged out.

The differences between elliptic and trigonometric functions lead to fundamental complications in performing the averaging. First, elliptic functions depend on two parameters (argument and modulus) rather than simply one argument. The modulus (varying with L and T) determines the period of the elliptic function: it is not a constant value. Second, the addition rule for elliptic functions does not lead to a simple rule for expressing powers of elliptic functions in terms of elliptic functions with integer multiples of the argument. Such a rule for trigonometric functions leads to the set being closed under integration: this is not the case for elliptic functions. Thus, the averaging procedure can introduce new (non-elliptic) functions at each order of the computation beginning with the second (Coppola 1989).

Chernous'ko (1963) used elliptic averaging to investigate the effects of gravitational torques on asymmetric artificial satellites. Akulenko *et al.* (1982) used elliptic averaging to investigate the motion of a rigid body in a resistive medium about a fixed point. A detailed discussion of elliptic averaging appears in Coppola and Rand (1990) and Coppola (1989). Because of the computational complexities involving elliptic functions, the averaging procedure has been implemented using an Elliptic Function Processor (EFP) written in Mathematica. The EFP allows one to integrate, differentiate and simplify elliptic functions symbolically in much the same manner as for trigonometric functions.

In particular, we are concerned with torques of the form $M = G\omega$ where G is a 3×3 matrix. This form can be used to model linear feedback control systems or torques resulting from internal moving parts (Thomson 1986). The scaling requires that ω be sufficiently large compared to M so that the rigid-body motion is a 'fast' motion compared with the effects of disturbances.

The slow-flow equations further decouple so that the averaged system is completely determined from the slow flow on the momentum sphere (i.e. a nonlinear ODE for the modulus) and two subsequent integrations. For particular values of the elements of G, the averaged system predicts the existence of limit cycles.

Interestingly, a limit cycle predicted in ω-space implies the existence of one on the momentum sphere, but the converse is not true. Thus, the energy may change monotonically while the orientation (i.e. coning) repeats cyclically. This is an important prediction in the design of a linear feedback controller since a limit cycle might prevent a desired reorientation of a satellite and adversely affect its rotational energy.

REFERENCES

Akulenko, L. D., Leshchenko, D. D. and Chernous'ko F. L. (1982) Fast motion of a heavy rigid body about a fixed point in a resistive medium, *Izv. AN SSSR. MTT*, **17**(3), 5–13.

Chernous'ko, F. L. (1963) On the motion of a satellite about its center of mass under the action of gravitational moments, *PMM*, **3**, 474–483. English translation in *Appl. Math. Mech.*, **27** (3), 708–722.

Coppola, V. T. (1989) Averaging of Strongly Nonlinear Oscillators using Elliptic Functions, Ph.D. Dissertation, Cornell University.

Coppola, V. T. and Rand, R. H. (1990) Averaging using elliptic functions: approximation of limit cycles, *Acta. Mech.*, **81**, 125–142.

Hale, J. K. (1969) *Ordinary Differential Equations*, New York, Wiley.

Hughes, P. C. (1986) *Spacecraft Attitude Dynamics*, New York, Wiley.

Sanders, J. and Verhulst, F. (1985) *Averaging Methods in Nonlinear Dynamical Systems*, Applied Mathematical Sciences Vol. **59**, New York, Springer-Verlag.

Thomson, W. T. (1986) *Introduction to Space Dynamics*, Mineola, NY, Dover.

V. T. Coppola, *Department of Aerospace Engineering University of Michigan, Ann Arbor, MI 48109–2140, USA.*

A COMBINED APPROACH TO EVALUATE THE EFFECT OF MODELLING APPROXIMATIONS IN PREDICTING VESSEL CAPSIZING

J. M. Falzarano

This work describes a combined steady-state and transient approach to evaluating the qualitative and quantitative importance of modelling approximations on nonlinear ship rolling motion possibly leading to capsizing.

The subject of large-amplitude (nonlinear) ship rolling motion leading to capsizing is an important practical problem which has only recently received adequate attention. Prior to this recent interest, the state of the art in analysing such motion was to study the static roll-restoring moment in isolation or to study the linearized dynamics. Anyone even vaguely familiar with the physical problem will instantly realize that both of these approaches are hopelessly inadequate. Until recently the only way to analyse highly nonlinear systems was by numerical simulations; however, a simulation approach is also inadequate since only a finite number of simulations can be performed, yielding an incomplete understanding of the system's behaviour. In order to overcome these shortcomings, many researchers have used the techniques of nonlinear vibration (i.e. perturbation techniques) theory and dynamical systems theory. The perturbation method of approach suffers from limitations with respect to the size and complexity of the problem and the relative size of the nonlinearity. With the exception of some isolated studies on pitch/roll coupling and saturation, the perturbation approach has been restricted to single-degree-of-freedom studies. Furthermore, the fact that perturbation methods are in general restricted to moderate-amplitude steady-state response makes them of limited use in studying the large-amplitude response approaching capsizing.

In order to overcome these shortcomings, a numerical approach has been used which is not restricted in either the size of the problem or the relative size of the nonlinearity. This approach has been combined with a determination of the critical basin boundaries in the

Nonlinearity and Chaos in Engineering Dynamics
Edited by J. M. T. Thompson and S. R. Bishop,   © 1994 John Wiley & Sons Ltd

Poincaré map to yield a somewhat more complete understanding of the dynamical system than is otherwise possible (Falzarano 1990). Using this combined approach, it is possible to analyse additional significant aspects of more general dynamical systems than using either method in isolation. Although the individual approaches have been published previously by the author and his distinguished colleagues, this approach simultaneously applies steady-state bifurcation analysis and transient Poincaré map analysis to evaluate critical behaviour of coupled large-amplitude nonlinear roll motion. The focus is on critical behaviour to guide further analysis. This study should be important in evaluating the significance of various levels of modelling approximation in effecting the predicted qualitative and quantitative behaviour of the ship dynamical system. Moreover, this study should provide guidance to researchers and practitioners in this area concerning the relative importance of the various modelling approximations in accurately predicting the large-amplitude vessel roll motion. This guidance will suggest what parameters most strongly qualitatively and quantitatively effect the roll motion, so that researchers can focus their efforts in refining their prediction of these parameters and so that practioners can assess the accuracy of their aproximate predictions. Improved analysis techniques should 'bridge the gap' between unnecessarily simple approaches that miss critical physics and unnecessarily complicated and detailed models that require many CPU hours on the largest of supercomputers. Furthermore, the emphasis of this approach is to directly study physically meaningful quantities in convenient formats in lieu of specialized derived quantities which cause the analyst to lose sight of the original problem.

The approach outlined above will combine the numerical path-following approach previously applied (Falzarano *et al.* 1990, 1991) to study the coupled steady-state roll motion with the Poincaré mapping/invariant manifolds technique applied in Falzarano and Troesch (1990) and Falzarano *et al.* (1992). The steady-state analysis techniques and the transient techniques are being extended to consider even more general and strongly nonlinear systems. These two improved techniques combine into a consistent analysis scheme to provide improved insight and prediction capability for strongly nonlinear systems which may exhibit chaotic response.

This proposed approach of combining the steady-state and transient analysis procedures should be a fundamental and practical contribution to the field of nonlinear dynamical systems analysis; to this author's knowledge, this has not been done to date. Moreover, the approach is unique with respect to analysing the two sets of basin boundaries that separate the two steady-state resonance-type solutions, and the basin boundary that separates these bounded and the unbounded capsizing solutions. This type of analysis is similar to that undertaken previously (Falzarano 1990), but this analysis considers one wholly new type of basin-boundary type which should be of greater practical significance. Considering that the multivaluedness of the magnification curve is the nonlinear analogue to linear resonance this should be of practical importance in identifying a 'worst-case' periodic response which could guide further analysis and simulation studies of realistic randomly forced systems.

REFERENCES

Falzarano, J. M. (1990) Predicting Complicated Dynamics Leading to Vessel Capsizing, PhD Dissertation, Naval Architecture Department, The University of Michigan.

Falzarano, J. M. and Troesch, A. W. (1990) Application of modern geometric methods for dynamical systems to the problem of vessel capsizing with water-on-deck, *4th International Conference on the Stability of Ships and Ocean Vehicles, Naples, Italy, September 1990.*

Falzarano, J., Stiendl, A., Troesch, A. and Troger, H. (1990) Bifurcation analysis of a vessel slowly turning in waves, *4th International conference on the Stability of Ships and Ocean Vehicles, Naples, Italy, September 1990.*

Falzarano, J., Steindl, A., Troesch, A. and Troger, H. (1991) Rolling motion of ships treated as a bifurcation problem *Bifurcation and Chaos: Analysis, Algorithms and Applications,* R. Seydel *et al.* (eds.), Birkhäuser, Basel.

Falzarano, J., Steindl, A., Troesch, A. and Troger, H. (1991) Rolling motion of ships treated as a bifurcation problem *Bifurcation and Chaos: Analysis, Algorithms and Applications,* R. Seydel *et al.* (eds.), Basel, Birkhäuser.

J. M. Falzarano, *School of Naval Architecture and Marine Engineering, University of New Orleans, New Orleans, LA 70148, USA*

NONLINEAR BEHAVIOR OF HYDRAULIC ENGINE MOUNTS: THEORY AND EXPERIMENTS

A. G. Haddow

This note focuses on the design, analysis and performance of a class of passive engine mounts known as hydraulic engine mounts (HEM). A detailed description and thorough literature review of such devices can be found in Brach and Haddow (1992). In essence, the HEM is an isolator whose exterior is fabricated from a combination of rubber and metal and is filled with a fluid. A typical cross-section of such a device is shown in the figure. In this figure, the refinement of a decoupler and an inertial track has been added. Over a small range of motion the decoupler is free to move between its stops and the mount's characteristics are dominated by the properties of the rubber. However, when the decoupler bottoms against one of its stops, the fluid must be transferred between chambers A and B via the inertial track. Complex fluid interactions and resonances then occur which result in nonlinear dynamic characteristics which are both amplitude- and frequency-dependent.

None of the analysis techniques employed to date correctly allow for these nonlinear characteristics. Even when the isolation units are bench tested, the stiffness and damping coefficients are evaluated using algorithms which are valid only for linear systems. While the end product does operate, it is more a testament to the practical experience of the design engineer than to the soundness of the analysis techniques employed.

As a result of the linear approach that has been taken, industry is now encountering two major problems. These are:

1. The presence of 'chortle' which is described as an intermittent rattling and/or knocking sound audible in the passenger compartment. This phenomenon is little understood by the automotive industry but the author will present a combination of experimental results and theoretical analysis to shed light on this phenomenon. It may be shown to be a type of intermittent chaos (Bergé *et al.* 1986), resulting from impacting conditions internal to the mount.

2. Presently, inadequate testing procedures and signal analysis techniques are employed to properly document a mount's characteristics. Using standard techniques (and based on a

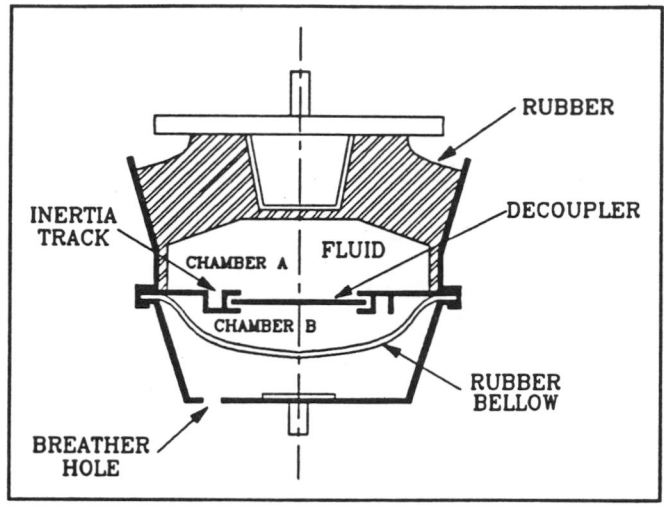

Figure A1   Cross-section of a typical hydraulic engine mount.

linear viewpoint) it is impossible to distinguish subtle differences between mounts. It is possible to apply newer techniques, such as wavelet transforms (Daubechies 1990) to alleviate this problem.

In addition to addressing the issues as itemized above, it is possible to tailor the mount characteristics by way of utilizing nonlinear phenomena to further improve the mount's performance, and also adapt the mount in order to use it as a nonlinear vibration absorber.

REFERENCES

Bergé, P., Pomeau, Y. and Vidal, C. (1986) *Order within Chaos*, Wiley-Interscience, New York.
Brach, R. M. and Haddow, A. G. (1992) Hydraulic engine mounts, Report to Delco Products, Mechanical Engineering Department, Michigan State University, East Lansing, Michigan, September.
Daubechies, I. (1990) The wavelet transform, time frequency localization and frequency analysis, *IEEE Trans. Inf. Theory*, **36** (5).

A. G. Haddow, *Department of Mechanical Engineering, Michigan State University, A231 Engineering Building, East Lansing, MI 48824–1226, USA.*

BIFURCATION ANALYSIS OF DYNAMIC SYSTEMS WITH CONTINUOUSLY PIECEWISE LINEARITY

H. Hu and W. Schiehlen

The periodic response and associated bifurcations of a harmonically driven oscillator with continuously piecewise linearity has received great attention during the past decade. The oscillator serves as a widely used model to describe various mechanical systems with clearances, elastic constraints etc. Even if the 'corners' of the piecewise-linear restoring force in the oscillator make the system non-smooth on the interface between two linear regions, the bifurcation analysis published up to now is still based on the theory of smooth dynamic systems (e.g. Natsiavas 1990; Kleczka *et al.* 1992). The validity of such an extention is an open problem.

To avoid the difficulty caused by the non-smoothness in bifurcation analysis, it is very natural to replace the non-smooth restoring force by a smoothed one under certain approximation. The smoothed model proposed here is based on a locally defined polynomial in the neighbourhood of the restoring-force corner. It is proved that the polynomial can be tangential to both sides of the corner in any required order of smoothness and that the maximal error can be less than any desired quantity. One may expect, therefore, that the bifurcation analysis to the smoothed system give approximate results for the original system. The numerical simulations did support the expectation in most cases, but sometimes new phenomena were observed, which cannot be explained using the theory of smoothed dynamic systems.

In order to understand the non-smoothness effect of the restoring force on the system behaviour, the analytical properties of the Poincaré map for a continuously piecewise-linear system are studied in detail. According to Filippov (1988), it is proved that the second derivative of a Poincaré map suddenly jumps on the interface between two linear regions in state space and that the jumping amplitude becomes very large if the slope of the restoring force has a high jump or the trajectory passes through the interface at low velocity. By means of the Lyapunov–Schmidt procedure, the Poincaré map of a dissipative system near a fixed point can be reduced to a truncated one-dimensional normal form with a control parameter λ:

$$p(y, \lambda) = p_y(0, 0)y + p_\lambda(0, 0)\lambda + \frac{1}{2} \begin{cases} p_{yy}(0^+, 0)y^2, & y \geqslant 0, \\ p_{yy}(0^-, 0)y^2, & y < 0. \end{cases} \tag{1}$$

Moreover, the normal form for the smoothed system can be approximately written as

$$q(y, \lambda) = q_y(0, 0)y + q_\lambda(0, 0)\lambda + \tfrac{1}{2} q_{yy}(0, 0)y^2 + \tfrac{1}{6} q_{yyy}(0, 0)y^3, \tag{2}$$

where

$$\left.\begin{aligned}
q_y(0, 0) &= p_y(0, 0), \\[4pt]
q_\lambda(0, 0) &= p_y(0, 0), \\
q_{yy}(0, 0) &= \tfrac{1}{2}(p_{yy}(0^+, 0) + p_{yy}(0^-, 0)), \\[6pt]
q_{yyy}(0, 0) &= \frac{9}{8\delta}(p_{yy}(0^+, 0) - p_{yy}(0^-, 0)),
\end{aligned}\right\} \tag{3}$$

and δ is the radius of the smoothed range.

Now the simplest case is considered, where the normal form of the smoothed system yields the condition of saddle–node bifurcation, i.e.

$$q_y(0, 0) = 1, \quad q_\lambda(0, 0) \neq 0, \quad q_{yy}(0, 0) \neq 0. \tag{4}$$

From equation (1), one finds that the fixed point undergoes a bifurcation due to $p_y(0, 0) = 1$, but only if the inequality $p_{yy}(0^+, 0)p_{yy}(0^-, 0) > 0$ is valid does the bifurcation look like a 'saddle–node type', i.e. a pair of fixed points with opposite stabilities disappears suddenly through the bifurcation. In the case of, $p_{yy}(0^-, 0)p_{yy}(0, 0) < 0$, the bifurcation looks like a 'hysteresis type with co-dimension 1', changing the number of fixed points and their stabilities not at all. But the non-zero codimension of the bifurcation usually results in a jump of the bifurcation curve in practice. This is a local bifurcation caused by the non-smoothness of the restoring force and has not been reported before to the authors' knowledge.

To verify the theoretical results, numerical simulations were performed for a piecewise-linear softening oscillator under harmonic excitation. Usually the smoothed system provides a good approximation to the original system both in response calculation and in bifurcation analysis. But if the original system undergoes a 'hysteresis-type' bifurcation, the standard bifurcation analysis of the smoothed system fails to offer correct information. In fact, the term y^3 in the normal-form equation (2) takes an important role in such a case, as the coefficient $q_{yyy}(0, 0)$ is very large.

All in all, the bifurcation analysis based on the smoothed system can be applied to the continuously piecewise-linear system, but with great care.

ACKNOWLEDGEMENT

Supported by Alexander von Humboldt Foundation, Germany

REFERENCES

Filippov, A. F. (1988) *Differential Equations with Discontinuous Righthand Sides*, Dordrecht, Kluwer, pp. 117–122.

Golubitsky, M. and Schaeffer, D. G. (1985) *Signularities and Groups in Bifurcation Theory*, **1**, 182–212.
Kleczka, M., Kreuzer, E. and Schiehlen, W. (1992) Local and global stability of a piecewise linear oscillator, *Philosophical Transactions of the Royal Society of London A* **338**, 533–546.
Natsiavas, S. (1990) Stability and bifurcation analysis for oscillators with motion limiting constraints, *Journal of Sound and Vibration*, **141** (1), 97–102.

H. Hu and W. Schiehlen, *Institut B für Mechanik, Universität Stuttgart, D-70550 Stuttgart, Germany*

[2]On leave from Nanjing University of Aeronautics and Astronautics, 210016 Nanjing, China

CHAOS AND TRANSIENT CHAOS IN NONLINEAR DYNAMICS

H. M. Isomäki and M. Franaszek

Transient chaos is related to the existence of a chaotic saddle or repeller, i.e. a non-attracting invariant set. The study of its hyperbolic properties has aroused a lot of interest recently, particularly because in the non-hyperbolic case the dynamics are especially anomalous and pathological (Lai *et al.* 1993).

In this note we discuss the anomalous chaotic transients of the bouncing-ball dynamics (Isomäki 1990), where the stable and unstable manifolds are found to be tangential in vast regions of the phase space (Franaszek and Isomäki 1991). Moreover, one believes that a chaotic repeller has discontinuities and holes transversal to the unstable direction, i.e. it is a double-Cantor structure. However, this need not always be rigorously true and consequently the transient chaos may be a complex problem of a non-hyperbolic repeller without characteristic holes.

The repeller is constructed by using the Kantz–Grassberger ensemble method. In this repeller we do not find any holes transversal to the unstable direction. In a careful check using two other methods as well, the Nusse–Yorke PIM triple method and the overlapping manifolds method, we find the same result. We show how the stable and unstable manifolds touch each other tangentially in vast regions of the phase plane. Moreover, these results are valid for wide ranges of the control parameter H (or λ) (Franaszek and Isomäki 1991). Because basin boundaries embody a chaotic saddle-type repeller a practical consequence is (Franaszek and Isomäki 1991) that the stable manifold is very dense with a capacity dimension close to 2 as shown in the figure. The calculated box dimension is ≈ 1.85 by using the uncertainty exponent method of Grebogi *et al.* Due to this high dimension, the computational difficulties increase drastically. For instance, by increasing the initial condition accuracy by huge 6–7 decades in the black region the yield of correct final states increases only by a factor of 10.

Nonlinearity and Chaos in Engineering Dynamics
Edited by J. M. T. Thompson and S. R. Bishop,   © 1994 John Wiley & Sons Ltd

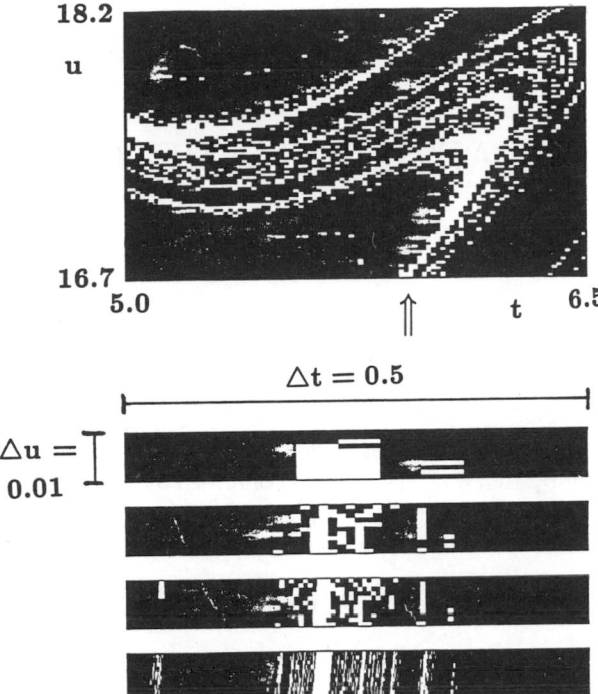

Figure A2   Basin for $H = 2.661$ ($\lambda = 4.95$). The black denotes the chaotic self-reanimating-mode basin and the white denotes the ZR 2-impact period-6 basin. The lower part shows the blow-up of a horizontal bar of 0.5×0.01 at the white throat of the upper figure marked with an arrow. The refined grids are from top to bottom 10, 50, 100 and 1000, respectively.

ACKNOWLEDGEMENTS

We would like to thank Tamás Tél (Eötvös Loránd University), Gert Eilenberger (IFF der KFA Jülich), Michael Thompson (University College London), Francis C. Moon (Cornell University), Olli-Pekka Piirilä (Helsinki University of Technology) and Jukka A. Ketoja (University of Helsinki) for illuminating discussions.

REFERENCES

Franaszek, M. and Isomäki, H. M. (1991) Anomalous chaotic transients and repellers of the bouncing-ball dynamics, *Phys. Rev. A8*, 4231–4236.

Isomäki, H. M. (1990) Fractal properties of the bouncing-ball dynamics. In: *Nonlinear Dynamics in Engineering Systems*, W. Schiehlen (ed.), Berlin, Springer, pp. 125–131.

Lai, Y.-C., Grebogi, C., Yorke, J. A. and Kan, I. (1993) *Nonlinearity*, **6**, 779–797.

H. M. Isomäki, *Faculty of Information Technology, Helsinki University of Technology, FIN-02150 Espoo 15, Finland*

M. Franaszek, *Department of Computer Science, Higher Educational School, PL-30-084 Kraków, Poland*

SHIP CAPSIZING AND CHAOS

M. Kan and H. Taguchi

A new mode of capsizing accompanied by period doubling bifurcations or chaos was observed in a model experiment by Kan et al. (1990a) and confirmed by Kan et al. (1993) this mode has been unfamiliar to most naval architects. Being stimulated by this new experimental observation, the following five capsize equations were examined:

$$\text{symmetric type: } d^2\phi/dt^2 + \kappa d\phi/dt + \phi - \phi^3 = B\cos(\Omega t + \delta), \tag{1}$$

$$\text{asymmetric type: } d^2\phi/dt^2 + \kappa d\phi/dt + \phi - \phi^3 = B_0 + B\cos(\Omega t + \delta), \tag{2}$$

$$\text{Mathieu N-type: } d^2\phi/dt^2 + \kappa d\phi/dt + (1 - e\sin\Omega t)(\phi - \phi^3) = B\cos\Omega t, \tag{3}$$

$$\text{Mathieu L-type: } d^2\phi/dt^2 + \kappa d\phi/dt + (1 - e\sin\Omega t)\phi - \phi^3 = B\cos\Omega t, \tag{4}$$

$$\text{Loll type: } d^2\phi/dt^2 + \kappa d\phi/dt - \phi(1 - \phi^2)(1 - r^2\phi^2) = B\cos\Omega t. \tag{5}$$

Bifurcation diagrams, time histories and phase portraits, chaotic attractors, fractal metamorphoses of the safe basin, and fractal capsize boundaries in the control parameter space were examined systematically for these five equations. Melnikov values were obtained analytically for equations (1), (3) and (4), and numerically for equations (2) and (5), and confirmed by magnifying the safe basins of the first beginning stage of fractal erosion. Approximate bifurcation analyses were also done to see the possibility of application to the safety criterion.

The typical time history of roll with the clear period-doubling bifurcation such as observed in the experiment was simulated better by the asymmetric type (2) and the Mathieu types (3) and (4). The difference between the period doubling bifurcation and chaos was not clear in the time history because the chaotic attractor of equations (1)–(4) formed only simple folded bands. However, it was confirmed that the period doubling bifurcations and/or chaos in the roll response should be regarded as a warning sign of the imminent danger of capsizing. It was also confirmed that only a few cases of time simulation, which was a usual practice, could not estimate the safety against capsizing in the vicinity of the capsize boundary both of the initial-value plane and the control-parameter space because of the fractal feature of the boundary (Kan and Taguchi 1990b, 1991a, b, c, 1992b, c).

Nonlinearity and Chaos in Engineering Dynamics
Edited by J. M. T. Thompson and S. R. Bishop,   © 1994 John Wiley & Sons Ltd

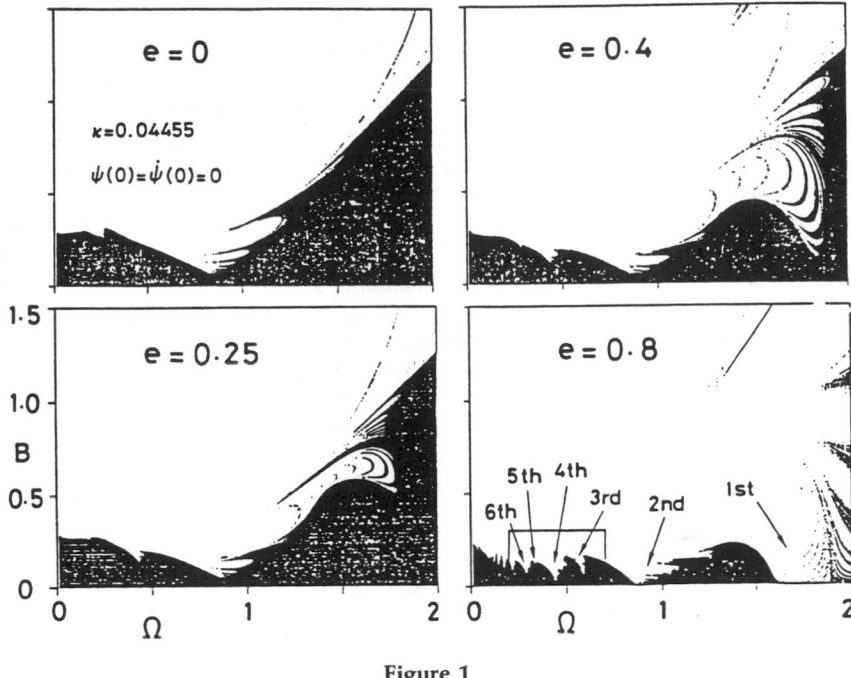

Figure 1

The loll-type capsize equation (5), which was applied to the ship with negative GM due to a flooded compartment or water on the deck, had several unusual characteristics such as clear chaotic behaviours of roll crossing the two-potential well, beautiful chaotic attractors, exterior or interior invasion region of the safe basin, and new fractal capsize boundaries in the control parameter space (Kan and Taguchi 1992a).

The Mathieu-type capsize equations (3) and (4), which should be applied to the ship running in quartering waves because the restoring term inevitably varies with the encounter period, had several extraordinary nonlinear phenomena such as various kinds of bifurcation diagrams and fractal development of the Mathieu's unstable region. The figure shows the variation of capsize boundary in the $\Omega-B$ plane for the Mathieu N-type capsize equation (3). As a parametric excitation increases, the Mathieu's first unstable region develop from the thick finger of the second unstable region and becomes more dangerous, and the nth unstable region becomes deeper and clearer (Kan and Taguchi 1992b, c).

A number of the references cited are to be published in English and future work will investigate an actual capsizing accident.

REFERENCES

Kan, M., Saruta, T., Taguchi, H. and Yasuno, M. (1990a) Capsizing of a ship in quartering seas (Part 1. Model experiments on mechanism of capsizing), *Journal of the Society of Naval Architects of Japan*, **167**, 81–90.

Kan, M. and Taguchi, H. (1990b) Capsizing of a ship in quartering seas (Part 2. Chaos and fractal in capsizing phenomenon), *Journal of the Society of Naval Architects of Japan*, **168**, 213–222.

Kan, M. and Taguchi, H. (1991a) Fractal of capsize boundary of a ship in control space, *Transactions of the West Japan Society of Naval Architects*, **81**, 143–151.

Kan, M. and Taguchi, H. (1991b) Capsizing of a ship in quartering seas (Part 3. Chaos and fractal in asymmetric capsize equation), *Journal of the Society of Naval Architects of Japan*, **169**, 1–13.

Kan, M. and Taguchi, H. (1991c) On the effect of the phase of exiciting force on fractal capsize boundaries of a ship, *Transactions of the West Japan Society of Naval Architects*, **82**, 69–81.

Kan, M. and Taguchi, H. (1992a) Chaos and fractals in loll type capsize equation, *Transactions of the West Japan Society of Naval Architects*, **83**, 131–149.

Kan, M. and Taguchi, H. (1992b) Capsizing of a ship in quartering seas (Part 4. Chaos and fractals in forced Mathieu type capsize equation), *Journal of the Society of Naval Archictects of Japan*, **171**, 229–244.

Kan, M. and Taguchi, H. (1992c) Effects of the variation of stability on fractal capsize boundaries of a ship, *Transactions of the West Japan Society of Naval Architects*, **84**, 1–14.

Kan, M., Saruta, T. and Taguchi, H. (1993) Capsizing of a ship in quartering seas (Parts 5. Comparative model experiments on mechanism of capsizing), *Journal of the Society of Naval Architects of Japan*, **173**, 133–145.

M. Kan and H. Taguchi, *Ship Research Institute, Ministry of Transport, 6-38-1 Shinkawa, Mitaka, Tokyo 181, Japan*

For the historical background to this field, see also the papers by Thompson (1989) and Thompson et al. (1990, 1992) referenced in Chapter 1.

BIFURCATIONS OF THE DOUBLE SCROLL CIRCUIT

T. Matsumoto, R. Tokunaga and M. Komuro

This paper reports several bifurcation mechanisms observed in the Double Scroll Circuit given in figure 1. The dynamics of the circuit is described by

$$C_1 \frac{dv_{C_1}}{dt} = G(v_{C_2} - v_{C_1}) - g(v_{C_1})$$

$$C_2 \frac{dv_{C_2}}{dt} = G(v_{C_1} - v_{C_2}) + i_L$$

$$L \frac{di_L}{dt} = -v_{C_2}.$$

where g(.) is the characteristic of the subcircuit N which is given in figure 2.

Theorem (Normal form) Consider a three-dimensional, three-region continuous piecewise-linear vector field $f : \mathbb{R}^3 \to \mathbb{R}^3$ with (i) $f(-x) = -f(x)$, and (ii) There is no plane or line parallel to the boundary planes which is invariant under the action of the linear vector field in the middle region. Let

$$U^\pm = \{x \in \mathbb{R}^3 : \pm \langle \alpha, x \rangle - 1 = 0\}$$

be the boundary planes. if f has eigenvalues μ_1, μ_2, μ_3 in the middle region, and v_1, v_2, v_3 in the outer regions, then f is uniquely determined up to a linear conjugacy as

$$f(x) = \begin{cases} Ax & (x \in R_0) \\ B(x - P) & (x \in R_+) \\ B(x + P) & (x \in R_-) \end{cases}$$

Nonlinearity and Chaos in Engineering Dynamics
Edited by J. M. T. Thompson and S. R. Bishop,   © 1994 John Wiley & Sons Ltd

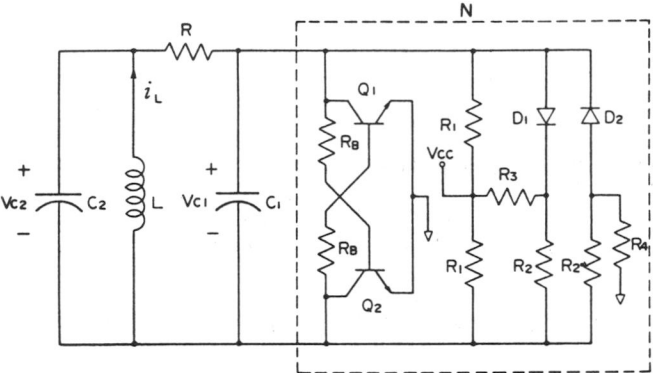

Figure 1 Double Scroll circuit (reprinted with permission from Matsumoto, Chua and Tokumasu, © IEEE).

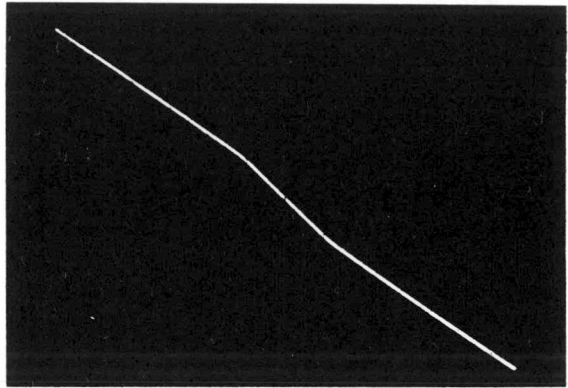

Figure 2 Characteristic of subcircuit N (reprinted with permission from Matsumoto, Chua and Tokumasu 1986 © IEEE).

where

$$R_{\pm} = \{x \in \mathbb{R}^3 : \pm \langle \alpha, x \rangle - 1 > 0\}, R_0 = \{x \in \mathbb{R}^3 : |\langle \alpha, x \rangle| \leqslant 1\}$$

$$\alpha = \begin{bmatrix} 1 \\ 0 \\ 0 \end{bmatrix}, \qquad p = \begin{bmatrix} c_1 \\ c_2 \\ c_3 \end{bmatrix}, \qquad P = \begin{bmatrix} 1 - \dfrac{a_3}{b_3} \\ \dfrac{c_1 a_3}{b_3} \\ \dfrac{c_2 a_3}{b_3} \end{bmatrix}$$

$$A = \begin{bmatrix} 0 & 1 & 0 \\ 0 & 0 & 1 \\ a_3 & a_2 & a_1 \end{bmatrix}, \qquad B = \begin{bmatrix} c_1 & 1 & 1 \\ c_2 & 0 & 1 \\ c_3 + a_3 & a_2 & a_1 \end{bmatrix} = A + p\alpha^T$$

$$a_1 = \mu_1 + \mu_2 + \mu_3, \quad a_2 = -(\mu_1\mu_2 + \mu_2\mu_3 + \mu_3\mu_1), \quad a_3 = \mu_1\mu_2\mu_3,$$

$$b_1 = \nu_1 + \nu_2 + \nu_3, \quad b_2 = -(\nu_1\nu_2 + \nu_2\nu_3 + \nu_2\nu_1), \quad b_3 = \nu_1\nu_2\nu_3,$$

$$c_1 = b_1 - a_1, \quad c_2 = b_2 - a_2 + c_1 a_1, \quad c_3 = b_3 - a_3 + a_2 c_1 + a_1 c_2.$$

For $x \in U^+$ let

$$t = \inf\{t' > 0 : |\langle \alpha, e^{-At'}x\rangle| = 1\} \tag{1}$$

$$s = \inf\{s' > 0 : |\langle \alpha, e^{Bs'}(x - P) + P\rangle| = 1\}. \tag{2}$$

Then, one can show that

$$x = [e_1\alpha^T e^{-At} + e_2\alpha^T x + e_3\alpha^T e^{Cs}]^{-1}h,$$

where $e_1 = (1, 0, 0)^T, e_2 = (0, 1, 0)^T, e_3 = (0, 0, 1)^T$ and $C = A^{-1}BA$.

The pair (t, s) is called the *return time coordinate* of x because there is a one-to-one correspondence between (t, s) and $x \in U^+$.

Theorem (Exact Periodicity Condition) Consider

$$e^{(Cs-At)}[e_1\alpha^T e^{-At} + e_2\alpha^T x + e_3\alpha^T e^{Cs}]^{-1}h = 0,$$

where (t, s) satisfies (1) and (2). Then an x with return time coordinate (t, s) is a periodic point. Using this formula one can derive various explicit bifurcation equations. Explicit homoclinicity/heteroclinicity condition can also be derived.

REFERENCES

Chua, L. O., Komuro, M., and Matsumoto, T. (1986) The double scroll family, *IEEE Trans. Circuits and Systems*, **CAS33**, 1073–1118.

Komuro, M. (1988) Normal forms of continuous piecewise-Linear vector field and chaotic attractors: Part I, *Japan J. Appl. Math*, **5**, 257–304.

Komuro, M. (1988). Normal forms of continuous piecewise-linear vector field and chaotic attractors: part II, *Japan J. Appl. Math.*, **5**, 503–549.

Komuro, M. (1992) Bifurcation equations of continuous piecewise-linear vector fields, *Japan J. Appl. Math.*, **9**, 269–312.

Komuro, M., Tokunaga, R., Matsumoto, T., Chua, L. O., and Hotta, A. (1991) A global bifurcation analysis of the double scroll circuit, *Int. J. Bifurcation and Chaos*, **1**, 139–182.

Matsumoto, T. (1987) Chaos in electronic circuits, *Proc. IEEE*, **75**, 1033–1055.

Matsumoto, T., Chua, L. O., and Ayaki, K. (1988) Reality of chaos in the double scroll circuit: a computer assisted proof, *IEEE Trans. Circuits and Systems*, **CAS35**, 909–925.

Matsumoto, T., Chua, L. O., and Komuro, M. (1985) The double scroll, *IEEE Trans. Circuits and Systems*, **CAS32**, 797–818.

Matsumoto, T., Chua, L. O., and Komuro, M. (1986) The double scroll bifurcations, *Int. J. Circuit Theory and Applications*, **14**, 117–146.

Matsumoto, T., Chua, L. O., and Komuro, M. (1987) Birth and death of the double scroll, *Physica*, **24**, 97–124.

Matsumoto, T., Chua, L. O., and Tokumasu, K. (1986) Double scroll via a two-transistor circuit, *IEEE Trans. Circuits and Systems*, **CAS33**, 828–835.

Tokunaga, R., Abe, Y., and Matsumoto, T. (1993) Observing codimension 2 bifurcations, *Chaos*, **3**, 63–72.

T. Matsumoto, *Dept. of Electrical Engineering, Waseda University, Ohkubo 3-4-1, Shinjuku-ku, Tokyo 169, Japan*

R. Tokunaga, *Institute of Electronics and Information Sciences, University of Tsukuba, Tsukuba 305, Japan*

M. Komuro, *Department of Mathematics, Nishi-Tokyo University, Yamanashi 606-01, Japan*

A GENERAL FORMULATION FOR THE NONLINEAR DYNAMICS OF DISCRETE ELASTIC SYSTEMS

C. E. N. Mazzilli

Considerable effort is being presently devoted to revealing the striking complexity of nonlinear dynamics of few-degree-of-freedom 'deterministic' systems (see Thompson and Stewart 1986). Yet, it should be reckoned that very little is being done to reduce the increasing gap between the achievements in the analysis of these 'simple' systems and of typical large-size engineering problems. In fact, nonlinear analysis of several-degree-of-freedom systems is still a vast unpopulated territory. Computational mechanics considers it from the point of view of developing efficient codes for the numerical analysis, as in Bathe (1982) and Belytschko and Hughes (1983). However, it does not seem to be aware of the importance of either adopting robust models or performing thorough parametric surveys, as remarked by Symonds and Yu (1985) and Fanelli (1990). Applied mechanics, on the other hand, is busy applying perturbation methods and searching for new geometrical, computational and analytical techniques for the analysis of nonlinear and chaotic phenomena in systems with few degrees of freedom. Extension to higher phase-space dimensions is unclear. Engineering analysis has therefore much to gain if the achievements of applied mechanics can be made accessible to computational mechanics. A valuable effort would be a research line starting with consistent and general formulations of analytical dynamics, having in mind a robust modelling. The next step would be its application to the finite-element method which would put within reach the analysis of complex engineering problems. Stationary basic solutions could then be surveyed. It would follow a modal analysis considering small perturbations around the chosen basic state, which would give evidence of the relevant modes to be kept in a low-dimension version of the large-size problem. Here, intuitive engineering reasoning – such as the consideration of the energy imparted to the selected modes and the possibility of internal and external resonances – together with computational techniques for location of invariant manifolds and their tangencies are required. Nonlinear parametric studies for the associated few-degree-of-freedom system would follow, supplying a valuable qualitative knowledge of alternative competing regimes.

Nonlinearity and Chaos in Engineering Dynamics
Edited by J. M. T. Thompson and S. R. Bishop,   © 1994 John Wiley & Sons Ltd

These studies would then be used to define the quantitative analysis to be performed in the original large-size system via the finite-element method. As a matter of fact, such a research line is already being pursued at LMC – Computational Mechanics Laboratory of Escola Politécnica, University of São Paulo. At this moment, a general consistent formulation of analytical dynamics – which is the subject of this note – is already available. Based upon it the ANDROS finite-element program was developed (Mazzilli and Brasil 1992). Discrete structural systems of geometrically nonlinear behaviour constituted by linear elastic material are initially considered. The equations of motion in explicit form are investigated, retaining nonlinearities in both the elastic and inertial forces. The formulation is already in adequate form for application to the finite-element method and is able to tackle conservative and non-conservative applied loads, as well as translation and rotation support excitations. The secant matrices of mass, equivalent damping – included the inertial damping – and stiffness, as well as the equivalent load vector, are characterized within the *global equations of motion*:

$$M^{rs}\ddot{Q}_s + D^{rs}\dot{Q}_s + K_S^{rs}Q_s = \mathscr{F}^r,$$

$$M^{rs} = F^{rsjm}\delta_{jm},$$

$$D^{rs} = \mu^{rs} + (F^{rsjm} - F^{srjm})a_j^i\ddot{a}_m^k\delta_{ik} + \left(F_{,t}^{rsjm} - \frac{1}{2}F_{,r}^{stjm}\right)\delta_{jm}\dot{Q}_t,$$

$$\mathscr{F}^r = P^r - G^{rjm}a_j^i\ddot{a}_m^k\delta_{ik} - H^{rj}a_j^i\ddot{S}^k\delta_{ik}.$$

Viscous damping (μ^{rs}) is included in D^{rs}, K_S^{rs} is the secant stiffness matrix and F^{rsjm}, G^{rjm} and H^{rj} are auxiliary functions of the generalized coordinates, defining the inertial effects. They can be obtained from the holonomic constraint equations implicit in the position vector of a generic material point:

$$R = [a_j^i(t)y^j(Q_1, Q_2, \dots, Q_n) + S^i(t)]e_i.$$

Support rotation is characterized by $a_j^i(t)$ and support translation by $S^i(t)$.

$$F^{rsjm} = \int_\Omega \frac{\partial y^j}{\partial Q_r}\frac{\partial y^m}{\partial Q_s}\rho d\Omega; \quad G^{rjm} = \int_\Omega \frac{\partial y^j}{\partial Q_r}y^m\rho d\Omega; \quad H^{rj} = \int_\Omega \frac{\partial y^j}{\partial Q_r}\rho d\Omega.$$

alternatively, the *incremental equations of motion* can be written:

$$M_T^{rs}\delta\ddot{Q}_s + D_T^{rs}\delta\dot{Q}_s + K_T^{rs}\delta Q_s = \delta\mathscr{P}^r,$$

$$M_T^{rs} = M^{rs},$$

$$D_T^{rs} = \mu^{rs} + (F^{rsjm} - F^{srjm})a_j^i a_m^k\delta_{ik} + (F_{,t}^{rsjm} + F_{,s}^{rtjm} - F_{,r}^{stjm})\delta_{jm}\dot{Q}_t,$$

$$K_T^{rs} = U_{,rs} + G_{,s}^{rjm}a_j^i\ddot{a}_m^k\delta_{ik} + H_{,s}^{rj}a_j^i\ddot{S}^k\delta_{ik} + F_{,s}^{rtjm}\delta_{jm}\ddot{Q}_t$$

$$+ (F_{,st}^{rujm} - \tfrac{1}{2}F_{,rs}^{utjm})\delta_{jm}\dot{Q}_u\dot{Q}_t + (F_{,s}^{rtjm} - F_{,s}^{trjm})a_j^i\ddot{a}_m^k\delta_{ik}\dot{Q}_t + \mu_{,s}^{rt}\dot{Q}_t,$$

$$\delta\mathscr{P}^r = \delta P^r - G^{rjm}(\delta a_j^i\ddot{a}_m^k + a_j^i\delta\ddot{a}_m^k)\delta_{ik} - H^{rj}(\delta a_j^i\ddot{S}^k + a_j^i\delta\ddot{S}^k)\delta_{ik}$$

$$- (F^{rsjm} - F^{srjm})(\delta a_j^i\dot{a}_m^k + a_j^i\delta\dot{a}_m^k)\delta_{ik}.$$

In the K_T^{rs} expression, $U_{,rs}$ stands for the partial derivative of the strain energy (U) with respect to Q_r and Q_s. Both the global and the incremental formulations may be conveniently adapted to consider elastoplastic behaviour.

REFERENCES

Bathe, K.-J. (1982) *Finite Element Procedures in Engineering Analysis,* Prentice-Hall, Englewood Cliffs, NJ.

Belytschko, T. and Hughes, T. J. R. (1983) *Computational Methods for Transient Analysis,* North-Holland.

Fanelli, M. (1991) La dinamica dei sistemi non lineari, *ENEL,* **3959**.

Mazzilli, C. E. N. and Brasil, R. M. L. R. F. (1992) ANDROS – A finite-element program for nonlinear dynamics. Boletim Técnico BT/PEF-9213, Escola Politécnica, University of São Paulo.

Symonds, P. S. and Yu, T. X. (1985) Counter-intuitive behavior in a problem of elastic-plastic beam dynamics, *ASME Journal of Applied Mechanics,* **52**(3), 517–522.

Thompson, J. M. T. and Stewart, H. B. *Nonlinear Dynamics and Chaos,* Chichester, Wiley.

C. E. N. Mazzilli, *Escola Politécnica da Universidade de São Paulo, CP 61548, CEP 05498, São Paulo, Brazil.*

LOCAL BIFURCATIONS AND MODAL INTERACTIONS OF TWO-DEGREE-OF-FREEDOM SELF-EXCITED OSCILLATORS

S. Natsiavas

Dynamical systems with self-excited characteristics appear in many engineering fields. Typical examples are found in the areas of flow-induced vibrations, rotor dynamics and metal cutting. As a first step towards understanding the dynamics of these systems – which at the same time produce approximate results predicting the major response features – the behaviour of the self-excited component is usually modelled by a van der Pol-type damper. The great majority of the previous work on this subject has focused on presenting and analysing response diagrams, obtained by direct integration or approximate analytical methods.

This study presents an analysis on the dynamics of two-degree-of-freedom oscillators including van der Pol- and Duffing-type nonlinearities. The equations of motion are first put in a convenient normalized form. Then, the method of multiple scales is applied and a set of averaged equations is derived for cases of primary external resonance. These equations are shown to admit two types of constant solutions. The first type involves the directly excited mode only. However, for some parameter values the nonlinear damping terms excite the second mode also, even when no internal resonance is present. Determination of the amplitudes and phases of both of these types of motion is reduced to finding the solution of cubic polynomials. Then, detailed stability and local bifurcation analyses of these solutions are carried over. Explicit conditions involving combinations of the system parameters leading to codimension-one, -two and -three bifurcations are presented. The emphasis is placed on presenting a formulation with general validity, providing global information on the dynamics of such systems. The main objective is to present analytical results forming the basis for bringing the system equations in suitable forms, allowing the direct utilization of results from recent theories and developments in the area of dynamical systems. Based on this approach, many aspects of the response can be analysed in detail. In addition, a stimulus is provided

Nonlinearity and Chaos in Engineering Dynamics
Edited by J. M. T. Thompson and S. R. Bishop,   © 1994 John Wiley & Sons Ltd

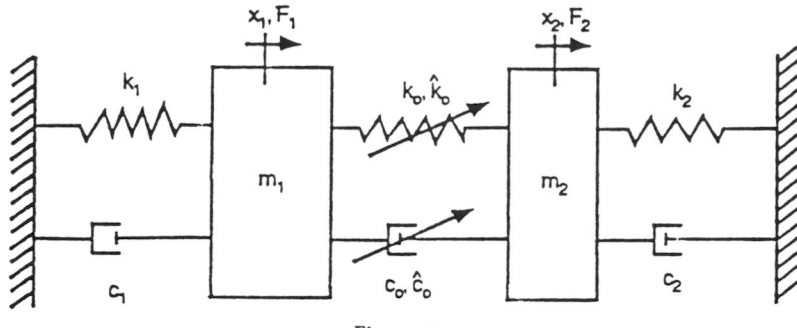

for more study of some dynamical aspects of great practical importance, which still remain unexplored mathematically.

The analysis is applied to a specific practical system, taken from the area of metal cutting (see figure). The first oscillator represents the workpiece, while the other oscillator represents the cutting-tool subsystem. The nonlinearity arises by the cutting process and is modelled by the spring and the damper connecting the two oscillators.

For the example system, comprehensive numerical results are obtained in a systematic manner. The averaged equations appear in the form of a codimension-eight periodically perturbed bifurcation. Here a three-parameter study is proved sufficient to reveal the major behavioural features. A sequence of bifurcation diagrams in the plane of frequency-detuning and forcing amplitude is first presented, by varying a third parameter. Besides saddle–node type and Hopf bifurcations, this study verified the existence of codimension-two bifurcations (asymmetric Takens–Bogdanov and Hopf/saddle–node) as well as the occurrence of a triple-zero bifurcation. The centre manifold around the last bifurcation is three-dimensional and the results confirm that the oscillators examined exhibit very rich dynamics for parameter values leading close to that bifurcation: namely, the results show the existence of resonance horns on the bifurcation planes and illustrate their interactions. Further information about the existence, coexistence and interaction of the single-mode and mixed-mode response is provided by presenting selected response diagrams in conjunction with results from direct integration. Among other things, these results demonstrate the existence of two-and three-frequency quasi-periodic response. In the latter case, a locking phenomenon can occur which cancels one of the three frequencies and leads to two-frequency motions. All these results are discussed and explained in terms of classical engineering terminology.

S. Natsiavas, *Department of Mechanical Engineering, Aristotle University, 54006 Thessaloniki, Greece*

A COMPUTER IMPLEMENTATION FOR THE INTERACTIVE STUDY OF CELL MAPPINGS

A. B. Nordmark

The technique of using cell mappings has become a widely used tool for the study of global behaviour in nonlinear oscillations.

The implementation of the cell-mapping method that is described here is distinguished in that it aims at being *interactive* and *flexible*. Flexibility is obtained by separating the cell-mapping analysis algorithms and the graphical presentation and interaction, which is independent of the particular system studied and not very computing-intensive, from the actual calculation of the cell mapping. The latter usually requires much numerical work, but it is also subject to considerable variation, depending on the system. The separate parts of the implementation communicate over a computer network when they are executed on separate machines. For systems defined by differential equations, the use of a massively parallel CM-200 enables the computation to complete in less than one minute, even if cell grids with several hundred thousand cells are used.

The implementation enables the user to perform a mapping analysis in a short time, and to have several windows representing different parameter values or different models active at the same time, thus facilitating comparison. An interactive technique that proves very powerful is to let the user have access to iterates of the cell mapping by clicking the pointer. The fact that iteration only consists of table look-ups makes it reasonable to have iterates continuously updated as the pointer is moved. In this interactive environment, the technique enables the user to quickly obtain information like the type of stability of periodic orbits or the location of saddle points.

The program can be used for the study of models of impacting systems. Such models, where the velocity changes discontinuously at the time of impact, are known to exhibit non-smooth bifurcations. One example is when a periodic oribit somewhere in the orbit is close to impacting, and through a parameter change acquires an additional, zero − velocity impact. This generally leads to a sudden loss of stability and is called a *grazing bifurcation*

Nonlinearity and Chaos in Engineering Dynamics
Edited by J. M. T. Thompson and S. R. Bishop,   © 1994 John Wiley & Sons Ltd

(C-bifurcation, touch-down bifurcation). If, on the other hand, the impact is modelled by a rapidly increasing but continuous force, the grazing bifurcation should be resolved into a series of ordinary smooth bifurcations. Such a model is the piecewise linear oscillator

$$\ddot{q} + \frac{2D}{W}\dot{q} + F(q) = A\left[\left(\frac{1}{W^2} - 1\right)\cos t - \frac{2D}{W}\sin t\right], \qquad (1)$$

where

$$F(q) = \begin{cases} \dfrac{1}{W^2}q, & q \geqslant -1, \\[2mm] \dfrac{M}{W^2}q + \dfrac{M-1}{W^2}, & q \leqslant -1, \end{cases} \qquad (2)$$

with M large.

If A is increased from 0, the non-impacting periodic solution becomes grazing at $A = 1$, and for still increasing values of A starts to undergo bifurcations. The bifurcations can be studied using a $t = 0 \bmod 2\pi$ type Poincaré mapping. It can be shown that a simple closed-form mapping can also be used to described the bifurcation sequence for large M, and the only parameters that have to be known are the eigenvalues of the periodic orbit at $A = 1$ and a sign.

For example, when $D = 0.4320$ and $W = 6.2574$, the eigenvalues are complex with sum $a_1 = 0.8$ and product $a_2 = 0.42$ and the sign is negative. The mapping that describes the dynamics is then

$$\begin{aligned} x &\leftarrow f(x, d) + a_1 x + y, \\ y &\leftarrow -a_2 x, \end{aligned} \qquad (3)$$

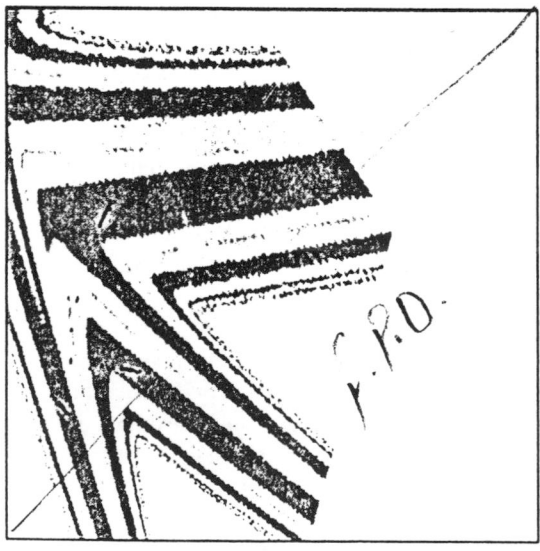

Figure 1   Basins of attraction for impact oscillator.

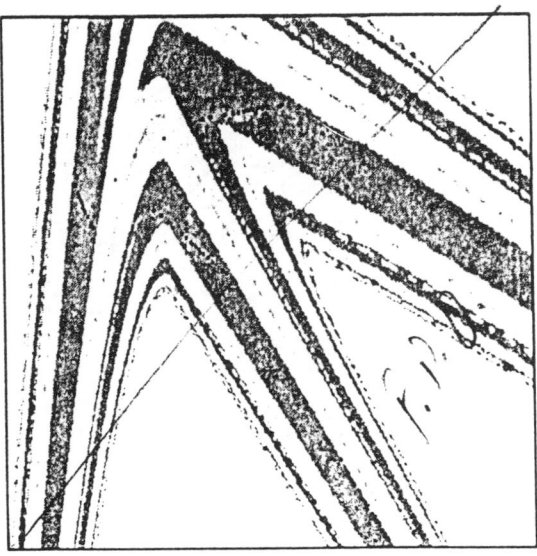

Figure 2

where

$$f(x, d) = \begin{cases} 0, & x \geq d, \\ (d - x)^{3/2}, & x \leq d. \end{cases} \qquad (4)$$

Here x and y are suitable local coordinates whose scales are of the order M^{-2}, and d is a bifurcation parameter proportional to $M^2(A - 1)$. In Figure 1, the basins of attraction for $A = 1.0020$ and $M = 400$ in (1) are shown, and in Figure 2 the basins for the corresponding value $d = 10$ in (3). The light areas make up the basin of a stable period-two solution and the darker areas belong to a chaotic solution. The similarities, apart from a linear change of coordinates, are readily seen.

REFERENCES

Faole, S. and Bishop, S. R. (1993) Bifurcations in impact oscillators. To appear in *Nonlinear Dynamics*.

Hsu, C. S. (1987) *Cell-to-Cell Mapping*, Applied Mathematical Sciences Vol. 64, Berlin, Springer-Verlag.

Ivanov, A. P. (1993) Impact oscillations: linear theory of stability and bifurcations. To appear in the *Journal of Sound and Vibration*.

Nordmark, A. B. (1991) Non-periodic motion caused by grazing incidence in an impact oscillator, *Journal of Sound and Vibration*, **145**, 279–297.

A. B. Nordmark, *Department of Mechanics, Royal Institute of Technology, S-100 44 Stockholm, Sweden*

FLEXURAL–TORSIONAL RESPONSE OF A CRACKED ROTATING SHAFT DURING PASSAGE THROUGH CRITICAL SPEED

R. H. Plaut

There has been extensive research on the vibrational behaviour of cracked shafts and the use of response characteristics to detect cracks (e.g. see the reference lists in Wauer (1990) and Collins et al. (1991)). Most of this previous work involved shafts rotating at constant angular speed, often focusing on the changes of the natural frequencies or modes. However, Wang et al. (1987) and Bently and Thomson (1987) noted that it is easier to detect cracks during a start-up or run-down process than at a steady speed. According to Nilsson (1982), cracks are usually more evident when a rotor passes through a resonance than under normal operation. Vibration monitoring during run-down is sometimes used in an attempt to detect cracks, but research on transient responses of cracked shafts has been limited.

In the present study, transient responses of a simply supported, rotating shaft with a disc are studied analytically. Euler–Bernoulli theory is applied. Torsional and flexural vibrations are analysed, and forces due to eccentricity, gravity, and internal and external damping are included. The shaft contains a single transverse crack that is assumed to be either completely open or completely closed at any given time, depending on the curvature of the shaft at the cross-section containing the crack. Results for this breathing crack are compared with those for a crack that is always open and with those for an uncracked shaft.

The governing equations of motion are bilinear. Galerkin's method is utilized with five-term approximations for the displacement functions, and the resulting equations are integrated numerically. Natural frequencies and critical speeds are determined for the unforced, undamped shaft with an open crack and with no crack. Then time histories of the response are computed when the shaft is accelerated or decelerated past the fundamental critical speed at a constant rate or exponentially. The maximum response is determined, and the effects of the acceleration and deceleration rate, crack depth, crack position along the shaft, and eccentricity angle (with respect to the crack face) are investigated.

Nonlinearity and Chaos in Engineering Dynamics
Edited by J. M. T. Thompson and S. R. Bishop,   © 1994 John Wiley & Sons Ltd

For the breathing crack, the problem is 'inherently nonlinear'. Natural frequencies and modes do not exist, no matter how small the motion. Another interesting aspect is the presence of both parametric and forcing excitations in the equations of motion for displacements in the rotating reference frame. Let $\phi(x, t)$ be the torsional displacement and $v(x, t)$, $w(x, t)$ be the transverse displacements. In the absence of the disc and the crack, the equations of motion are

$$\mu(v_{tt} - 2\dot{\theta}\omega_t - \dot{\theta}^2 v - \ddot{\theta}w) + d_e\mu(v_t - \dot{\theta}w) + EI_1(v_{xxxx} + d_i v_{txxxx})$$
$$- e\mu[(\dot{\theta}^2 + 2\dot{\theta}\dot{\phi}_t + \ddot{\theta}\phi)\cos\delta + (\ddot{\theta} + \phi_{tt} - \dot{\theta}^2\phi)\sin\delta] - \mu g\cos\gamma\sin\theta = 0; \qquad (1)$$

$$\mu(w_{tt} + 2\dot{\theta}v_t - \dot{\theta}^2 w + \ddot{\theta}v) + d_e\mu(w_t + \dot{\theta}v) + EI_2(w_{xxxx} + d_i w_{txxxx})$$
$$- e\mu[(\dot{\theta}^2 + 2\dot{\theta}\dot{\phi}_t + \ddot{\theta}\phi)\sin\delta - (\ddot{\theta} + \phi_{tt} - \dot{\theta}^2\phi)\cos\delta] - \mu g\cos\gamma\cos\theta = 0; \qquad (2)$$

$$\mu(r_1^2 + r_2^2)(\ddot{\theta} + \phi_{tt}) + d_T\mu(r_1^2 + r_2^2)(\dot{\theta} + \phi_t) - GJ(\phi_{xx} + d_i\phi_{txx})$$
$$- e\mu(v_{tt} - 2\dot{\theta}w_t - \dot{\theta}^2 v - \ddot{\theta}w)\sin\delta + e\mu(w_{tt} + 2\dot{\theta}v_t - \dot{\theta}^2 w + \ddot{\theta}v)\cos\delta$$
$$+ \mu eg\phi\cos\gamma\sin(\theta + \delta) - \mu eg\cos\gamma\cos(\theta + \delta) = 0. \qquad (3)$$

Subscripts x and t denote partial differentiation.

In equations (1)–(3), $\dot{\theta}(t)$ is the angular velocity of the shaft at $x = 0$, e is the eccentricity, δ is a measure of the eccentricity location, γ is the angle of the shaft with the horizontal plane, and d_e, d_i and d_T are damping coefficients. The crack introduces localized forces which change when the crack opens or closes. The transition condition depends on the curvature of the shaft at the crack location. Due to this bilinear behaviour in the stiffness and damping, and to the combined parametric and forcing excitations, the response during passage through a critical speed exhibits a number of interesting features.

One of the most surprising features is the effect of the eccentricity angle δ on the response. For certain ranges of δ, the maximum response of the cracked shaft during passage through a critical speed may be smaller than the response of the uncracked shaft, despite a decrease in the stiffness of the system. This is true both for a breathing crack and for a crack that remains open during the motion, and is of significance in the use of vibration response to detect cracks. Since the problem involves gyroscopic forces, inherently nonlinear behaviour, combined parametric and forcing excitations, and non-stationary angular velocity, the response is not simple or easily predictable. This investigation demonstrates the basic characteristics of the dynamics of the system, with an important application to the prevention of failures in rotating machinery.

REFERENCES

Bently, D. E. and Thomson, A. S. (1987) Detection of cracks in rotors, *Incipient Failure Detection Conference, Philadelphia, March 10–12.*

Collins, K. R., Plaut, R. H. and Wauer, J. (1991) Detection of cracks in rotating Timoshenko shafts using axial impulses, *J. Vib. Acous.*, **113**, 74–78.

Nilsson, L. R. K. (1982) On the vibrational behavior of a cracked rotor-bearing system, *Proc. Int. Conf. on Rotordynamic Problems in Power Plants, IFToMM, Rome*, 515–524.

Wang, Y. B., Duan, Z. S., Huang, X. L. and Wen, B. C. (1987) The responses of the simple rotor with surface transverse crack. In: *Rotating Machinery Dynamics*, A. Muszynska and J. C. Simonis (eds.), DE-Vol. 2, New York, ASME, pp. 595–600.

Wauer, J. (1990) On the dynamics of cracked rotors: a literature survey, *Appl. Mech. Rev.*, **43**, 13–17.

R. H. Plaut, *The Charles E. Via Jr. Department of Civil Engineering, Virginia Polytechnic and State University, Blacksburg, VA 24061-0105, USA*

EFFECT OF SMALL RANDOM DISTURBANCE ON THE 'PROTECTION THICKNESS' OF ATTRACTORS OF NONLINEAR DYNAMIC SYSTEMS

J. Q. Sun
Thomas Lord Research Center, USA

The stable steady-state solutions, often referred to as attractors, of a nonlinear dynamic system have domains of attraction. The domain of attraction protects an attractor in the sense that if a disturbance is injected into the system, the system will oscillate around the attractor as long as the disturbed motion of the system remains in the domain of attraction. The domain of attraction thus provides a good measure of reliability of the attractor against disturbances. The analytical study of variation of the domain of attraction with respect to system parameters and random disturbance is a difficult problem. However, this study is well suited to the cell-to-cell mapping method, and we have attacked the problem by using this method. This note presents a numerical study of the effect of small random disturbances on the 'protection thickness' of nonlinear dynamic systems *having multiple steady-state solutions*.

The domains of attraction were first viewed as the 'protecting' region of attractors by Hsu and Chiu (1987). The erosion of the domains of attraction was related to the safety criterion against ship capsize by Thompson (1990) and Lansbury *et al.* (1992). The latter study added significant engineering meaning to the domains of attraction. Since a real system is often exposed to some random disturbances, it is important to study the effect of random disturbances on the domains of attraction. This note and the early work by Sun and Hsu (1991) follow this line.

The 'protection thickness' of an attractor is defined here as the minimum distance between the attractor and the outer boundary of its domain of attraction, or in the GCM terminology, the minimum distance between the persistent group and the set of multi-domicile cells. We examine how the 'protection thickness' changes with the disturbance strength.

Nonlinearity and Chaos in Engineering Dynamics
Edited by J. M. T. Thompson and S. R. Bishop,   © 1994 John Wiley & Sons Ltd

Example 1. Consider a 2-dimensional nonlinear mapping that describes the motion of a hinged bar subjected to a periodic impact excitation (Hsu 1987):

$$x_1(n+1) = x_1(n) - C_1\alpha(n)\sin[x_1(n)] + C_1x_2(n),$$
$$x_2(n+1) = -D_1\alpha(n)\sin[x_1(n)] + D_1x_2(n), \tag{1}$$

where $C_1 = (1 - e^{-2\mu})/(2\mu)$, $D_1 = e^{-2\mu}$, μ is a damping factor, and $\alpha(n)$ is related to the forcing amplitude at the nth cycle. Assume that $\alpha(n)$ is contaminated by small random disturbances and can be modelled as a random process having an independent uniform distribution for each n. In Hsu (1987) it was found that when $\mu = 0.1\,\pi$ and the excitation is deterministic, the system may have a P-2 and two P-1 solutions for some values of the forcing amplitude. Note that for a P-K solution represented by a persistent group consisting of K disjoint subsets of cells in the state space, a minimum distance can be computed between each subset of the persistent group and the set of multi-domicile cells. Naturally, the minimum of these K distances is by definition the 'protection thickness' of the P-K solution.

Figure 1 presents the variation of the 'protection thickness' of the P-1 and P-2 solutions of system (1) with respect to a deterministic forcing amplitude and the standard deviation of a random forcing amplitude, respectively. Because of the symmetry, two minimum distances for the P-2 solution are equal and the 'protection thickness' for the two P-1 solutions is the same.

Example 2 Consider a nonlinear Duffing oscillator defined by

$$\ddot{x} + k\dot{x} + \alpha x + x^3 = B\cos t + w(t) \tag{2}$$

where $w(t)$ is a zero-mean Gaussian random process such that $E[w(t)w(t+\tau)] = \sigma^2\delta(\tau)$, $k = 0.25$ and $B = 8.5$. When $\alpha = -0.02$ or 0.005 and $\sigma = 0$, the system has a strange attractor and a P-3 harmonic motion. Figure 2 shows the variation of the 'protection thickness' of these two solutions with respect to the noise strength σ.

Observations and conclusions. We have seen that K minimum distances of a P-K solution vanish simultaneously as the system parameter or the disturbance strength changes. Note that

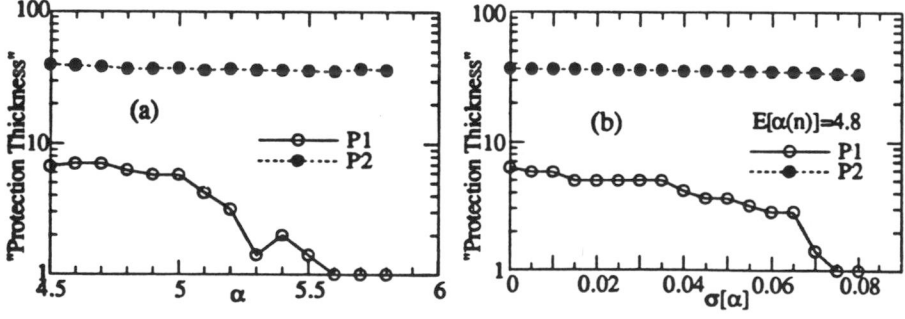

Figure 1   Variation of the 'protection thickness' of P-1 and P-2 solutions of system (1): (a) with a parameter α and (b) with the standard deviation of α when α is random. The 'protection thickness' presented here is scaled by the cell size.

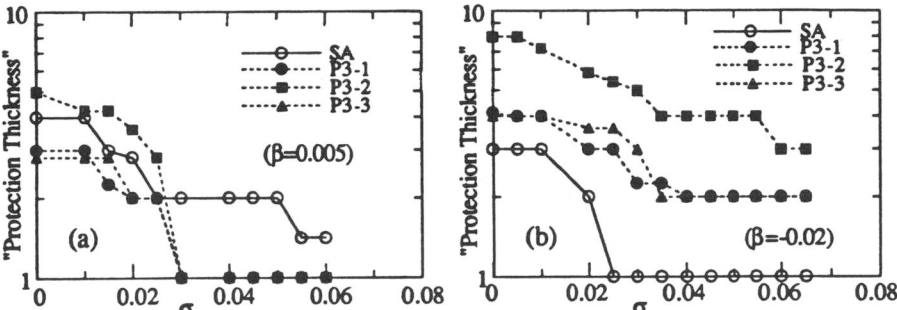

Figure 2 Variation of the 'protection thickness' of the strange attractor (SA) and the P-3 solution of system (2) with σ.

vanishing of the 'protection thickness' indicates the disappearance of an attractor. We have also observed that the 'weakly protected' solution usually disappears before the 'deeply protected' solution. The degree of protection or the reliability of an attractor against disturbances is well measured by the 'protection thickness'. The results of this study show that the 'protection thickness' concept can be useful as a means to detect the in-coming disappearance of attractors.

REFERENCES

Hsu, C. S. (1987) *Cell-to-Cell Mapping: A Method of Global Analysis for Nonlinear Systems*, New York, Springer-Verlag.

Hsu, C. S. and Chiu, H. M. (1987) Global analysis of a system with multiple responses including a strange attractor, *J. of Sound and Vib*, **114**, 203–218.

Lansbury, A. N. *et al.* (1992) Basin erosion in the twin-well Duffing oscillator: two distinct bifurcation scenarios, *Int. J. of Bif. and Chaos*, **2**, 505–532.

Sun, J. Q. and Hsu, C. S. (1991) Effects of small random uncertainties on non-linear systems studied by the generalized cell mapping method, *J. of Sound and Vib.* **147**, 185–201.

Thompson, J. M. T. (1990) Loss of engineering integrity due to the erosion of absolute and transient basin boundaries, *IUTAM Symposium: Nonlinear Dynamics in Engineering System, Stuttgart, Germany*, W. Schielen (ed), Berlin, Springer, pp. 313–320.

J. Q. Sun, *Lord Corporation, Thomas Lord Research Center, 405 Gregson Drive, Cary, NC 27511, USA.*

PARAMETRIC IDENTIFICATION OF NONLINEAR DYNAMIC SYSTEMS, WITH APPLICATION TO AN AIRCRAFT LANDING-GEAR DAMPER

G. Verbeek, A. de Kraker and D. H. van Campen

Eindhoven University of Technology, The Netherlands

For the experimental analysis of linear dynamic systems, it is common to employ shaker test procedures for the parameter identification or certification of a postulated model. The identification of nonlinear dynamic systems, however, is mostly performed by using various transient measurement signals, which are defined in order to resemble the actual or future trajectories of the system under consideration. In landing-gear manufacturing, for example, airworthiness authorities have defined a series of so-called drop-test procedures that will produce transient signals which resemble the operational conditions of the system during a landing. The drop-test facility could be used for identification or optimization of prototypes. Another approach for developing a design or certification tool analogous to the experimental analysis of linear systems, is the application of periodic loads in order to measure the corresponding periodic equilibrium solutions and/or outputs.

The identification method used here will be briefly addressed and consists of a combination of a Bayesian estimator (Bard 1974) and a periodic equilibrium solution technique based on time discretization (Fey 1992) or shooting. The Bayesian estimator utilizes the assumed deterministic mathematical model, the measured data, the assumed independent normally distributed residuals and prior knowledge concerning the parameters. The estimation problem can be reduced to a maximization problem (Verbeek 1991):

$$\Phi = \frac{1}{2}\left(-\sum_{\mu=1}^{n} \ln(\det V_{\mu}) - \sum_{\mu=1}^{n} e_{\mu}^{T} V_{\mu}^{-1} e_{\mu} - \ln \prod_{\alpha=1}^{l'} \sigma_{\alpha}^{2} - \sum_{\alpha=1}^{l} \frac{1}{\sigma_{\alpha}^{2}}(\theta_{\alpha} - \bar{\theta}_{\alpha})^{2} \right). \tag{1}$$

Nonlinearity and Chaos in Engineering Dynamics
Edited by J. M. T. Thompson and S. R. Bishop,   © 1994 John Wiley & Sons Ltd

In equation (1) n and l stand for the number of measurements and parameters respectively, e_μ are the residuals, V_μ are the covariance matrices, and σ_α and $\bar{\theta}_\alpha$ contain the prior normally distributed knowledge on the parameters θ_α. This maximization problem can be solved for the optimal parameters by modified Newton–Gauss iteration following the suggestions of Hendriks (1991).

This technique may be illustrated for axial measurements on an F16 nose landing-gear damper under periodic excitation. The experimental set-up for these tests will be discussed, as well as the choice of the experiments. First, the commonly used simple one-degree-of-freedom (1-DOF) dynamic model for this damper is postulated (Batill and Bacarro 1988):

$$\theta_1(\ddot{u}+g) + \theta_2|\dot{u}|\dot{u} + \theta_3\left[\frac{1}{1-\theta_4 u}\right]^{\theta_5} + \theta_6\arctan(\theta_7\dot{u}) + \theta_8 = F_{exc}, \tag{2}$$

with output equation

$$\hat{y} = u + \theta_9, \tag{3}$$

in which u stands for the displacement and the θ_i's are the parameters. The gas spring force is based on the assumption of polytropic ideal gas behaviour and the friction force is modelled as 'continuous' coulomb friction. As in numerical simulations of these experiments, good results can be obtained with this model for identifications on a single frequency and amplitude, but the predictive power in other parts of the parameter or state space appears to be very low, because the assumption of polytropic gas behaviour is not met.

Re-evaluation of the measurement data leads to an improved 2- or 3-DOF model for the damper, which consists of the equation of motion, the first law of thermodynamics, and a constitutive relation for the pressure p, the displacement u and the temperature T, regarding solubility of gas in oil:

$$\theta_1(\ddot{u}+g) + \theta_2|\dot{u}|\dot{u} + \theta_3 p + (\theta_4 + \theta_5 p)\sin(\theta_6\arctan(\theta_7\dot{u})) + \theta_8 = F_{exc},$$

$$\dot{T} + \theta_9(T-T_0) + \frac{\theta_{10}T\dot{u}}{1-\theta_{11}u} = 0, \tag{4}$$

$$(1 + \theta_{12}u + \theta_{13}T)p + \theta_{14}T = 0.$$

In equations (4) the second and third terms in the energy balance stand for stationary heat flow through the damper wall and a heat source due to compression. The output equations are

$$\hat{y} = \begin{pmatrix} u + \theta_{15} \\ p + \theta_{16} \end{pmatrix}. \tag{5}$$

The new identification method described above is applied to both models for this landing-gear damper. For describing the observed physical phenomena, it can be concluded from the results that the thermodynamic model is superior to the commonly used one.

REFERENCES

Bard, Y. (1974) *Nonlinear Parameter Estimation*, New York, Academic Press.

Batill, S. M. and Bacarro, J. M. (1988) Modeling and identification of nonlinear dynamic systems with application to aircraft landing gear, *Structures, Structural Dynamics and Materials conference, 29th. Williamsburg, VA April 18–20*.

Fey, R. H. B. (1992) Steady-state behaviour of reduced dynamic systems with local nonlinearities. Ph.D. thesis, Eindhoven University of Technology.

Hendriks, M. A. N. H. (1991) Identification of the mechanical behaviour of solid materials. Ph.D. thesis, Eindhoven University of Technology.

Verbeek, G. (1991) Identification of mechanical systems by means of periodic excitation, *Euromech Symposium, 280. Lyon, October 28–31.* Balkema, Rotterdam.

G. Verbeek, A. de Kraker and D. H. van Campen, *Faculty of Mechanical Engineering, Section of Computational and Experimental Mechanics, Eindhoven University of Technology, PO Box 513, 5600 MB Eindhoven, The Netherlands.*

Appendix II Addresses of Participants

Prof. H. D. I. Abarbanel

INLS, Mail Code 0402
U. C. San Diego
La Jolla
CA 92093
USA

Dr. S. A. Al-Athel

President,
King Abdulaziz City for
Science and Technology,
PO BOX 6068, Riyadh 11442
SAUDI ARABIA

Dr. J. Anderson

Aerospace Engineering Dept.
University of Glasgow
Glasgow
G12 8QQ
UK

Dr. J. Angles

EDF-DER
Department MMN
1 av. du Gal de Gaulle
92141 Clamart
FRANCE

Prof. S. T. Ariaratnam

Dept of Civil Engineering
University of Waterloo
Waterloo
Ontario, N2L 3G1
CANADA

Prof. A. Bajaj

School of Mechanical Engineering
Purdue University

West Lafayette
IN 47907-1288
USA

Prof. V. V. Beletsky

Institut B für Mechanik
TU-München
Postfach 202420
8000 München 2
GERMANY

Dr. A. K. Belyaev

Institut f. Technische Mechanik
Technische Universität Braunschweig
Postfach 3329
D-W-3300, Braunschweig
GERMANY

Prof. F. Benedettini

Dipartimento Strutture
Università di L'Aquila 67040 Monteluco Di Roio
L'Aquila
ITALY

Dr. S. R. Bishop

Centre for Nonlinear Dynamics
University College London
Gower Street
London, WC1E 6BT
UK

Ing. F. Bontempi

Dip. Ingegneria Strutturale
Politecnico di Milano
P. zza Leonardo da Vinci 32
20133 Milano
ITALY

Prof. J. Brindley

Dept. of Applied Mathematics
Leeds University
Leeds
LS2 9JT
UK

Dr. D. S. Broomhead

Defence Research Associates
St. Andrews Road
Malvern, Worcs.
WR14 3PS
UK

Prof. D. H. van Campen

Faculty of Mechanical Engineering
Eindhoven University of Technology
Den Dolech 2, PO Box 513
NL-5600 MB Eindhoven
NETHERLANDS

Dr. C. Cheng

Elliger Hohe 11
Wohnung A-3
53177 Bonn
GERMANY

Prof. V. T. Coppola

Dept. of Aerospace Engineering
The University of Michigan
Ann Arbor
MI 48109-2140
USA

Prof. J. P. Cusumano

Dept. Engineering Science
Pennsylvania State University
227 Hammond Building
University Park
PA 16802, USA

Prof. H. G. Davies

Mechanical Engineering Department
University of New Brunswick
P.O. Box 4400
Fredericton, N.B.
CANADA E3B 5A3

Dr. M. E. Davies

Centre for Nonlinear Dynamics
University College London

Gower Street
London, WC1E 6BT
UK

Prof. J. Dorning

Thornton Hall (Reactor Facility)
University of Virginia
Charlottesville, VA 22903-2242
USA

Prof. E. H. Dowell

Duke University
School of Engineering
305 Teer Building, Box 90271
Durham, NC 27708-0271
USA

Prof. J. M. Falzarano

Dept. of Naval Architecture
and Marine Engineering
University of New Orleans
New Orleans, LA 70148
USA

Dr. R. H. B. Fey

TNO Building & Construction
Research
Postbus 29
2600 AA Delft
NETHERLANDS

Dr. S. Foale

Centre for Nonlinear Dynamics
University College London
Gower Street
London, WC1E 6BT
UK

Dr. A. H. Foster

University of Maryland Hospital
22 S. Greene St.
Rm N4W94
Baltimore, Maryland 21201
USA

Dr. Ch. Glocker

Lehrstuhl B für Mechanik
TU München
Postfach 202420
D-80290 München
GERMANY

Prof. V. Gontar
IGCS
Ben Gurion University of the Negev
P.O.B. 1025
Beer Sheva 84110
ISRAEL

Prof. C. Grebogi
Dept. of Mathematics
University of Maryland
College Park
MD 20742
USA

Prof. A. G. Haddow
Dept. of Mechanical Engineering
Michigan State University
East Lansing
MI 48824
USA

Prof. K. Hedrih
Mechanical Engineering Faculty
University of Nis
ul. Beogradska 14
18000-NIS

Prof. S. J. Hogan
Dept. of Engineering Maths
University Walk
Bristol
BS8 1TW
UK

Mr. P. G. Holborn
Centre for Nonlinear Dynamics
University College London
Gower Street
London, WC1E 6BT
UK

Dr. N. Hollingworth
Kluwer Academic Publishers
Spuiboulevard 50/PO Box 17
3300 AA Dordrecht
NETHERLANDS

Prof. P. Holmes
Dept. Theor. and Applied Mechanics
Cornell University
Thurston Hall

Ithaca, NY 14853
USA

Dr. H. Hu
Institute B of Mechanics
University of Stuttgart
Pfaffenwaldring 9
D-7000 Stuttgart 80
GERMANY

Dr. R. A. Ibrahim
Dept. of Mechanical Engineering
Wayne State University
Detroit
MI 48202
USA

Dr. H. M. Isomäki
Faculty of Information Technology
Helsinki University of Technology
Main Building Y329
Otakaari 1, SF-02150, Espoo 15
FINLAND

Dr. K. Janicki
Inst. of Fundamental Research
Swietokrzyska 21
00-049 Warsaw
POLAND

Dr. M. Kan
Ship Research Institute
Ministry of Transport
6-38-1 Shinkawa
Mitaka, Tokyo 181
JAPAN

Dr. T. Kapitaniak
Division of Control and Dynamics
Technical University of Lódź
Stefanowskiego 1/15
90-924 / Lódź
POLAND

Dr. W. Kleczka
Meerestechnik II-Strukturmechanik
TU Hamburg-Harburg
Eissendorfer Strasse 42
D-2100 Hamburg 90
GERMANY

Prof. A. N. Kounadis

Civil Engineering Department
National Tech. University of Athens
Patision 42
GR-10682 Athens
GREECE

Prof. W. F. Langford

Dept. of Mathematics & Statistics
University of Guelph
Guelph
Ontario N1G 2W1
CANADA

Dr. A. N. Lansbury

Dept. of Physics
Brunel University
Uxbridge
Middlesex
UK

Prof. P. V. E. McClintock

Dept. of Physics
Lancaster University
Lancaster
LA1 4YB
UK

Mr. F. A. McRobie

Centre for Nonlinear Dynamics
University College London
Gower Street
London, WC1E 6BT
UK

Prof. T. Matsumoto

Dept. of Electrical Engineering
Waseda University
Ohkubo 3-4-1
Shinjuku-ku, Tokyo 169
JAPAN

Prof. C. E. N. Mazzilli

Av. Prof. Mello Moraes, 2373
Dep. Eng. Estrutura e Fundacoes
EPUSP
05508-900 São Paulo
BRAZIL

Prof. F. C. Moon

204 Upson Hall
Mechanical & Aerospace
Engineering
Cornell University
Ithaca
NY 14853, USA

Prof. N. Sri Namachchivaya

Nonlinear Systems Group
Dept. of Aeronautical Engineering
University of Illinois
Urbana, IL 61801
USA

Prof. S. Narayanan

Machine Dynamics Laboratory
Dept. of Applied Mechanics
Indian Institute of Technology
Madras, IITPO 600 036
INDIA

Prof. S. Natsiavas

Dept. of Mech. and Aero. Engr.
Arizona State University
Tempe
AZ 85287
USA

Prof. A. H. Nayfeh

Dept. Engineering Science
VPI&SU
Blacksburg
Virginia 24061
USA

Dr. A. B. Nordmark

Department of Mechanics
Royal Institute of Technology
S-100 44 Stockholm
SWEDEN

Dr. F. Peterka

Institute of Thermomechanics
Czeck Academy of Sciences
Dolejskova 5
182 00 Prague 8
CZECH REPUBLIC

Prof. F. Pfeiffer

Lehrstuhl B für Mechanik
TU-München
Postfach 202420
D-8000 München 2
GERMANY

Prof. R. H. Plaut

Dept. of Civil Engineering
Virginia Tech
Blacksburg
VA 24061-0105
USA

Prof. G. Rega

Dip. Ingegneria Dell Strutture
Acque e Terreno
Università di L'Aquila
I-67040 I'Aquila (Monteluco Roio)
ITALY

Dr. S. Schaub

Institute B of Mechanics
University of Stuttgart
D-70550 Stuttgart
GERMANY

Prof. W. Schiehlen

Institut B für Mechanik
Universität Stuttgart
Pfaffenwaldring 9
W-7000 Stuttgart 80
GERMANY

Dr. C. Semler

Dept. of Mechanical Engineering
McGill University
Montreal
Quebec
CANADA H3A 2K6

Prof. S. D. Sharma

Dept. of Marine Technology
Building BK
D-47048 Duisburg
GERMANY

Prof. S. W. Shaw

Dept. Mechanical Engineering &
Applied Mechanics
University of Michigan
2250 GG Brown
Ann Arbor, MI 48109-2125, USA

Dr. E. Slivsgaard

Laboratory of Applied Mathematical
Physics and MIDIT
The Technical University of

Denmark
Bldg. 303, DK-2800 Lyngby
DENMARK

Dr. M. S. Soliman

Dept. of Mechanical Engineering
Queen Mary and Westfield College
Mile End Road
London, E1 4NS
UK

Dr. J. Stark

Centre for Nonlinear Dynamics
University College London
Gower Street
London, WC1E 6BT
UK

Dr. N. D. Stein

Dept. of Physics
Lancaster University
Lancaster
LA1 4YB
UK

Dr. A. Steindl

Institut für Mechanik
Technische Universität Wien
Wiedner Haupstrasse 8-10/325
A-1040 Wien
AUSTRIA

Dr. G. Stépán

Dept. of Mechanics
Technical University of Budapest
H-1521 Budapest
HUNGARY

Dr. N. G. Stocks

Dept. of Engineering
University of Warwick
Coventry
CV4 7AL
UK

Dr. J. Q. Sun

Lord Corporation
405 Gregson Drive
Cary
North Carolina 27511-7900
USA

Prof. W. Szemplínska-Stupnicka

Inst. of Fundamental Research
Polish Academy of Sciences
ul. Świtokrzyska 21
PL-00-049 Warsaw
POLAND

Prof. J. M. T. Thompson

Centre for Nonlinear Dynamics
University College London
Gower Street
London, WC1E 6BT
UK

Prof. H. Troger

Institut für Mechanik
Technische Universität Wien
Wiedner Haupstrasse 8-10/325
A-1040 Wien
AUSTRIA

Prof. H. True

MIDIT
The Technical University of
Denmark
DK-2800 Lyngby
DENMARK

Dr. G. Verbeek

Faculty of Mechanical Engineering
Eindhoven University of Technology
P.O. Box 513, WH -1.127
5600 MB Eindhoven
NETHERLANDS

Prof. L. N. Virgin

Dept. Mechanical Engineering
School of Engineering
Duke University
Durham, North Carolina 27708-0300
USA

SYMPOSIUM PHOTOGRAPH

Abarbanel, H. D. I.	(80)	Ellis, M. D.	(3)	Lansbury, A. N.	(60)	Semler, C.	(63)
Anderson, J.	(53)	Falzarano, J. M.	(44)	Lighthill, Sir James	(87)	Sharma, S. D.	(88)
Ariaratnam, S. T.	(91)	Fey, R. H. B.	(13)	Lady Lighthill	(86)	Shaw, S. W.	(66)
Mrs Ariaratnam	(90)	Foale, S	(9)	MacMaster, A. G.	(6)	Slivsgaard, E.	(29)
Bajaj, A. K.	(43)	Foster, A. H.	(62)	Malhotra, N.	(71)	Soliman, M. S.	(46)
Beletsky, V. V.	(78)	Glocker, Ch.	(22)	Matsumoto, T.	(69)	de Souza, J. R.	(47)
Belyaev, A. K.	(24)	Gontar, V.	(50)	McClintock, P.	(18)	Mrs de Souza	(23)
Benedettini, F.	(27)	Grebogi, C.	(79)	McRobie, F. A.	(14)	Stark, J.	(2)
Bishop, S. R.	(72)	Haddow, A. G.	(75)	Moon, F. C.	(76)	Mrs Stark	(49)
Brindley, J.	(64)	Hedrih, K.	(35)	Sri Namachchivaya, N.	(70)	Steindl, A.	(56)
Broomhead, D. S.	(4)	Holborn, P. G.	(8)	Mrs Sri Namachchivaya	(68)	Stépán, G.	(34)
van Campen, D. H.	(55)	Holmes, P.	(61)	Narayanan, S.	(67)	Stocks, N. G.	(73)
Mrs van Campen	(31)	Hu, H.	(19)	Natsiavas, S.	(45)	Sun, J. Q.	(17)
Chastell, P. R.	(5)	Ibrahim, R. A.	(40)	Nayfeh, A. H.	(77)	Sz-Stupnicka, W.	(38)
Cheng, C.	(16)	Isomäki, H. M.	(58)	Nordmark, A. B.	(28)	Tan, N. H.	(74)
Coppola, V. T.	(42)	Janicki, K.	(51)	Peterka, F.	(59)	Thompson, J. M. T.	(83)
Croll, J. G. A.	(1)	Kan, M.	(20)	Pfeiffer, F.	(52)	Mrs Thompson	(36)
Cusumano, J. P.	(26)	Mrs Kan	(37)	Plaut, R. H.	(21)	Helen Thompson	(15)
Davies, H. G.	(32)	Kapitaniak, T.	(30)	Rega, G.	(25)	Thompson, M. G.	(7)
Davies, M. E.	(10)	Kleczka, W.	(41)	Schaub, S.	(39)	Troger, H.	(54)
Dorning, J.	(48)	Kounadis, A. N.	(89)	Schiehlen, W.	(84)	Verbeek, G.	(11)
Dowell, E. H.	(81)	Kovács, Z.	(33)	Mrs Schiehlen	(85)	Virgin, L. N.	(65)
Mrs Dowell	(82)	Langford, W. F.	(57)	Selman, C. G.	(12)		

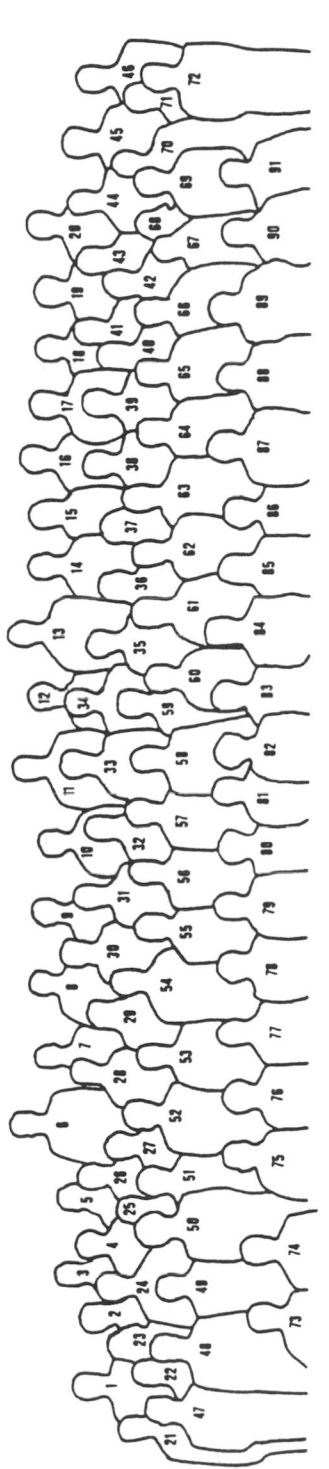

AUTHOR INDEX

SUBJECT INDEX